Building
Scientific
Apparatus

Building
Scientific
Apparatus

A Practical Guide to Design and Construction

John H. Moore ▪ Christopher C. Davis ▪ Michael A. Coplan

University of Maryland

1983

ADDISON-WESLEY PUBLISHING COMPANY
Advanced Book Program / World Science Division

London ▪ Amsterdam ▪ Don Mills, Ontario ▪ Sydney ▪ Tokyo

Figures 4.17, 4.18, 4.19(a), 4.45, and 4.119 are reprinted with permission from *Handbook of Lasers*, R. J. Pressley, Ed., CRC Press, Cleveland, 1971. Copyright The Chemical Rubber Co., CRC Press, Inc.

Illustrations by JAMES S. KEMPTON

Library of Congress Cataloging in Publication Data

Moore, John H., 1941–
  Building scientific apparatus.

  Includes index.
    1. Scientific apparatus and instruments—
Handbooks, manuals, etc. 2. Instrument
manufacture—Handbooks, manuals, etc.
I. Davis, Christopher C., 1944–   . II. Coplan,
Michael A., 1938–   . III. Title.
Q185.M66   1983   681'.75   82-8726

ISBN 0–201–05532–5   AACR2

BCDEFGHIJ-HA-8987654

To Our Families

TABLE OF CONTENTS

2 WORKING WITH GLASS

3 VACUUM TECHNOLOGY

4   OPTICS

Over the past 40 years, the technology associated with scientific research has grown dramatically. To be successful, research scientists or technicians must acquire technical skills that are not taught in undergraduate lectures or labs. They must become familiar with a wide range of commercially produced equipment and learn to use it effectively. In addition, they must be able to design apparatus that cannot be purchased, and often must build it as well.

Although many books are devoted exclusively to such topics as electronics, optics, vacuum technology, and mechanical design, research workers, under pressure to obtain results, often must choose to strike out on their own rather than taking the time to sift through this voluminous literature. We have written this text with the conviction that an overview of these fields is needed—an overview that will permit the practicing scientist to capitalize on new and unfamiliar technologies without having to take the time to acquire an engineering education in each field. We also believe that a student who is beginning experimental research can benefit from studying the laboratory technologies systematically rather than depending entirely on irregular instruction from a research supervisor and informal interaction with other students. We intend this book to serve both as an introductory text for the beginning researcher and as a shelf reference for the experienced scientist.

This volume owes much to the spirit of the classic laboratory text, *Procedures in Experimental Physics*, written over 40 years ago by John Strong. The emphasis in that text is on the construction of apparatus, but we recognize that today it is often more efficient for experimentalists to rely on others to fabricate parts of their apparatus rather than making their own. In successive chapters we deal with mechanical, vacuum, optical, electron-optical, and electronic devices. In each chapter we review the physical principles that must be understood in order to use these devices intelligently. We then describe the operation of particular instruments. When instruments are commercially available, we provide information to help practicing scientists determine which model best fits their needs. In the case of apparatus that must be custom-built, we deal with the engineering principles that should be mastered in order to carry out the design, and we provide engineering data to assist in this task. We also describe construction procedures for some apparatus that may be fabricated in the laboratory.

Throughout the text, we make specific reference to manufacturers of instruments, components, and materials. Our choice of particular suppliers is based upon our own experience, which must be considered rather narrow in view of the large industry that has grown up to serve the modern technician and scientist. The enterprising reader can easily find many sources of scientific equipment other than those mentioned in the text.

The choice of units poses a problem. In many areas, metric units are being adopted only slowly. In some places, both metric and English units are used, the choice depending upon the country in which a manufacturer is located. In discussing each class of apparatus, we use the system of units most often employed by the specialist.

CHAPTER 1

MECHANICAL DESIGN

Every scientific apparatus, even a device that is fundamentally electronic or optical in nature, requires a mechanical structure. The design of this structure determines to a large extent the usefulness of the apparatus, and thus a successful scientist must acquire many of the skills of the mechanical engineer in order to proceed rapidly with an experimental investigation.

The designer of research apparatus must strike a balance between the makeshift and the permanent. Too little initial consideration of the expected performance of a machine may frustrate all attempts to get data. Too much time spent planning can also be an error, since the performance of a research apparatus is not entirely predictable. A new machine must be built and operated before all the shortcomings in its design are apparent.

The function of a machine should be specified in some detail before design work begins. One must be realistic in specifying the job of a particular device. The introduction of too much flexibility can hamper a machine in the performance of its primary function.

The beginning is also the time to consider the problem of assembly and disassembly, since research equipment rarely functions properly at first and often must be repeatedly taken apart and reassembled. Sometimes it is useful to allow space in an initial design for anticipated modifications.

Make a habit of studying the design and operation of research apparatus. Learn to visualize in three dimensions the size and positions of the parts of an instrument in relation to one another.

Before beginning a design learn what has been done before. It is a good idea to build and maintain a library of commercial catalogs in order to be familiar with what is available from outside sources. Too many scientific designers waste time and money on the reinvention of the wheel and the screw. Use nonstandard parts only when their advantages justify the great cost of one-off construction in comparison with mass production. Consider modifications of a design that will permit the use of standardized parts. Become aware of the available range of commercial services. In most big cities, such operations as casting, plating, and heat treating are performed inexpensively by specialty job shops. In many cases it is cheaper to have others provide these services rather than attempt them yourself. The addresses of a few of the thousands of suppliers of useful services, as well as manufacturers of useful materials, are listed at the end of this chapter.

In the following sections we discuss the properties of materials and the means of joining materials to create a machine. The physical principles of mechanical design are presented. These deal primarily with controlling the motion of one part of a machine with respect to another, both where motion is desirable and where it is not. There are also sections on machine tools and on mechanical drawing. The former is mainly intended to provide enough information to enable the scientist to make intelligent use of the services of a machine shop. The latter is detailed enough to allow effective communication with people in the shop.

1

1.1 TOOLS AND SHOP PROCESSES

A scientist must be able to make proper use of hand tools in order to assemble and modify his research apparatus. He or she should also be capable of performing elementary operations with a drill press, lathe, and milling machine in order to make or modify simple components quickly. Even when a scientist works with instruments that are fabricated and maintained by research technicians and machinists, an elementary knowledge of machine-tool operations will allow the design of apparatus that can be constructed with efficiency and at reasonable cost. The following is intended to familiarize the reader with the capabilities of various tools. Skill with machine tools is best acquired under the supervision of a competent machinist.

1.1.1 Hand Tools

A selection of hand tools for the laboratory is given in Table 1.1. A research scientist in physics or chemistry will have frequent use for most of these tools, and if possible should have the entire set in the lab. The tool set outlined in Table 1.1 is not too expensive for any scientist to have on hand.

It is wise for a laboratory scientist to adopt a craftsmanlike attitude toward tools. Far less time is required to find and use the proper tool for a job than will be required to repair the damage resulting from using the wrong one.

1.1.2 Machines for Making Holes

Holes up to about 1-in. diameter are made using a *twist drill* in a drill press. Larger holes are usually made by enlarging a drilled hole using a *cutting tool in a boring bar* (Figure 1.1) in a lathe or jig boring machine.

Twist drills are available in fractional inch sizes and metric sizes as well as in a numbered series of sizes at intervals of only a few thousandths of an inch. Standard drill sizes are given in Table 1.2. The included angle at

Table 1.1   TOOL SET FOR LABORATORY USE

Screwdrivers:
  No. 1, 2, and 3 drivers for slotted-head screws
  No. 1, 2, and 3 Phillips screwdrivers
  Set of Allen drivers for $\frac{1}{16}-\frac{1}{4}$-in. socket-head screws
  Set of nut drivers for $\frac{1}{8}-\frac{1}{2}$-in. hex-head screws and nuts
  Set of jeweler's screwdrivers

Wrenches:
  Set of combination box and open-end wrenches, $\frac{3}{8}-1$ in. in $\frac{1}{16}$-in. increments
  Set of socket wrenches, $\frac{3}{8}-1$ in. in $\frac{1}{16}$-in. increments, with $\frac{3}{8}$-in. square ratchet driver
  6-in., 8-in., and 12-in. adjustable wrenches
  Pipe wrench

Pliers:
  Slip-joint pliers
  Channel-locking pliers
  Large and small needle-nose pliers
  Large and small diagonal cutters
  Small flush cutters
  Hemostats

Hammers:
  2-oz and 6-oz ball-peen hammers
  Soft-faced hammer with plastic or rubber inserts

Files:
  6-in. and 8-in. second-cut flat, half-round, and round files with handles
  6-in. and 8-in. smooth-cut flat, half-round, and round files with handles

Miscellaneous:
  Forceps
  Sheet-metal shears
  Hacksaw
  Tubing cutter
  Spring-loaded center punch
  Scriber
  Small machinist's square
  Steel scale
  Divider
  Dial caliper, Vernier caliper, or micrometer
  4-40, 6-32, 8-32, 10-24, and $\frac{1}{4}$-20 taps
  Electric hand drill or small drill press
  Drills, $\frac{1}{16}-\frac{1}{2}$ in. in $\frac{1}{32}$-in. increments, in drill index
  Drills, Nos. 1–60, in drill index
  Small bench vise

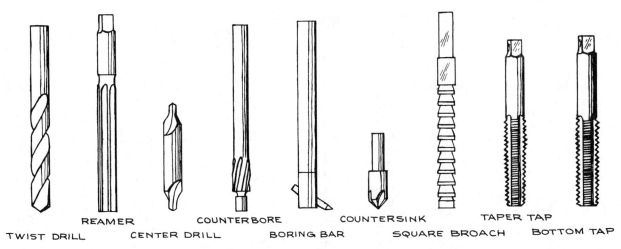

REAMER   COUNTERBORE   COUNTERSINK   TAPER TAP

TWIST DRILL   CENTER DRILL   BORING BAR   SQUARE BROACH   BOTTOM TAP

Figure 1.1 Tools for making and shaping holes.

the point of a drill is 118°. A designer should always choose a hole size that can be drilled with a standard size drill, and the shape of the bottom of a blind hole should be taken to be that left by a standard drill unless another shape is absolutely necessary.

If many holes of the same size are to be drilled, it may be worthwhile to alter the drill point to provide the best performance in the material that is being drilled. In very hard materials the included angle of the point should be increased to as much as 140°. For soft materials such as plastic or fiber it should be decreased to about 90°. Many shops maintain a set of drills with points specially ground for drilling in brass. The included angle of such a drill is 118°, but the cutting edge is ground so that its face is parallel to the axis of the drill in order to prevent the drill from digging in.

The location of a drilled hole can be determined to within about 0.010 in. by scribing two intersecting lines and center-punching the intersection. Locational accuracy of 0.001 in. can be achieved with a jig borer. A drill tends to produce a hole that is out of round and oversize by as much as 0.005 in. This is particularly true when drilling material that is so thin that the point breaks out on the under side before the shoulder enters the upper side. To prevent this problem the work is clamped to a backup block of similar material. A drill point also tends

to deviate from a straight line. This runout can amount to 0.005 in. for a $\frac{1}{4}$-in. drill making a 1-in.-deep hole. However, a $\frac{1}{2}$-in. drill would only run out 0.001 in. in this distance.

Before drilling, the location of the hole should be center-punched and the work should be securely clamped to the drill-press table. The drill should enter perpendicular to the work surface. When drilling curved or canted surfaces, it is desirable to mill a flat, perpendicular to the hole axis at the location of the hole, before drilling.

The speed at which the drill turns is determined by the maximum allowable surface speed at the outer edge of the bit. For aluminum the speed at the outer edge should be about 250 surface feet per minute (sfpm), for brass about 200 sfpm, and for steel 50–100 sfpm. While drilling the bit should be cooled and lubricated by flooding with soluble cutting oil or other suitable cutting fluid, although brass or aluminum can be drilled without cutting oil if necessary.

A drilled hole that must be round and straight to close tolerances is drilled slightly undersize and then reamed using a tool such as is shown in Figure 1.1. *Reamers* are available in $\frac{1}{64}$-in. increments for $\frac{1}{8}$- to $\frac{3}{4}$-in. diameter and in $\frac{1}{16}$-in. increments to 3 in. The diameter tolerance on a reamed hole can be 0.001 in. or better.

Table 1.2 TWIST-DRILL SIZES

Size	Diameter (inches)	Size	Diameter (inches)	Size	Diameter (inches)	Size	Diameter (inches)
80	.0135	53	.0595	26	.1470	A	.234
79	.0145	52	.0635	25	.1495	B	.238
78	.0160	51	.0670	24	.1520	C	.242
77	.0180	50	.0700	23	.1540	D	.246
76	.0200	49	.0730	22	.1570	E	.250
75	.0210	48	.0760	21	.1590	F	.257
74	.0225	47	.0785	20	.1610	G	.261
73	.0240	46	.0810	19	.1660	H	.266
72	.0250	45	.0820	18	.1695	I	.272
71	.0260	44	.0860	17	.1730	J	.277
70	.0280	43	.0890	16	.1770	K	.281
69	.0292	42	.0935	15	.1800	L	.290
68	.0310	41	.0960	14	.1820	M	.295
67	.0320	40	.0980	13	.1850	N	.302
66	.0330	39	.0995	12	.1890	O	.316
65	.0350	38	.1015	11	.1910	P	.323
64	.0360	37	.1040	10	.1935	Q	.332
63	.0370	36	.1065	9	.1960	R	.339
62	.0380	35	.1100	8	.1990	S	.348
61	.0390	34	.1110	7	.2010	T	.358
60	.0400	33	.1130	6	.2040	U	.368
59	.0410	32	.1160	5	.2055	V	.377
58	.0420	31	.1200	4	.2090	W	.386
57	.0430	30	.1285	3	.2130	X	.397
56	.0465	29	.1360	2	.2210	Y	.404
55	.0520	28	.1405	1	.2280	Z	.413
54	.0550	27	.1440				

Note: Sizes designated by common fractions are available in $\frac{1}{64}$-in. increments in diameters from $\frac{1}{64}$ to $1\frac{3}{4}$ in., in $\frac{1}{32}$-in. increments in diameters from $1\frac{3}{4}$ to $2\frac{1}{4}$ in., and in $\frac{1}{16}$-in. increments in diameters from $2\frac{1}{4}$ to $3\frac{1}{2}$ in.

A drilled hole can be threaded with a *tap*. Threads in larger holes may be cut on the lathe.

A bolt can have its head recessed by enlarging the entrance of the bolt hole with a *counterbore* (shown in Figure 1.1).

A drilled hole can be slotted or made square or hexagonal by shaping the hole with a broach (Figure 1.1). The *broach* is a cutting tool that is pulled or pushed through a hole by a broaching machine. A designer should determine whether or not a broaching machine and broaches are available before specifying a shaped hole.

1.1.3 The Lathe

A lathe (Figure 1.2) is used to produce cylindrical or conical surfaces and other surfaces of revolution. The work to be turned is grasped by a *chuck*, which is rotated by the driving mechanism within the lathe *headstock*. Long pieces are supported at the free end by a center mounted in the *tailstock*. A cutting tool held atop the lathe carriage is brought against the work as it turns. As shown in Figure 1.2, the *tool holder* is mounted to the *compound rest*, which is mounted to a rotatable table atop the *cross-feed*, which in turn rests on the

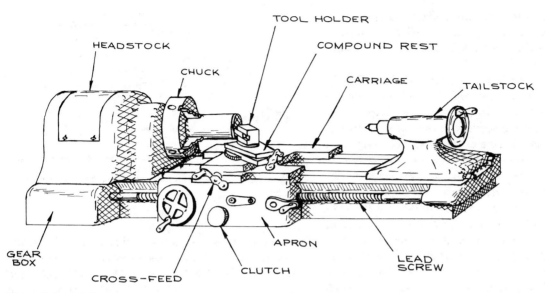

Figure 1.2 A lathe.

carriage. The carriage can be moved parallel to the axis of rotation along slides or ways on the lathe bed. A cylindrical surface is produced by moving the carriage up or down the ways, as in the first cut illustrated in Figure 1.3. Driving the cross-feed produces a face perpendicular to the axis of rotation. A conical surface is produced by driving the tool with the compound-rest screw.

Most lathes have a *lead screw* along the side of the lathe bed. This screw is driven in synchronization with the rotating chuck by the motor drive of the lathe. A groove running the length of the lead screw can be engaged by a clutch in the carriage *apron* to provide power to drive either the carriage or the cross-feed in order to produce a long uniform cut. The threads of the lead screw can be engaged by a split nut in the apron to provide uniform motion of the carriage for cutting threads.

A variety of attachments are available for securing work to the spindle in the headstock. Most convenient is the three-jaw chuck. All three jaws are moved inward and outward by a single control so that a cylinder placed in the chuck is automatically centered to an

accuracy of about 0.002 in. A four-jaw chuck with independently controlled jaws is used for grasping workpieces that are not cylindrical or for holding a cylindrical piece off center. Large irregular work can be bolted to a face plate that is attached to the lathe spindle. Small round pieces can be grasped in a *collet chuck.* A collet is a slotted tube with an inner diameter

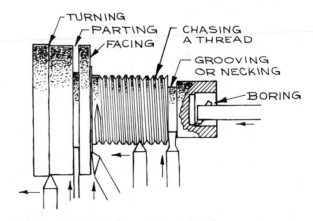

Figure 1.3 Common cuts made on a lathe.

of the same size as the work and a slightly tapered outer surface. The work is clamped in the collet by a mechanism that draws the collet into a sleeve mounted in the lathe spindle.

The quality of work produced in a lathe is largely determined by the cutting tool. The efficiency of the tool bit used in a lathe depends upon the shape of the cutting edge and the placement of the tool with respect to the workpiece. A cutting tool must be shaped to provide a good compromise between sharpness and strength. The sharpness of the cutting edge is determined by the *rake angles* indicated in Figure 1.4. The indicated *relief angles* are required to prevent the non-cutting edges and surfaces of the tool from interfering with the work. A tool bit is ground to shape on a grinding wheel and after grinding the cutting edge should be honed on a medium oilstone. Placement of the tool in relation to the workpiece is illustrated in Figure 1.5.

As in drilling, the cutting speed for turning in a lathe depends upon the hardness of the material being machined. Using tool bits of high-speed tool steel, acceptable cutting speeds are: 300 sfpm for aluminum, 200 sfpm for brass, 50–100 sfpm for steel. Stellite cutting tools are about three times faster, and modern carbide

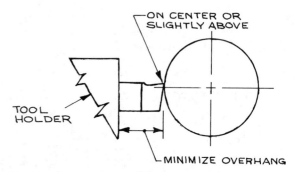

Figure 1.5 Location of a tool with respect to the work in a lathe.

cutters five to ten times faster, than tool-steel bits; they also produce a cleaner, more precise cut. Typically, a cut should be 0.003–0.010 in. deep, although much deeper cuts are permissible for rough work if the lathe and workpiece can withstand the stress.

Holes in the center of a workpiece may be drilled by placing a twist drill in the tailstock and driving the drill into the rotating work with the handwheel drive of the tailstock. The hole should first be located with a *center drill* (Figure 1.1), or the drill point will wander off center.

Tolerances of 0.005 in. can be maintained with ease when machining parts in a lathe. Diameters accurate to within 0.0005 in. can be obtained by a skilled operator at the expense of considerable time. Any modern lathe will maintain a straightness tolerance of 0.005 in./ft provided the workpiece is stiff enough not to spring away from the cutting tool.

1.1.4 Milling Machines

Milling, as a machine-tool operation, is the converse of lathe turning. In milling the workpiece is brought into contact with a rotating cutter. Milling machines are used to produce plane surfaces and grooves or channels in plane surfaces.

Typical milling cutters are illustrated in Figure 1.6. A *plain milling cutter* has teeth only on the periphery and is used for milling flat surfaces. A *side milling cutter* has cutting edges on the periphery and either one or both

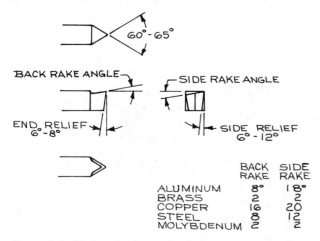

Figure 1.4 Tool angles for a right-cutting round-nose tool. A right-cutting tool has its cutting edge on the right when viewed from the point end.

PLAIN MILLING
CUTTER

SIDE MILLING
CUTTER

END MILL

FLY CUTTER

Figure 1.6 Typical milling cutters.

UP MILLING

CLIMB MILLING

ends so that it can be used to mill a channel or groove. An *end mill* is rotated about its long axis and has cutters on both the end and sides. There are also a number of specially shaped cutters for milling dovetail slots, T-slots, and Woodruff keyslots. A *fly cutter* is another useful milling tool. It consists of a cylinder with a single, movable cutting edge and is used for cutting round holes and milling large flat surfaces. Radial saws are also used in milling machines for cutting narrow grooves and for parting off.

There are two main types of milling machine. The *plain miller* has a horizontal shaft, or *arbor*, on which a cutter is mounted. The work is attached to a movable bed below the cutter. The other type of milling machine is the *vertical mill* which has a vertical spindle located over the bed. Milling cutters can be mounted to an arbor in the spindle of the vertical mill, or the arbor can be replaced by a collet chuck for grasping an end mill. Motion of the mill bed in three dimensions is controlled by handwheels. Big machines may have power-driven beds. An essential accessory for a vertical mill is a rotating table so that the workpiece can be rotated under the cutter for cutting circular grooves and for milling a radius at the intersection of two surfaces.

The two possible cutting operations are illustrated in Figure 1.6. *Climb milling*, in which the cutting edge enters the work from above, has the advantage of producing a cleaner cut. Also, climb milling tends to

hold the work flat and deposits chips behind the direction of the cut. There is however a danger of pulling the work into the cutter and damaging both the work and the tool. *Up milling* is preferred when the work cannot be securely mounted and when using older, less rigid machines.

Cutter speeds should be about 500 sfpm for aluminum, 100 sfpm for brass, and 50 sfpm for steel.

Dimensional accuracy better than about 0.005 in. is difficult to achieve in a milling operation, although flatness and squareness of high precision are easily maintained. Both the mill operator and the designer specifying a milled surface should be aware that milled parts tend to curl after they are unclamped from the mill bed. This problem is particularly acute with thin pieces of metal. It can be alleviated somewhat if cuts are taken alternately on one side and then the other, finishing up with a light cut on each side.

1.1.5 Grinders

Grinders are used for the most accurate work and to produce the smoothest surface attainable in most machine shops. A grinding machine is similar to a plain milling machine except that a grinding wheel rather than a milling cutter is mounted on the rotating arbor. In most machines the work is clamped magnetically to the table and the table is raised until the work touches the grinding wheel. The table is automatically moved back and forth under the wheel at a fairly rapid rate. Many lathes incorporate, as an accessory, a grinder that can be mounted to the compound rest in place of the tool holder, so that cylindrical and conical surfaces can be finished by grinding.

For instrument applications it is possible to produce very flat parallel surfaces even if a surface grinder is not accessible. Simply clamp a drive motor with a grinding wheel above a surface plate at a height such that the wheel just touches the work resting on the surface plate. Then slide the work along the surface under the grinder until no point on the work is left to touch the wheel. Invert the workpiece, shim it up 0.00l in. with brass shim stock, and finish the opposite side. Repeat the process, adding shims each time the work is inverted.

Even the hardest steel can be ground, but grinding is seldom used to remove more than a few thousandths of an inch of material. For complicated pieces to be made

of hardened bronze or steel, it is advantageous to soften the stock material by annealing, machine it slightly oversize with ordinary cutting tools, reharden the material, and then grind the critical surfaces.

Tolerances of 0.0001 in. can be routinely maintained in grinding operations. The average variation of a ground surface should not exceed 50 microinches, and a surface roughness of less than 10 microinches (rms) is possible.

1.1.6   Tools for Working Sheet Metal

Most machine shops are equipped with the tools necessary for making panels, brackets, and rectangular and cylindrical boxes of sheet metal. The basic sheet-metal processes are illustrated in Figure 1.7.

Sheet metal is cut in a guillotine *shear*. Shears are designed for making long straight cuts or for cutting out inside corners. A typical shear can make a cut several feet in length in sheet metal of up to $\frac{1}{16}$ in. in thickness.

Figure 1.7   Basic sheet-metal shop processes.

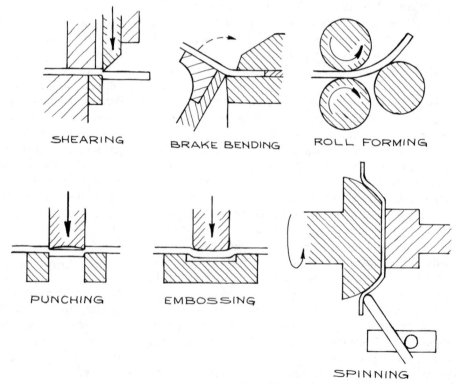

SHEARING

BRAKE BENDING

ROLL FORMING

PUNCHING

EMBOSSING

SPINNING

A *sheet-metal brake* is used to bend sheet stock. A typical instrument-shop brake can accommodate sheet stock 2 to 4 ft wide and up to $\frac{1}{16}$ in. thick. The minimum bend radius is equal to the thickness of the sheet metal. Dimensional tolerances of 0.03 in. can be maintained.

Sheet can be formed into a simple curved surface on a *sheet-metal roll*. The roll consists of three long parallel rollers, one above and two below. Sheet is passed between the upper roller and the two lower rollers. The upper roller is driven. The distance between the upper roller and the lower rollers is adjustable and determines the radius of the curve that is formed.

Holes can be punched in sheet metal. A sheet-metal punch consists of a *punch*, a *guide bushing*, and a *die*. The punch is the male part. The cross section of the punch determines the shape and size of the hole. The punch is a close fit into the die, so that sheet metal placed between the two is sheared by the edge of the punch as it is driven into the die. Round and square punches are available in standard sizes. A punch-and-die set to make a nonstandard hole can be fabricated, but the cost is justified only if a large number of identical holes are required.

Sheet metal can be embossed with a *stamp and die*, which is similar to a punch and die except that the die is somewhat larger than the stamp, so that the metal is formed into the die rather than sheared off at the edge.

Sheet metal can be formed into surfaces that are figures of revolution by *spinning*. The desired shape is first turned in hard wood in a lathe. A circular sheet-metal blank is then clamped against the wooden form by a rubber-faced rotating center mounted in the lathe tailstock. Then as the wooden form is rotated the sheet metal is gradually formed over the surface of the wood by pressing against the sheet with a blunt wooden or brass tool. Spinning requires few special tools and is economical for one-off production.

1.1.7 Casting

Sand casting is the most common process used for the production of a small number of cast parts. Although few instrument shops are equipped to do casting, most competent machinists can make the required wooden patterns, which can then be sent to a foundry for casting in iron, brass, or aluminum alloy. There are both mechanical and economical advantages to producing some complicated parts by casting rather than by building them up from machined pieces. Very complicated shapes can be produced economically because the patternmaker works in wood rather than metal. Castings can be made very rigid by the inclusion of appropriate gussets and flanges, which can only be produced and attached with difficulty in built-up work. Most parts on an instrument require only a few accurately located surfaces. In this case the part can be cast and then only critical surfaces machined.

A designer must understand the sand-casting process in order to design parts that can be produced by this method. A sand-casting mold is usually made in two parts as shown in Figure 1.8. The lower mold box, called the *drag*, is filled with sand, and the wooden

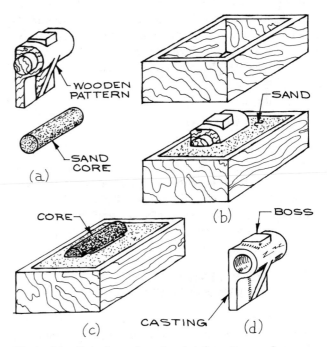

Figure 1.8 Steps in sand casting: (*a*) the pattern and core; (*b*) making an impression in the lower mold box; (*c*) inserting the core in the impression; (*d*) the finished casting.

pattern is pressed into the sand. The sand is leveled and dusted with dry parting sand. Then the upper box, or *cope*, is positioned over the lower and packed with sand. The two boxes are separated, the pattern is removed, and a filling hole, or *sprue*, is cut in the upper mold. A deep hole can be cast by placing a sand *core* in the impression, as shown in Figure 1.8(*c*). Note that the pattern has lugs, called *core prints*, which create a depression to support the core. The two halves of the mold are then clamped together and filled with molten metal.

It is obvious that a part to be cast must include no involuted surfaces above or below the parting plane, so that the pattern can be withdrawn from the mold without damaging the impression. It is also good practice to taper all protrusions that are perpendicular to the parting plane to facilitate withdrawal of the pattern from the mold. A taper or *draft* of 1–2° is adequate. Surfaces to be machined should be raised to allow for the removal of metal. That is the purpose of the *boss* shown in Figure 1.8(*d*). When possible, all parts of the casting should be of the same thickness to avoid stresses that may build up as the molten metal solidifies. All corners and edges should be radiused. With care, a tolerance of 0.03 in. can be maintained.

1.2   MATERIALS

Before discussing the properties of the materials available to the designer of scientific apparatus, we shall first define the parameters used to specify these properties. In Table 1.3 the properties of many useful materials are tabulated in terms of these parameters.

1.2.1   Parameters to Specify Properties of Materials

The strength and elasticity of a metal are best understood in terms of a stress-strain curve such as is shown in Figure 1.9. *Stress* is the force applied to the material per unit of cross-sectional area. This may be a stretching, compressing, or shearing force. *Strain* is the deflection

per unit length which occurs in response to a stress. Most metals deform in a similar way under compression or elongation, but their response to shear is different. When a stress is applied to a metal the initial strain is *elastic* and the metal will return to its original dimensions if the stress is removed. Beyond a certain stress however, *plastic* strain occurs and the metal is permanently deformed.

The *tensile strength* or ultimate strength of a metal is the stress applied at the maximum of the stress-strain curve. The metal is very much deformed at this point, so that in most cases it is impractical to work at such a high load.

A more important parameter for design work is the *yield strength*. This is the stress required to produce a stated, small plastic strain in the metal—usually 0.2% permanent deformation. For some materials the *elastic limit* is specified. This is the maximum stress which the material can withstand without a permanent deformation occurring.

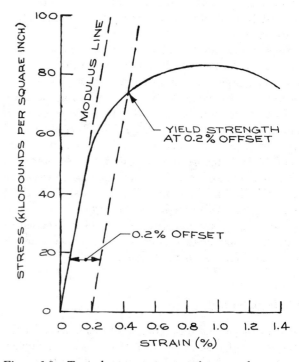

Figure 1.9   Typical stress-strain curve for a metal.

Table 1.3 TYPICAL PROPERTIES OF SOME RESEARCH MATERIALS

Material	Density (lb/in.^{3})	Yield Strength (10^3 psi)	Tensile Strength (10^3 psi)	Modulus of Elasticity (10^6 psi)	Shear Modulus (10^6 psi)	Hardness[a]	Coefficient of Thermal Expansion (10^{-6} °C^{-1})	Comments
Ferrous Metals								
Cast gray iron (ASTM 30)	0.253	25	30	13	5.2	200 BHN	11	
1015 steel (hot-finished)	0.284	27	50	29	11	100 BHN	15	Low carbon (0.15% C)
1030 steel (hot-finished)	0.284	37	68	29	11	137 BHN	15	Medium carbon (0.30% C)
1050 steel (hot-finished)	0.284	50	90	29	11	180 BHN	15	High carbon (0.50% C)
Type 302 stainless steel (annealed)	0.286	37	90	28	10	150 BHN	17	Austenitic
Type 304 stainless steel (annealed)	0.286	35	85	28	10	150 BHN	17	Austenitic
Type 304 stainless steel (cold-worked)	0.286	75	110	28	10	240 BHN	17	Austenitic
Type 316 stainless steel (cold-worked)	0.286	60	90	28	10	190 BHN	16	Austenitic
Aluminum Alloys								
1100-0	0.100	5	13	10	4	23 BHN	23	99% Al
1100-H14	0.100	17	18	10	4	32 BHN	23	99% Al, strain-hardened
2024-T4	0.098	47	68	10.6	4	120 BHN	23	3.8% Cu, 1.2% Mg, 0.3% Mn
2024-T4 (200°C)	0.098	12	18					
6061-T6	0.100	40	45	10	4	95 BHN	23	0.15% Cu, 0.8% Mg, 0.4% Si
7075-T6	0.101	73	83	10.4	4	150 BHN	23	5.1% Zn, 2.1% Mg, 1.2% Cu
Copper Alloys								
Yellow brass (annealed)	0.306	14	46	15	6		20	65% Cu, 35% Zn
Yellow brass (cold-worked)	0.306	60	74	15	6	400 BHN	20	65% Cu, 35% Zn
Cartridge brass ($\frac{1}{2}$ hard)	0.308	52	70	16	6	400 BHN	20	70% Cu, 30% Zn
Beryllium copper (precipitation-hardened)	0.297	140	175	19	7	380 BHN	17	98% Cu, 2% Be
Plastics								
Phenolics	0.049		7.5	1		125 R	81	Bakelite®, Formica®
Polyethylene (low density)	0.033		2	0.025		10 R	180	
Polyethylene (high density)	0.034		4	0.12		40 R	216	
Polyamide	0.040	11.8		0.410		118 R	90	Nylon®
Polymethylmethacrylate	0.043		8	0.42		220 R	72	Lucite®, Plexiglas®
Polytetrafluoroethylene	0.077		2.5	0.060		70 R	99	Teflon®
Polychlorotrifluoroethylene	0.076		6	0.25		110 R	70	Kel-F®
Wood								
Douglas fir (air-dried)	0.011	6.3	0.34[b]	1.4			6, 35[c]	Typical of softwoods
Oregon white oak (air-dried)	0.026	6.6	0.83[b]	2.3			5, 55[c]	Typical of hardwoods

Note: See Section 1.2.1 for explanation of terms.

[a] BHN = Brinell hardness number; R = Rockwell hardness.

[b] Tensile strength perpendicular to grain.

The slope of the straight-line portion of the stress-strain curve is called the *modulus of elasticity E*. It is a measure of the stiffness of the material. It is valuable to note that E is about the same for all grades of steel (about 30×10^6 psi) and about the same for all aluminum alloys (about 10×10^6 psi), regardless of the strength or hardness of the alloy. The effect of elastic deformation is specified by *Poisson's ratio* μ, which is the ratio of transverse contraction per unit dimension of a bar of unit cross section to its elongation per unit length, when subjected to a tensile stress. For most metals $\mu \simeq 0.3$.

The *hardness* of a material is a measure of its resistance to indentation and is determined from the force required to drive a standard indenter into the surface of the material or from the depth of penetration of an indenter under a standardized force. The two most common hardness scales are the Brinell hardness number (BHN) and the Rockwell C scale. Type-304 stainless steel is about 150 BHN or about 0 Rockwell C. A file is about 600 BHN or 60 Rockwell C.

A designer must consider the machinability of a material before specifying its use in the fabrication of a part that must be lathe-turned or milled. In general, the harder and stronger a material, the more difficult it is to machine. On the other hand, very soft materials such as copper and some nearly pure aluminums are also difficult to machine because the metal tends to adhere to the cutting tool and produce a ragged cut. Some metals are alloyed with other elements to improve their machinability. Free-machining steels and brass contain a small percentage of lead or sulfur. These additives do not usually affect the mechanical properties of the metal, but since they have a relatively high vapor pressure, their outgassing at high temperature can pose a problem in some applications.

1.2.2   Heat Treating and Cold Working

The properties of many metals and metal alloys may be considerably changed by heat treating or cold working, which changes the chemical or mechanical nature of the granular structure of the metal. The two classes of heat treatment are quenching and annealing.[1] In *quenching*, the metal is heated above a transition point and then quickly cooled in order to freeze in the granular properties possessed by the metal above the transition. The cooling is accomplished by plunging the heated part into water or oil. Quenching is usually performed to harden a metal. Hardened metals may be softened by *annealing*, wherein the metal is heated above the transition temperature and then slowly cooled. It is frequently desirable to anneal hardened metals before machining and reharden them after working. *Tempering* is an intermediate heat treatment wherein previously hardened metal is reheated to a temperature below the transition point in order to relieve stresses and then cooled at a rate that preserves the desired properties of the hardened material.

Cold working consists of rolling or otherwise plastically deforming a metal to reduce the grain size. Not all metals benefit from cold working, but for some the strength is greatly increased. Because of the annealing effect, the strength and hardness derived from cold working begin to disappear as a metal is heated. This occurs above 250°C for steel and above 125°C for aluminum. In rolling or spinning operations, some metals will work-harden to such an extent that the material must be periodically annealed during fabrication to retain its workability. Cold working reduces the toughness of metal. The surface of a metal part can be work-hardened, without modifying the internal structure, by peening or shot blasting and by some rolling operations. The strength of metal stock may depend upon the method of manufacture. For example, sheet metal and metal wire are usually much stronger than the bulk metal because of the work hardening produced by rolling or drawing.

The surface of a metal part may be chemically modified and then heat-treated to increase its hardness and strength. This process is called *case hardening*. The many methods of case hardening are usually named for the chemical that is added to the surface. *Carburizing*, *cyaniding*, and *nitriding* are common processes for case-hardening steel. Carburizing of low-carbon steel is most conveniently performed in the lab or shop. The part to be hardened is packed in bone charcoal or Kasenit in an open metal box and heated in a furnace to 900°C (salmon-red heat). The time in the furnace, from 15 minutes to several hours, determines the depth of the case. Steel will absorb carbon to a maximum depth of 0.015 in. The part is removed from the furnace and promptly quenched in water. The quenching hardens the case, but the core remains tough and ductile.

1.2.3   Iron and Steel

For instrument work, iron is used primarily as a casting material. The advantages of cast iron are its hardness and its high degree of internal damping. Cast iron is harder than most steels and is used for sliding parts where resistance to wear is important. Vibrations set up in an iron casting are damped about ten times as fast as in steel. Thus the frame of a delicate optical or electrical instrument is often cast of iron.

The two main classes of cast iron are grey iron and white iron. Grey iron is inexpensive and easy to machine. Class-30 grey iron has a Brinell hardness of about 200 BHN. It is, however, rather brittle and is only about half as strong as steel. White iron is stronger and harder than grey iron, and thus it is very wear-resistant. It is very difficult to machine. The machinability and ductility of cast irons can be considerably improved by various heat treatments.

There are more than 10,000 different types of steel. We shall of course only be able to discuss the properties of a few. Steels are classed as cast or wrought steels depending on the method of manufacture. *Wrought steels* are produced by rolling and are almost the only kind used for one-off machine work.

Steels are also classified as either carbon steels or alloy steels. *Carbon steels* are specified by a four digit number. Plain carbon steels are specified as 10*XX*, where *XX* is a two-digit number that indicates the carbon content in hundredths of a percent. In general the strength and hardness of a carbon steel increases with carbon content, but these properties also depend upon the nature of the heat treatment and cold working that the metal has received.

Steels with a carbon content in the 0.10–0.50% range are referred to as *low* and *medium-carbon steels* or mild steels. These steels are supplied rolled or drawn and are most suitable for machine work.

High-carbon steels are difficult to machine and weld, although their machinability is improved if they are first annealed by heating to 750°C (orange heat) and cooling slowly in air. They may be returned to their hard state by reheating to 750°C and quenching in water; however, this process inevitably produces some distortion.

The most common *alloy steels* are the types known as *stainless steels*. The main constituents of stainless steels are iron, chromium, and nickel. They are classified as ferritic, martensitic, or austenitic, depending on the nickel content. The high-nickel or austenitic alloys comprise the 200 and 300 series of stainless steels. 300-series stainless steels are most useful for instrument construction because of their superior toughness, ductility, and corrosion resistance. These austenitic stainless steels are

sensitive to heat treatment and cold working, although the heat-treatment processes are rather more complicated than for plain carbon steels. For example, quenching from 1000°C leaves these steels soft. Cold working, however, can more than double the strength of the annealed alloy. The heat associated with welding of austenitic stainless steels can result in carbide precipitation at the grain boundaries in the vicinity of a weld, leaving the weld subject to attack by corrosive agents. The carbides can be returned to solution by annealing the steel after welding. There are also some special stainless steels that do not suffer from carbide precipitation. In general, stainless steels are more expensive than plain carbon steels, but their superior qualities often offset the extra cost.

Type 303 is the most machinable of the 300-series alloys. Type 316 has the greatest resistance to heat and corrosion. All of these alloys are fairly nonmagnetic, but type 304 is the most nonmagnetic. Machining and cold working tends to increase the magnetism of these alloys. Residual magnetism can be relieved by heating to 1100°C (safely above the temperature range of 450–900°C where carbide precipitation occurs) and quenching in water. Thin pieces that might distort excessively may be quenched in an air blast.

In some instances the dimensions of a particular component of an instrument must be stable in spite of temperature variations. Several iron-nickel alloys have been developed to satisfy this requirement. Two of the most useful are Invar and Super-Invar (from Carpenter Technology), which have thermal coefficients of 1.5×10^{-6} and 0.36×10^{-6} K^{-1}, respectively, between 0 and 100°C. By comparison, the coefficient for stainless steel is about 1.7×10^{-5} K^{-1} in this temperature range. These materials can only be used where thermal stability is an absolute necessity, as they are very expensive and only available in small quantities.

1.2.4 Copper and Copper Alloys

Copper is a soft, malleable metal. Of all metals (short of the noble metals) it is the best electrical and thermal conductor. For instrument applications, oxygen-free high-conductivity (OFHC) copper is preferred because of its high purity and excellent conductivity and because it is not subject to hydrogen embrittlement. Ordinary copper becomes brittle and porous when exposed to hydrogen at elevated temperatures such as may occur in welding or brazing.

Alloys of copper are classified as either brasses or bronzes. *Brass* is basically an alloy of copper and zinc. Generally, brass is much easier to machine than steel but not quite so strong, and its electrical and thermal coefficients are about half those of pure copper. Its properties vary considerably with the proportions of copper to zinc, and also with the addition of small amounts of other elements. Brass parts are readily joined by soldering or brazing. Brass is expensive, but, because it is easy to work with, it is well suited to instrument construction, where the cost of fabrication far outweighs the cost of materials.

For general machine work, yellow brass (35% Zn) is most suitable. Free-machining brass is similar except that it contains 0.5–3% lead. Cartridge brass (30% Zn) is very ductile and is suitable for the manufacture of parts by drawing and spinning and for rivets. It work-hardens during cold forming and may require periodic annealing to 600°C. Stresses that have built up in a cold-formed part during manufacture can be relieved by heating to 250°C for an hour.

The name *bronze* originally meant an alloy of copper and tin, but the term now also refers to alloys of copper and many other elements, such as aluminum, silicon, beryllium, or phosphorus. Bronzes, generally harder and stronger than brasses, often are of special value to the designer of scientific apparatus.

Beryllium-copper or beryllium bronze is a versatile material whose properties can be considerably modified by heat treatment. In its soft form it has excellent formability and resistance to fatigue failure. Careful heat treating will produce a material that is harder and stronger than most steels. It is useful for the manufacture of springs and parts that must be corrosion-resistant. Also it is nearly impossible to strike a spark on beryllium bronze; this makes it useful for parts to be used in an explosive atmosphere. Phosphor bronze cannot be heat-treated, but it is very formable and has great resistance to fatigue. It is used for bellows and springs. Aluminum bronze is highly resistant to corrosion and is

strong and tough. It is therefore used for many marine applications.

Bronze stock can be manufactured by sintering. Sintered bronze is produced by compacting and heating a bronze powder at a temperature below the melting point. It is quite porous in this form and can be used as a filter or can be impregnated with lubricant and used as a bearing.

1.2.5 Aluminum Alloys

Aluminum alloys are valued for their light weight, good electrical and thermal properties, and excellent machinability. Aluminum is about one-third as dense as steel. Its electrical conductivity is 60% that of copper, but it is a better conductor per unit weight.

Aluminum is resistant to most corrosive agents except strong alkalis, owing to an oxide film which forms over the surface. This layer is hard and tenacious and forms nearly instantaneously on a freshly exposed surface. Because the oxide layer is electrically insulating, aluminum parts can in some applications accumulate a surface charge. This problem can be overcome by copper- or gold-plating critical surfaces. The oxide layer makes plating difficult, but reliable plating can be carried out by specialty plating shops.

Aluminum parts cannot be soldered or brazed, but they can be joined by heliarc welding. Welds in aluminum tend to be porous, and welding weakens the metal in the vicinity of the weld because of the rapid conduction of heat into the surrounding metal during the operation.

Aluminum stock is produced in cast or wrought form. Wrought aluminum, produced by rolling, is used for most machine work. Aluminum alloys are specified by a four-digit number that indicates the composition of the alloy, followed by a suffix, beginning with a *T* or *H*, that specifies the state of heat treatment or work hardness. The heat-treatment scale extends from T0 through T10. T0 indicates that the material is dead soft, and T10 that it is fully tempered. 1100-T0 is a soft, formable alloy that is nearly pure aluminum. It is used for extrusions. 2024-T4 or -T6 is as strong as annealed mild steel and is used for general machine work. 6061-T6 is not quite so strong as 2024-T4, but is more easily cold-worked. Its machinability is excellent. The strongest readily available aluminum alloy is 7075-T6. It is easily machined; however, its corrosion resistance is inferior to that of other alloys.

Aluminum is available as plate, bar, and round and square tube stock. Aluminum jig plate is useful for instrument construction. This is rolled plate that has been fly-cut by the manufacturer to a high degree of flatness.

Aluminum may be anodized to give it a pleasing appearance and to improve its corrosion resistance. A number of proprietary formulas, such as Birchwood-Casey Aluma Black, are available to blacken aluminum to reduce its reflectivity.

1.2.6   Other Metals

Monel and inconel are nickel alloys with properties that may be valuable to the instrument designer. These alloys are somewhat stronger and more corrosion-resistant than the stainless steels discussed above, and more importantly, they retain these properties up to 1200°C.

Magnesium alloys are useful because their density is only 0.23 times that of steel. Magnesium is very stiff per unit weight, and is completely nonmagnetic. It is very easy to machine, although there is some fire hazard. Magnesium chips can be ignited, but not the bulk material. The strength of magnesium alloys is considerably reduced above 100°C. The cost of these alloys is at least 10 times that of steel.

Titanium is about half as dense as steel and is much stronger than most steels. It is very resistant to corrosion and completely nonmagnetic. Titanium is useful for extreme-temperature service, since it maintains its strength and corrosion resistance from −250 to 600°C. Titanium is very expensive and is also fairly difficult to machine.

Molybdenum is useful because of its very low, uniform surface potential. It is used for fabricating electrodes and electron-optical elements. Molybdenum is brittle and must be machined slowly using very sharp tools.

Tungsten is the most dense of the readily available elements. Its density is 2.5 times that of steel. Compact counterweights in scientific mechanisms are often made of this material. Tungsten is also used for extreme high-temperature service, such as torch nozzles and plasma electrodes.

1.2.7   Plastics

Plastics are classified as either thermoplastic or thermosetting. Within limits, *thermoplastics* can be softened by heating and they return to their original state upon cooling. *Thermosetting plastics* undergo a chemical change when heated during manufacture, and they cannot be resoftened.

Phenol-formaldehyde plastics, or phenolics, are the most widely used thermosetting plastics. Phenolics are hard, light in weight, and resistant to heat and chemical attack. They make excellent electrical insulators. Phenolics are usually molded to shape in manufacture. Because of their tendency to chip, they are difficult to machine. Bakelite is a phenolic.

Representative thermoplastics are polyethylene, nylon, Delrin, Plexiglas, Teflon, and Kel-F.

Polyethylene is inexpensive, machinable, and very resistant to attack by most chemicals. It is soft and not very strong. Its physical properties vary with the molecular weight and the extent of chain branching within the polymer. It is produced in high-density and low-density forms, which are whitish and translucent. It is also manufactured in a blackened form, which is not degraded by ultraviolet radiation as is the white form. Polyethylene softens at 90°C and melts near 110°C. It can be welded with a hot wire or a stream of hot air from a heat gun. It can be cast, although it shrinks a great deal upon solidifying. Care must be exercised when heating polyethylene because it is burnable. Polyethylene is an excellent electrical insulator, with a dielectric strength of 1000 volts per mil (0.001 in.).

Nylon, a polyamide, is strong, tough, and very resistant to fatigue failure. It retains its mechanical strength up to ≃120°C. It does not cold-flow, and is self-lubricating and thus is useful for all types of bearing surfaces. Nylon is available as rod and sheet in both a white and a black form. Nylon balls and gears are also readily available. There is no fire hazard with nylon, since it is self-extinguishing. It is hygroscopic, and its volume increases and strength decreases somewhat when moisture is absorbed.

Delrin is a polyacetal resin with properties similar to hard grades of nylon in most respects, except that it is not hygroscopic.

All plastics are electrical insulators, although the hygroscopic ones are liable to surface leakage currents. Du Pont produces a polyimide film known as Kapton that is widely used as an electrical insulator. This material has a high dielectric strength in thin films. It is strong and tough, resists chemicals and radiation, and is a better thermal conductor than most plastics. It is used as a wire insulation and in sheets as an insulator between layers of windings in magnets.

Plexiglas, Lucite, and Perspex are all trade names for polymethylmethacrylate. These materials, also known as acrylic plastics, are strong, hard, and transparent. Because they are machinable, heat-formable, and shatter-resistant, polymethylmethacrylates are used in many applications as replacements for glass.

Thermoforming of acrylic sheet into curved surfaces is a fairly simple operation. Begin by producing a wooden or metal pattern of the desired shape. Rest the sheet on top of the pattern, warm the plastic to about 130°C, and permit gravity to pull the softened sheet down over the form. The forming may be done in an oven, or the plastic may be heated with infrared lamps or with hot air from a heat gun. It is necessary to apply the heat slowly and uniformly.

Polymethylmethacrylate dissolves in a number of organic solvents. Parts can be cemented by soaking the edges to be joined in acetone or methylene chloride for about a minute and then clamping the softened edges together until the solvent has evaporated.

General Electric manufactures a polycarbonate plastic known as Lexan. It is transparent and machinable, it has excellent vacuum properties, and is so tough that it is nearly unbreakable. Lexan is one of the most useful plastics for instrument construction.

Teflon is a fluorocarbon polymer. It is only slightly stronger and harder than polyethylene, but it is useful at sustained temperatures as high as 250°C and as low as −200°C. Teflon is resistant to all chemicals, and nothing will stick to it. In most applications it is not so good as nylon as a bearing surface, since it tends to cold-flow away from the area where pressure is applied. A thin film of Teflon, such as is found on modern cookware, provides dry lubrication and prevents dust and moisture adhesion. These films can be applied by specialty shops that serve the food industry. Teflon is about ten times as expensive as nylon.

Kel-F is a fluorochlorocarbon polymer. Its chemical resistance is similar to Teflon, but it is much stronger and harder. Like Teflon, Kel-F does not absorb water.

A number of rubberlike polymeric materials, known as elastomers, find application in scientific apparatus. Materials of interest include Buna-N, Viton-A, and RTV. Buna-N is a synthetic rubber. It can be used at sustained temperatures up to 80°C without suffering permanent deformation under compression. It is available as sheet or in block form. It is useful as a gasket material and for vibration isolation. Viton-A is a fluorocarbon elastomer similar in appearance and mechanical properties to Buna-N. Useful up to 250°C, it tends to take a set at higher temperatures. Unlike Buna-N, Viton-A shows no tendency to absorb water or cleaning solvents. RTV, a self-curing silicone rubber produced by General Electric, comes in a semiliquid form and cures to a rubber by reaction with moisture in the air. It is useful as a sealant, a potting compound for electronic apparatus, and an adhesive. It can be cast in a plastic mold.

1.2.8   Glasses and Ceramics

Borosilicate glasses such as Corning Pyrex 7740 or Kimble KG-33 are used extensively in the lab. These glasses are strong, hard, and chemically inert, and they retain these properties to 500°C. As will be discussed in the succeeding chapter, they are conveniently worked with a natural-gas–oxygen flame.

Quartz or fused silica has a number of unique properties. It has the smallest known coefficient of thermal expansion of any pure material. The coefficient for quartz is 1×10^{-6} K^{-1}. By comparison, the thermal expansion coefficient of the borosilicate glasses is 3×10^{-6} K^{-1}. Quartz can be drawn into long, thin fibers.

The internal damping in these fibers is very low, and their stress-strain relation for twisting is linear to the breaking point. Springs of quartz fibers provide the best possible realization of Hooke's law. Torsion springs and coil springs made of a quartz fiber are used in delicate laboratory balances. Quartz retains its excellent mechanical properties up to 800°C. It softens at 1500°C and must be worked with a hydrogen-oxygen flame.

Alumina (Al_2O_3) is the most durable and heat-resistant of the readily available ceramics. On the Vickers hardness scale, which is similar to the Brinell scale, alumina is 2800 VHN. Only diamond and a few other exotic metal carbides are harder. The compressive strength of alumina exceeds 300,000 psi, although its tensile strength and shear strength are somewhat less than those of steel. Alumina is brittle and, because of its hardness, can only be worked by grinding with a diamond-grit wheel. Alumina rod, tube, thermocouple wells, and electrical insulators are available. These shapes are produced by molding a slurry of alumina powder and various binders to the desired shape and then firing at high temperature to produce the hard ceramic. One particularly useful product is alumina rod that has been centerless-ground to a roundness tolerance of better than 0.001 in. and a straightness of $\frac{1}{32}$ in. per ft. Polycrystalline as well as single-crystal (sapphire) balls are available that have been ground to a roundness tolerance of 0.000025 in. They are surprisingly inexpensive: balls of 0.125–0.500-in. diameter cost only a few dollars. These balls are used as electrical insulators and bearings, and in valves and flowmeters.

Corning produces a machinable ceramic known as MACOR, and Aremco produces machinable ceramics called AREMCOLOX. These materials are not nearly so strong or hard as alumina, but can be machined to intricate shapes that find wide application in instrument work. They can be drilled and tapped and can be turned and milled with conventional high-speed steel tools. They resist chipping and are insensitive to thermal shock. Machinable ceramics have a dielectric strength of 1000–2000 volts per mil (0.001 in.) and thus are excellent electrical insulators.

Aluminum silicate, or lava, is another useful, machinable ceramic. Blocks of grade-A lava are available in the unfired condition. This material is readily machined into intricate shapes, which are subsequently fired at 1050°C for half an hour. The fired ceramic should be cooled at about 150 K/hour to prevent cracking. It shrinks slightly as it cools.

1.3 JOINING MATERIALS

The parts of an apparatus may be permanently joined by soldering, brazing, welding, or riveting, or they may be joined by demountable fasteners such as screws, pins, or retaining rings. We shall discuss both types of joints.

1.3.1 Soldering

In soldering and brazing, two metal surfaces are joined by an alloy that is applied in the molten state to the joint. The melting point of the alloy is below that of the base metal, that is, the metal of the parts to be joined. The two methods differ in the composition of the fusible alloy and the degree to which it is necessary to heat the base metal. For brazing, the base metal is heated to the melting temperature of the alloy, while this is not absolutely required for soldering. Solder is usually an alloy of tin and lead or tin and silver that melts at a relatively low temperature—in the 200–300°C range. Brazing was originally done with a brass alloy that melts at about 900°C. There are now many different brazing alloys with melting points from 200°C to more than 2000°C.

For routine soldering of copper or brass parts, 50-50 lead-tin solder (50% lead, 50% tin) is used. This material is used where great strength and cleanliness are not required and when it is most convenient to work at low temperatures. The shear strength of a lead-tin soldered joint is about 5000 psi. This strength is considerably reduced at temperatures above 100°C.

The first step in soldering a joint is to clean the surfaces to be joined with fine emery paper or steel wool. The surfaces should then be covered with soldering flux. When heated, the flux cleanses the surfaces so that the molten solder will wet them. A sal ammoniac (NH_4Cl) solution is used as a flux for lead-tin solders.

Other proprietary formulas are available for more exotic solders. The parts are then assembled and heated. A natural-gas–air flame or propane-air flame is most convenient. The assembly may also be heated on an electric or gas hot plate. Flame heating should be done indirectly if possible. Play the flame on the metal near the joint, particularly on the heavier of the two pieces. Periodically test the temperature of the two pieces by touching each with the solder. When each piece is hot enough to melt the solder, remove the flame and apply solder to the joint. The solder should flow by capillary action into and through the joint. Sufficient solder should be applied to fill the joint and to produce a buildup of solder, called a *fillet* (Figure 1.10), at the juncture of the two parts. The fillet radius should be $\frac{1}{16}-\frac{1}{8}$ in. A gap of 0.003–0.006 in. is desirable between surfaces to be soldered. In some cases it may be necessary to insert a piece of brass shim stock to maintain this clearance. The lap between soldered surfaces should be at least four times the thickness of the thinner part.

Steel and stainless steel as well as brass and copper can be soldered with silver-alloy solders. Staintin 157PA from Eutectic Welding Alloys is particularly useful. This is a slurry of silver-tin alloy and flux. The slurry is painted on the surfaces to be joined and indirectly heated until the solder flows. The alloy melts at 200°C and can produce a joint with a tensile strength of 15,000 psi. It does not contain lead, zinc, antimony, cadmium, or other volatile metals, and thus is suitable for high-vacuum applications.

1.3.2 Brazing

There are hundreds of different brazing alloys.[2] The choice of an alloy depends upon the metals to be joined and the application of the brazed part. For routine brazing of steels, borate-flux-coated brass brazing rod is available from Sears, Roebuck and other hardware suppliers. The tensile strength of joints made with brass brazing alloy is of the order of 50,000 psi. Silver brazing alloys are preferred for most laboratory work. Silver alloys are lower-melting than brass alloys and can be used on brass as well as steel parts. The tensile strength

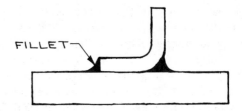

Figure 1.10 A soldered joint.

of silver brazed joints is in the 30,000–50,000-psi range, depending upon the alloy and the joint design.

Handy and Harmon Easy-Flo 45 is a good general-purpose silver brazing alloy. It melts at 620°C and thus can be conveniently worked with the same gas-oxygen torch used for laboratory glassblowing. A fluoride flux, such as Handy Flux, is used with this and most other silver alloys. Easy-Flo 45 contains 45% Ag, 15% Cu, 16% Zn, and 24% Cd. The cadmium in this alloy, and many others as well, represents a serious health hazard. Brazing should always be carried out in a well-ventilated area. The presence of the volatile metals cadmium and zinc in Easy-Flo 45 makes this alloy unsuitable for brazing parts that will be used at elevated temperatures or in high-vacuum apparatus.

When volatile metals cannot be tolerated, the silver-copper eutectic mixture (72% Ag, 28% Cu) makes a suitable brazing alloy for copper, stainless steel, or Kovar. EutecRod 1806 (Eutectic Welding Alloys) and Silvaloy 301 (American Platinum and Silver) are silver-copper eutectics. This alloy melts at 780°C. A lower-melting alloy with much the same properties as the silver-copper eutectic alloy can be made by adding indium to the silver-copper mixture. Incusil 10 (Western Gold and Platinum) is a silver-copper-indium mixture that melts at 730°C. Incusil 15 melts at 685°C.

Torch brazing is the simplest and most convenient brazing operation. An oxyacetylene torch is required for brass brazing alloys. A natural-gas–oxygen flame can be used for lower-melting alloys. The flame must be non-oxidizing. The flame should have a well-defined, blue inner cone about $\frac{1}{4}$ in. long. The oxyacetylene flame should have a greenish feather at the tip of the inner cone. A fireproof working surface of transite or firebrick is also required. Most fluxes contain sodium, which strongly emits yellow light when heated. A pair of didymium eyeglasses, such as are used in glassblowing, will filter out the sodium yellow lines and make the work much more visible while brazing.

Parts to be brazed should, if possible, be designed to be self-locating. A clearance between parts of 0.001–0.003 in. will assure maximum strength in the finished joint. Surfaces to be wetted by the brazing alloy must be clean.

Coat the joint surfaces with flux, then assemble the parts and clamp them together. Preheat the base metals in the area of the joint nearly to the melting temperature of the brazing alloy. The flux should melt from its crystalline form to a glassy fluid as the brazing temperature is approached. Then start at one end of the joint and heat a small area. Keep the flame moving over the work. Apply the brazing rod to the heated area and melt off a bead of metal. The rod should melt in contact with the work. Do not apply the flame directly to the brazing rod. With the flame moving in a circular or zigzag pattern, chase the puddle of molten alloy along the joint. The alloy should completely fill the joint and leave a fillet at the juncture of the parts being joined. It is sometimes convenient to preplace snippets of brazing wire in or near the joint before heating rather than applying the alloy after the work is heated.

Copper, Kovar, monel, nickel, steel, and stainless steel may be brazed without flux if the metal is heated in a hydrogen atmosphere. The hydrogen serves as a flux by reducing surface oxides to produce a clean surface that will be wetted by the molten brazing alloy. Alloys containing chromium must be brazed in dry hydrogen. The surface oxide of chromium found on stainless steel is reduced by dry hydrogen at temperatures above 1050°C. Because hydrogen brazing is carried out in a furnace, there is usually less distortion of the work than in torch brazing. The work is also stress-relieved to some extent. Hydrogen-brazed parts are clean and bright over their entire surface.

A setup for hydrogen brazing is illustrated in Figure 1.11. The container may be Pyrex, quartz, stainless steel, or monel. Porous ceramics such as firebrick are unsatisfactory, since they absorb moisture. The work can be heated by r.f. induction or by radiation from a heating element surrounding the work. All the air in the container must be displaced by hydrogen and a slow flow of hydrogen maintained during the brazing operation. It is of course necessary that the hydrogen overflow be exhausted in a safe manner. A wire-mesh plug in the exhaust line will reduce the possibility of a flashback into the oven if the exhaust is ignited.

For brazing stainless steel, the hydrogen must be free of water vapor and oxygen. The hydrogen dew point should be below −50°C. Tank-grade dry hydrogen is usually satisfactory. As a precaution the hydrogen may be passed through a catalytic oxidizer, such as a Baker De-Oxo unit, to convert oxygen to water, and then through a water-vapor trap filled with dry Linde 13X molecular sieve. The molecular sieve is dried by heating

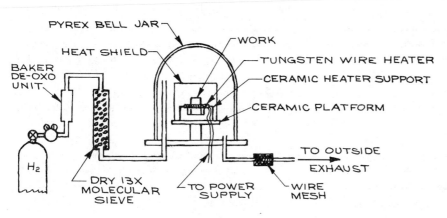

Figure 1.11 Apparatus for hydrogen brazing.

to 150°C for an hour while dry nitrogen or argon flows backwards through the trap. The trap can be heated with heating tape wrapped around the outside.

The filler metal used in hydrogen brazing should melt at a temperature higher than that required for the hydrogen to effectively clean the surface of the base metal. OFHC copper has a melting point of 1083°C and is thus a very good filler metal for brazing steel, stainless steel, Kovar, and nickel. If low-melting alloys such as the silver-copper eutectic are used, a small amount of flux is required. Brazing alloys that contain volatile metals should never be used for any hydrogen brazing.

It is unwise to heat ordinary electrolytic copper in a hydrogen atmosphere. Oxides in the metal are reduced to water that is trapped within the intergranular structure. This water can cause outgassing problems in vacuum applications, and if the copper is heated, the expanding water vapor will produce internal cracks (hydrogen embrittlement). For instrument work, OFHC copper should be used for all copper parts that are to be brazed.

Parts to be hydrogen-brazed should be chemically clean. Avoid fingerprints, since inorganic salts from the skin are not affected by hydrogen and will remain as stains on the finished work. Assemble the parts and place snippets of filler-metal wire near the joints to be brazed. A rough calculation of the volume to be filled will determine the amount of wire to be used. When the filler metal melts, it is drawn into the nearby joint by capillary action. If the joints to be brazed are not visible from outside the furnace, place a piece of brazing metal on top of the work where it can be observed by the operator, in order to determine when the brazing temperature has been reached. If the walls of the brazing apparatus are opaque, a thermocouple will be required for this purpose.

Many commercial heat-treating shops operate hydrogen or ammonia furnaces. In an ammonia furnace the ammonia dissociates to H_2 and N_2 and provides a sufficiently dry atmosphere for hydrogen brazing of stainless steel. It is worthwhile to ascertain the availability and cost of the services of a commercial shop before embarking on the construction of a hydrogen furnace or

before rejecting the advantages of hydrogen brazing because of its complexity.

1.3.3 Welding

In welding, parts are fused together by heating above the melting temperature of the base metal. Sometimes a filler metal of the same type as the base metal is used. Fusion temperatures are attained with a torch, with an electric arc, or by ohmic heating at a point of electrical contact. For most instrument work the preferred method is arc welding under an argon atmosphere using a nonconsumable electrode. This method is referred to as tungsten–inert-gas (TIG) welding or heliarc welding. All metals can be welded, although refractory metals must be welded under an atmosphere that is completely free of oxygen. Arc welding requires special skills and equipment, but most instruments shops are prepared to do TIG welding of steel, stainless steel, and aluminum on a routine basis.

Welded joints are strongest and most free of strain if the rate of heat dissipation is the same to each part of the work during the welding operation. When a thick piece of metal is to be joined to a thin piece, it is common practice to cut a groove in the thick piece, near the joint, to limit the rate of heat loss into the heavier piece during welding. The design of work to be welded is discussed in Section 3.6.2 (see Figure 3.30).

The effect of the high temperatures involved must be considered when specifying a welded joint. Some distortion of welded parts is inevitable. Provision should be made for remachining critical surfaces after welding. The state of heat treatment of the base metal is affected by welding. Hardened steels will be softened in the vicinity of a weld. Stainless steels may corrode because of carbide precipitation at a welded joint.

Spot welding or resistance welding is a good technique for joining metal sheets or wires. The joint is heated by the brief passage of a large electric current through a spot contact between the pieces to be joined. Steel, nickel, and Kovar as well as refractory metals such as tungsten and molybdenum can be spot-welded. Copper and the precious metals cannot usually be spot-welded, because their electrical resistance is so low that a current dissipates little power. Small spot-welding units with hand-held electrodes are available from commercial sources and are quite useful in the lab.

1.3.4 Threaded Fasteners

Threaded fasteners are used to join parts that must be frequently disassembled. A thread on the outside of a cylinder, such as the thread on a bolt, is referred to as an external or male thread. The thread in a nut or a tapped hole is referred to as an internal or female thread. The terminology used to specify a screw thread is illustrated in Figure 1.12. The *pitch* is the distance between successive crests of the thread. The pitch in inches is the reciprocal of the number of threads per inch (tpi). The *major diameter* is the largest diameter of either an external or an internal thread. The *minor diameter* is the smallest diameter. In the United States, Britain, and Canada the form of a thread is specified by the Unified Standard. Both the *crest* and *root* of these threads are flat, as shown in Figure 1.12, or else slightly rounded. The *thread angle* is always 60°.

There are two thread series commonly used for instrument work. The *coarse-thread series*, designated UNC for Unified National Coarse, is for general use. Coarse threads provide maximum strength. The *fine-thread series*, designated UNF, is for use on parts subject to shock or vibration, since a tightened fine-

thread nut and bolt are less likely to shake loose than a coarse-thread nut and bolt. The UNF thread is also used where fine adjustment is necessary.

The fit of threaded fasteners is specified by tolerances designated as 1A, 2A, 3A, and 5A for external threads and 1B, 2B, 3B, and 5B for internal threads. The fit of 2A and 2B threads is adequate for most applications, and such threads are usually provided if a tolerance is not specified. Many types of machine screws are available only with 2A threads. The 2A and 2B fits allow sufficient clearance for plating. 1A and 1B fits leave sufficient clearance that dirty and scratched parts can be easily assembled. 3A and 3B fits are for very precise work. 5A and 5B are interference fits such as are used on studs that are to be installed semipermanently.

The specification of threaded parts is illustrated by the following examples:

1. An externally threaded part with a nominal major diameter of $\frac{3}{8}$ in., a coarse thread of 16 threads per inch, and a 2A tolerance is

$$\frac{3}{8}\text{-16 UNC-2A.}$$

2. An internally threaded part with a nominal major diameter of $\frac{5}{8}$ in., a fine thread of 18 tpi, and a 2B tolerance is designated

$$\frac{5}{8}\text{-18 UNF-2B.}$$

3. Major diameters less than $\frac{1}{4}$ in. are specifed by a gauge number; thus a thread with a nominal major diameter of 0.164 in. and a coarse thread of 32 tpi is designated

$$\text{8-32 UNC.}$$

The specifications of the UNC and UNF thread forms are listed in Table 1.4.

Pipes and pipe fittings are threaded together. Pipe threads are tapered so that a seal is formed when an externally threaded pipe is screwed into an internally threaded fitting. The American Standard Pipe Thread,

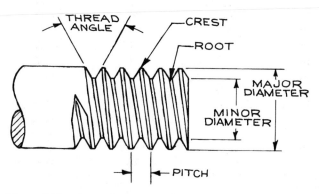

Figure 1.12 Screw-thread terminology.

designated NPT for National Pipe Thread, has a taper of 1 in 16. The diameter of a pipe thread is specified by stating the nominal internal diameter of a pipe that will accept that thread on its outside. For example, a pipe with a nominal internal diameter of $\frac{1}{4}$ in. and a standard thread of 18 tpi is designated

$$\frac{1}{4}\text{-}18 \text{ NPT}$$

or simply

$$\frac{1}{4}\text{-NPT.}$$

The American Standard Pipe Thread specifications are listed in Table 1.5.

The common forms of machine screws are illustrated in Figure 1.13. *Hex-head* cap screws are ordinarily available in sizes larger than $\frac{1}{4}$ in. They have a large bearing surface and thus cause less damage to the surface under the head than other types of screws. A large torque can be applied to a hex-head screw, since it is tightened with a wrench.

Slotted-head screws are available with *round, flat,* and *fillister* heads. The flat head is countersunk so that the top of the head is flush. Fillister screws are preferred to round-head screws, since the square shoulders of the head provide better support for the blade of a screwdriver.

Socket-head cap screws with a hexagonal recess are preferred for instrument work. These are also known as *Allen* screws. L-shaped, straight, and ball-pointed hex drivers are available that permit Allen screws to be

Table 1.4 AMERICAN STANDARD UNIFIED AND AMERICAN NATIONAL THREADS

Size (nominal diameter)	Coarse (NC, UNC)		Fine (NF, UNF)	
	Threads per Inch	Tap Drill[a]	Threads per Inch	Tap Drill[a]
0 (0.060)			80	$\frac{3}{64}$
1 (0.073)	64	No. 53	72	No. 53
2 (0.086)	56	No. 50	64	No. 50
3 (0.099)	48	No. 47	56	No. 45
4 (0.112)	40	No. 43	48	No. 42
5 (0.125)	40	No. 38	44	No. 37
6 (0.138)	32	No. 36	40	No. 33
8 (0.164)	32	No. 29	36	No. 29
10 (0.190)	24	No. 25	32	No. 21
12 (0.216)	24	No. 16	28	No. 14
$\frac{1}{4}$	20	No. 7	28	No. 3
$\frac{5}{16}$	18	Let. F	24	Let. I
$\frac{3}{8}$	16	$\frac{5}{16}$	24	Let. Q
$\frac{7}{16}$	14	Let. U	20	$\frac{25}{64}$
$\frac{1}{2}$	13	$\frac{27}{64}$	20	$\frac{29}{64}$
$\frac{9}{16}$	12	$\frac{31}{64}$	18	$\frac{33}{64}$
$\frac{5}{8}$	11	$\frac{17}{32}$	18	$\frac{37}{64}$
$\frac{3}{4}$	10	$\frac{21}{32}$	16	$\frac{11}{16}$
$\frac{7}{8}$	9	$\frac{49}{64}$	14	$\frac{13}{16}$
1	8	$\frac{7}{8}$	12	$\frac{59}{64}$

Note: ASA B1.1-1960.
[a] For approximately 75% thread depth.

Table 1.5   AMERICAN STANDARD TAPER
PIPE THREADS

Nominal Pipe Size	Actual O.D. of Pipe	Threads per Inch	Normal Length of Engagement by Hand	Length of Effective Thread
$\frac{1}{8}$	0.405	27	0.180	0.260
$\frac{1}{4}$	0.540	18	0.200	0.401
$\frac{3}{8}$	0.675	18	0.240	0.408
$\frac{1}{2}$	0.840	14	0.320	0.534
$\frac{3}{4}$	1.050	14	0.340	0.546
1	1.315	$11\frac{1}{2}$	0.400	0.682
$1\frac{1}{4}$	1.660	$11\frac{1}{2}$	0.420	0.707
$1\frac{1}{2}$	1.900	$11\frac{1}{2}$	0.420	0.724
2	2.375	$11\frac{1}{2}$	0.436	0.756
$2\frac{1}{2}$	2.875	8	0.682	1.136
3	3.500	8	0.766	1.200

Note: ASA B2.1-1960.

installed in locations inaccessible to a wrench or screwdriver. An Allen screw can only be driven by a wrench of the correct size, so the socket does not wear so fast as the slot in a slotted-head screw. These screws have a relatively small bearing surface under the head and thus should be used with a washer.

Setscrews are used to fix one part in relation to another. They are often used to secure a hub to a shaft. In this application it is wise to put a flat on the shaft where the setscrew is to bear; otherwise the screw may mar the shaft, making it impossible for it to be withdrawn from the hub. Setscrews should not be used to lock a hub to a hollow shaft, since the force exerted by the screw will deform the shaft. In general, setscrews are suitable only for the transmission of small torque, and their use should be avoided if possible.

Machine screws shorter than 2 in. are threaded their entire length. Longer screws are only threaded for part of their length. Screws are usually only available with class 2A coarse or fine threads.

The type of head on a screw is usually designated by an abbreviation such as "HEX HD CP SCR" or "FILL HD MACH SCR." For example, a socket-head screw $1\frac{1}{2}$ in. long with an 8-32 thread is designated

$$8\text{-}32 \text{ UNC} \times 1\tfrac{1}{2} \text{SOC HD CAP SCR.}$$

The torque T required to produce a tension load F in a bolt of diameter D is

$$T = CDF,$$

where the coefficient C depends upon the state of lubrication of the threads. In general, C may be taken as 0.2. If the threads are oiled or coated with molybdenum disulfide, a value of 0.15 may be more accurate. If the threads are very clean, C may be as large as 0.4. The tension load on the bolt is equal to the compressive force exerted by the underside of the bolt head.

Steel bolts meeting SAE specifications are identified by markings on the head:

SAE grades 0, 1, 2:   no mark

SAE grade 3:

SAE grade 5:

SAE grade 6:

The strength of the bolt increases with the SAE grade number. Most common steel bolts are SAE grade 2. In

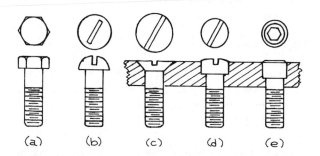

Figure 1.13   Common machine screws: (*a*) hex head; (*b*) round head; (*c*) flat head; (*d*) fillister head; (*e*) socket head.

sizes up to 1 in., these bolts have a yield strength of about 50,000 psi. SAE grade 5 bolts have a yield strength of about 80,000 psi.

When a bolt is to be tightened to a specified torque, the threads should be first seated by an initial tightening. It should then be loosened and retightened to the computed torque with a torque wrench. The torque corresponding to the yield strength should not be exceeded during this operation.

Because a bolt is usually made of a fairly hard material, excessive wear may result if it is frequently screwed into a threaded (tapped) hole in a soft material such as plastic or aluminum. Threaded inserts are made to alleviate this problem. These inserts are tightly wound helices of stainless steel or phosphor-bronze wire that have a diamond-shaped cross section. The insert is placed in a tapped oversize hole and the mating bolt is screwed into the insert. Special taps are required to prepare a hole for the insert, and a special tool is required to drive the insert into the hole.

To obtain maximum load-carrying strength, a steel bolt engaging an internal thread in a steel part should enter the thread to a distance equal to at least one bolt diameter. For a steel bolt entering an internal thread in brass or aluminum, the length of engagement should be closer to two bolt diameters.

1.3.5 Rivets

Rivets are used to permanently join sheet-metal or sheet-plastic parts together. They are frequently used when some degree of flexibility is desired in a joint, as when joining the ends of a belt to give a continuous loop. The most common rivet shapes are shown in Figure 1.14. Rivets are made of soft copper, aluminum, or steel. To join two pieces, a hole, slightly larger than the body of the rivet, is drilled or punched in each piece. A rivet is inserted through the holes, and a head is formed on the plain end of the rivet using a hammer or, preferably, a riveting machine. The hammering action swells the body of the rivet to fill the hole.

"Pop" rivets, illustrated in Figure 1.14, are useful in the lab. These can be installed without access to the back side of the joint. The mandrel is grasped by a

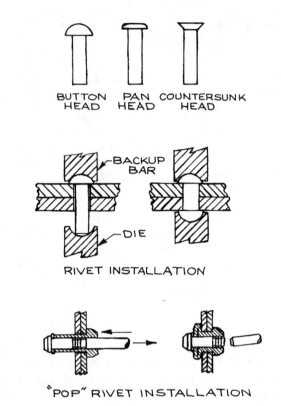

Figure 1.14 Rivets and riveted joints.

special rivet gun, the rivet is inserted into the hole, and the mandrel is pulled back until it breaks. The head of the mandrel rolls the stem of the rivet over to form a head, and the remaining broken portion of the mandrel seals the center hole of the rivet.

1.3.6 Pins

Pins such as are shown in Figure 1.15 are used to precisely locate one part with respect to another or to fix a point of rotation. To install a pin, the two pieces to be joined are clamped together and the hole for the pin is drilled and reamed.

Straight pins are made of steel that has been hardened and ground to a diameter tolerance of 0.0001 in. They require a hole that has been reamed to a close tolerance, since they are intended to be pressed into place. If a

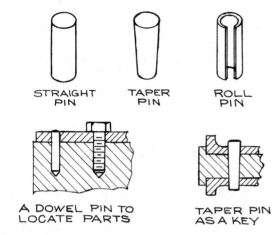

Figure 1.15 Pins.

small end threaded are drawn into place with a nut, which then secures the installed pin.

A *shoulder screw* is used as a pivot pin. These screws are hardened, and the shoulder is ground to a diameter tolerance of 0.0001 in.

1.3.7 Retaining Rings

A retaining ring serves as a removable shoulder on a shaft or in a hole to position parts assembled on the shaft or in the hole. An axially assembled external retaining ring is expanded slightly with a special pair of pliers, then slipped over the end of a shaft and allowed to spring shut in a groove on the shaft. An axially assembled internal ring is compressed, inserted in a hole and permitted to spring open into a groove. A radially assembled external retaining ring is forced onto a shaft from the side. It springs open as it slides over the diameter and then closes around the shaft.

Some retaining rings have a beveled edge, as shown in Figure 1.16, so that the spring action of the ring on the edge of its groove produces an axial load to take up unwanted clearances between assembled parts. This scheme is frequently used to take up the end play in a ball bearing. Another solution to end-play take-up is the use of a bowed retaining ring as shown in Figure 1.16.

straight pin is to serve as a pivot or if it is to locate a part that is to be removable, the hole in the movable part is reamed slightly oversize.

A *roll pin* has the advantage of ease of installation and removal and does not require a precision-reamed hole. The spring action of the walls of the pin hold it in place.

Taper pins are installed in holes that are shaped with a special reamer. The taper is 0.250 in. per ft. Plain taper pins are driven into place. Taper pins with the

Figure 1.16 Retaining rings and retaining-ring installations.

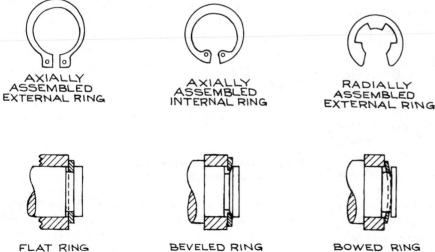

1.3.8   Adhesives

Epoxy resins are the most universally applicable adhesives for the laboratory. They are available in a variety of strengths and hardnesses. Epoxies that are either thermally or electrically conducting are also available. Epoxy adhesives consist of a resin and a hardener, which are mixed just prior to use. The mixing proportions and curing schedule must be controlled to achieve the desired properties. Tra-Con and Devcon manufacture a variety of epoxy adhesives that are prepackaged in small quantities in the correct proportions.

Epoxies will adhere to metals, glass, and some plastics. Hard, smooth surfaces should be roughened prior to the application of the adhesive. Sandblasting is a convenient method. Parts of a surface that are not to receive the adhesive can be masked with tape during this operation. To obtain maximum strength in an epoxied joint, the gap filled by the epoxy should be 0.002–0.006 in. wide. To maintain this gap between flat smooth surfaces, shims can be inserted to hold them apart.

Self-curing silicone rubber such as General Electric RTV can be used as an adhesive. RTV is chemically stable and will stick to most surfaces. The cured rubber is not mechanically strong; however, this can be an advantage when making a joint that may occasionally have to be broken.

Cyanoacrylate contact adhesives have found wide use in instrument construction. Eastman 910 and Techni-Tool Permabond are representative of this type of adhesive. These adhesives are monomers that polymerize rapidly when pressed into a thin film between two surfaces. They will adhere to most materials, including metal, rubber, and nylon. Cyanoacrylate adhesives are not void-filling and only work to bond surfaces where contours are well matched. An adhesive film of about 0.001 in. gives best results. A firm set is achieved in about a minute, and maximum strength is usually reached within a day. When joining metals and plastics, a shear strength in excess of 1000 psi is possible.

Sauereisen manufactures a line of ceramic cements. These are inorganic materials in powder, paste, or liquid form that may be used to join pieces of ceramic or to bond ceramic to metal. Some are designed to be used as electrically insulating coatings and others may be used for casting small ceramic parts. The tensile strength of these materials is only of the order of 500 psi, but the materials are serviceable up to at least 1100°C.

1.4   MECHANICAL DRAWING

Mechanical drawing is the language of the scientific designer. Initial ideas for a design are best expressed in terms of simple, full-scale drawings that develop in complexity and detail as the design matures. The construction of an apparatus is realized through communications to shop personnel in the form of working drawings of each part of the device. Success for the designer depends in large part on his command of the language.

1.4.1   Drawing Tools

A scientist who spends any more than five or ten percent of his time on apparatus design should acquire the set of tools listed in Table 1.6. The drawing board should be located in a well-lighted corner of an office or laboratory. Light from a north-facing window is most desirable. The drawing tools should be stored nearby. A scientist who does a considerable amount of work at the drawing board should consider acquiring a regular drafting table and draftsman's stool. A drafting machine for the drawing board instead of a T-square will improve efficiency considerably. It is also helpful to cover the drawing surface with graph sheet ruled in 0.1-in. squares or with a ruled surface sheet such as K&E Laminene.

1.4.2   Basic Principles of Mechanical Drawing

If mechanical drawing is the designer's language, then the line is the alphabet of this language. Some of the basic lines used in pencil drawing are illustrated in the

Table 1.6   DRAWING TOOLS FOR SCIENTIFIC DESIGN WORK

Semiautomatic drawing pencils with 4H, 2H, and H leads
Mechanical pencil pointer
Sandpaper pad
Erasing equipment:
    Ruby pencil eraser
    Artgum cleaning eraser
    Erasing shield
    Dust brush
Triangles:
    8 in. 45° triangle
    10 in. 30–60° triangle
T-square or drafting machine (preferred)
Engineer's scales:
    Scale with divisions in fractions of an inch
    Scale with divisions in tenths of an inch
Drawing board (at least 30 in. wide)
Drafting tape
Set of drawing instruments:
    Large (6-in.) bow compass with fixed needle-point leg and removable pencil and pen legs and extender beam
    Small ($3\frac{1}{2}$-in.) bow compass with fixed needle-point leg and removable pencil and pen legs
    Large friction-joint dividers
    Pen holder
100%-rag tracing paper:
    11×17 in. engineer's form
    17×22 in. engineer's form
    30-in.-wide roll

The above may be augmented by:
    Set of ship's curves
    Set of technical fountain pens (Koh-i-Noor, Rapidograph, or Leroy Nos. 1–4)
    Leroy lettering instrument and lettering templates
    Circle and ellipse templates

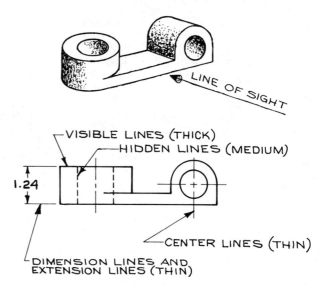

Figure 1.17   Some of the lines used in pencil drawing, exemplified in a side view of the object shown at the top.

plane view shown in Figure 1.17. Three line widths are employed: *thick lines* for visible outlines, short breaks, and cutting planes (explained below); *medium lines* for hidden outlines; *thin lines* for center, extension, dimension, and long break lines. H-grade lead may be used for heavy lines and lettering, 2H for medium and light lines, 4H for layout.

The shape of an object is described by *orthographic projection.* This is the view that one obtains when an object is far from the eye so that there is no apparent perspective. There are six principal views of an object, as illustrated in Figure 1.18. In America the relative position of the views must always be as shown there. Thus the view seen from the right side of an object is placed to the right of the front view. Of course, for clarity or convenience, the draftsman is initially free to choose any side of the object as the "front." As indicated in the illustration, not all views are required to completely describe an object. The draftsman should present only those views that are necessary for a complete description.

In some cases an *auxiliary view* will simplify a presentation. For example, a view normal to an inclined surface may be easier to draw and more informative than one of the six principal views. Such a case is illustrated in Figure 1.19. The relationship of the auxiliary view to the front view must be as shown. That is, the auxiliary view of an inclined surface must be placed in the direction of the normal to the surface.

Often the internal structure of an object cannot be clearly delineated with hidden edges indicated by dashed lines. In this case it is useful to show the view that would be seen if the object were cut open. Such a view

Figure 1.18 The six principal views of an object. The views not required to describe the object in this particular case have been crossed out.

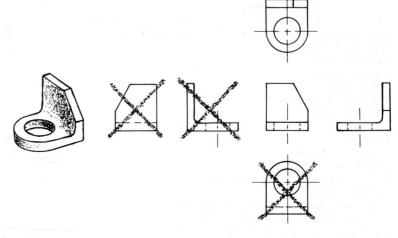

is called a *section*. Several examples of full sections are shown in Figure 1.20. The *cutting plane* in the section is represented by cross-hatching with fine lines to suggest a saw cut. The cross-hatch lines should always be at an oblique angle to the heavy lines that indicate the outline of the section. Cross-hatch lines on two adjacent pieces should be in opposite directions. The cutting plane in a view perpendicular to the section is shown by a heavy line of alternating long and two short dashes. Arrows on the cutting-plane line indicate the direction of sight from which the section is viewed. Sections may be identified with letters at each end of the cutting-plane line.

It is often convenient to have the cutting plane change directions so that it passes through several features that are desired to be shown in section. Angled surfaces are then revolved into a common plane and offset surfaces are projected onto a common plane to give an *aligned section* as illustrated in Figures 1.21 and 1.22, rather than a true projection of the cutting plane.

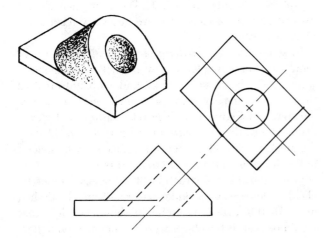

Figure 1.19 Effective use of an auxiliary view.

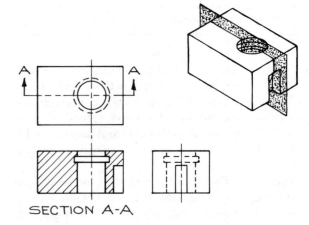

SECTION A-A

Figure 1.20 Full sections cut along the indicated planes.

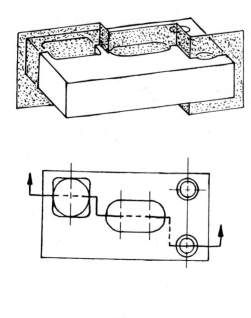

Figure 1.21   An aligned section.

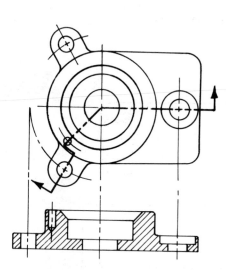

Figure 1.22   An aligned section.

Other convenient sectioning techniques include the *half section* of a symmetrical object as shown in Figure 1.23 and the *revolved section* of a long bar or spoke as shown in Figure 1.24.

Many details, such as screw threads, rivets, springs, and welds, are so tedious to draw, and appear so often, that they are designated by symbols rather than faithful representations. Symbols for thread are illustrated in Figure 1.25.

Drawings of parts too long to be conveniently fitted on a piece of drawing paper can be represented as

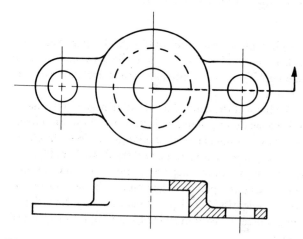

Figure 1.23   A half section.

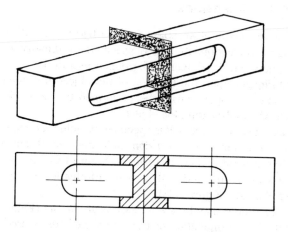

Figure 1.24   A revolved section.

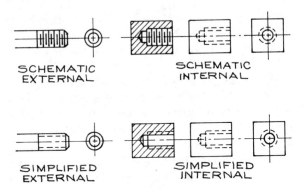

Figure 1.25   Thread symbols.

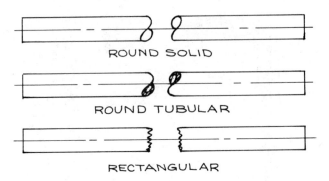

Figure 1.26   Conventional breaks.

though a piece from the middle of the part had been broken out and discarded and the two ends moved together. Three conventional breaks are illustrated in Figure 1.26.

Mechanical drawings include dimensions, tolerances, and special descriptions and instructions in written form. Lettering on mechanical drawings is done freehand using only uppercase letters. Letters about $\frac{1}{4}$ in. high are usually appropriate. Lettering must be uniform and legible. Lightly penciled guidelines should always be used. A scientist should plan to spend some time developing a neat, consistent style of lettering for mechanical drawing.

1.4.3   Dimensions

After the shape of an object is described by an orthographic projection, its size is specified by dimensions and notes on the drawing. The dimensions to be specified depend upon the function of the object and upon the machine operations to be performed by the workman when fabricating the object.

Distance on a drawing is specified by either a dimension or a note. A *dimension* indicates the distance between points, edges, or surfaces (Figure 1.27). A *note* is a written instruction that gives information on the size of a part (Figure 1.28).

Dimensions may be given in *series* or *parallel* (Figure 1.29). If a combination of these two methods is used, care must be taken to ensure that the size of the object is not overdetermined. An example of an overdimen-

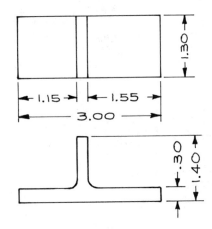

Figure 1.27   Dimensions.

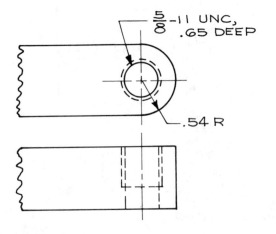

Figure 1.28   A note.

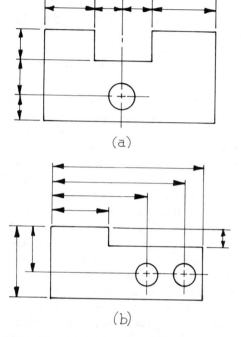

Figure 1.29 Dimensions: (*a*) series dimensions; (*b*) parallel dimensions.

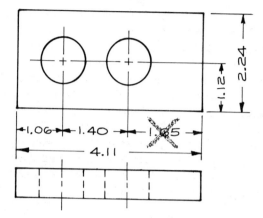

Figure 1.30 An overdimensioned drawing. One of the dimensions is superfluous.

sioned drawing is given in Figure 1.30. Dimensions should not be duplicated, and a drawing should include no more dimensions than are required. When several features of a drawing are obviously identical, it is only necessary to dimension one of them.

Dimensions are often specified with respect to a *datum*. This is a single feature whose location is assumed to be exact. When choosing a datum, consider the functional importance of locating other features of an object relative to the datum, as well as the ease with which the workman can locate the datum itself.

If possible, all dimensions should be placed outside of the view with extension lines leading out from the view. Dimensions within the view are permissible to eliminate long extension lines and if the dimension does not obscure some detail of the view. The dimension closest to the outside of a view should be about $\frac{1}{2}$ in. from the edge. Successive dimensions should be placed at $\frac{3}{8}$- to $\frac{1}{2}$-in. intervals. Dimension lines never cross another line. Dimension values should be staggered as in Figure 1.31 rather than stacked vertically or horizontally. Dimension

values are oriented so that they can be read from the bottom or the right side of the drawing. Notice that the extension lines do not touch the outline of the view, but rather stop about $\frac{1}{16}$ in. short.

The units of a dimension are not ordinarily specified on a drawing. In America all dimensions are assumed to be in inches. In Europe they are assumed to be in millimeters.

Notes are used to specify the size of standard parts and parts too small to be dimensioned. Notes are also used to specify a feature that is derived from a standard

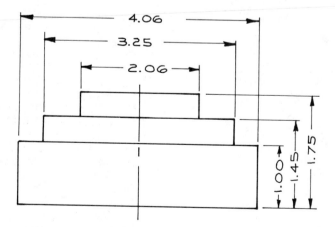

Figure 1.31 Placement of dimensions and dimension values.

operation such as drilling, reaming, tapping, or spotfacing. Notes such as "TAP 10-32 UNF, .75 DEEP" or "$\frac{1}{2}$ DRILL, SPOTFACE .75D×.10 DEEP," which refer to a specific feature, are accompanied by a leader emanating from the beginning or end of the note and terminating in an arrowhead that indicates the location of the feature (Figure 1.28). General notes, such as "REMOVE BURRS" or "ALL FILLETS $\frac{1}{4}$ R," which refer to the whole drawing, do not require a leader.

Many abbreviations are used in writing dimensions and notes. Common abbreviations used on mechanical drawings are given in Table 1.7.

1.4.4   Tolerances

One of the most important parts of design work is the selection of manufacturing tolerances so that a designed part fulfills its function and yet can be fabricated with a minimum of effort. The designer must carefully and thoughtfully specify the function of each part of an apparatus in order to arrive at realistic tolerances on each dimension of a drawing. It is important to appreciate the capabilities of the machines that will be used in manufacture and the time and effort required to maintain a given tolerance. As indicated in Figure 1.32, the cost of production is about inversely related to the tolerances.

Usually only one or two tolerance specifications will apply to most of the dimensions on a drawing. It is then convenient to encode the tolerance specification in the dimension rather than affix a tolerance to each dimension. In instrument design most dimensions are written as decimals. The tolerance on a dimension can then be indicated by the number of decimal places of the dimension. Dimensions expressed to two decimal places have one tolerance specification, to three decimal places another, and so on. The specifications are given in a note. An example of this system is given in Figure 1.33.

Fractions may be used to give the size of some features derived from standardized operations, such as drilling or threading. Usually a tolerance specification is unnecessary in these cases, since both the designer and workman should know what precision can be expected.

Table 1.7   ABBREVIATIONS USED ON MECHANICAL DRAWINGS

Word	Abbr.	Word	Abbr.	Word	Abbr.
Bearing	BRG	Fillister	FIL	Right hand	RH
Bolt circle	BC	Grind	GRD	Rivet(ed)	RIV
Bracket	BRKT	Groove	GRV	Round	RD
Broach(ed)	BRO	Ground	GRD	Root mean square	RMS
Bushing	BUSH	Head	HD	Screw	SCR
Cap screw	CAP SCR	Inside diameter	ID	Socket	SOC
Center line	CL	Key	K	Space(d)	SP
Chamfer	CHAM	Keyway	KWY	Spot-face(d)	SF
Circle	CIR	Left hand	LH	Square	SQ
Circumference	CIRC	Long	LG	Stainless	STN
Concentric	CONC	Maximum	MAX	Steel	STL
Counterbore	CBORE	Minimum	MIN	Straight	STR
Counterdrill	CDRILL	Not to scale	NTS	Surface	SUR
Countersink	CSK	Number	NO	Taper(ed)	TPR
Cross section	XSECT	Opposite	OPP	Thread(ed)	THD
Diameter	DIA or D	Outside diameter	OD	Tolerance	TOL
Drawing	DWG	Pipe tap	PT	Typical	TYP
Drill(ed)	DR	Pipe thread	PT	Vacuum	VAC
Each	EA	Press	PRS	Washer	WASH
Equal(ly)	EQ	Punch	PCH	With	W/
Fillet	FIL	Reference line	REF	Without	W/O

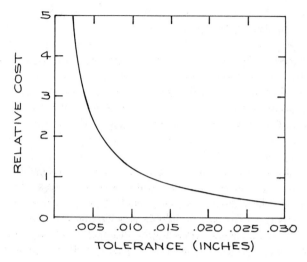

Figure 1.32 Approximate relation between tolerance and the cost of production.

There are two methods of expressly stating the tolerance on a dimension. The allowable variation, plus and minus, may be stated after the dimension. Alternatively, the limits of the dimension can be stated without giving

a tolerance at all. Examples of these two methods are given in Figure 1.34.

When the absolute size of two mating parts is not important but the clearance between them is critical, specify the desired *fit*. Fits are classified as *sliding fits, clearance locational fits, transition fits, interference fits,* and *force fits*. The uses of the various classes of fits and tables of the tolerances required to give such fits are given in standard texts on mechanical drawing.

Beware of an undesirable accumulation of tolerances. In series dimensioning, the tolerance on the distance between two points separated by two or more dimensions is equal to the sum of the tolerances for each of the intervening dimensions. In Figure 1.29(*a*) the tolerance on the lateral location of the hole with respect to the right side of the object is given by the sum of two tolerances. For parallel dimensions, the location of each feature relative to the datum depends on only one tolerance; however, the difference in the distance between two locations depends upon two tolerances. In Figure 1.29(*b*), the tolerance on the distance between the two holes is given by the sum of the tolerances on the dimensions which locate the holes.

As much as possible, dimensions and tolerances should

Figure 1.33 A decimal system for specifying tolerances.

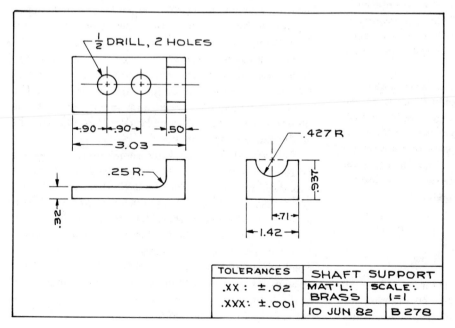

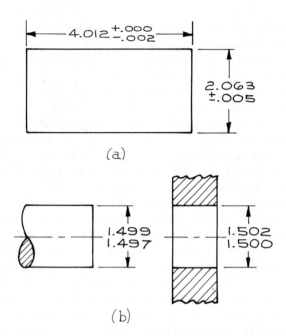

Figure 1.34   Two methods of stating the tolerance on a dimension: (*a*) as the allowable variation on the dimension; (*b*) in terms of the allowable limits of the dimension.

be chosen so that materials can be used in their stock sizes and shapes. Do not call for unnecessary finishing. For example, the area around a hole in a casting or other rough piece of metal can be spotfaced to provide a bearing surface for a bolt if the quality of the rest of the surface is not critical.

The dimensions and tolerances on a finished drawing should be carefully checked. Imagine yourself as the machinist who must use the drawing to make a part. Proceed mentally through each step of the fabrication to see that all necessary dimensions appear on the drawing and that the specified dimensions are easy to use. Also satisfy yourself that all desired tolerances are within the limits of the required machine-tool operations.

1.4.5 From Design to Working Drawings

The first step in designing an instrument is to draw the entire assembly. Work to scale. Start by drawing the largest parts of the instrument, and then add smaller parts. Most dimensions, tolerances, and fine details may be omitted at this point.

Pay attention to the manner in which pieces fit together. Will the apparatus be convenient to assemble and disassemble? Are some parts much stronger or weaker than required? Does a strong part depend upon a weak one for location or support? Are all shapes as simple as possible? Are standard shapes of materials and standard bolts, couplings, bearings, and shafts used wherever possible?

Time spent at the drawing board mentally assembling and disassembling a device will save hours of frustration in the laboratory. Careful consideration of production techniques can save hundreds of dollars in the shop.

When a full-size orthographic drawing of the apparatus is complete, it is wise to discuss it with the people who will do the work.

After the design has reached its final form, the designer must make a working drawing of each piece of the apparatus. A working drawing is a fully dimensioned and toleranced drawing to be used in the shop. Work to scale if possible. Leave the design drawing attached to the board and place a piece of tracing paper over the part to be drawn. This system speeds work and helps prevent errors when transferring information from the design drawing to the working drawing. To check a working drawing of a particular part it is also useful to lay it over the drawing of each mating part to see that they fit together properly.

The critical test of a set of working drawings is their usefulness. All necessary communication between scientist and machinist should appear on the drawings. The machinist should not require oral instructions, and he should not be required to make decisions affecting the performance of an instrument.

The original drawings should remain in the possession of the designer. Copies of the working drawings are submitted to the shop. Pencil drawings on tracing paper are usually reproduced by the diazo-dry process. This is a contact printing method. The original drawing is placed over a special light-sensitive paper and exposed to ultraviolet light. The sensitized paper is then developed by ammonia vapors, which produce a blue-line copy of the original. Machines such as the Ozalid Copier perform these operations automatically. Most shops are equipped with a print copier.

In the above discussion we have assumed that the scientific designer has the services of a model shop. This is the case in industrial laboratories and in many university laboratories. However, a designer should follow the same procedure when he fabricates his own apparatus. All questions of sizes, tolerances, and fits must be answered before construction. When a scientist attempts to make these decisions as he proceeds with construction, the results are inevitably poor.

1.5   PHYSICAL PRINCIPLES OF MECHANICAL DESIGN

No material is perfectly rigid. When any member of a machine is subjected to a force, no matter how small, it will bend or twist to some extent. A member subjected to a force that varies in time will vibrate. A designer must appreciate the extent of deflection of mechanical parts under load.

1.5.1   Bending of a Beam or Shaft

When a beam bends, one side of the beam experiences a tensile load and the other a compressive load. Consider a point in the flexed beam in Figure 1.35. The stress at this point depends upon its distance, c, from the centroid of the beam in the direction of the applied force. The *centroid* is the center of gravity of the cross section of the beam [see Figure 1.35(b)]. The stress is given by

$$s = \frac{Mc}{I},$$

where M is the bending moment at the point of interest

and I is the centroidal moment of inertia of the section of the beam. Clearly the stress in a flexed beam is greatest at the outer surface of the beam.

The *centroidal moment of inertia* of the section (more correctly known as the second moment of the area of the cross section) is taken about an axis that passes through the centroid and in the direction perpendicular to the applied force. For the rectangular section in Figure 1.35(b)

$$I = \int_{-B/2}^{B/2} \int_{-H/2}^{H/2} h^2 \, dh \, db = \frac{BH^3}{12},$$

where B is the width of the section and H is the height. The centroidal moments of inertia of some common symmetrical sections are given in Figure 1.36.

The *bending moment* is proportional to the curvature produced in a beam by the applied force:

$$M = EI \frac{d^2y}{dx^2},$$

where y is the deflection produced by the force and E is the modulus of elasticity of the beam. For the example given in Figure 1.35, assuming the weight of the shaft can be ignored, the bending moment is simply

$$M = -F(L - x).$$

The bending moments for this and other common systems are given in Figure 1.37.

The deflection of a stressed beam depends upon the modulus of elasticity of the material. Expressions for the deflection of both point-loaded and uniformly loaded beams are given in Figure 1.37.

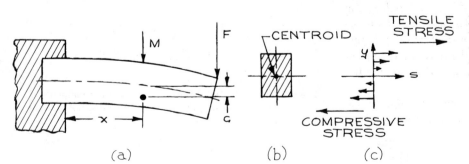

Figure 1.35   A flexed beam: (a) the beam bending under a load; (b) a cross section of the beam showing the centroid; (c) the distribution of shear forces along a vertical line through the centroid of the beam.

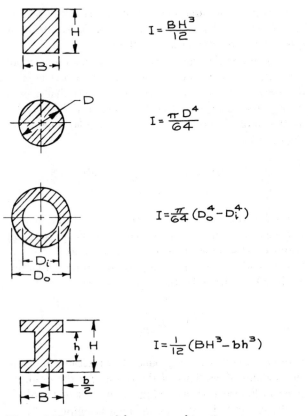

Figure 1.36   Centroidal moments of inertia.

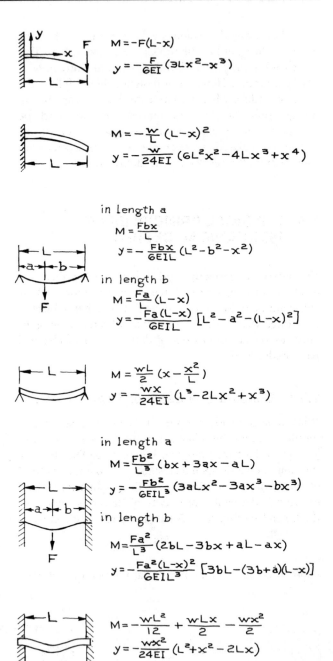

Figure 1.37   Bending formulae: M = bending moment; E = modulus of elasticity; I = centroidal moment of inertia; w = weight per unit length.

If the bending moment on a beam is sufficiently large, it will be permanently deformed. A conservative relation between the bending moment at the yield point, M_y, and the yield strength of the material, s_y, is given by $M_y = s_y I / c_{max}$. Actually, the yield strength for a beam in bending is greater than for a beam in tension, since the inner fibers of the material must be loaded to their yield point to a considerable depth before the beam is permanently deformed. To determine the maximum allowable force on a beam or shaft one simply sets M_y equal to the expression for the maximum value of the bending moment on the beam and solves for the force.

The preceding discussion assumes that the tensile strength of a material is the same as the compressive strength. For materials such as cast iron, where this is not true, the bending equations are much more complicated. The foregoing also ignores shear stresses in a loaded beam. For very short beams, shear stresses become important and the shear strength of the material must be considered.[3]

1.5.2 Twisting of a Shaft

The stress at a point in a round shaft subjected to a torsional load is

$$s = \frac{Tc}{J}, \qquad J = \frac{\pi}{32}\left(D_o^4 - D_i^4\right),$$

where c is the distance of the point of interest from the center of the shaft and T is the applied torque. J is the centroidal polar moment of inertia of the section of the shaft, and D_o and D_i are the outer and inner diameters of the shaft.

The total angle of twist in a shaft of length L is

$$\theta = \frac{57LT}{GJ} \text{ degrees,}$$

where G is the modulus of elasticity in shear or *shear modulus* of the material of the shaft. For metals, the shear modulus is about $\frac{1}{3}$ the elastic modulus.

1.5.3 Stress Relief

An abrupt change in the cross section of a shaft produces a concentration of stresses as shown in Figure 1.38(a). Stresses are increased at steps, grooves, keyways, holes, dents, and scratches. A stress concentration is an area where there is a large gradient in the stress. Material tends to fail because of the shear forces at a stress concentration. Anyone who has ever broken a bolt is familiar with this effect: the bolt invariably fails just where the head joins the shaft. Methods of relieving stresses at a step in a shaft are suggested in Figure 1.38(b), (c), and (d).

There are a number of means of detecting stress in a part. One of the most useful is the *photoelastic method*.[4]

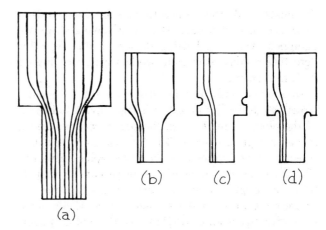

(a) (b) (c) (d)

Figure 1.38 (*a*) Stress concentrations in a shaft. Lines indicate surfaces of constant stress. Parts (*b*), (*c*), and (*d*) show methods of relieving stresses at a step in a shaft; (*c*) and (*d*) are useful when a shoulder is required to locate a bearing.

In this technique a transparent plastic model is stressed in the same way as the element of interest without necessarily duplicating the magnitude of the stress. The model is illuminated with polarized, monochromatic light and viewed through a polarizer. As shown in Figure 1.39, the stress distribution appears as fringes in

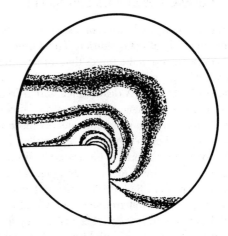

Figure 1.39 Stress distribution in a plastic model as observed by the photoelastic method.

the image, and stress concentrations are characterized by closely spaced fringes. Most glassblowing shops are equipped with a polarizing device of this type for detecting residual stresses in worked glass (see Figure 2.16, next chapter).

In a similar fashion, fabricated parts can be nondestructively tested by means of holography. A hologram of the object is made; then the part is stressed and a second hologram is made. The two holograms are then superposed and placed in the holographic projector. The resulting three-dimensional image consists of patterns of fringes, which are densest in the areas of greatest distortion in the stressed object. Holographic nondestructive testing (HNDT) is carried out as a commercial service by materials-testing laboratories such as Jodon Engineering.

A particularly simple and inexpensive means of locating stress concentrations in a working part has been developed by Magnaflux. The part to be tested is sprayed with Stresscoat, which forms a hard, brittle layer over the surface. The part is returned to service for a time, and any flexing of the part causes the surface coating to crack. The part is then removed and dipped in a dye, which permeates the cracks to reveal the areas of greatest strain.

1.5.4 Vibration of Beams and Shafts

In many instruments, vibration of a supporting beam or shaft can adversely affect operation. The amplitude of a vibration depends upon the forces that initiate and drive the vibration. All mechanical elements have natural frequencies of vibration. A disturbance with a frequency near this natural frequency will be more effective in causing a vibration than one which occurs more rapidly or more slowly. The response of an oscillator to the frequency of a driving force is illustrated in Figure 1.40.

There are two ways for the designer to prevent destructive vibrations. The critical element can be designed so that its natural frequency of vibration is far removed from the frequency of a disturbing force, or the disturbing force can be inhibited in the vicinity of the natural frequency of the critical element. The latter

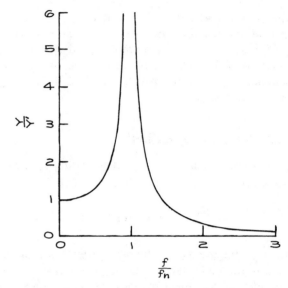

Figure 1.40 Response of a machine element to a periodically varying force. Y' is the amplitude of vibration of the element, and f_n is its natural frequency of vibration. Y and f are the amplitude and frequency of the driver.

scheme is called damping. Very low-frequency vibrations can be damped by a hydraulic or friction shock absorber such as is employed on automobile suspensions. High frequencies can be damped by coupling the objectionable source of vibration to its surroundings through a member that has a very low natural frequency of vibration. Mounts made of rubber or cork are very effective for absorbing high-frequency vibrations and sudden shocks, since these materials possess a low modulus of elasticity.

The natural frequency of vibration (in Hz) of a shaft or beam with one or more concentrated masses attached is given by

$$f_n^2 = \frac{\dfrac{a}{4\pi^2} \displaystyle\int_0^L \dfrac{M^2}{EI}\,dx}{\displaystyle\int_0^L a\rho A Y^2\,dx + \sum_i F_i Y_i^2},$$

where

$M = M(x) = maximum$ bending moment at x, i.e., the amplitude of the bending moment induced by the forces on the shaft at x,

$a =$ inertial acceleration experienced by the shaft and its weights (for a nonrotating shaft in the gravitational field of the earth, $a = g$),

$E =$ modulus of elasticity of shaft material,

$\rho =$ mass density of the shaft ($g\rho$ is the weight density of the shaft),

$A =$ cross-sectional area of the shaft,

$Y = y(x) =$ maximum deflection of the shaft at x, i.e., the amplitude of vibration at x,

$F_i =$ force exerted by the ith mass (for a nonrotating shaft in the earth's field, this is the weight of the ith mass),

$Y_i =$ maximum deflection at the location of the ith mass.

For a system similar to one of the static deflection cases treated in Figure 1.37, one need only substitute the appropriate expressions for the bending moment and deflections and solve the resulting equation to determine the natural frequency. Consider for example a beam of negligible mass, which is supported at one end and supports a mass of weight F at its free end. This is the case illustrated in Figure 1.35. The natural frequency of vibration is

$$f_n^2 = \frac{\dfrac{g}{4\pi^2} \displaystyle\int_0^L \dfrac{(-Fx)^2}{EI}\,dx}{F\left[-\dfrac{F}{GEI}(-2L^3)\right]^2},$$

$$f_n = \frac{1}{2\pi}\left(\frac{3gEI}{FL^3}\right)^{1/2}\ \text{Hz}.$$

In general, the shape of the deflection curve is not known. However, the assumption of some reasonable deflection curve usually will provide a useful result. In most cases one can assume a sine curve for a shaft supported at both ends, or a parabola for a shaft supported at only one end. Alternatively, one can employ one of the rather complicated graphical methods outlined in many engineering texts on vibration.[5]

Consider once again Figure 1.35. A parabolic deflection curve with the correct limit properties is given by

$$y = -\frac{y_0}{L^2}(L - x)^2,$$

where y_0 is the maximum deflection of the free end. The bending moment is then

$$M = EI\frac{d^2y}{dx^2}$$

$$= -2EI\left(\frac{y_0}{L^2}\right).$$

Substituting in the equation for the natural frequency gives

$$f_n^2 = \frac{\dfrac{g}{4\pi^2}\displaystyle\int_0^L \dfrac{\left[-2EI(y_0/L^2)\right]^2}{EI}\,dx}{F(-y_0)^2},$$

$$f_n = \frac{1}{\pi}\left(\frac{gEI}{FL^3}\right)^{1/2}\ \text{Hz},$$

in reasonable agreement with the exact solution derived above.

For the special case of a heavy shaft or beam with a centrally placed load, a correct solution can be derived by assuming the shaft to be weightless and adding half the weight of the shaft to the concentrated load.

1.5.5 Shaft Whirl and Vibration

A shaft and rotor can never be perfectly balanced, and the axis of rotation can never be precisely located by the shaft bearings. As a shaft rotates, centrifugal force on the unbalanced mass deflects the shaft and causes it to *whirl* around its axis of rotation. This whirling motion appears as a vibration to a stationary observer, and can be treated as such. At the natural frequency of the shaft, the whirl becomes violent and can destroy the shaft or its bearings. This critical condition occurs when the shaft speed (in rps) equals the natural frequency (in Hz) of the shaft. Thus one need only compute the natural frequency of a shaft and rotor assembly in order to determine the critical speed.

When using the equation for the natural frequency of a shaft to determine the critical speed of rotation, it is not necessary to determine the centrifugal load. This is because the load F always appears as the ratio F/a and

$$\frac{F}{a} = \frac{W}{g} = m,$$

where W is the weight of the element which is exerting the load.

The critical speed for a rotor on a weightless shaft supported on thin bearings is

$$f_c = \frac{0.276}{ab}\left(\frac{gEIL}{W}\right)^{1/2} \text{ rps},$$

where W is the weight of the rotor, L is the length of the shaft, and a and b are the distances from the rotor to the bearings. If the ends of the shaft are rigidly supported in long bearings the critical speed is

$$f_c = \frac{0.276L}{ab}\left(\frac{gEIL}{abW}\right)^{1/2} \text{ rps}.$$

Notice that f_c can be increased by placing the rotor near one end of the shaft so that a or b is small. In this case the rotor acts as a gyro tending to stiffen the shaft.

For an unloaded shaft on thin bearings the critical speed is

$$f_c = \frac{1.57}{L^2}\left(\frac{EI}{\rho A}\right)^{1/2} \text{ rps},$$

and for an unloaded shaft on long bearings

$$f_c = \frac{3.57}{L^2}\left(\frac{EI}{\rho A}\right)^{1/2} \text{ rps},$$

where ρ is the mass density ($g\rho$ is the weight density).

Once again, the special case of a centrally placed rotor on a heavy shaft can be treated by adding half the weight of the shaft to the rotor and then assuming the shaft to be weightless.

The designer must manipulate the dimensions and material of a shaft and rotor, and the location of the rotor, so that its critical speed does not coincide with the design speed of an assembly. It is usually sufficient for these two speeds to differ by a factor of $\sqrt{2}$. The most desirable situation is for the critical speed to exceed the design speed. If this is not possible, the drive motor for the rotating assembly should be so powerful that the shaft accelerates very rapidly through the region of the critical speed. The drive motor will require an excess of power because considerable power will be lost to vibration at the critical speed.

At high speeds there is the possibility of encountering shaft whirl at modes above the fundamental. A shaft with a number of weights has as many different modes of vibration as it has weights. An unloaded shaft has a critical speed at every integral multiple of the fundamental.

For rotational speeds far away from a critical speed a rotating assembly can be considered to be rigid. Nevertheless, imbalance can lead to inertial (centrifugal) forces that may damage the shaft or its bearings. If the unbalanced mass lies within a plane perpendicular to the axis of rotation, as for the case of a thin rotor on a shaft, the rotor can be statically balanced. The condition for *static balance* is that the axis of rotation pass through the center of gravity. In practice this can be accomplished by placing the shaft with its rotor across a pair of knife-edge rails. The shaft will roll or rock until coming to rest with the unbalanced inertial force pointed radically downwards. Weight is then added or removed along the upward pointing radius until balance is achieved.

If there are unbalanced masses acting at different points along the axis, then dynamic balancing is required. The conditions for *dynamic balance* are that the axis of rotation pass through the center of gravity and that the axis of rotation coincide with a principal axis of the inertial force. Dynamic balancing is difficult without specialized equipment. However, it is done inexpensively and accurately by commercial firms such as Balance Technology.

1.6   CONSTRAINED MOTION

In addition to a rigid support structure, most machines include some moving parts. The motion of these parts must usually be constrained in some fashion to obtain the required function. Most often the desired motion is translation in a plane or along a line, or rotation about one or more axes. The design of the constraining mechanism for a moving part must meet dimensional, strength, and wear-resistance specifications.

To achieve constrained motion the designer must either create a support structure for the moving part that will give the desired alignment or make use of a standard machine element that is produced commercially. In the former case, it is necessary for the designer to understand the principles and limitations of geometric design. In the latter, he or she must be familiar with the range and precision of commercial bearings, shafts, sliders, and so on. The economics of the choice should not be ignored.

1.6.1   Kinematic Design

A rigid body has six independent degrees of freedom of motion. These are usually taken to be the translational motions along three orthogonal axes and the rotational motions about these axes. Any motion can be described as a linear combination of these six motions. The position of a body at any time is defined by six coordinates: three position coordinates and three angle coordinates. Motion is constrained and the number of degrees of freedom is reduced if any one of these coordinates is fixed or if some linear combination of these coordinates is fixed. For example, a rigid object will be constrained to move in a plane if one of its position coordinates is fixed. Kinematic design for constrained motion consists of constraining a point on a body for each degree of freedom to be constrained.[6]

In general, one degree of freedom can be removed by constraining a point on a body to remain in contact with a reference surface. It must be kept in mind that this degree of freedom does not necessarily correspond to a motion of one of the six coordinates arbitrarily chosen to define the position of a body. The lost degree of freedom may be some linear combination of the reference motions. The following examples illustrate this principle of kinematic design:

1. Consider a sphere constrained to maintain point contact with each of two reference planes: a ball resting in a V-groove. The ball is free to rotate about any axis but it can only translate along a line parallel to the intersection of the planes. The ball has lost two degrees of freedom.

2. A body that is constrained to maintain three points of contact with a plane surface has only three degrees of freedom: translation in two directions parallel to the plane and rotation about an axis perpendicular to the plane.

3. A body in contact at three points with a cylindrical surface has three degrees of freedom: translation in the direction of the cylinder axis, revolution about the cylinder axis, and rotation about an axis perpendicular to the plane defined by the three points of contact.

A complication exists if two contact points are degenerate in the sense that the function of the two points can be served by a single point of contact. Consider a four-legged table *vis-à-vis* a three-legged table. One of the four legs provides a degenerate point of contact with the floor. The fourth leg may provide some much-needed stability, but it also introduces uncertainty in the location of the table top unless all four legs are of precisely the same length. The object of kinematic design is to permit motion to be constrained without depending upon precision of manufacture. Of course, the sort of considerations that lead to the production of four-legged tables often dictates a semikinematic or degenerate kinematic design for other devices as well.

Examples of kinematic design are illustrated in Figures 1.41–1.43.

The ball feet on the carriage shown in Figure 1.41 provide five points of contact between the carriage and its base. Thus the carriage is only free to slide in one direction. One problem with this design is that it is

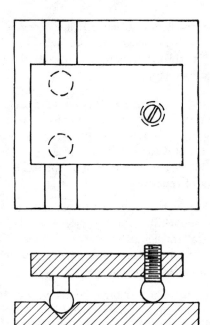

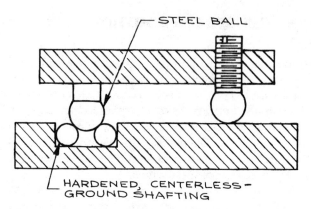

Figure 1.42 An improved version of the design shown in Figure 1.41.

Figure 1.41 Kinematic design that constrains a carriage to move in a straight line.

difficult to mill a V-groove with smooth surfaces. The surface quality of the groove can be improved by lapping with carborundum, using a V-shaped brass lap that fits in the groove.

An improved version of the design in Figure 1.41 is shown in Figure 1.42. The V-groove has been replaced by a pair of steel rods, and the ball feet of the carriage have been replaced by rolling balls. This design is both more precise and more economical than the previous one. Stainless-steel shafting that has been hardened and centerless-ground to a diameter tolerance of ± 0.00005 in. is available at a cost of less than a dollar a foot. Stainless-steel balls with a roundness tolerance of ± 0.00005 in. are also inexpensive. The channel that contains the rods must be milled, but the surface quality in this channel is not critical. In a milling operation, the sides of the channel can easily be kept straight and parallel to within 0.0001 in./ft.

The three grooves in the platform shown in Figure 1.43 allow precise relocation of a three-legged table after

it has been removed. The ball feet of the table make six points of contact with the platform, so that there are no degrees of freedom.

A semikinematic design to give motion about a single axis is shown in Figure 1.44. If only three balls are used, there are six points of contact with the lower groove: one more than is required for a true kinematic arrangement. This design provides smooth rotary motion, particularly if the grooves are cut as shown,[7] so that the balls roll rather than slide as they would in a symmetrical groove. The circular grooves are cut on a lathe, and much better surface quality can be obtained than for the straight, milled grooves mentioned above.

The difficulty with the design of Figure 1.44 is that the location of the axis of rotation is indeterminate to the extent that the radii of curvature of the two grooves are not equal. If the radii of the grooves are not the same, the centerline of the lower circular groove may not coincide with the centerline of the upper groove, and the axis of rotation can be anywhere between the two centers. This problem can be somewhat alleviated if the bearing is "run in" under load, so that the balls wear a shallow indentation in the faces of the grooves and thus establish a preferred axis of rotation.

To obtain rotation about a fixed axis it is generally less expensive and more accurate to use commercially produced ball bearings. Properly installed, a good ball bearing, costing only a few dollars, will maintain a

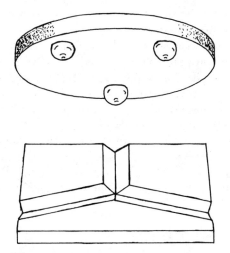

Figure 1.43 Kinematic design that permits an accurately located part to be removed and replaced in the same position.

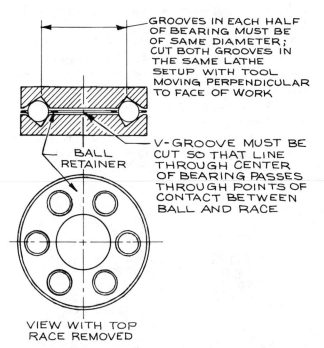

GROOVES IN EACH HALF
OF BEARING MUST BE
OF SAME DIAMETER;
CUT BOTH GROOVES IN
THE SAME LATHE
SETUP WITH TOOL
MOVING PERPENDICULAR
TO FACE OF WORK

BALL
RETAINER

V-GROOVE MUST BE
CUT SO THAT LINE
THROUGH CENTER
OF BEARING PASSES
THROUGH POINTS OF
CONTACT BETWEEN
BALL AND RACE

VIEW WITH TOP
RACE REMOVED

Figure 1.44 Semikinematic design that allows rotation about a single axis. If the grooves are cut as indicated, the balls will always be in rolling contact with the grooves.

tolerance of ± 0.0001 in. on the location of a rotating part.

The design of Figure 1.44 can be used when a bearing must be constructed of exotic materials. For example, in high-vacuum work, ball bearings with ceramic or sapphire balls and stainless-steel races are used because steel balls in a steel race tend to cold-weld when they are very clean, as they would be for a vacuum application. In practice, more than three balls can be used. Since this is a semikinematic design, the addition of more balls will increase the bearing strength without significantly decreasing accuracy.

1.6.2 Plain Bearings

A bearing is a stationary element that locates and carries the load of a moving part. Bearings can be divided into two categories depending upon whether there is sliding or rolling contact between the moving and stationary parts. Sliding-contact bearings are called *plain bearings*. Rolling bearings will be discussed in the next section.

Plain bearings may be designed to carry a radial load, an axial load, or both. Different types are illustrated in Figure 1.45. A radial bearing consists of a cylindrical shaft or *journal* rotating or sliding within a shell, which is the *bearing* proper. The entire assembly is referred to as a *journal bearing*. An axial bearing consists of a flat bearing surface, like a washer, against which the end of the shaft rests. These are called *thrust bearings*. A journal bearing may incorporate a *flanged journal*, in

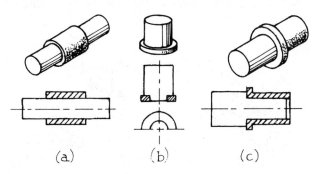

(a)   (b)   (c)

Figure 1.45 Different types of plain bearings: (*a*) a journal bearing; (*b*) a thrust bearing; (*c*) a flanged journal bearing.

which case it will support a radial load as well as an axial load.

A journal is usually hardened steel or stainless steel. Precision-ground shafts and shafts with precision-ground journals are available in diameters from $\frac{1}{32}$ to 1 in. Bearing shells of bronze or oil-impregnated bronze are available to fit. Nylon bearings for light loads and oil-free applications are also available. Commercial shafting and bearings are manufactured to provide a clearance of 0.0002–0.0010 in.

Many different methods of lubrication can be employed instead of oil impregnation. The inner surface of the bearing can be grooved, and oil or grease can be forced into the groove through a hole in the shell. If oil is objectionable, a groove on the inner surface of the bearing can be packed with molybdenum disulfide or other dry lubricant.

A plain bearing is installed by pressing it into a hole in the supporting structure. An interference of about 0.001 in. is desirable for bearings up to an inch in diameter. That is, the outer diameter of the bearing shell should be about 0.001 in. larger than the hole into which it is pressed. If the interference is too great, the inner diameter of the bearing may become significantly reduced.

A variety of *bearing housings* and *pillow blocks* (Figure 1.46) are available. These mountings are bored to accept standard bearings and in many instances the bearing is premounted. These mounts replace precision-bored bearing mounts.

It is usually necessary to provide axial location for a shaft in a journal bearing. This can be accomplished with a retaining ring in a groove on the shaft or a collar secured by a setscrew.

Plain bearings run smoothly and quietly, and have a high load-carrying ability. Properly installed and lubricated, they have a very long life. Because of the close clearances between parts, they are not easily fouled by dirt in their environment. However, they are limited to low-speed operation. Speeds in excess of a few hundred rpm are not practical without forced lubrication. The primary disadvantage of plain bearings is their high starting friction, although, when properly installed and lubricated, their running friction can be very low.

1.6.3 Ball Bearings

The rolling element in a rolling contact bearing may be a ball, cylinder, or cone. Ball bearings are used for light loads and high speeds. Roller bearings, which employ cylindrical or conical rollers, are suitable for very heavy loads and are not often used for instrument work. We shall discuss only ball bearings.

As with plain bearings, there are both radial and thrust ball bearings (Figure 1.47). A *radial ball bearing* consists of an inner and outer *race* with a row of balls between. The grooves in each race have a radius slightly larger than the radius of the balls so that there is only point contact between the balls and the race. The balls are separated by a *retainer* which prevents the balls from rubbing against one another and keeps them uni-

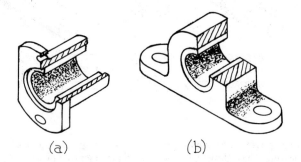

Figure 1.46 (*a*) A bearing housing; (*b*) bearing mounted in a a pillow block.

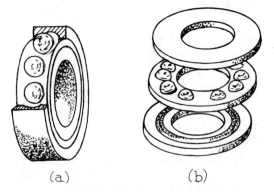

Figure 1.47 Ball bearings: (*a*) a radial ball bearing; (*b*) a thrust ball bearing.

formly spaced around the bearing. A radial ball bearing can tolerate a substantial thrust load, but for pure axial loads a thrust bearing should be used. A *thrust bearing* is similar to an axial bearing except that it has upper and lower races rather than inner and outer.

Ball bearings are made of steel or stainless steel. They are graded 1, 3, 5, 7 or 9 depending on manufacturing tolerances. Grades 7 and 9 have ground races and are made to the closest tolerances. They cost little more than lower-grade ones, and they should be specified for instrument applications.

Proper installation is required to obtain good performance from a ball bearing. The rotating race should be given a firm interference fit, and the stationary race given a light "push fit" to permit some rotational creep. This slight movement of the stationary race helps prevent the maximum load from always bearing on the same spot.

Press fitting changes the internal clearances in a bearing. Bearing manufacturers specify the amount of interference that should be used. As shown in Figure 1.48, the press arbor used to drive a bearing onto a shaft or into a housing should be designed so that the thrust is not transmitted through the balls. Never hammer a bearing into place.

The surface quality and diameter tolerance of the shaft that is to be fitted into a bearing, or the hole that is to house a bearing, must be of the same quality as the bearing. For high-quality bearings the mounting surfaces should be ground. Fortunately for the builder of scientific apparatus, centerless-ground precision shafting is available to fit all standard bearings. Bearing mounts and pillow blocks with premounted bearings are similarly available to eliminate the need for precision machine work when installing ball bearings.

Ball bearings are designed with both radial and axial clearances. This play is intended to allow for axial misalignment and for dimensional changes that occur upon installation or because of thermal stresses. For the best locational precision and to obtain smooth, vibrationless operation, a ball bearing should be *preloaded* to remove most of this play. A preloading force that displaces one race axially with respect to the other will remove both radial and axial play by causing the balls to roll up the sides of their grooves. The use of a shim to take up the play in a bearing is illustrated in Figure 1.49. For light-duty applications, the end play can also be taken up by installing a spring washer (Section 1.6.5) instead of a shim. With proper installation, a ball bearing will locate a revolving axis to within a few ten-thousandths of an inch.

Bearings must be protected from effects that will damage the race surface on which the balls roll. Ball bearings are most likely to fail because of large static loads, which produce an indentation in the race. Such a

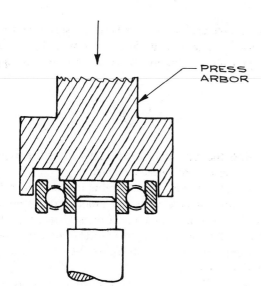

Figure 1.48 Installation of a bearing. The press arbor should bear on the race that is being fitted.

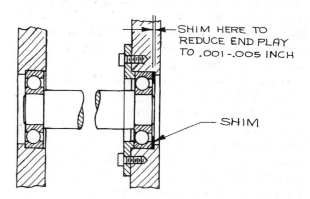

Figure 1.49 Installation of a shim to remove play in a shaft mounted in ball bearings.

dent is called a *brinell*. Dynamic loads are distributed around the race and are thus less likely to cause damage; however, a hard vibration can cause brinelling in the form of a series of dents or waves on the surface of a race. Of course, a bearing is also damaged by the introduction of foreign matter which abrades or corrodes the bearing surfaces.

Bearings should be lubricated with petroleum oils or greases. For high speeds and light loads the lightest, finest grades of machine oil can be used. Ball bearings require very little oil. Lubrication is sufficient if there is enough oil to produce an observable meniscus at the point where each ball contacts a race.

Cleanliness is important. In a dirty environment, bearings with built-in side shields should be used. It is probably wise to use enclosed bearings in all instrument applications to keep the bearings clean and to prevent oil from contaminating the environment.

The chief advantages of ball bearings are their low starting friction and very low running friction. They are well suited to high-speed, low-load operation. Relative to plain bearings, ball bearings are noisy and occupy a large volume. The cost of quality ball bearings is so low that economic considerations are usually not important in choosing between rolling bearings and plain bearings for instrument use.

1.6.4 Linear-Motion Bearings

A linear-motion bearing may be of either the sliding or rolling type. A plain journal bearing can be used to locate a shaft that is to move axially. A V-shaped or dovetail groove sliding over a mating rail can serve as a bearing between a heavily loaded, slowly moving carriage and a stationary platform. This is the type of bearing used between the carriage and bed of a lathe.

Recently, linear ball bearings have become commercially available. In a bearing for use with an axially moving shaft, the balls that carry the load between the outer race and the shaft move in grooves that run parallel to the axis of the shaft. The balls are recirculated through a return track when they roll to the end of a groove. Linear-motion ball bearings will locate a shaft to within ± 0.0002 in. of a reference axis. Their

cost is comparable to conventional rotating ball bearings. Complete roller-slide assemblies are also available. These employ balls rolling in V-grooves. Roller slides will maintain straight-line motion to within 0.0002 in. per inch of travel.

1.6.5 Springs

In many instances it is desirable for a motion to be constrained by a flexible element such as a spring. Springs are used to hold two parts in contact when zero clearance is required, to absorb shock loads, to damp vibrations, and to measure forces.

A spring is characterized by the ratio of the magnitude of an applied force to the resulting deflection. This is the *spring rate*

$$k = \frac{F}{\delta}.$$

As is the case for any flexible system, an assembly consisting of a spring and an *attached* load has a natural frequency of vibration

$$f_n = \frac{1}{2\pi}\left(\frac{k}{m}\right)^{1/2},$$

where m includes the mass of both the spring and any load that is affixed to it. Since $m = W/g = F/a$, we have upon substituting for the spring rate in this equation

$$f_n = \frac{1}{2\pi}\left(\frac{g}{\delta_{st}}\right)^{1/2},$$

where δ_{st} is the static deflection produced by the weight of the spring plus the attached load.

In most applications it is desirable to choose a spring that will not resonate with any other part of the apparatus in which it is installed. If the spring is expected to damp a vibratory motion, its natural frequency should differ from that of the disturbance by more than an order of magnitude.

A spring will exert an uneven force when it is subjected to a periodically varying load whose frequency is close to the natural frequency of the spring. When the vibration of the spring is in phase with the load, the reactive force of the spring will be less than its static force for any given deflection. When the spring is out of phase with the load, it will exert a greater force than expected. This phenomenon is called *surge*. Surge can be reduced or eliminated by using two springs with different natural frequencies. In the case of helical springs, they can be placed one inside the other.

In instrument work, helical springs are most often used. *Helical compression, extension,* and *torsion* springs are illustrated in Figure 1.50. The number of coils in a helical spring must be sufficient to ensure that the spring wire remains within its elastic limit when the spring is at maximum deflection. The number of coils in a compression spring also determines the minimum length, which is realized when successive coils come into contact. A long compression spring may buckle under stress. This tendency is discouraged if the ends of the spring are square. It can be prevented by placing a rod through the center or by installing the spring in a hole. In general, a compression spring must be supported by some means if its length exceeds its diameter by more than a factor of five.

The spring rate of a helical spring made of round wire is

$$k = \frac{Gd^4}{8D^3N},$$

where G is the shear modulus of the spring material (about one-third of the elastic modulus), d the diameter of the wire, N the number of coils, and D the mean diameter (the average of the inner and outer diameters

of the spring). The natural frequencies for free vibration are

$$f_n = \frac{nd}{4\pi D^2 N}\left(\frac{G}{2\rho}\right)^{1/2} \text{Hz},$$

where ρ is the mass density of the spring wire and n is an integer. For steel wire

$$f_n = \frac{14000\,nd}{ND^2} \text{Hz},$$

when d and D are in inches.

A variety of steels and bronzes are used for spring manufacture. High-carbon steel wire works well. If necessary the spring can be wound in the annealed state and hardened after forming. Music wire, or piano wire, is one of the best materials for one-off construction of small springs. It is available in diameters of 0.004 to 0.103 in. This wire is very strong and hard, because of the drawing process used in its production, and does not need to be hardened after forming. Type-302 stainless-steel wire is useful for springs that are subject to a corrosive environment. Springs of beryllium-copper wire are especially useful in applications where spring deflection is used to gauge a force, since this material maintains a linear stress-strain relation almost to the point of permanent deformation. As mentioned earlier, quartz fiber springs can also be used in these applications.

Helical coil springs are conveniently formed by winding wire on a mandrel, which is rotated in a lathe. The wire must be kept under tension as it is pulled onto the mandrel, and when this tension is released the formed spring will expand. Thus the mandrel must be somewhat smaller than the desired inner diameter of the finished spring. The production of a small number of springs of a given size and spring rate is probably best carried out by cut and try.

Commercially manufactured springs are readily available. They are convenient to use because such properties as the spring rate and free length are specified by the supplier.

There are hundreds of possible spring configurations; however, *disc* springs are the only form that we shall mention other than coil springs. A disc spring, also

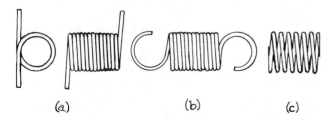

(a)   (b)   (c)

Figure 1.50 Springs: (*a*) helical compression spring; (*b*) extension spring; (*c*) torsion spring.

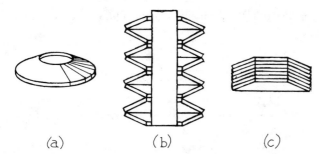

Figure 1.51   Belleville spring washers: (*a*) a disc spring; (*b*) stacked disc springs; (*c*) disc springs stacked in parallel.

known as a Belleville spring washer, is a cone-shaped disc with a hole in the center [Figure 1.51(*a*)]. When loaded, the cone flattens. This is a very stiff spring and thus can absorb a large amount of energy per unit length. A spring of any desired travel can be created by stacking disc springs as in Figure 1.51(*b*). They must be aligned by a rod passing through their centers or else by stacking them in a hole slightly larger than the outer diameter of the discs. If the discs are stacked in parallel as shown in Figure 1.51(*c*), they will provide a great deal of damping owing to friction between the faces of the discs.

CITED REFERENCES

1. *Metals Handbook*, *Heat Treating*, *Cleaning and Finishing*, Vol. 2, 8th edition, American Society for Metals, Metals Park, Ohio, 1964.
2. A comprehensive list of brazing alloys has been compiled by W. H. Kohl, in *Handbook of Vacuum Physics*, Vol. 3, *Technology*, A. H. Beck, Ed., Pergamon Press, Elmsford, N.Y., 1964; *Materials and Techniques for Electron Tubes*, Reinhold, New York, 1959.
3. R. J. Roark, *Formulas for Stress and Strain*, 4th edition, McGraw-Hill, New York, 1965.
4. J. W. Dally and W. F. Riley, *Experimental Stress Analysis*, McGraw-Hill, New York, 1965, pp. 165–221.
5. H. H. Mobie and F. W. Ocvirk, *Mechanisms and Dynamics of Machinery*, 2nd edition, Wiley, New York, 1963, pp. 489–491.
6. A useful review of kinematic design is given by J. E. Furse, J. Phys. E., 14, 264, 1981.
7. Carl Pelander of the JILA Instrument Shop of the University of Colorado is responsible for this groove design.

GENERAL REFERENCES

Design of Moving and Rotating Machinery

R. M. Phelan, *Dynamics of Machinery*, McGraw-Hill, New York, 1967.

Machine-Shop Manual

J. L. Feirer and E. A. Tatro, *Machine Tool Metalworking*, McGraw-Hill, New York, 1961.

Mechanical-Drawing Texts

T. E. French and C. J. Vierck, *Engineering Drawing and Graphic Technology*, 11th edition, McGraw-Hill, New York, 1972; (entitled *A Manual of Engineering Drawing* in its 1st–10th editions).

F. E. Giesecke, A. Mitchell, H. C. Spencer, and I. L. Hill, *Technical Drawing*, 5th edition, Macmillan, New York, 1967.

Practical Mechanical-Engineering Texts

A. D. Deutschman, W. J. Michels, and C. E. Wilson, *Machine Design*, Macmillan, New York, 1975.

V. M. Faires, *Design of Machine Elements*, 4th edition, Macmillan, New York, 1965.

R. E. Parr, *Principles of Mechanical Design*, McGraw-Hill, New York, 1970.

R. M. Phelan, *Fundamentals of Mechanical Design*, 3rd edition, McGraw-Hill, New York, 1970.

Properties of Materials

A. J. Moses, *The Practicing Scientist's Handbook*, Van Nostrand Reinhold, New York, 1978.

MANUFACTURERS AND SUPPLIERS

Adhesives

Devcon Corp.
Danvers, MA 01923

Sauereisen Cements Co.
Blawnox Station
Pittsburgh, PA 15238

Techni-Tool, Inc.
5 Apollo Rd.
Plymouth Meeting, PA 19462

Tra-Con, Inc.
55-T North St.
Medford, MA 02155

Ceramics

Aremco Products, Inc.
P.O. Box 429
Ossining, NY 10562

Corning Glass Works
Corning, NY 14830

Industrial Tectonics, Inc. (balls)
P.O. Box 1128
Ann Arbor, MI 48106

McDanel Refractory Porcelain Co. (rod, tube)
510 Ninth Ave.
Beaver Falls, PA 15010

Drafting Supplies

K&E (Keuffel & Esser Co.)
20 Whippany Rd.
Morristown, NJ 07960

Glass

Corning Glass Works
Corning, NY 14830

Kimble Glass
P.O. Box 1035
Toledo, OH 43666

Linear Ball Bearings

GM, Saginaw Steering Gear Div.
Saginaw, MI 40605

Linear Rotary Bearings, Inc.
59 New York Ave.
Westbury, NY 11590

Thomason Industries, Inc.
Manhasset, NY 11030

Turnomat Div.
455 Adirondack St.
Rochester, NY 14606

Pins, Especially Spring Pins and Roll Pins

Driv-Lok, Inc.
1140 Park Ave.
Sycamore, IL 60178

Elastic Stop Nut Corp. of America
2330 Vauxhall Rd.
Union, NJ 07083

Esna Rollpin
2330 Vauxhall Rd.
Union, NJ 07083

Precision Shafts, Balls, Bearings, and Mounts

Ace Plastic Company (nonmetallic balls)
91-30 Van Wyck Expwy.
Jamaica, NY 11435

W. M. Berg, Inc.
499 Ocean Ave.
East Rockaway, NY 11518

Industrial Tectonics, Inc. (balls)
P.O. Box 1128
Ann Arbor, MI 48106

New Hampshire Ball Bearings, Inc.
Astro Division
Laconia, NH 03246

Northfield Precision Instrument Corp.
4400 Austin Blvd.
Island Park, NY 11558

PIC
P.O. Box 335, Benrus Center
Ridgefield, CT 06877

Product Components (nonmetallic balls)
30 Lorraine Ave.
Mt. Vernon, NY 10553

RMB Miniature Bearings
4 Westchester Plaza
Elmsford, NY 10523

Specialty Ball Co.
P.O. Box 1128
Ann Arbor, MI 48106

Retaining Rings

Truarc Retaining Rings Div.
Waldes Kohinoor, Inc.
47-16 Austed Place
Long Island City, NY 11101

Soldering, Brazing, and Welding Supplies

American Platinum and Silver
Engelhard Industries, Inc.
70 Wood Ave.
Iselin, NJ 08830

Eutectic Welding Alloys Corp.
40-40 172nd St.
Flushing, NY 11358

Handy and Harmon
850 Third Ave.
New York, NY 10022

Western Gold and Platinum Co.
477 Harbor Blvd.
Belmont, CA 94002

Spot Welders

Ewald Instruments Corp.
Route 7-T
Kent, CT 06757

Unitek Corp.
Weldmatic Div.
1820 South Myrtle
Monrovia, CA 91016

Springs

Associated Spring Corporation
Wallace Barnes Division
Bristol, CT 06010

W.M. Berg, Inc.
499 Ocean Ave.
East Rockaway, NY 11518

Lee Spring Company, Inc.
30 Main St.
Brooklyn, NY 11201

PIC
P.O. Box 335
Benrus Center
Ridgefield, CT 06877

Threaded Fasteners and Inserts

Heli-Coil Products
1564 Shelter Rock Lane
Danbury, CT 06810

Tridair Industries
3000 W. Lomita Blvd.
Torrance, CA 90505

Miscellaneous Manufacturers and Suppliers

Balance Technology
120 Enterprise Dr.
Ann Arbor, MI 48103

Birchwood-Casey
7900 South Fuller Rd.
Eden Prairie, MN 55343

Carpenter Technology Corporation
150 W. Bern St.
Reading, PA 19603

Jodon Engineering Associates, Inc.
145 Enterprise Dr.
Ann Arbor, MI 48103

Magnaflux Corp.
6300 W. Lawrence Ave.
Chicago, IL 60656

CHAPTER 2

WORKING WITH GLASS

Glass has been called the miraculous material. The ubiquity of glass in the modern laboratory certainly confirms this. Because glass is chemically inert, most containers are made of it. Glass is transparent to many forms of radiation, and its transmission properties can be varied by controlling its composition; hence all sorts of windows and lenses are made of glass. Because glass can be polished to a high degree and is dimensionally stable, most mirrors are supported on glass surfaces. Glass is strong and stiff and is often used as a structural material. Considering its mechanical rigidity and density, it is a reasonably good thermal insulator. It is an excellent electrical insulator. Perhaps the greatest virtue of this material is that most glasses are inexpensive and can be cut and shaped in the laboratory without using expensive tools.

Thirty-five years ago, most glass laboratory apparatus was produced by the scientist or technician *in situ* by blowing molten glass or by grinding, cutting, and polishing hard glass. Today the glass industry has grown to such an extent that nearly all components of a glass apparatus are available from commercial sources at prices as low as a few dollars per pound. These include all sorts of containers, chemical labware, vacuum-system components, mirrors, windows, and lenses. It is often only necessary for laboratory scientists to acquaint themselves with the range of components available and to acquire the skills needed to assemble an apparatus from these components.

2.1 PROPERTIES OF GLASSES

The chemical composition of glass is infinitely variable, and so therefore are the thermal, electrical, mechanical, and chemical properties of glass. Furthermore, glass is a fluid that retains a memory of its past history. It is possible however to review the general properties of glass and to specify the properties of glasses of a particular composition and method of manufacture.

2.1.1 Chemical Composition and Chemical Properties of Some Laboratory Glasses

The chief constituent of any commercial glass is silica (SiO_2). All laboratory ware is at least three-quarters silica, with various other oxides added to obtain certain thermal properties or chemical resistance.

The least expensive and, until the last quarter century, the most common glass used for laboratory ware is known as *soda-lime* glass or *soft* glass. This glass typically contains 70–80% silica, 5–10% soda (Na_2O), 5–10% potash (K_2O), and 10% lime (CaO). The chief advantage of this glass is that it can be softened in a natural-gas–air flame.

Soda-lime glass has been largely replaced by *borosilicate* glass for the manufacture of labware. In this glass the alkali found in soft glass is replaced by B_2O_3 and

51

alumina (Al_2O_3). Borosilicate glass is superior to soda-lime glass in its resistance to chemical attack and thermal or mechanical shock. It softens at a higher temperature than soft glass, however, and is more difficult to work. Borosilicate glasses of many different compositions are manufactured for various laboratory applications, but far and away the most common laboratory glass is the borosilicate glass designated by Corning as *Pyrex 7740* and by Kimble as *Kimax KG-33*. The composition is: SiO_2, 80.5%; B_2O_3, 12.9%; Na_2O, 3.8%; Al_2O_3, 2.2%; K_2O, 0.4%.

Glassware of extremely good chemical resistance is produced from borosilicate glass by heat treatment, which causes the glass to separate into two phases—one high in silica and the other rich in alkali and boric oxides. This second phase is then leached out with acid, and the remaining phase is heated to give a clear consolidated glass that is nearly pure silica. This glass is known as *96% silica glass* and is designated by Corning as Vycor No. 7900.

Glass composed only of silica is known as *vitreous* or *fused silica* or simply as "quartz." Because of its refractory properties and chemical durability, this material would be the most desirable glass were it not for the high cost of making it and the extremely high temperatures required to work it. The relative cost of quartz is decreasing to the extent that the market for 96% silica glass is rapidly disappearing.

Most laboratory glasses are transparent to light, making visual distinction between the various glass compositions impossible. For this reason it is important to label glass materials before storing and to avoid mixing different kinds of glass. When necessary it is possible to distinguish different glasses from one another by differences in thermal or optical properties. The gas-air flame of a Bunsen burner will soften soda-lime glass but not borosilicate glass. A natural-gas–oxygen flame is required to soften borosilicate glass, but this flame will not affect fused silica. Fused silica must be raised to a white heat in an oxyhydrogen flame before it will soften. Glasses of different composition cannot generally be successfully fused together because of differences in the amount of expansion and contraction on heating and cooling. An unknown piece of glass may be compared with a known piece by placing the two side by side with their ends coincident. The two ends are softened together in a flame and pressed together with tweezers. Then the fused ends are reheated and drawn out into a long fiber about $\frac{1}{2}$ mm in diameter and permitted to cool. If the fiber remains straight, the two pieces have the same coefficient of expansion. If the fiber curves, the two pieces are of different composition.

A measurement of the refractive index is usually a sensitive and reliable test of glass composition. A piece of glass placed in a liquid of exactly the same refractive index will become invisible. For example, a test solution for Pyrex 7740 can be made of 16 parts by volume of methanol in 84 parts benzene. This test solution should be kept in a tightly covered container so that its composition does not change as a result of evaporation.

As can be judged from the extreme chemical environments to which glasses are routinely subjected, glass is indeed resistant to chemical action. However, one need only observe windows clouded by the action of rainfall in a polluted atmosphere or glassware permanently stained by laboratory chemicals to confirm that glass is not entirely impervious to chemical attack. Glass is attacked most readily by alkaline solutions, and all types are affected about equally. Water has an effect. The soft glasses are most susceptible, borosilicate glass is only slightly affected, and 96% silica hardly at all, since most of its soluble components are leached out in manufacture. Acids attack glass more readily than water, although again borosilicate glass is more resistant to acids than soda glass, and 96% silica is more resistant still.

2.1.2 Thermal Properties of Laboratory Glasses

One of the most important properties of glass is its very low coefficient of thermal expansion. It is this property that permits glass to be formed at high temperatures in a molten state and then cooled without changing shape or breaking. The coefficients of linear expansion of several types of glass are given in Table 2.1. As can be seen from these data, borosilicate glass is much more resistant to thermal shock than soft glass. Fused silica is

Table 2.1   THERMAL PROPERTIES OF GLASS

Glass	Linear Expansion Coefficient $(\text{cm}\,\text{cm}^{-1}\,\text{K}^{-1})$	Strain Point (°C)	Annealing Point (°C)	Softening Point (°C)
Soda-lime (typical)	$8\text{--}10 \times 10^{-6}$	500	550	700
Pyrex 7740 (borosilicate)	3.3×10^{-6}	510	555	820
Vycor 7900 (96% silica)	0.75×10^{-6}	820	910	1500
Fused Silica	0.55×10^{-6}	900	1150	1650

so stable that a white-hot piece can be immersed in liquid air without fracturing.

Glass does not have a melting point. Instead, the working properties of a glass are specified by particular points on its viscosity-temperature curve. The *softening point* is approximately the temperature at which a glass can be observed to flow under its own weight. The *annealing point* is the temperature at which internal stresses can be relieved (annealed) in a few minutes. The *strain point* is the temperature below which glass can be quickly cooled without introducing additional stress.

The specific heat of glass is about $0.2\ \text{cal}\,\text{g}^{-1}\,\text{K}^{-1}$, and the thermal conductivity of glass is about $0.0002\ \text{cal}\,\text{cm}\,\text{sec}^{-1}\,\text{cm}^{-2}\,\text{K}^{-1}$. For reference, this thermal conductivity is about an order of magnitude less than that of graphite, two orders less than that of metal, and about an order of magnitude greater than that of wood.

2.1.3   Optical Properties of Laboratory Glassware

In many experiments light must be transmitted through the wall of a glass container. The transmission as a function of wavelength depends upon the composition of the glass. The composition of the soda-lime glasses varies considerably, and the optical properties of each

piece should be determined by spectroscopic analysis before it is used in an experiment requiring light transmission. Spectrophotometric curves for some lab glasses of well-defined composition are given in Figure 2.1. As can be seen, fused silica is transparent over the widest wavelength range.

2.1.4   Mechanical Properties of Glass

The tensile, compressive, and shear strength of a piece of glass depends upon its shape and history and upon the time over which it is loaded. Glass appears to be much stronger in compression than tension. This is in part due to the fact that a glass surface can be made very smooth in order to uniformly distribute a compressive load. Values of 60,000–180,000 psi are quoted for the compression strength.

Glass is nearly perfectly elastic. Even when it breaks it does so without any plastic deformation. Springs made of glass or quartz fibers behave almost ideally. The modulus of elasticity of glass is high and of course depends upon composition. Young's modulus for Pyrex 7740 is 9.14×10^6 psi at 0°C and 9.40×10^6 psi at 100°C. For fused silica the corresponding values are 10.47×10^6 and 10.67×10^6 psi.

Glass fibers are very strong. A quartz fiber 0.01 in. in diameter has a tensile strength of 50,000 psi; a fiber

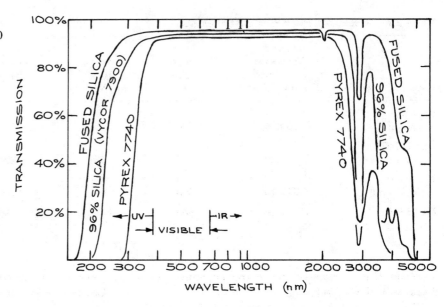

Figure 2.1 Optical-transmission curves for Pyrex 7740, Vycor 7900 (96% silica), and fused silica.

0.001 in. in diameter a strength of at least 200,000 psi; and a fiber 0.0001 in. in diameter a strength in excess of 1,000,000 psi.

2.2 LABORATORY COMPONENTS AVAILABLE IN GLASS

The list of laboratory apparatus components produced commercially in glass is nearly endless. The laboratory scientist should make a careful survey of lab suppliers' literature before embarking on the design and construction of a glass apparatus. Very often a complex device can be assembled in the lab entirely from inexpensive components without requiring the services of a skilled glassblower. Some of the most frequently used glass components are described below.

2.2.1 Tubing and Rod

Most laboratory supply houses carry a wide range of tubing and rod of Pyrex 7740 or Kimax KG-33 in 4-ft lengths at a cost of only a few dollars per pound. Outer

diameters between 3 and 178 mm are readily available. The standard wall thickness ranges from 0.5 mm for the smallest tubing to 3.5 mm for the largest. Heavy-wall tubing with wall thickness ranging from 2 to 10 mm is also available. In addition, heavy-wall tubing of very small bore diameter (0.5–4 mm), known as capillary tubing, is available. Finally, solid rod of 3- to 30-mm diameter is a standard item.

In addition to borosilicate-glass stock, tubing and rod of soft glass and quartz are generally available. The soft-glass stock is usually intended for student use. Tubing or rod of quartz costs eight to ten times more than Pyrex 7740.

2.2.2 Demountable Joints

Laboratory apparatus can be quickly assembled if components are connected with joints of the type illustrated in Figure 2.2.

A gas- or liquid-tight seal between two close-fitting pieces of glass can be achieved if the mating surfaces are lightly coated with a viscous lubricant before assembly. A number of low-vapor-pressure lubricants such as Apiezon vacuum grease or Dow-Corning silicone vacuum grease are especially formulated for this purpose. The taper joint and the ball-and-socket joint in Figure 2.2(*a*) and (*b*) are assembled in this manner. The mating surfaces must fit together very well. In general this requires that they be lapped together. This is accomplished by coating the surfaces with a fine abrasive such as Carborundum in water, fitting the pieces together, and rotating one with respect to the other. The rotation should not be continuous, but rather after half a turn or so the pieces should be pulled apart and then assembled again so as to redistribute the abrasive. Fortunately, taper and ball-and-socket joints with ground mating surfaces are now commercially produced in standard sizes with sufficient precision that lapping and grinding in the lab is seldom necessary.

The standard taper (indicated $\mathfrak{I}$) for ground joints is 1:10. Standard-taper joints on tubing with outer diameters between 8 and 50 mm are available. These joints are identified by figures that indicate the diameter of

the large end of the taper and the length of the ground zone in millimeters. For example, $\mathfrak{I}$ 10/30 indicates a ground zone 10 mm in diameter at the large end and 30 mm in length. Standard-taper ground joints are available in borosilicate glass, quartz, and type-303 stainless steel. These components are interchangeable, and thus it is often convenient to make a transition from glass to quartz or from glass to metal with a taper joint.

Ball-and-socket ground joints do not seal as reliably as taper joints. However, their design permits misalignment and even some slight motion between joined parts. This type of joint must be secured with a suitable clamp. Ball-and-socket joints are designated by a two-number code (i.e., 28/15). The first number gives the diameter of the ball in millimeters, the second number the inner diameter of the tubing to which the ball and socket are attached. Ball-and-socket joints of standard sizes are commercially available in borosilicate glass, quartz, and stainless steel.

The O-ring joint illustrated in Figure 2.2(*c*) is rapidly replacing the ground joint as a means of making demountable joints in glass vacuum apparatus. These joints require no grease. Furthermore, the mating pieces are sexless, since each member of the joint is grooved to a depth of less than half the thickness of the O-ring. To date, these joints are only available in borosilicate glass.

"Quick" connects of the type shown in Figure 2.2(*d*) are also sealed with an O-ring. They are suited to joining either glass or metal tubing to a metal container. For example, quick connects are often used to join glass ion-gauge tubes to metal vacuum apparatus.

2.2.3 Valves and Stopcocks

A glass stopcock of one of the types illustrated in Figures 2.3(*a*)–(*c*) has traditionally been used for controlling fluid flow in glass apparatus. These consist of a hollow tapered body and a plug with a hole bored through it that can be aligned by rotation with inlet and outlet ports in the body. The simplest and least expensive version uses a glass plug with a ground surface that mates with a ground surface on the interior of the stopcock body. The surface of the plug must be lightly lubricated with stopcock grease to ensure a good seal

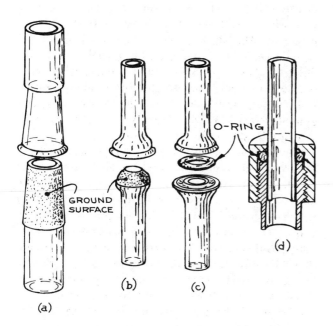

Figure 2.2 Demountable joints: (*a*) standard-taper joint; (*b*) ball and socket; (*c*) O-ring joint; (*d*) a quick connect. (*a*), (*b*), and (*c*) are used for attaching glass components to one another; (*d*) is used for joining a glass tube to a metal component.

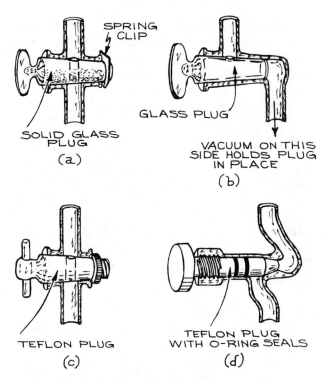

SOLID GLASS PLUG
(a)

SPRING CLIP

GLASS PLUG

VACUUM ON THIS SIDE HOLDS PLUG IN PLACE
(b)

TEFLON PLUG
(c)

TEFLON PLUG WITH O-RING SEALS
(d)

Figure 2.3 Stopcocks and valves for controlling fluid flow: (*a*) glass stopcock with solid glass plug; (*b*) high-vacuum stopcock with vacuum cup and hollow plug; (*c*) glass stopcock with Teflon plug; (*d*) threaded glass vacuum valve with O-ring-sealed Teflon stem.

and freedom of rotation. A stopcock of this type may be used to control liquid flow or gas flow at pressures down to a few millitorr. For use at pressures above 1 atm the end of the plug may be fitted with a spring-loaded clip or collar, which prevents its being blown out. Stopcocks of this type are now available with a Teflon plug. These are well suited for use with liquids. No lubrication is required, and the plug seldom freezes in the body of the stopcock.

In glass vacuum systems, the stopcock illustrated in Figure 2.3(*b*) is far more reliable than the simple solid plug design. The plug is hollow, and the small end of the body is closed off by a vacuum cup so that the interior of the stopcock can be evacuated. The plug is then held securely in place by atmospheric pressure.

Valves with threaded borosilicate glass bodies and threaded Teflon plugs [Lab-Crest, by Fisher and Porter: Figure 2.3(*d*)] are rapidly replacing conventional stopcocks for most applications. These valves are no more expensive than good vacuum stopcocks. They require no lubrication, so that only glass and Teflon are exposed to the interior of the system. They are suitable for use with most liquids and gases and may be used at pressures from 10^{-6} torr to 15 atm.

2.2.4 Graded Glass Seals and Glass-to-Metal Seals

In general, glasses of different composition cannot be fused together because differences in the amount of contraction upon cooling produce destructive stresses. In practice two glasses can be successfully joined if their coefficients of thermal expansion differ by no more than about $1 \times 10^{-6}\,\mathrm{K}^{-1}$. The problem of joining two glasses with significantly different coefficients of thermal expansion is solved by interposing several layers of glass, each with a coefficient only slightly different from its neighbors. This stack of glass when fused together is called a *graded seal*. Tubing with graded seals joining different glasses is manufactured commercially. Borosilicate-to-soft-glass seals are available in tube sizes from 7 to 20 mm. Seals graded from borosilicate glass (for example Pyrex 7740) to quartz (Vycor 7913) are available in tube sizes from 7 to 51 mm.

Glass tubing can with care be joined to metal tubing. Ideally the glass and the metal should have the same coefficient of expansion. It has been demonstrated, however, that glass tubing can be joined to metal tubing that has a very different coefficient provided that the end of the metal tubing has been machined to a thin, sharp, feathered edge.[1] The stresses of thermal expansion and contraction can then be taken up by stretching of the metal. The stresses created by joining glass to metal can also be reduced by interposing glasses whose coefficient of thermal expansion is intermediate between that of the metal and the final glass.

Direct tube seals between borosilicate glass and copper or stainless steel are available in diameters from $\frac{1}{4}$ to 2 in. Graded glass seals between borosilicate glass and Kovar, an iron-nickel-cobalt alloy, are available in sizes from $\frac{1}{8}$ to 3 in. These graded seals are robust, and far and away the most common glass-to-metal seal. Kovar is easily silver-soldered or welded to most steels and brasses.

2.3 LABORATORY GLASSBLOWING SKILLS

Considering the range of components now available, it is usually only necessary for the laboratory worker to be able to join commercially produced components to assemble a complete apparatus. The required skills consist primarily of joining tubes in series or at right angles, cutting and sealing off tubing, and making simple bends. It is easiest and often necessary to make a joint between a piece of tubing held in the hand and a piece attached to an apparatus or otherwise rigidly clamped. In the following sections only the simplest glassblowing operations with borosilicate glass are described. There are certainly many more elegant and complex manipulations that can be carried out by a skilled glassblower using only hand tools. Workers who find a need for these operations or discover in themselves a flair for glassblowing should consult a more complete text on the subject.

2.3.1 The Glassblower's Tools

The necessary equipment for laboratory glasswork is illustrated in Figure 2.4. The basic tool is of course a torch. The National type-3A blowpipe with #2, #3 or #4 tip is the standard in the trade. A torch stand is essential to hold the lighted torch when both hands are required to manipulate a piece of glass. A source of natural gas and low-pressure oxygen is required, along with two lengths of flexible rubber or Tygon tubing for connecting the gas supplies to the torch. For working

---

Many of the simplified methods described in Section 2.3 were developed by John Trembly of the University of Maryland for use in his remarkably successful introductory glassblowing classes.

quartz a hydrogen-oxygen flame will be required. A rigid ring stand with appropriate asbestos-covered clamps is needed to support glass being worked upon. The torch and stand should be set up on a bench top covered by several square feet of asbestos board such as Transite. A number of small hand tools, inexpensively obtained from a scientific supply house, are required. These include a torch lighter, a wax pencil, a glass-cutting knife, a swivel, a mouthpiece (an old tobacco-pipe bit will do), glassblowing tweezers, a collection of corks and one-hole stoppers, and a variety of reaming tools. These reaming tools consist of flat triangular pieces of brass fitted with wooden file handles, and tapered round or octagonal carbon rods in various sizes. Didymium eyeglasses are necessary to filter out the intense sodium D-line emission produced when glass is exposed to the flame. Without these it is impossible to see hot glass as it is worked. Clip-on didymium lenses are available for these who wear prescription eyeglasses.

A little preliminary practice with the hand torch is advisable. The gas inlet may be connected directly to a low-pressure natural-gas outlet in the lab. The outlet valve is turned completely open and gas flow is controlled by the valve in the body of the torch. Oxygen is usually obtained from a high-pressure cylinder. A pressure regulator for the oxygen tank is essential. The outlet pressure should be between 6 and 10 psi. Once again, the flow of oxygen gas is controlled by the valve in the body of the hand torch. The proper procedure is to open the gas control and ignite the gas first. The gas should be adjusted to give a flame 2 to 3 inches long before the oxygen is introduced. If the flame blows out, the oxygen valve should be closed and the gas permitted to flow for a few moments in order to purge the torch of the explosive gas-oxygen mixture before reignition is attempted.

With a #3 torch tip it is possible to produce a range of flames suitable for work on tubing from a few millimeters up to 50 mm in diameter. Opening the gas valve wide and adding only a little oxygen gives a large, bushy yellow flame. This is a cool reducing flame suitable for preheating large pieces of glass and for annealing finished work. At the opposite extreme a sharp blue oxidizing flame can be obtained by restricting the gas flow. For work on fine tubing and for cutting tubing it

Figure 2.4 Glassblower's equipment.

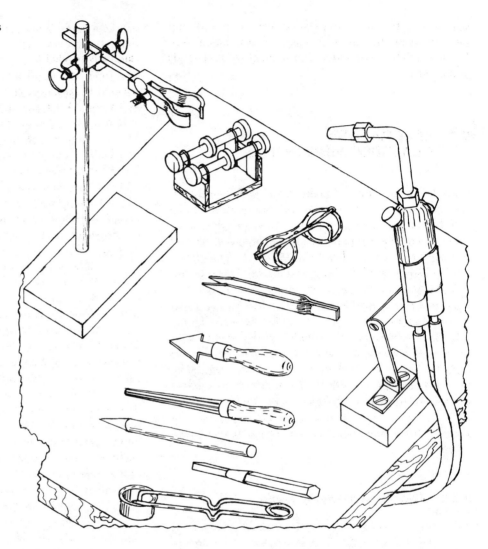

is possible to make a flame no more than an inch long and $\frac{1}{8}$ in. in diameter. The hottest flame is made with an excess of oxygen. The hottest part of the flame is the tip of the blue inner cone.

Some practice will be required to master even the simple operations described below. The neophyte should obtain the necessary tools and a supply of clean dry borosilicate glass tubing of 8–12-mm diameter. About ten hours of practice is required before one can profitably embark on any serious apparatus construction.

2.3.2 Cutting Glass Tubing

Tubing up to about 12-mm diameter can be broken cleanly by hand. The location of the desired break may be indicated by marking the tubing with a wax pencil. The glass is then scored with a glass knife or the edge of a triangular file as shown in Figure 2.5. The scratch should be perpendicular to the axis of the tube and need only be a few millimeters in length. Use considerable pressure and make only one stroke. Do not saw at the

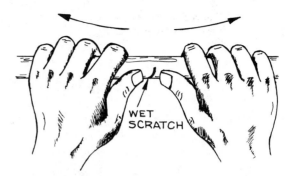

Figure 2.6 Breaking glass tubing. The tubing is bent and simultaneously pulled apart.

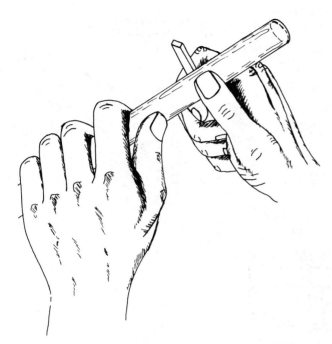

Figure 2.5 Scoring glass tubing in preparation for breaking. Use only one stroke with considerable force.

removed and the hot zone is brushed with a wet pipe cleaner. The thermal shock produced should result in a clean break.

The edge of the glass at a fresh break is sharp and fragile. After cutting, the end of a tube can be *fire-polished* to relieve this hazardous condition. This is accomplished simply by heating the end of the tube in the flame until the glass is soft enough that surface tension rounds and thickens the sharp edge.

2.3.3 Pulling Points

Points are elongations on the end of a tube produced by heating the glass and stretching it. These points serve as

glass. Wet the scratch and then, holding the tubing as shown in Figure 2.6 with the scratch toward you, pull the ends apart. Applying a force that tends to bend the ends of the tube away from you, while simultaneously pulling, will facilitate a clean break.

Tubing of 12- to 25-mm diameter may be cut by cracking it with a flame. First score the glass around approximately one third of its circumference with the glass knife. Then wet the scratch and touch the end of the scratch with a very fine sharp flame that is oriented tangentially to the circumference as shown in Figure 2.7.

Large-diameter tubing can be parted by cracking it using a resistively heated wire held in a yoke as illustrated in Figure 2.8. The tubing is placed on the resistance wire with one end butted against a stop. The wire is heated red-hot by passing current through it, and the tube is rotated to heat a narrow zone around its circumference. After a moment the tube is quickly

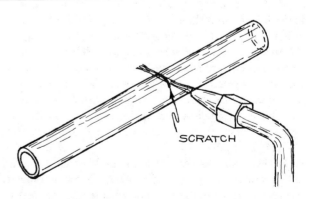

Figure 2.7 Cracking tubing with a flame.

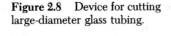

Figure 2.8 Device for cutting large-diameter glass tubing.

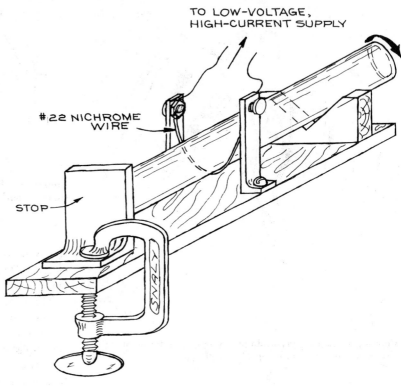

handles for manipulating short pieces of tubing. Pulling a point is the first step in closing the end of a tube.

The procedure for pulling points depends upon whether the tubing can be rotated or not. If the tubing is free, then place the torch in its holder with the flame pointed away from you and hold the glass in both hands as illustrated in Figure 2.9. Adjust the torch to give a fairly full, neutral flame. Then, rotating the glass at a uniform rate, pass it through the flame to heat a zone about as wide as the tube diameter. When the glass softens, remove it from the flame and pull the heated section to a length of about 8 inches. Then quickly heat the center of the stretched section and pull the glass in two, leaving a small bead at the end of each point. It is important that the points be on the axis of the original tube. This requires that the circumference of the tube be heated uniformly and that the ends of the tube be pulled straight away from one another. Some practice is required to synchronize the motion of the hands so that both ends of the tubing are rotated in such a way that

the tube does not twist or bend when it becomes pliable in the flame. Beginners tend to overheat glass, thus exacerbating this problem of misalignment. When glass is sufficiently plastic for proper manipulation, it is still rigid enough to help support the end sections.

If one end of a tube is attached, so that the tube cannot be rotated, then it becomes necessary to swing the torch around the tubing rather than rotate the glass before the stationary torch. To achieve a smooth motion with the torch so that the tube is heated uniformly around its circumference requires practice. The unattached end of the tubing is supported with the free hand until the glass is sufficiently hot to be pulled.

2.3.4 Sealing off Tubing: The Test-Tube End

Tubing is closed off by forming a hemispherical bubble similar to that found at the end of a test tube. In order for the test-tube end to form a strong closure, it must be

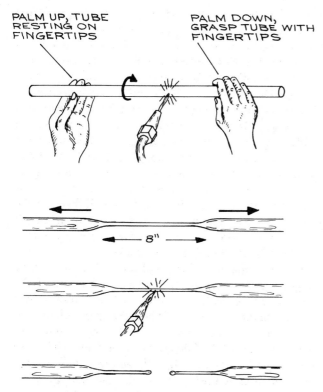

Figure 2.9 Pulling points.

smooth and of a uniform thickness. The hemispherical shape is formed by closing off the tubing and then softening the glass in the flame and blowing into the tube.

Begin by pulling a point on the tubing that is to be closed off. If the tubing is not rigidly mounted, place it in a clamp secured to a ring stand on the workbench. If possible the work should be at chest height with the point upward. Press a one-hole stopper, with a piece of soft rubber tubing attached, into the open end of the tube as shown in Figure 2.10(a). Warm the glass at the base of the point with a soft bushy flame. Adjust the gas-oxygen mixture to the torch to give a fairly small flame with a well-defined, blue inner cone. Heat the point close to the shoulder where it joins the tube. Swing the torch around the work so that the glass is heated uniformly. When the glass melts, pull the point off the tube. Only a small bead of glass should be left behind. A large blob of glass will result in a closure that is too thick in relation to the wall of the tubing. A test-tube end that is thick in the middle and thin at the sides will develop destructive strains, since it does not cool at a uniform rate. The key is to heat only a narrow zone of the point near the shoulder and to perform the whole operation rather quickly. If a large globule of

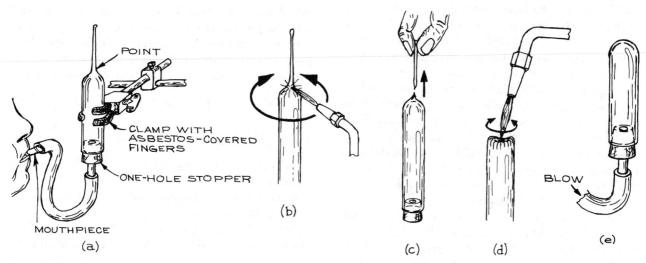

Figure 2.10 Blowing a test-tube end.

glass is left after pulling off the point, the excess must be removed. With a sharp flame, heat the globule until it melts. Remove the flame, touch the molten bead with the end of a piece of glass rod ("cane"), and quickly pull the rod away. Excess glass will be pulled away from the soft bead into a fiber which can be broken or melted off. The test-tube end is completed by heating the closed-off end as far as the shoulder with a somewhat larger flame until it is soft and the glass has flowed into a fairly uniform thickness. The flame should be directed downward at the end of the tube with a circular motion. Remove the flame and gently blow the end into a hemispherical shape. If a flat end is desired, reheat the round end and gently blow while pressing a flat carbon block against the end.

2.3.5 Making a T-Seal

A tube is joined at a right angle to the side of another tube with a *T-seal*. To make a T-seal, close one end of a tube with a cork and attach the blow tube to the other. If possible, mount the tube horizontally at chest height. With a sharp flame heat a spot on the top side of the tube. Using a circular motion of the torch, soften an area about the size of the cross section of the tube that is to be joined. Then move the flame away and gently blow out a hemispherical bulge as shown in Figure 2.11(*b*). Reheat the bulge and blow out a thin-walled bubble. Finally, while blowing to pressurize the inside of the tube, touch the bubble with the tip of the flame so that a hole is blown through it. Now heat the edge of this hole so that surface tension pulls the glass back to the edges of the original bubble. The result should be a round opening slightly smaller than the diameter of the tube that is to be joined, as shown in Figure 2.11(*f*). If the opening is too small or if the edge is irregular, the hole can be reamed and shaped by softening the glass and shaping the hole with a tapered carbon. A tool for this job can be made by sharpening a $\frac{1}{4}$-in.-diameter carbon rod in a pencil sharpener.

With a cork or a point, close off the end of the tube that is to be sealed to the horizontal tube. Then, holding this tube above the hole and tipped slightly back, simultaneously heat the edges of the tube and the hole until they become soft and tacky. Bring the tube into contact with the back edge of the hole, remove the flame, and tip the tube forward until it rests squarely on the hole. The tacky edges should join together to produce an airtight seal. Blow to check the seal. Small holes can be closed by heating the seal all around until the glass is soft and tipping the vertical tube in the direction of the leak. It also is possible to "stitch" a gap closed with cane. This is done by pulling a point on a piece of glass rod. Then warm the gap which is to be repaired, heat the point of the cane until it is molten and deposit the tiny blob of molten glass thus produced into the gap. The seal at this point is quite fragile because the glass is too thick around the seal. Cracks will develop if the glass is permitted to cool before the seal is blown out in order to reduce the wall thickness.

Glassblowing around the seal proceeds in four steps. Begin with a sharp flame parallel to the axis of the horizontal tube. Swing the torch back and forth to heat one quadrant of the seal. When the glass softens, remove the flame and blow gently. Timing is important here. Blowing while the glass is in the flame will cause the thinnest parts to bulge. If, however, one waits a moment after removing the flame, the thinnest parts will cool and blowing will preferentially thin out the thick portions of the seal, yielding a more uniform wall thickness. The joint between the vertical and horizontal tubes should then be a smooth curve of glass. During the blowing operation the vertical tube must be held with the free hand to prevent its tipping or sagging into the horizontal tube. Next repeat the heating and blowing operation on the side diametrically opposite. Finally, in a third and fourth step, blow out the remaining two quadrants.

2.3.6 Making a Straight Seal

A *straight seal* joins two tubes coaxially. There are two simple procedures for making such a seal.

If the tubes are quite different in diameter, close off the larger one with a test-tube end and mount it vertically with the closed end up. Then blow a bubble of the appropriate size in the test-tube end and proceed as in making a T-seal.

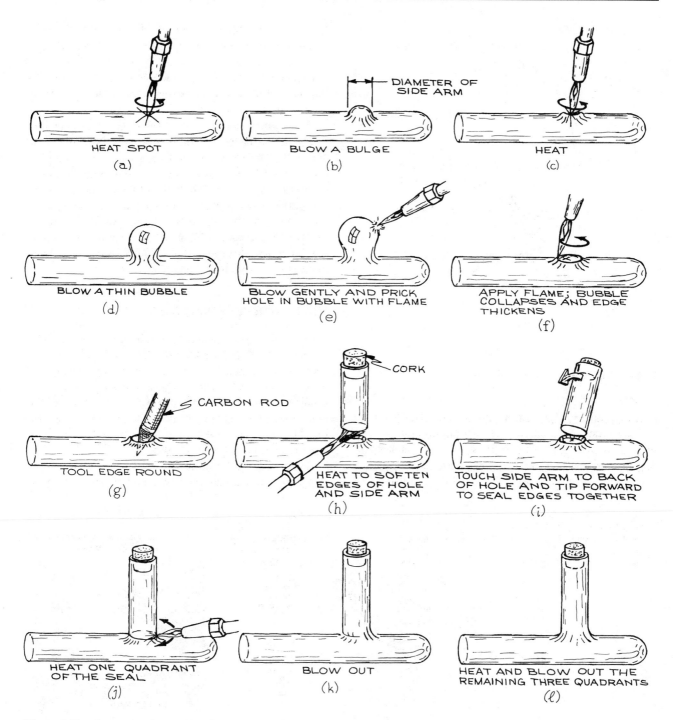

Figure 2.11   Steps in making a T-seal.

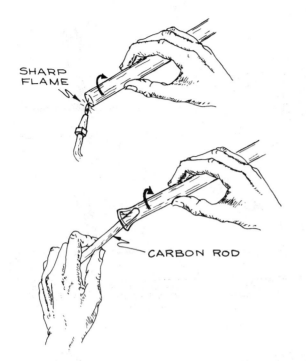

SHARP
FLAME

CARBON ROD

Figure 2.12   Flaring the end of a tube.

Alternatively, the smaller tube can be flared out to the diameter of the larger one. The flaring operation is illustrated in Figure 2.12. With the torch in its stand, heat the edge of the tube by rotating it in a sharp flame. Remove it from the flame, and while continuing to rotate the tube, insert a tapered carbon and gently bring the soft edge of the glass up against the taper. Do not attempt to produce a flare of the desired diameter in one rotation of the tube. It will not come out round. The flaring operation is much easier if the tube is placed on a set of glassblower's rollers.

As shown in Figure 2.13(a), the larger tube should be mounted vertically if possible. Suspend the smaller tube with the flare just above the larger one. With the upper tube tipped back slightly, heat the edges of the larger tube and the rim of the flare until the glass is tacky. Bring them into contact at the back of the seal, remove the flame, and tip the upper piece forward to join the two pieces. Proceed to blow out the seal in four steps as for the T-seal. Minimize the distance above and below the seal over which the glass is heated.

2.3.7   Making a Ring Seal

A ring seal is used for sealing a tube inside a larger tube as in Figure 2.14(f), or for passing a tube through the wall of a container. There are many procedures for making such seals. The simplest involves bulging the smaller-diameter tube sufficiently for it to close off the hole through which the tube must pass. A bulge in a tube is known as a *maria*. Making a maria is a freehand operation and requires some practice.

Begin with the torch in its holder and the gas supply adjusted to give a long, narrow flame. Then rotating the

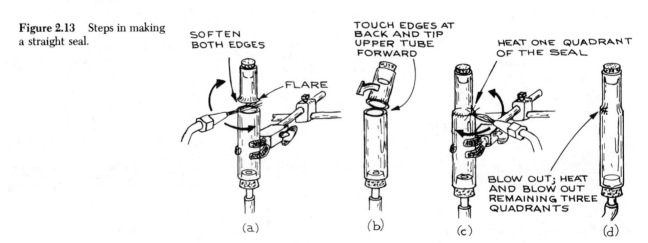

Figure 2.13   Steps in making a straight seal.

SOFTEN
BOTH EDGES

FLARE

TOUCH EDGES AT
BACK AND TIP
UPPER TUBE
FORWARD

HEAT ONE QUADRANT
OF THE SEAL

BLOW OUT; HEAT
AND BLOW OUT
REMAINING THREE
QUADRANTS

(a)   (b)   (c)   (d)

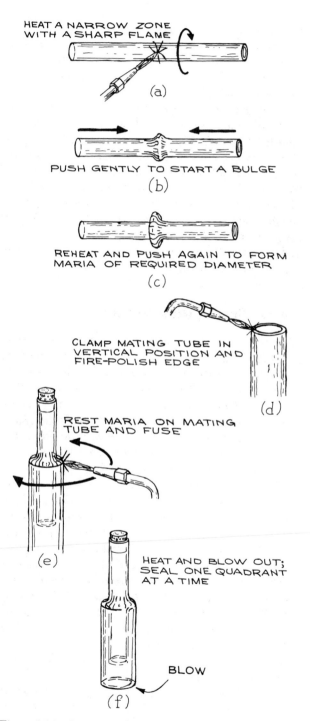

HEAT A NARROW ZONE WITH A SHARP FLAME

(a)

PUSH GENTLY TO START A BULGE

(b)

REHEAT AND PUSH AGAIN TO FORM MARIA OF REQUIRED DIAMETER

(c)

CLAMP MATING TUBE IN VERTICAL POSITION AND FIRE-POLISH EDGE

(d)

REST MARIA ON MATING TUBE AND FUSE

(e)

HEAT AND BLOW OUT; SEAL ONE QUADRANT AT A TIME

BLOW

(f)

Figure 2.14 Steps in making a ring seal.

tubing in both hands, heat a very narrow zone at the desired location. As soon as the glass softens, drop the tube out of the flame, and, with your elbows braced on the bench top or against your sides, push the ends of the tube toward one another. If the tube has been heated uniformly around and the ends are held coaxial, a bulge should develop around the tube. If the heating is not uniform, the tube will bend. If the glass is too soft, the ends of the tube are liable to move out of alignment and an eccentric bulge will result. When a uniform bulge is attained, reheat it, remove it from the flame, and push again to increase the diameter of the maria. The bulge should be solid glass, If too wide a zone is heated, the maria will be hollow and dangerously fragile.

To make a coaxial ring seal between two tubes, form on the smaller tube a maria of the same diameter as the larger. Mount the larger tube vertically, fire-polish the top edge, and rest the smaller tube on top as in Figure 2.14(e). With a small flame, heat the junction uniformly between the maria and the edge of the larger tube. The weight of the smaller tube should cause it to settle and become fused to the larger one. Finally blow out the seal in four steps, as in making a conventional seal. Exercise care to prevent the wall of the outer tube from sagging into the inner tube. If the inner tube is not centered, it can be realigned by softening the seal with a cool, bushy flame and manipulating its protruding end.

2.3.8 Bending Glass Tubing

Bends are best made beginning with the tubing vertical [Figure 2.15(a)]. Close off the upper end with a cork or a point and attach the blowtube to the other end. While supporting the upper end of the tube with the free hand, apply a large bushy flame at the location of the desired bend. Heat the glass uniformly over a length equal to about four times the diameter of the tube. When the glass becomes pliable, remove the flame and quickly bend the upper end over to the desired angle. As soon as the bend is completed, blow to remove any wrinkles which may have developed and to restore the tubing to its original diameter. Some reheating may be necessary.

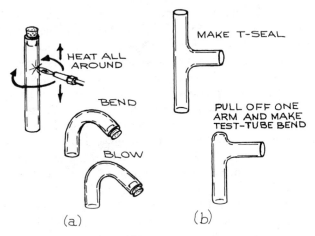

Figure 2.15   Bending tubing.

can be used to observe strain in glass in order to test the effectiveness of an anneal. A polariscope can be purchased from a laboratory supplier or can be constructed with high-extinction sheet polarizers as shown in Figure 2.16. The polarizers are crossed so that unstrained glass appears dark when placed between them. Areas of strain rotate the plane of polarization and appear bright as in Figure 2.17. A large stationary piece of glasswork, such as a vacuum manifold, can be tested by illuminating it from behind with vertically polarized light and then viewing the work through Polaroid sun glasses (which transmit horizontally polarized light). A suitable source can be constructed by placing a 100-watt lamp behind a diffuser in a ventilated, open top box, and covering the box with a sheet of polarizer sandwiched between glass plates.

A sharp right-angle bend can be produced by first making a T and then pulling off one running end of the T close to the seal [Figure 2.15(*b*)]. Blow out a test-tube end where the tubing is pulled off.

2.3.9   Annealing

Even when a seal is carefully blown out to give a uniform wall thickness, harmful stresses are inevitably frozen into the glass. Worked glass must be annealed by heating to a temperature above the annealing point (see Table 2.1) and then slowly cooling to the strain point. In a large professional glass shop, special ovens are employed for this operation. In the lab, however, glass must be annealed in the flame of the torch. After a piece of glass has been blown, it is heated uniformly over a wide area surrounding the work with a large, slightly yellow, brush flame. For Pyrex 7740 the glass should be brought to an incipient red heat. Care must be taken to prevent the glass from becoming so hot that it begins to sag under its own weight. The heated work is then cooled by slowly reducing the oxygen to the torch until the flame becomes sooty. When soot begins to adhere to the glass the flame can be removed.

Areas in glass which are under stress will rotate the plane of polarization of transmitted light. A *polariscope*

2.3.10   Sealing Glass to Metal

Glass-to-metal seals are fragile and difficult to produce in the lab. The simplest means of attaching metal tubing to a glass system or of passing a metallic electrode through a glass wall is to make use of a commercially produced glass-to-metal seal [Figure 3.24(*b*)]. If necessary a direct seal between a variety of metals and glasses can be produced with careful attention to design and fabrication.[2] Of these, the tungsten-to-Pyrex seal is most easily and reliably fabricated in the lab.

Although the linear coefficient of expansion of tungsten is about 50% greater than that of 7740 borosilicate glass, it is possible to make a strong, vacuum-tight seal between Pyrex and tungsten wire or rod of up to 0.050 in diameter.[3] Since the glass will wet tungsten oxide but not tungsten, it is necessary that the metal surface be properly prepared. Tungsten wire or rod usually has a longitudinal grain by virtue of the method of manufacture. The metal stock should be cut to length by grinding to prevent splintering. To prevent gas from leaking through the seal along the length of the wire it is good practice to fuse the ends of the wire in an electric arc. Alternatively, a copper or nickel wire can be welded to each end of the tungsten wire. Tungsten electrodes with copper or nickel wires attached are also available from commercial sources. Holding the wire in a pin vise, heat

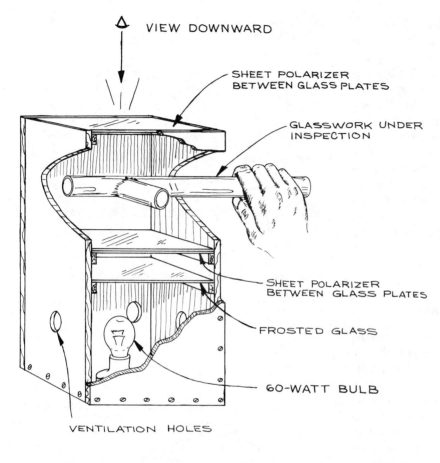

VIEW DOWNWARD

SHEET POLARIZER
BETWEEN GLASS PLATES

GLASSWORK UNDER
INSPECTION

SHEET POLARIZER
BETWEEN GLASS PLATES

FROSTED GLASS

60-WATT BULB

VENTILATION HOLES

Figure 2.16 A polariscope.

the area to be sealed to a dull red and immediately rub the hot metal with a stick of sodium nitrite. The salt combines with the tungsten oxide to leave a clean metal surface. Rinse the metal with tap water to remove the salt and then rinse with distilled water. The metal must then be reoxidized in an oxygen-rich flame to produce an oxide layer of the proper thickness. If the layer is too thick it will pull away from the metal; if it is too thin it may all dissolve into the glass. The clean tungsten should be gently heated only until the metal appears blue-green when cool. A piece of 7740 glass, 2 cm long with a wall thickness of about 1 mm and an inner diameter slightly larger than the tungsten wire, is slipped over the wire and slid along to the sealing position. Be careful not to scratch the oxide layer. The glass tube is fused to the metal using a small sharp flame. Melting

Figure 2.17 The appearance of strain in unannealed glass as seen in the polariscope.

should proceed from one end to the other with continuous rotation to prevent the trapping of air. If the metal is properly wetted by the glass, the wire will appear oversize where it passes through the seal. A copper or amber color after the seal is cool indicates a good seal. The beaded tungsten rod is then sealed into the apparatus, as in making a ring seal (Section 2.3.7). If necessary, a shoulder can be built up around the middle of the bead by fusing on a winding of 1-mm glass rod.

A tungsten-to-Pyrex seal makes an excellent electrical feedthrough. Wires can be brazed or spot-welded to either end of the tungsten rod. Since tungsten is not a good electrical conductor, care must be taken to ensure that electrical currents do not overheat the feedthrough. The safe current-carrying capacity of the tungsten in this application is about 5000 A in.$^{-2}$.

2.3.11 Grinding and Drilling Glass

In many instances the ends of a tube must be ground square or at an angle, or a tube must be ground to a specific length. The ends of a laser tube, for example, must be cut and ground to Brewster's angle before optical windows are glued in place. Grinding a plane surface is performed in a water slurry of abrasive on a smooth iron plate. It is done with a circular motion while applying moderate pressure. The work should be lifted from the surface frequently in order to redistribute the abrasive. If the angle of the ground surface relative to the workpiece is not critical, the work can be grasped in the hand; otherwise, a wooden jig, such as is shown in Figure 2.18, may be fabricated to orient the work with respect to the grinding surface.

Carborundum or silicon carbide grit of #30, 60, 120, or 220 mesh is used for coarse grinding, and emery of #220, 400, or 600 mesh is used for fine grinding. In any grinding operation, begin with the grit required to remove the largest irregularities and proceed in sequence through the finer grits. Before changing from one grit to the next finer, the work and the grinding surface must be scrupulously washed so that coarse abrasive does not become mixed with the next finer grade. Before changing from one grit to the next, it is

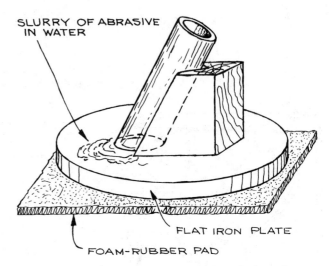

Figure 2.18   A wooden jig used in grinding the face of a piece of tubing.

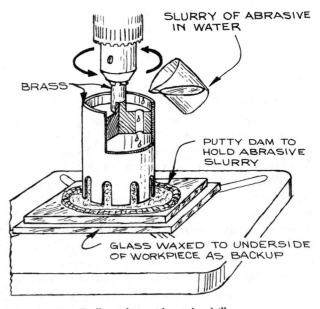

Figure 2.19   Drilling glass with a tube drill.

useful to inspect the ground surface under a magnifying glass. If the abrasive in use has completed its job, the scratch marks left by the abrasive should be of a fairly uniform width.

Holes can be cut in glass, and round pieces of glass can be cut from larger pieces, with a slotted brass tube drill in a drill press (Figure 2.19). These drills are easily fabricated in sizes from $\frac{1}{4}$ in. to several inches in diameter. Either flat or curved glass can be drilled. Cutting is accomplished by continuously feeding the drill with a slurry of #60 or #120 mesh abrasive. It is usually most convenient to form a dam of soft wax or putty around the hole location to hold the slurry of abrasive around the drill. Work slowly with a light touch at a drill speed of 50–150 rpm. The tool should be backed out frequently to permit the abrasive to flow under. Because grinding action occurs along both the leading edge and the side of the drill tube, the finished hole will be somewhat larger than the drill body. Frequently the edges of the hole will chip where the drill breaks through on the under side of the work. This can be prevented when drilling flat glass by backing up the workpiece with a second piece of glass waxed onto the underside.

Glass can be sawn as well as drilled. Many glass shops use a powered, diamond-grit cutting wheel for precision cutting. In the absence of such a machine, it is possible to cut glass in much the same manner as it is drilled. A hacksaw with a strip of 20-gauge copper as a blade is used. The work is clamped in a miter box and sawn with a light touch while feeding an abrasive slurry to the blade.

CITED REFERENCES

1. W. G. Housekeeper, Trans, A.I.E.E., 42, 870, 1923.
2. F. Rosebury, *Handbook of Electron Tube and Vacuum Techniques*, Addison-Wesley, Reading, Mass., 1965, pp. 54–66.
3. J. Strong, *Procedures in Experimental Physics*, Prentice-Hall, Englewood Cliffs, N.J., 1938, p. 24; G. W. Green, *The Design and Construction of Small Vacuum Systems*, Chapman and Hall, London, 1968, pp. 100–102.

GENERAL REFERENCES

W. E. Barr and J. V. Anhorn, *Scientific and Industrial Glass Blowing and Laboratory Techniques*, Instruments Publishing, Pittsburgh, 1959.

Corning Glass Works, *Laboratory Glass Blowing with Corning's Glasses*, Publ. No. B-72, Corning Glass Works, Corning, N.Y..

J. E. Hammesfahr and C. L. Strong, *Creative Glass Blowing*, Freeman, San Francisco, 1968.

W. Morey, *The Properties of Glass*, Reinhold, New York, 1954.

F. Rosebury, *Handbook of Electron Tube and Vacuum Techniques*, Addison-Wesley, Reading, Mass., 1965.

MANUFACTURERS AND SUPPLIERS

Ace Glass, Inc.
P.O. Box 688
1430 Northwest Blvd.
Vineland, NJ 08360

Corning Glass Works
Corning, NY 14830

Fischer and Porter Co.
Warminster, PA

Kimble Glass, Owens-Illinois, Inc.
P.O. Box 1035
Toledo, OH 43666

Labglass
Northwest Blvd.
Vineland, NJ 08360

CHAPTER 3

VACUUM TECHNOLOGY

In the modern laboratory, there are many occasions when a gas-filled container must be emptied. Evacuation may simply be the first step in creating a new gaseous environment. In a distillation process, there may be a continuing requirement to remove gas as it evolves. Often it is necessary to evacuate a container to prevent air from contaminating a clean surface or interfering with a chemical reaction. Beams of atomic particles must be handled *in vacuo* to prevent loss of momentum through collisions with air molecules. A vacuum system is an essential part of laboratory instruments such as the mass spectrometer and the electron microscope. Many forms of radiation are absorbed by air and thus can propagate over large distances only in a vacuum. Far-IR, far-UV, and X-ray spectrometers are operated within vacuum containers. Simple vacuum systems are used for vacuum dehydration and freeze-drying. Nuclear particle accelerators and thermonuclear devices require huge, sophisticated vacuum systems.

3.1 GASES

The pressure and composition of residual gases in a vacuum system vary considerably with its design and history. For some applications a residual gas density of tens of billions of molecules per cubic centimeter is tolerable. In other cases no more than a few hundred thousand molecules per cubic centimeter constitutes an acceptable vacuum: "One man's vacuum is another man's sewer."[1] It is necessary to understand the nature of a vacuum and of vacuum apparatus to know what can and cannot be done, to understand what is possible within economic constraints, and to choose components that are compatible with each other as well as with one's needs.

3.1.1 The Nature of the Residual Gases in a Vacuum System

The pressure below one atmosphere is loosely divided into vacuum categories. The pressure ranges and number densities corresponding to these categories are listed in Table 3.1. As a point of reference for vacuum work it is useful to remember that the number density at 1 mtorr is about 3.5×10^{13} cm^{-3}, and at 1 Pa about 2.7×10^{14} cm^{-3}, and that the number density is proportional to the pressure.

The composition of gas in a vacuum system is modified as the system is evacuated because the efficiency of a vacuum pump is different for different gases. At low pressures molecules desorbed from the walls make up the residual gas. Initially, the bulk of the gas leaving the walls is water vapor; at very low pressures, in a container that has been baked, it is hydrogen.

Table 3.1   AIR AT 20°C

	Pressure (torr)[a]	Number Density (cm^{-3})	Mean Free Path (cm)	Surface Collision Frequency (cm^{-2}sec^{-1})	Times for Monolayer Formation[b] (sec)
One atmosphere	760	2.7×10^{19}	7×10^{-6}	3×10^{23}	3.3×10^{-9}
Lower limit of:					
Rough vacuum	10^{-3}	3.5×10^{13}	5	4×10^{17}	2.5×10^{-3}
High vacuum	10^{-6}	3.5×10^{10}	5×10^3	4×10^{14}	2.5
Very high vacuum	10^{-9}	3.5×10^7	5×10^6	4×10^{11}	2.5×10^3
Ultrahigh vacuum	0				

[a]1 torr = 132 Pa.
[b]Assuming unit adhesion efficiency and a molecular diameter of 3×10^{-8} cm.

3.1.2 Kinetic Theory

In order to understand mass flow and heat flow in a vacuum system it is necessary to appreciate the immense change in freedom of movement experienced by a gas molecule as the pressure decreases.

The average velocity of a molecule can be deduced from the Maxwell-Boltzmann velocity distribution law:

$$\bar{v} = \left(\frac{8kT}{\pi m} \right)^{1/2}.$$

For an air molecule (molecular weight of about 30) at 20°C,

$$\bar{v} \simeq 5 \times 10^4 \, \text{cm sec}^{-1} = \tfrac{1}{2} \, \text{km sec}^{-1}.$$

Each second a molecule sweeps out a volume with a diameter twice that of the molecule and a length equal to the distance traveled by the molecule in a second. As shown in Figure 3.1, this molecule collides with any of its neighbors whose center lies within the swept volume. The number of collisions per second is equal to the number of neighbors within the swept volume. On the average, the number of collisions per second (Z) is the number density of molecules (n) times the volume swept by a molecule of velocity $\bar{v}$ and diameter ξ. More precisely,

$$Z = \sqrt{2} \, n\pi\xi^2\bar{v},$$

where the $\sqrt{2}$ accounts for the relative motion of the molecules. The time between collisions is the reciprocal of this *collision frequency*, and the average distance between collisions, or *mean free path*, is

$$\lambda = \bar{v}Z^{-1}$$
$$= \frac{1}{\sqrt{2} \, n\pi\xi^2}.$$

The mean free path is inversely proportional to the pressure. For an N_2 or O_2 molecule, $\xi \simeq 3 \times 10^{-8}$ cm. The number density at 1 mtorr is about 3.5×10^{13} cm^{-3}. Thus the mean free path in air at 1 mtorr is about 5 cm. Recalling the relationship between λ and P,

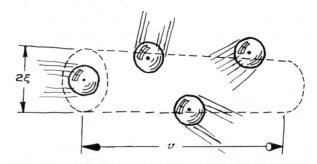

Figure 3.1   The volume swept in one second by a molecule of diameter ξ and velocity $\bar{v}$ (cm sec^{-1}). The molecule will collide with any of its neighbors whose center lies within the volume.

a valuable rule of thumb is that for air at $20°C$

$$\lambda \simeq \frac{5}{P(\text{mtorr})} \text{ cm.}$$

3.1.3 Surface Collisions

The frequency of collisions of molecules within a container with the surface of the container, per unit area, is

$$Z_{\text{surface}} = \frac{n\bar{v}}{4} \quad (\text{sec}^{-1}\text{cm}^{-2}).$$

The sticking probability for most air molecules on a clean surface at room temperature is between 0.1 and 1.0. For water the sticking probability is about unity for most surfaces. Assuming unit sticking probability and a molecular diameter $\xi = 3 \times 10^{-8}$ cm, the time required to form a monolayer of adsorbed air molecules at $20°C$ is

$$t = \frac{2.5 \times 10^{-6}}{P(\text{torr})} \text{ sec.}$$

Thus to maintain a clean surface for a useful period of time may require a gas pressure over the surface less than 10^{-9} torr.

3.1.4 Bulk Behavior versus Molecular Behavior

The pressure of gas within a vacuum system may vary over ten or more orders of magnitude as the system is evacuated from atmospheric pressure to the lowest attainable pressure. At high pressures, when the mean free path is much smaller than the dimensions of the vacuum container, gas behavior is dominated by intermolecular interactions. These interactions, resulting in viscous forces, ensure good communication among all regions of the gas. At high pressures, a gas behaves as a homogeneous fluid. When the gas density is decreased to the extent that the mean free path is much larger than the container, molecules rattle around like the balls on a billiard table (not a pool table, where the ball density can be high), and gas behavior is determined by the random motion of the molecules as they bounce from wall to wall.

In the *viscous-flow* region where the mean free path is relatively small, gas flow improves with increasing pressure because gas molecules tend to queue up and push on their neighbors in front of them. Gas flow is impeded by turbulence and by viscous drag at the walls of the pipe that is conducting the gas. In the viscous region the coefficients of viscosity and thermal conductivity are independent of pressure.

When the mean free path far exceeds the dimensions of the container, the process of gas flow is called *molecular flow*. In this region momentum transfer occurs between molecules and the wall of a container, but molecules seldom encounter one another. A gas is not characterized by a viscosity. Gas flows from a region of high pressure to one of low pressure simply because the number of molecules leaving a unit of volume is proportional to the number of molecules within that volume. Gas flow is a statistical process. At very low pressure a molecule does not linger at a surface for a sufficient time to reach thermal equilibrium. Thus thermal conductivity at low pressures is a function of gas density, and the coefficient of thermal conductivity depends upon the pressure and upon the condition of the surface.

3.2 GAS FLOW

3.2.1 Parameters for Specifying Gas Flow

Before discussing vacuum apparatus it is necessary to define the parameters used by vacuum engineers to characterize gas flow.

The volume rate of flow through an aperture or across a cross section of a tube is defined as the *pumping speed* at that point:

$$S \equiv \frac{dV}{dt} \quad [\text{liter sec}^{-1}; \text{ ft}^3\text{min}^{-1} (\text{cfm})].$$

The capacity of a vacuum pump is specified by the

speed measured at its inlet:

$$S_P \equiv \frac{dV}{dt} \text{ (at pump inlet)}.$$

The mass rate of flow through a vacuum system is proportional to the *throughput*:

$$Q \equiv PS \qquad (\text{torr liter sec}^{-1}; \text{ torr cfm}).$$

To determine the throughput it is necessary that the pressure and speed be measured at the same place, since these quantities vary throughout the system.

The ability of a tube to transmit gas is characterized by its *conductance C*. The definition of conductance is analogous to Ohm's law for electrical circuitry. The throughput of a tube depends upon the conductance of the tube and the driving force, which in this case is the pressure drop across the tube:

$$Q = (P_1 - P_2)C \qquad (P_1 > P_2).$$

Notice that conductance has the same units as pumping speed.

3.2.2   Network Equations

A complicated network of tubes can be reduced to a single equivalent conductance for the purpose of analysis. In analogy to electrical theory once again, a number of tubes in series can be replaced by a single equivalent tube with conductance C_{series} given by

$$\frac{1}{C_{\text{series}}} = \frac{1}{C_1} + \frac{1}{C_2} + \frac{1}{C_3} + \cdots.$$

A parallel network can be replaced by an equivalent conductor with conductance

$$C_{\text{parallel}} = C_1 + C_2 + C_3 + \cdots.$$

3.2.3   The Master Equation

By use of the network equations given above, an entire vacuum system can be reduced to a single equivalent conductance leading from a gas source to a pump as

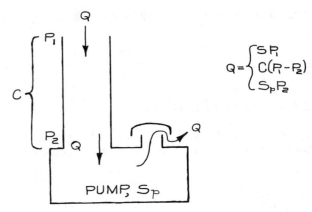

Figure 3.2   Analytical representation of a vacuum system.

shown in Figure 3.2. The net speed of the system is

$$S = \frac{Q}{P_1},$$

and the speed at the pump inlet is

$$S_p = \frac{Q}{P_2}.$$

Notice that the throughput is the same at every point in this system, since there is only one gas source. The throughput of the conductance is

$$Q = (P_1 - P_2)C.$$

Upon substitution for P_1 and P_2 from the previous two equations, this equation after some rearrangement becomes

$$\frac{1}{S} = \frac{1}{S_p} + \frac{1}{C}.$$

This is the master equation that relates the net speed of a system to the capacity of the pump and the conductors leading to the pump. From this equation we see that both the conductance of a system and the speed of the pump exceed the net speed of the system. In vacuum-system design, economics should be considered. The cost of a vacuum pump is usually greater than the cost of constructing a tube leading from a vacuum container to the pump. In order to achieve a given net speed for a system, it is usually most economical to design a system whose equivalent conductance exceeds the required net speed by a factor of three or four so that the speed of the pump need only exceed the net speed by a small amount.

3.2.4 Conductance Formulae

In the viscous flow region the conductance of a tube depends upon the gas pressure and viscosity. For a tube of circular cross section the conductance for air at 20°C is, in the absence of turbulence, is

$$C = 180 \frac{D^4}{L} P_{av} \text{ liter sec}^{-1},$$

when the diameter D and length L of the tube are in centimeters, and the average pressure P_{av} is in torr. This equation is useful for determining the dimensions of roughing and forepump lines. Elbows, bends and joints in a tube will reduce the conductance, but since conductors are usually overdesigned, it is sufficient to consider only the simple cylindrical tubes in a network and ignore the junctions between these tubes.

In the molecular-flow region conductance is independent of pressure. For air at 20°C the conductance of a tube of circular cross section is

$$C = 12 \frac{D^3}{L} \text{ liter sec}^{-1},$$

when the diameter and length are measured in centimeters. This equation is useful in determining the dimensions of a high-vacuum chamber and the size of the tubes leading from the chamber to a diffusion pump or ion pump. As an example of poor design consider the case of a vacuum chamber connected to a 100-liter sec^{-1} diffusion pump by a tube 2.5 cm in diameter and 10 cm long. The net speed of the pump and connecting tube is

$$S = \left(\frac{1}{100} + \frac{1}{19} \right)^{-1} = 16 \text{ liter sec}^{-1}.$$

The pump in this case is being strangled by the connecting tube. Consider also the pressure drop across the

tube. Since the throughput is a constant, the pressure at the inlet of the tube P_1 is related to the pressure at the inlet of the pump P_2 by

$$SP_1 = S_p P_2;$$

thus

$$\frac{P_1}{P_2} = \frac{S_p}{S} = 6.$$

This result indicates the importance of proper location of a pressure gauge. In this case a gauge at the mouth of the pump would indicate a pressure six times lower than the pressure in the vacuum container.

In the molecular-flow region the conductance of an aperture for a gas of molecular weight M is

$$C = 3.7 \left(\frac{T}{M} \right)^{1/2} A \quad \text{liter sec}^{-1},$$

where A is the area in cm^2. This equation is useful for determining the rate of gas flow into a vacuum chamber through an aperture in a gas-filled collision chamber or ion source.[2]

There is a *transition region* between the viscous and molecular flow regions where the mean free path approximately matches the dimensions of the container. Any mathematical description of this region is difficult. In most vacuum systems the transition region is encountered only briefly during pumpdown from atmosphere to high vacuum. The formulae for conductance in the molecular flow region will give a conservative estimate of the conductance in the transition region.

3.2.5 Pumpdown Time

The time required to pump a chamber of volume V from pressure P_0 to P is

$$t = 2.3 \frac{V}{S} \ln \frac{P_0}{P},$$

assuming a constant net pumping speed and assuming no additional gas is admitted to the system. A typical high-vacuum system is rough-pumped to about 5×10^{-2}

torr with a mechanical pump and then pumped to very low pressures with a diffusion pump or some other pump that works effectively in the molecular flow region. The pumpdown calculation is performed in two steps corresponding to these two operations.

3.2.6 Outgassing

The rate of evacuation of a chamber at pressures below 10^{-6} torr is not controlled by the pumping speed of a system but rather by the rate of evolution of gas from the walls of the chamber. Stainless steel outgasses at about 10^{-7} torr liter sec^{-1} cm^{-2} after an hour of pumping. More than a day of pumping may be required to reduce this rate below 10^{-8} torr liter sec^{-1} cm^{-2}. The rate of outgassing from metal surfaces can be considerably reduced by baking to drive out adsorbed gases. Plastics and elastomers outgas at as much as 10^{-5} torr liter sec^{-1} cm^{-2} after an hour of pumping, and as a result such materials are not suitable for use at pressures below 10^{-7} torr.

Consider for example the problem of maintaining a base pressure of 10^{-7} torr in a stainless-steel chamber. If the outgassing rate is 10^{-8} torr liter sec^{-1} cm^{-2}, a pumping speed of 0.1 liter sec^{-1} for every cm^2 of surface is required. For a container 30 cm in diameter and 30 cm high, a pumping speed of 400 liter sec^{-1} is needed. This speed is typical of a diffusion pump with a nominal 4-in. inlet port.

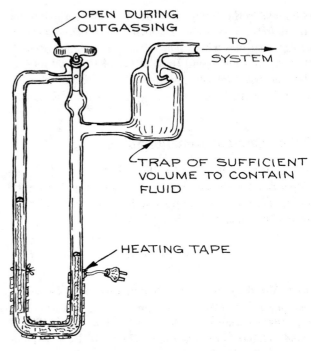

OIL DENSITIES (20° C)

di-n-butyl phthalate	1.044 gm cm^{-3}
Octoil	0.983
Octoil-S	0.912
DC-704	1.07
DC-705	1.09

Figure 3.3 Oil manometer. The trap is intended to contain the oil in the event that the low-pressure arm of the manometer is accidentally opened to the atmosphere.

3.3 PRESSURE MEASUREMENT

The pressure within a vacuum system may vary over ten or more orders of magnitude. No one gauge will operate over this great range, and thus most systems are equipped with several different gauges.

3.3.1 Mechanical Gauges

The simplest pressure gauges are hydrostatic gauges such as the oil or mercury manometer. A closed-end U-tube manometer filled with mercury is useful down to

1 torr. Diffusion-pump oils such as Octoil, di-*n*-butyl phthalate, or the Dow Corning silicone oils DC704 or DC705 may also be used as manometer fluids. Oil-filled manometers may be used down to 0.1 torr because the oil density is much less than that of mercury. As shown in Figure 3.3, an oil manometer should be made of glass tubing of at least 1-cm diameter in order to reduce the effects of capillary depression. To prevent oil adhesion to the walls, the manometer should be cleaned with dilute HF before filling. Because air is somewhat soluble in oil, a heater is usually provided to outgas the oil prior to use.[3] Several manometer designs have been developed to increase the surface area of oil exposed to the

vacuum in order to hasten outgassing.[4] Both oil and mercury manometers will contaminate a vacuum system with vapor of the working fluid. If this is a problem, a cold trap is placed between the gauge and the system to condense the offending vapor. Mercury in a manometer can be isolated and protected from chemical attack by floating a few drops of silicone oil on top of the mercury column.

The McLeod gauge shown in Figure 3.4 is a sophisticated hydrostatic gauge and is sensitive to much lower pressures than a simple U-tube manometer. In a McLeod gauge a sample of gas is trapped and compressed by a known amount (typically 1000:1) by displacing the gas with mercury from the original volume into a much smaller volume. The pressure of the compressed gas is measured with a mercury manometer and the original pressure determined by use of the general gas law. Of course, it is not possible to use a McLeod gauge to measure the pressure of a gas that condenses upon compression. With care a high-quality McLeod gauge will provide accuracy of a few percent down to about 10^{-4} torr. Some gauges can be used at even lower pressures, but it is necessary to consider the error caused by the pumping action of mercury vapor streaming out of the gauge.[5] The useful pressure range of any one McLeod gauge is limited by geometrical constraints to about four orders of magnitude in pressure.

The McLeod gauge is constructed of glass and is easily broken. If the mercury is raised into the gas bulb too quickly, it can acquire sufficient momentum to shatter the glass. When a gauge is broken, the external air pressure frequently drives a quantity of mercury into the vacuum system. For this reason it is wise to place a ballast bulb of sufficient volume to contain the mercury charge of the gauge between the gauge and the system.

A number of different gauges depend upon the flexure of a metal tube or diaphragm as a measure of pressure. In the Bourdon gauge, a thin-wall, curved tube, closed at one end, is attached to the vacuum system. Pressure changes cause a change in curvature of the tube. A mechanical linkage to the tube drives a needle, which gives a pressure reading on a curved scale. Another type of mechanical gauge contains a chamber divided in two by a thin metal diaphragm. The volume on one side of the diaphragm is sealed, while the volume on the other side is attached to the system. A variation in pressure on

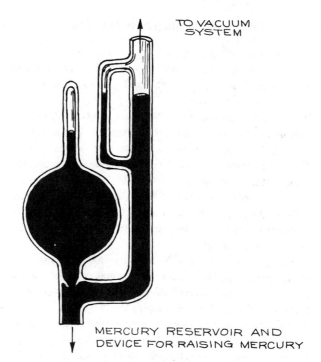

Figure 3.4   McLeod gauge.

one side relative to the other causes the diaphragm to flex, and this movement is sensed by a system of gears and levers, which drives a needle on the face of the gauge. The precision of these gauges is limited by hysteresis caused by friction in the linkage. To overcome this friction it is helpful to gently tap the gauge before making a reading. Unlike the liquid manometer, mechanical gauges are not absolute gauges. The pressure scale and zero location on these gauges must be calibrated against a McLeod gauge or U-tube manometer. Mechanical gauges are useful down to 1 torr. They offer the advantage of being insensitive to the chemical or physical nature of the gas. Excellent mechanical gauges are manufactured by Wallace & Tiernan and by Leybold-Heraeus.

A recent development of the diaphragm gauge is the capacitance manometer, wherein the diaphragm is one plate of an electrical capacitor. A change in the diaphragm position results in a change in capacitance, which is detected by a sensitive capacitance bridge. The MKS Baratron capacitance manometer is claimed to be accurate to a few percent at 10^{-4} torr.

3.3.2 Thermal-Conductivity Gauges

The thermal conductivity of a gas decreases from some constant value above 1 torr to essentially zero at about 10^{-3} torr. This change in thermal conductivity is used as an indication of pressure in the Pirani gauge and the thermocouple gauge. In both gauges a low-temperature filament is heated by a constant current. The temperature of the filament depends on the rate of heat loss to the surrounding gas. In the thermocouple gauge, the temperature is determined from the e.m.f. produced by a thermocouple in contact with the filament (Figure 3.5). In the Pirani gauge, a change in temperature of the filament results in a change in resistivity that is detected by a sensitive bridge.

The pressure indicated by a thermocouple gauge or a Pirani gauge depends upon the thermal conductivity of the gas. They are usually calibrated by the manufacturer for use with air. For other gases these gauges must be recalibrated point by point over their entire range, since thermal conductivity is a nonlinear function of pressure. In routine use a thermal-conductivity gauge can only be expected to be accurate to within a factor of two. This accuracy is adequate when the gauge is used to sense the foreline pressure of a diffusion pump or to determine whether the pressure in a system is sufficiently low to begin diffusion-pumping. The principal advantages of these gauges are their ease of use, ruggedness, and low cost.

3.3.3 Ionization Gauges

In the region of molecular flow, pressure is usually measured with an ion gauge. In this type of gauge, gas molecules are ionized by electron impact and the resulting positive ions are collected at a negatively biased electrode. The current to this electrode is a function of pressure. There are several types of ion gauges, differing primarily in the mechanism of electron production. The most common gauge is the thermionic or hot-cathode ionization gauge, shown schematically in Figure 3.6. Electrons from an electrically heated filament are accelerated through the gas toward a positively biased grid. Ions are collected at a central wire, and the positive ion

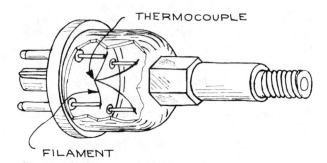

Figure 3.5 Thermocouple gauge.

current is measured by a sensitive electrometer. These gauges are useful in the 10^{-3}- to 10^{-9}-torr range. The indicated pressure depends upon the ionization cross section of the gas. When a gauge calibrated for air is used with hydrogen, it will read low by a factor of 3. With helium, it will be low by a factor of 8.

Ion gauges have the sometimes useful and sometimes detrimental property of functioning as pumps. This pumping action results from two mechanisms. Ions accelerated to and imbedded in the collector are effectively removed from the system. In addition, metal

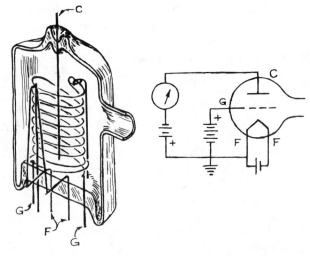

Figure 3.6 Bayard-Alpert type of thermionic ionization gauge.

evaporated from the filament deposits on the walls to produce a clean, chemically active surface that adsorbs nitrogen, oxygen, and water. At pressures below 10^{-7} torr this pumping action creates a pressure gradient, which causes the gauge to indicate a pressure different from the pressure in the system to which it is attached. On the other hand, it is possible to pump a small system solely with an ion gauge. Typically the pumping speed of an ion gauge is 0.2 liter sec^{-1} for N_2 when it is operated with an electron-emission current of 10 mA.

At low pressures the accuracy of an ion gauge is improved if the electrode surfaces and gauge walls are periodically outgassed by heating. With most commercial gauges this degassing is accomplished by heating the grid to incandescence by passing an a.c. current through it.

Other types of ionization gauges include the cold-cathode gauge (Penning gauge) and the Alphatron. In the first type a magnetically confined discharge is contained between two electrodes at room temperature. Molecules in the discharge are ionized and collected to produce a measurable current. These gauges are useful between 10^{-2} and 10^{-6} torr. In the Alphatron gauge the ionizing electrons are produced by a radium source. The Alphatron gauge operates between atmosphere and 10^{-3} torr.

3.3.4 Mass Spectrometers

Mass spectrometers may be used to determine the partial pressures of residual gases in a system and to detect leaks. Residual-gas analyzers (RGAs) may be of the magnetic-deflection, quadrupole, time-of-flight, r.f., or cycloidal type, but most commercial RGAs are of the first two types. Typically these devices cover the range of 1 to 200 amu and can detect partial pressures as low as 5×10^{-13} torr. Fixed-focus mass spectrometers set to detect helium are widely used as sensitive leak detectors. Leaks are located by monitoring the helium concentration within a vacuum system while the exterior of the system is probed with a small jet of helium. The detection limit of these devices is of the order of 10^{-10} torr liter sec^{-1}.

The pressure ranges in which the various gauges are useful are shown in Figure 3.7.

3.4 VACUUM PUMPS

Vacuum pumps, like pressure gauges, operate in a limited pressure range. In general, a pump that operates in the viscous flow region will not operate in the molec-

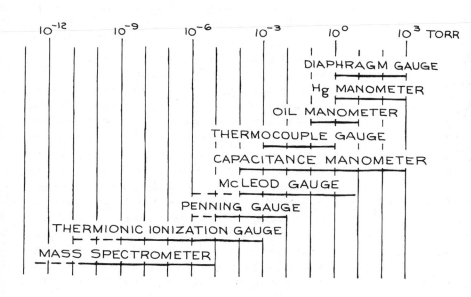

Figure 3.7 Operational pressure ranges of common gauges.

ular flow region and vice versa. The useful range of a pump is also limited by the vapor pressure of the materials of construction and the working fluids within the pump.

3.4.1 Mechanical Pumps

The pump most commonly used for attaining pressures down to a few millitorr is the oil-sealed rotary pump shown schematically in Figure 3.8. In this pump a rotor turns off-center within a cylindrical stator. The interior of the pump is divided into two volumes by spring-loaded vanes attached to the rotor. Gas from the pump inlet enters one of these volumes and is compressed and forced through a one-way valve to the exhaust. The seal between the vanes and the stator is maintained by a thin film of oil. The oil used in these pumps is a good-quality lubricating oil from which the high-vapor-

pressure fraction has been removed. These pumps are also made in a two-stage version in which two pumps with rotors on a common shaft operate in series. Rotary pumps to be used for pumping condensable vapors are provided with a gas ballast. This is a valve that admits air to the compressed gas just prior to the exhaust cycle. This additional air causes the exhaust valve to open before the pressures of condensable vapors exceed their vapor pressure and thus prevents these vapors from condensing inside the pump.

Oil-sealed rotary pumps will operate well for years if the inner surfaces do not rust and the oil maintains its lubricating properties. It is wise to leave these pumps operating continuously so that the oil stays warm and dry. For storage a pump should be filled with new oil and the ports sealed. In use, an increase in the lowest attainable pressure (the base pressure) indicates that the oil has been contaminated with volatile materials. The dirty oil should be drained while the pump is warm. The pump should be filled with new oil, run for several minutes, drained, and refilled.

Rotary pumps are available with capacities of 1 to 500 liter sec^{-1}. A single-stage pump is useful down to 50 mtorr, and a two-stage pump to 5 mtorr. Typical performance of a single-stage pump is indicated by the pumping-speed curve in Figure 3.8. With a two-stage pump, a base pressure of 10^{-4} torr can be achieved after a long pumping time if the back diffusion of oil vapor from the pump is suppressed by use of a sorption or liquid-air trap on the pump inlet. This is a simple scheme for evacuating small spectrometers and Dewar flasks or other vacuum-type thermal insulators.

The exhaust gases from an oil-sealed mechanical vacuum pump contain a mist of fine droplets of oil. This oil smoke is especially dense when the inlet pressure is in the 200- to 600-torr range. The oil droplets are extremely small, usually less than 5 microns. Over the course of time, this oil settles on the pump and its surroundings and collects dirt and grime. Furthermore, breathing the finely dispersed oil may injure the operator's lungs. Most pump manufacturers market filters to cope with this problem, and their use is recommended. An excellent system of coalescing-type filters is manufactured by Balston, Inc. These filters cause the droplets to accumulate into large drops, which run off into a

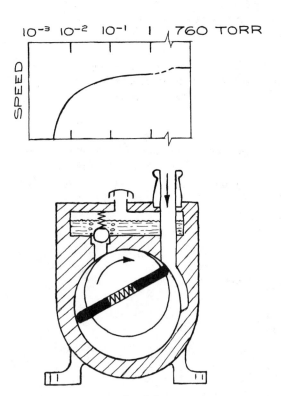

Figure 3.8 Two-vane, oil-sealed rotary pump.

sump at the bottom of the filter housing. Fittings are available to install these filters on most pumps. An alternative solution to the pump-exhaust problem, particularly when pumping toxic gases, is to vent the pump into a fume hood in the lab. Exhaust lines can be made of PVC drain pipe available from plumbing suppliers.

In order to achieve high pumping speeds in the 10- to 10^{-2}-torr region, a *Roots blower* can be used in series with an oil-sealed rotary pump. As illustrated in Figure 3.9, these pumps consist of a pair of counterrotating, two-lobed rotors on parallel shafts. Rotational speeds are about 3000 rpm. There is a clearance of a few thousandths of an inch between the rotors themselves and between the rotors and the housing. The roughing pump in series is required because there is no oil present in the Roots blower to attain a seal at high pressure. A Roots blower provides a compression ratio of the order of 10 : 1, and hence the speed required of the backing pump is correspondingly lower than that of the blower. Roots pumps are available with displacements of a few hundred to a few thousand liters per second.

There are now commercially available mechanical pumps, called *molecular-drag pumps*, which operate in the molecular flow regime. In these pumps one or more balanced rotors turn at 20,000 to 50,000 rpm within a slotted stator. The edge speed of the rotors approaches molecular velocities. When a molecule strikes a rotor, a significant component of momentum is transferred to the molecule in the direction of rotation. This transferred momentum causes molecules to move from the pump inlet toward the exhaust. The most common version of this pump is the *turbomolecular pump*, which, like a turbine, has a series of rotors with oblique radial slots turning between radially slotted stators. These pumps achieve a compression ratio of up to 10^6 : 1 provided the outlet pressure is kept below 100 mtorr. This requirement means that the turbomolecular pump must be run in a series with a conventional rotary pump, which is referred to as a *backing pump* or *forepump*. The forepump, pumping directly through the turbopump, can be used for initial roughing of a system from atmosphere to 10^{-1} torr. In this capacity the rotary pump is referred to as a *roughing pump*. Turbomolecular pumps are available with capacities of a few hundred to 10,000 liter sec^{-1}. They have the ad-

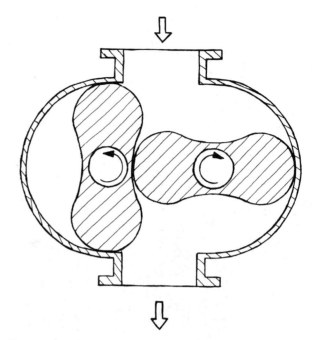

Figure 3.9   Roots blower.

vantage over diffusion pumps of providing an oil-free and mercury-free vacuum. Their main disadvantage is cost. For comparable pumping speeds, turbomolecular pumps cost about ten times more than diffusion pumps.

3.4.2 Vapor Diffusion Pumps

In a diffusion pump, gas molecules are moved from inlet to outlet by momentum transfer from a directed stream of oil or mercury vapor. As shown in Figure 3.10, the working fluid is evaporated in an electrically heated boiler at the bottom of the pump. Vapor is conducted upward through a tower above the boiler to a nozzle or an array of nozzles from which the vapor is emitted in a jet directed downward and outward toward the pump walls. The walls of the pump are cooled so that molecules of the working fluid vapor condense before their motion is randomized by repeated collisions. The pump walls are usually water-cooled, although in some small pumps they are air-cooled. The condensate runs down the pump wall to return to the boiler.

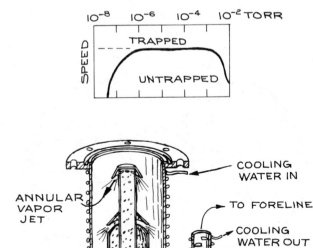

Figure 3.10 Diffusion pump.

As indicated by the pumping-speed curve in Figure 3.10, the pumping action of a diffusion pump begins to fail when the inlet pressure increases to the point where the mean free path is less than the distance from the vapor-jet nozzle to the wall. When this occurs the net downward momentum of vapor molecules is lost and the vapor begins to diffuse upward into the vacuum system. Diffusion pumping of a system may be initiated at 50 to 100 mtorr, but the system pressure should quickly fall below 1 mtorr or the system may become significantly contaminated with the vapor of the working fluid. Oil diffusion pumps can run against an outlet pressure of 300 to 500 mtorr, and mercury pumps can tolerate an outlet pressure of a few torr. Thus these pumps must be operated in series with a mechanical forepump. The pump speed is insensitive to foreline pressure up to some critical pressure. If this critical pressure is exceeded, the pump is said to stall. Stalling is a disaster, because hot pump-fluid vapor is flushed backwards through the pump into the system.

Oil diffusion pumps use low-vapor-pressure hydrocarbon or silicone oil as the working fluid. Hydrocarbon oils are subject to cracking and will oxidize if exposed to air when hot. Silicone oils are much less subject to chemical reaction and may be exposed to air when hot. Silicone oils are poor lubricants and thus should be used with a foreline trap that prevents their entering the backing pump. Many hydrocarbon oils have the advantage of being much less expensive than silicone oils. The absolute lowest pressure attainable with an untrapped diffusion pump is the room-temperature vapor pressure of the working fluid. The vapor pressures of some common pump oils are given in Table 3.2. For operation at pressures below 10^{-7} torr, and for a reasonably oil-free vacuum environment, a vapor trap must be placed immediately above a diffusion pump. Vapor traps tie up oil molecules by adsorption or condensation.

Mercury as a pump fluid has the advantage of being chemically inert and is used in systems where hydrocarbon contamination is unacceptable. Mercury diffusion pumps can tolerate inlet pressures ten times greater than the maximum inlet pressure tolerated by oil pumps. In addition the critical foreline pressure for a mercury pump is much higher. However, the room-temperature vapor pressure of mercury is about 10^{-3} torr, and thus an inlet cold trap is required to condense mercury vapor in order to achieve system pressures below 10^{-3} torr. Mercury amalgamates with many metals and alloys, particularly brass. Mercury vapor is toxic, and care must be taken to properly trap and vent the backing-pump exhaust.

To attain pressures in the 10^{-3}- to 10^{-9}-torr region, diffusion pumps are the simplest and least expensive route. Pumps with speeds of 50 to 50,000 liter sec^{-1} and with nominal inlet port diameters of 1 to 35 in. are available. Diffusion pumps are constructed of Pyrex glass, mild steel, or stainless steel. The jets and towers of pumps with steel barrels are aluminum. The choice of glass or steel usually depends upon the material that is used to construct the vacuum container, although it is also relatively simple to mate glass to steel with an elastomer gasket or through a metal-to-glass seal. For those systems that are intended to handle reactive gases, glass is the preferred material. However, glass pumps are available only in small sizes.

Table 3.2   PROPERTIES OF DIFFUSION-PUMP FLUIDS

Name (chemical composition)	Boiling Point at 1 Torr (°C)	Approximate Room-Temperature Vapor Pressure (torr)
(Mercury)	120	2×10^{-3}
(Di-n-butyl phthalate)	140	2×10^{-5}
Octoil (di-2-ethyl hexyl phthalate)	200	3×10^{-7}
Octoil-S (di-2-ethyl hexyl sebacate)	210	3×10^{-8}
Convoil-10 (saturated hydrocarbon)	150	10^{-4}
Convoil-20 (saturated hydrocarbon)	190	5×10^{-7}
Convalex-10 (polyphenyl ether)	280	3×10^{-9}
Neovac Sy (alkyldiphenyl ether)	240	10^{-8}
D.C. 702 (silicone)	180	5×10^{-7}
D.C. 704 (silicone)	210	6×10^{-8}
D.C. 705 (silicone)	250	10^{-9}

3.4.3 Sorption Pumps, Getter Pumps, Cryopumps, and Ion Pumps

A variety of vacuum pumps remove gas from a system by chemically or physically tying up molecules on a surface or by trapping them in the interior of a solid. Two of the principal advantages of these pumps are that they require no backing pump and they contain no fluids to contaminate the vacuum.

The simplest of this class of pumps is the *sorption pump*, illustrated in Figure 3.11. The sorbent material is activated charcoal or one of the synthetic zeolite materials known as molecular sieves. These materials are effective sorbents partly because of their huge surface area, which is of the order of thousands of square meters per gram. The most common molecular sieves are Linde 5A or 13X. The number in the sieve code specifies the pore size: 5A is preferred for air pumping, while 13X is used for trapping hydrocarbons. All of these materials will pump water and hydrocarbon vapors at room temperature, but they must be cooled to liquid-nitrogen temperature to absorb air. Sorbent materials do not trap hydrogen or helium at liquid-nitrogen temperature. In some cases, therefore, the pressure of hydrogen or helium in a system will establish the lowest attainable pressure. Sorption pumps must be provided with a poppet valve because the sorbent material releases all of its absorbed air as it warms to room temperature.

The sorbent is initially activated by baking to 300°C. After several pumping cycles the pores of the sorbent material will become clogged with water and the ef-

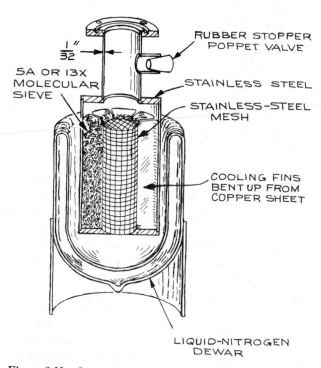

Figure 3.11   Sorption pump.

ficiency of the pump will deteriorate. Water is removed by baking the pump to 300°C with the poppet valve open and then cooling to room temperature with the valve closed so that moisture from the room is not reabsorbed. Baking can be accomplished by wrapping the pump with heating tape and covering with fiberglass insulation. Custom-made heating mantles can be obtained inexpensively from the Glas-Col Company.

In a well-designed pump, 50 g of dry Linde 5A will pump a 1-liter volume from atmosphere to less than 10^{-2} torr in about 20 minutes. The pump must be designed to provide good thermal contact between the sieve and the coolant, or the maximum pumping speed will not be achieved. In addition, the inlet tube should be as thin as possible to minimize heat conduction into the pump. Pressures below 10^{-6} may be attained in a system that has been roughed down. A simple way to achieve low pressures with sorption pumping is to use several pumps with appropriate valving so that one pump is used as a rough pump and successive pumps are used at lower pressures.

Clean surfaces of refractory metals such as titanium, molybdenum, tantalum, or zirconium will pump N_2, O_2, CO_2, H_2O, and CO by chemisorption. This process is called *gettering*. In a *getter pump*, the active metal surface is produced *in vacuo*. As in Figure 3.12, a simple pump can be made by wrapping titanium wire around a tungsten heater filament contained in a glass bulb. The filament is electrically heated to evaporate the titanium, which in turn condenses on the walls of the bulb. This pump operates effectively between 10^{-3} and 10^{-11} torr. The titanium surface must be renewed at a rate roughly equal to the rate at which a monolayer of gas is adsorbed. Down to 10^{-7} torr the filament is continuously heated. Below 10^{-7} torr the titanium need only be deposited periodically. The capacity of a titanium sublimation pump is about 30 torr liter per gram of Ti. A forepump is not required, but a mechanical or sorption roughing pump is needed to reach a starting pressure of 10^{-2} to 10^{-3} torr.

Saturated hydrocarbons and rare gases are not adsorbed at room temperature. However, if a getter is cooled to the temperature of liquid nitrogen, hydrocarbons and argon are adsorbed. A titanium sublimation pump can be used to achieve an ultrahigh vacuum if it

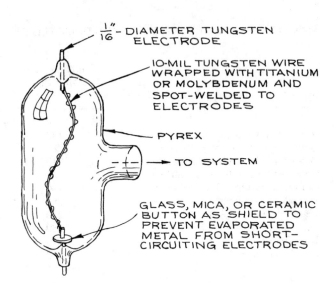

Figure 3.12   Simple titanium sublimation pump.

is used in conjunction with a small ion pump that removes rare gases, or if, as will be described in a later section, it is used to pump a system that has been purged of inert gases prior to evacuation.

Any surface will act as a pump for a gas that condenses at the temperature of the surface. A pump that relies primarily upon condensation on a cold surface is called a *cryopump*. Commercial versions of these pumps incorporate a closed-circuit helium refrigerator to cool the active surfaces. Working temperatures are typically below 20 K. These pumps usually employ two stages. In the first stage a metal surface is maintained at 30 to 50 K to trap water vapor, carbon dioxide, and the major components of air. The second stage, maintained at 10 to 20 K, is coated with a cryosorbent material such as charcoal to provide pumping of neon, hydrogen, and helium. Cryopumps provide high pumping speed for the easily condensed gases. Their performance with helium depends critically upon the quality and recent history of the cryosorbent surface.

An *ion pump* combines getter pumping with the pumping action exhibited by an ionization gauge. Within these pumps a magnetically confined discharge is maintained between a stainless-steel anode and a titanium cathode. The discharge is initiated by field emission when a potential of about 7 kV is placed across the electrodes. After the discharge is struck, a current-limited power supply maintains the discharge rate at 0.2 to 1.5 A, depending on the size of the pump. Inert-gas molecules, and other molecules as well, are ionized in the discharge and accelerated into the cathode with sufficient kinetic energy that they are permanently buried. Active gases are chemisorbed by titanium that has been sputtered off the cathode by ion bombardment and deposited on the anode. Ion pumps operate between 10^{-2} and 10^{-11} torr. Pumps with speeds from one to many thousands of liters per second are available.

Ion pumps require no cooling water or backing pump. They continue pumping by getter action even if the power fails. Ion pumps do not introduce hydrocarbon or mercury vapors into the vacuum. The positive-ion current to the cathode of an ion pump is a function of pressure; thus the pump serves as its own pressure gauge. Ion pumps cannot be used where stray electric and magnetic fields are unacceptable. The primary disadvantage of ion pumps is their cost. The price of an ion pump with its control unit is nearly an order of magnitude greater than for a comparable diffusion pump. However, on small ultrahigh-vacuum systems, an ion pump may compete economically with other pumps, since the pump obviates the need for a pressure gauge. The cost of a 1- to 5-liter sec^{-1} ion pump with a control unit that has a pressure readout is little more than the cost of an ion gauge and controller.

3.5 VACUUM HARDWARE

3.5.1 Materials

Borosilicate glasses such as Pyrex or Kimax glass are particularly well suited for the construction of small laboratory vacuum systems. These glasses are chemically inert and have a low coefficient of thermal expansion. Because of the plasticity of the material, complicated shapes are easily formed. Glass vacuum systems can be constructed and modified *in situ* by a moderately competent glassblower. The finished product does not have to be cleaned after working.

Hard glass tubing and glass vacuum accessories are inexpensive. Stopcocks and Teflon-sealed vacuum valves; ball joints, taper joints, and O-ring-sealed joints; traps; and diffusion pumps are all easily obtained at low cost. In most cases only the glassblowing ability to make straight butt joints and T-seals is required to make a complete system from standard glass accessories.

Glass pipe and pipe fittings are manufactured for the chemical industry. A variety of standard shapes such as elbows, tees, and crosses are available in pipe diameters of $\frac{1}{2}$ to 6 in. Glass process pipe with Teflon seals can be used to make inexpensive vacuum equipment for use down to 10^{-8} torr. Couplings are also available for joining glass pipe to metal pipe and to standard metal flanges and pipe fittings, and it is therefore easy to make a system that combines glass and metal vacuum accessories.

Glass and metal parts can be mated through a graded glass seal. Typically a graded seal appears to be simply a section of glass tubing butt-sealed to a section of Kovar metal tube. In fact, the glass tubing consists of a series of short pieces of glass whose coefficients of thermal expansion vary in small increments from that of hard glass to that of the metal. Graded seals are useful for joining glass accessories such as ion-gauge tubes to metal systems. Large, high-throughput glass stopcocks are not readily available. However, by using graded seals, large metal vacuum valves can be inserted into a glass vacuum line. Inexpensive graded seals in sizes up to 2 in. in diameter are available from commercial sources.

Brass and copper are useful vacuum materials. Brass has the advantage of being easily machined, and brass parts can be joined by either soft solder or silver solder. Unfortunately, brass contains a large percentage of zinc, whose volatility limits the use of brass to pressures above about 10^{-6} torr. Heating brass causes it to lose zinc quite rapidly. The vapor pressure of zinc is 10^{-5} torr at 200°C, and 2×10^{-3} torr at 300°C. Forelines for diffusion pumps are conveniently constructed of brass or copper tubing and standard plumbing elbows and tees. Ordinary copper water pipe is acceptable for forelines and other rough vacuum applications; however oxygen-free high-conductivity (OFHC) copper should be used for high temperature and high vacuum work. Copper can be heliarc-welded by a skilled technician.

The most desirable metal for the construction of high-vacuum apparatus is type-304 stainless steel. This material is strong, reasonably easy to machine, bakeable, and easy to clean after fabrication. Stainless-steel parts may be brazed or silver-soldered. Low-melting (230°C) silver-tin solder such as StainTin 157 PA (Eutectic Welding Alloys Co.) is useful for joining stainless-steel parts in the laboratory. However, it is best if stainless parts are fused together by arc welding using a nonconsumable tungsten electrode in an argon atmosphere. This process, known commonly as heliarc fusion welding or tungsten–inert-gas (TIG) welding, produces a very strong joint, and, because no flux or welding rod is used, such a joint is easily cleaned after welding. Most machine shops are prepared to do heliarc welding on a routine basis.

Type-304 stainless is widely used in the milk- and food-processing industry. As a result many stock shapes are commercially available at low cost. Elbows, tees, crosses, Y's, and many other fittings in sizes up to at least 6-in. diameter may be purchased from the Ladish Company or Alloy Products Company.

Many stainless steels are nearly nonmagnetic. The field produced by a piece of type 304 after machining is on the order of 10–30 milligauss at a distance of 1 cm. This magnetism can be reduced to 1 milligauss by quickly heating the material to 1100°C after machining and quenching in water.

Aluminum alloys, particularly the 6000 series, are used for vacuum apparatus. Aluminum has the advantage over stainless steel of being lighter and stiffer per unit weight, and much easier to machine. It is completely nonmagnetic. The chief disadvantages of aluminum stem from its porosity and the oxide layer that covers the surface. The rate of outgassing from aluminum is five to ten times greater than for stainless steel. Welds in aluminum are not as reliable as welds in stainless, and they tend to outgas volatile materials that are occluded in the weld. However, welded aluminum vacuum containers can be used down to about 10^{-7} torr. The hard oxide layer that forms instantly on a clean aluminum surface is an electrical insulator. This insulating surface tends to collect an electrical charge, which may be undesirable in some applications. Stray fields resulting from this charge may be eliminated by having critical aluminum parts copper- or gold-plated after fabrication.

Many plastics may be used at pressures down to 10^{-7} or 10^{-8} torr. The use of these materials is limited to varying degrees because they outgas air and plasticizers, and because they cannot be heated to high temperatures. Fluorocarbon polymers such as Teflon, Kel-F, and Viton-A have relatively low outgassing rates. Teflon can withstand temperatures up to 250°C, and its outgassing rate falls well below 10^{-8} torr liter sec^{-1} cm^{-2} after an initial pumpdown of a day at 100°C. Unfortunately, Teflon is relatively soft, and it cold-flows under mechanical pressure. Nylon, Delrin, and Vespel (polyimide) are harder than Teflon, and because they are self-lubricating they are useful as bearing surfaces. Delrin is preferred to nylon, since nylon is quite hygroscopic and outgasses water vapor after each exposure to air. Polyimide is somewhat hygroscopic as well, but may be baked to 250–300°C. Windows in vacuum systems may be made of acrylic plastic such as Plexiglas or Lucite. General Electric's Lexan is a very strong, tough, and machinable plastic well suited for use as a structural material or as an electrical insulator.

A variety of low-vapor-pressure sealers and adhesives have vacuum applications. Apiezon M grease and Dow-Corning silicone high-vacuum grease are used to seal stopcocks and ground-glass taper joints and, in some instances, as low-speed lubricants. Apiezon W black wax is useful for sealing windows to the ends of glass or metal tubes. This material melts at 60°C and has a room-temperature vapor pressure of about 10^{-6} torr. Glyptal is another low-vapor-pressure sealer and adhesive material.

Epoxy resins are particularly useful. Epoxy cements consist of a resin and a catalytic hardener, which are combined immediately before use. The proportions of resin and hardener and subsequent curing must be carefully controlled to prevent excessive outgassing from the hardened material. Small epoxy kits with resin and catalyst prepackaged in the correct proportions are available (e.g., from Tracon). Epoxy formulations with a range of flexibilities can be obtained, as well as ones with high electrical or thermal conductivity.

3.5.2 Demountable Vacuum Connections

Vacuum systems require detachable joints for convenience in assembling and servicing. There are a tremendous variety of vacuum connections, but demountable parts are most commonly joined by mating flanges or pipe threads that are sealed with some elastic material.

Pipe threads may be reliably sealed with Teflon thread dope that is sold in hardware stores for sealing threads in water pipes. This sealer consists of a thin Teflon tape that is stretched over the male thread before assembly. The tape must be replaced upon reassembly.

For pressures down to about 10^{-7} torr, vacuum connections are usually sealed with rubber O-rings. Several O-ring-sealed joints are illustrated in Figure 3.13. O-rings are circular gaskets with a round cross section. They are available in hundreds of sizes from 0.125-in. i.d. (inner diameter) with a cord diameter of 0.070 in. (nominally $\frac{1}{16}$ in.) to 2-ft i.d. with a cord diameter of 0.275 in. (nominally $\frac{1}{4}$ in.). Very large rings may be made from lengths of cord stock with the ends butted together and glued with Eastman 910 adhesive. O-rings are made of a variety of elastomers. The most common are Buna-N, a synthetic rubber, and Viton-A, a fluorocarbon polymer. Buna-N may be heated to 80°C and will not take a set after long periods of compression. Unfortunately this material outgasses badly, particularly

after exposure to cleaning solvents. Viton-A has a low rate of outgassing and will withstand temperatures up to 250°C. Viton-A will take a set after baking. Generally, the advantages of Viton-A O-rings offset their higher cost.

O-ring-sealed flanges and "quick connects" are easily fabricated, and they are also available ready-made. Mating flanges have a groove that contains the ring after the flanges have been pulled into contact with one another. Usually, one flange is flat and the groove is cut into the mating flange, but flanges can be made sexless by cutting a groove of half the required depth in both flanges. The cross section of the groove should be about 10% greater than the cross section of the O-ring cord, since rubber is deformable but incompressible. The groove depth should be about 70% of the cord diameter for static seals and 80% of the cord diameter for dynamic seals. The i.d. of the groove should match the O-ring i.d. It is sometimes convenient to undercut the sides of the groove slightly to give a closed dovetail cross section rather than a rectangular cross section, so that the ring is retained in the groove during assembly.

For rings up to a foot in diameter, the cord diameter should be $\frac{1}{8}$ in. or less. Choose a ring to fit in a groove that is as close to the flange i.d. as possible, in order to minimize the amount of gas trapped in the narrow space between the mated flanges. The bolts or clamps that pull the flanges together should be as close to the groove as possible to prevent flange distortion.

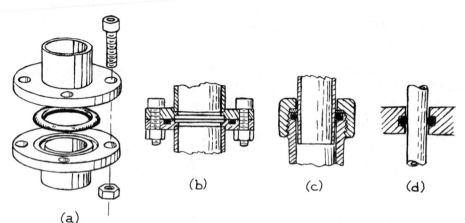

Figure 3.13 O-ring-sealed vacuum connections: (*a*) an exploded view of an O-ring-sealed flange joint; (*b*) a flange joint; (*c*) a quick connect; (*d*) a rotating-shaft seal.

(a) (b) (c) (d)

For static seals, O-rings should be used dry. Before assembly, the groove should be cleaned and the ring wiped free of mold powder with a lintless cloth. O-rings should not be cleaned with solvents. For rotating seals a very light film of vacuum grease on the ring will prevent abrasion. Occasionally a film of grease may be required on a static O-ring to help make a seal on an irregular surface.

Metal sealing materials are required for ultrahigh-vacuum (UHV) work. Elastomers are unacceptable for this because of their high vapor pressure and because they cannot be baked to high temperatures. The most common UHV seal consists of a flat OFHC copper gasket trapped between knife edges on the faces of a pair of flanges as shown in Figure 3.14. These flanges and gaskets are available from a number of manufacturers of vacuum hardware. A reliable UHV seal can be made by using a gold wire O-ring in a groove in the same manner as elastomer O-rings are used. The O-ring is made by butt-welding the ends of an appropriate length of gold wire and then annealing the whole ring by heating it to a dull red with a cold flame and permitting it to cool in air. After fabrication the ring must be handled carefully because it is easily stretched. Metal seals can be only used once, since they are permanently deformed during installation. The metal is recoverable and may be recycled.

3.5.3 Valves

A wide variety of valves are available commercially for use in glass systems and in high-vacuum and ultrahigh-vacuum metal systems. Generally, vacuum valves are sufficiently complex that it is uneconomical for the laboratory scientist to undertake their fabrication.

The two most common glass valves are illustrated in Figure 3.15. One (*a*) is a vacuum stopcock consisting of a tapered glass plug which is fitted into a glass body by lapping. The mating parts are sealed with a film of high-vacuum grease. As shown, these stopcocks are designed so that atmospheric pressure forces the plug into the body of the valve. The other, more modern type of glass valve (*b*) has a Teflon plug threaded into a glass body. The plug is sealed to the body with an O-ring. It

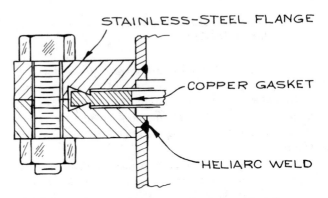

Figure 3.14   Detail of a bakeable, ultrahigh-vacuum flange seal.

is generally advisable to apply a light film of vacuum grease to the O-ring and the threads of these valves.

Several metal vacuum valves are illustrated in Figure 3.16. The bellows-sealed valve in Figure 3.16(*a*) is usually constructed of brass or stainless steel and is available in sizes suitable for use on tubing of $\frac{3}{8}$- (outer diameter), to $1\frac{5}{8}$-in. o.d. The drive screw that moves the sealing plate is contained within a bronze or stainless-steel bellows that maintains a vacuum seal while the plate moves from the open to the closed position. This type of valve is commonly used in the foreline of a diffusion pump, but it may also be used in small high-

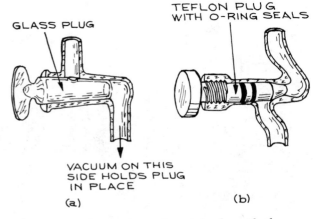

Figure 3.15   Glass vacuum valves: (*a*) valve with glass stopcock; (*b*) valve with Teflon plug.

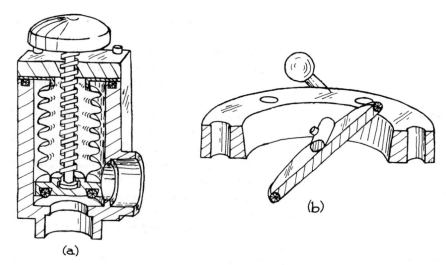

(a)

(b)

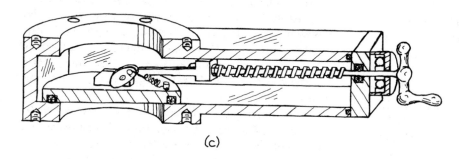

(c)

Figure 3.16 Metal high-vacuum valves: (*a*) bellows-sealed foreline valve; (*b*) quarter-swing gate valve; (*c*) sliding-gate valve.

vacuum systems. Stainless-steel, bellows-sealed valves with Viton O-rings can be heated to 200°C. Those with polyimide O-rings can be baked at 300°C.

The gate valves in Figure 3.16(*b*) and (*c*) are most often used to isolate a diffusion pump or ion pump from a high-vacuum chamber. These valves have very high conductance because of their low profile and because the sealing plate or gate does not significantly obstruct the valve aperture when the valve is open. Gate valves will seal against atmospheric pressure in either direction, but when used to isolate a pump they are usually installed so that atmospheric pressure in the chamber tends to hold them closed. The actuator may be either bellows-sealed or O-ring-sealed. The bellows seal is preferred. The body of a gate valve is cast from aluminum or stainless steel. They are available with nominal aper-

tures of 2 to 10 in. to match the inlet apertures of most diffusion pumps.

3.5.4 Mechanical Motion in the Vacuum System

It is frequently necessary to transmit linear or rotary motion through the wall of a vacuum container. A simple and inexpensive linear-motion feedthrough can be made from the actuator mechanism of a small bellows-sealed vacuum valve. As shown in Figure 3.17, the valve body is truncated above the seat, a mounting flange is brazed to the body, and a fixture is brazed or screwed to the valve plate for attaching the mechanism to be driven inside the vacuum. Linear motion can be

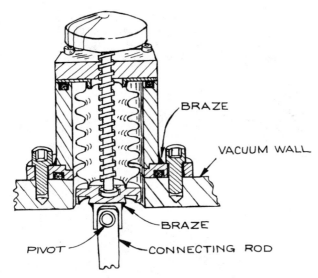

Figure 3.17 A bellows-sealed valve [Figure 3.16 (a)] converted to a linear-motion feedthrough.

converted to rotary motion inside the vacuum by applying the linear motion to a crank through a pivoting connecting rod. This scheme will provide up to 180° of rotary motion. Full 360° rotation may be achieved by means of a system of springs that carry the crank over center as shown in Figure 3.18.

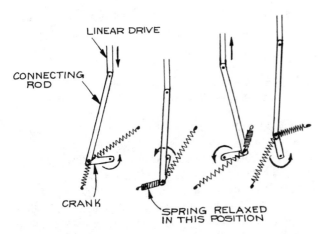

Figure 3.18 Spring arrangement to permit 360° rotation of a crank driven by a linear motion.

Rotary motion at speeds up to 100 rpm may be transmitted through an O-ring-sealed shaft as in Figure 3.13(d). These seals are inexpensive but may fail if the O-ring becomes abraded. In addition, lubricants and the elastomer O-ring material are exposed to the vacuum.

Rotary motion may be transmitted through a bellows-sealed wobble drive of the type shown in Figure 3.19. These drives are available commercially.

Moving parts can be magnetically coupled through a vacuum wall. All that is required is that a magnet be attached to the driving element and that the driven element be made of a magnetic material such as iron or nickel. Of course, the vacuum wall must be made of a nonmagnetic material such as Pyrex, brass, or type-304 stainless steel.

Metal surfaces become very clean *in vacuo*, particularly after baking, and metals in close contact tend to cold-weld. Because of this, unlubricated bearing surfaces within a vacuum system often become very rough after only a little use. There are a number of methods of improving bearing performance in a vacuum system without introducing high-vapor-pressure oils into the vacuum.

The tendency for a bearing to gall is reduced if the two mating bearing surfaces are made of different metals. For example, a steel shaft rotating without lubrication in a brass or bronze journal will hold up better than in a steel bushing. A solid lubricant may be applied to one of the bearing surfaces. Silver, lead-indium, and molybdenum disulfide have been used for this purpose. Graphite does not lubricate in a vacuum. MoS_2 is probably best. The lubricant should be burnished into the bearing surface. The part to be lubricated is placed in a lathe. As the part turns, the lubricant is applied and rubbed into the surface with the rounded end of a hardwood stick. By this means, the lubricant is forced into the pores. After burnishing, the surface should be wiped free of loose lubricant.

One component of a bearing may be fabricated of a self-lubricating material such as nylon, Delrin, Teflon, or polyimide. Teflon is good for this purpose, but its propensity to cold-flow will cause the bearing to become sloppy with time. Polyimide can be used in ultrahigh vacuum after baking to 250–300°C.

Brown, Sowinski, and Pertel[6] have overcome the

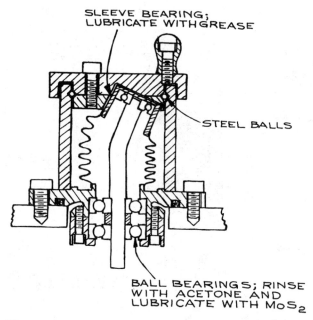

Figure 3.19 A bellows-sealed, wobble-drive, rotary-motion feedthrough.

leach clean the phenolic by boiling the bearings in chloroform-acetone, and then vacuum-dry at 100°C. The bearings are then lubricated by impregnating the phenolic with Du Pont Krytox 143AZ, a fluorinated hydrocarbon. Impregnation is accomplished by immersing the bearing in the lubricant and heating to 100°C at a pressure of 1 torr or less until air stops bubbling out of the phenolic. After this process the bearing must be wiped almost dry of lubricant. The Texwipe Company makes ultraclean foam cubes for this process. The amount of lubricant in the bearing is determined by weighing before and after impregnation. About 25 mg of Krytox is required for an R4 ($\frac{1}{4}$-in.) bearing, and about 50 mg is needed for an R8 ($\frac{1}{2}$-in.) bearing. This lubrication process should be carried out in a very clean environment, and the bearing should be inspected under a microscope for cleanliness before the side shields are replaced. Bearings treated in this manner have been run at speeds up to 60,000 rpm *in vacuo*.

3.5.5 Traps and Baffles

Traps are used in vacuum systems to intercept condensable vapors by means of chemisorption or physical condensation. Most high-vacuum systems have a trap in the foreline to prevent mechanical pump oil from backstreaming from the forepump to the diffusion pump. In addition, a trap is usually placed between a diffusion pump and a vacuum chamber to pump water vapor and to remove diffusion-pump fluid vapors that migrate backwards from the pump toward the chamber.

Foreline traps are similar in design to the molecular-sieve sorption pumps previously described, except, of course, that a trap must have both an inlet and an outlet. Two simple traps filled with 13X molecular sieve[8] are illustrated in Figure 3.20. The molecular sieve must initially be activated by baking to 300°C for several hours. In use the sieve material is regenerated at intervals of about a month by baking at 100°C to drive out absorbed oil and water. The baking may be done at atmospheric pressure or under a rough vacuum. If the baking is done in atmosphere, the sieve should be permitted to cool down from 100°C in vacuum to prevent reabsorption of water. If the baking is done *in*

cold-flow problem in the design of a drive screw for use in a vacuum. Both the screw and its nut are steel, and lubrication is accomplished by placing a Teflon key in a slot cut in the side of the screw. Teflon is then continuously wiped onto the screw threads as the screw turns.

For very precise location of rotating parts and for high rotational speeds, ball bearings are required. Precision, very low-speed ball bearings that are bakeable can be made using sapphire balls running in a stainless-steel race. Inexpensive precision sapphire balls are available from Industrial Tectonics. A ball retainer is required to prevent the balls from rubbing against one another. As explained in Section 1.6.1, the race must be designed so that the balls rotate without slipping against the race surface. It is helpful to burnish the race with MoS_2.

For high-speed applications a fluid lubricant is necessary. The following process has been developed at NASA-Goddard Space Flight Center.[7] Purchase high-quality stainless-steel ball bearings with side shields and phenolic ball retainers, such as the New Hampshire PPT series or Barden SST3 series. Remove the shields and

Figure 3.20 Two designs for a molecular-sieve foreline trap.

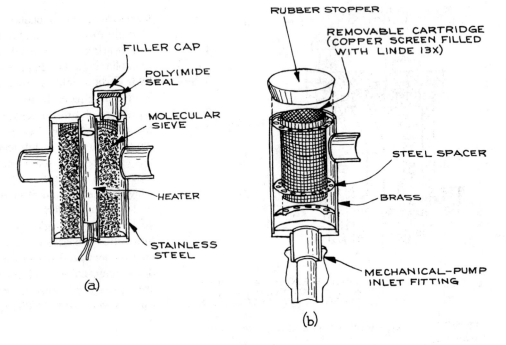

situ, as would be the case for the trap in Figure 3.20(*a*), the trap should be isolated from the diffusion pump with a valve, and the line between the trap and the mechanical pump should be warmed to prevent desorbed vapors from condensing in the foreline. Also, the gas ballast of the pump should be opened while the sieve is baking to prevent water vapor from condensing in the pump.

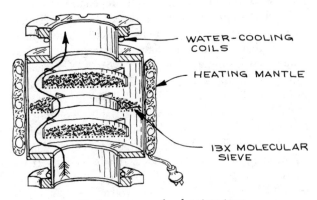

Figure 3.21 High-vacuum molecular-sieve trap.

The lowest pressure attainable with a two-stage mechanical pump is largely determined by the vapor pressure of the oil in the pump. By using a molecular-sieve trap in series with a mechanical pump to remove oil vapor it is possible to achieve pressures as low as 10^{-4} torr in a small system without using a diffusion pump.

A molecular sieve may also be used in a high-vacuum trap over the inlet of an oil diffusion pump. As shown in Figure 3.21, these traps are designed to be optically opaque so that a molecule cannot pass through the trap in a straight line. This precaution is necessary because this type of trap is intended for use at pressures where the mean free path of a molecule is very long. To ensure high conductance, the inlet and outlet ports should have the same cross-sectional area as the inlet of the attached pump. Also, the cross section perpendicular to the flow path through the trap (indicated by an arrow in Figure 3.21) should be at least as large as that of the inlet and outlet ports. The molecular sieve is activated by baking under vacuum to 300°C for at least six hours. The trap may be heated with heating tape covered with fiber-glass batting for insulation. Alternatively, a custom-made

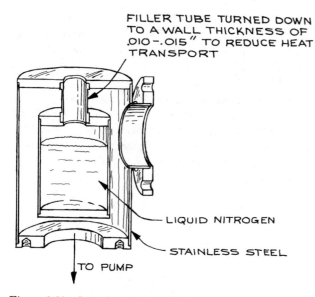

FILLER TUBE TURNED DOWN TO A WALL THICKNESS OF .010 - .015" TO REDUCE HEAT TRANSPORT

LIQUID NITROGEN

STAINLESS STEEL

TO PUMP

Figure 3.22 Liquid-nitrogen-cooled trap.

mantle may be purchased from the Glas-Col Apparatus Company. The flanges of the trap are water-cooled to prevent overheating the seals.

The maintenance of a high-vacuum trap is different from that of a foreline trap. A high-vacuum molecular-sieve trap is placed above an oil diffusion pump, and a gate valve is located above the trap to permit isolation of the trap and pump stack from the vacuum chamber. After the initial bakeout the pump is run continuously so that the trap is always under vacuum. The isolation valve is closed whenever it is necessary to open the chamber to the atmosphere, and the chamber is rough-pumped before reopening the isolation valve. It is unwise to attempt to regenerate the molecular sieve by baking. The initial bake of the sieve results in the evolution of water vapor, but subsequent baking will drive out absorbed pump-fluid vapor. This oil vapor will condense on the bottom of the isolation-valve gate and will be exposed to the vacuum whenever the valve is open. After the initial bakeout, the molecular sieve will trap oil vapor effectively for a period of at least six months if it is not exposed to moist air during that period. When the sieve becomes clogged with absorbed vapor, the base pressure attainable with a typical pump-and-trap combination will begin to rise. At this time the

molecular-sieve charge should be replaced. If it is absolutely necessary to stop the diffusion pump, the pump and trap should be filled with argon or dry nitrogen and isolated from the atmosphere in order to preserve the molecular sieve.

Liquid-nitrogen-cooled traps and baffles are frequently used with either oil or mercury diffusion pumps to condense backstreaming pump fluid. A metal cold trap is illustrated in Figure 3.22. These are very effective traps, but they have the disadvantage of requiring regular refilling with coolant. An alternative is to fill the trap with a low-melting liquid such as isopropyl alcohol and refrigerate the liquid with an immersion cooler. A temperature of $-40°C$ is usually adequate. When a cold trap is permitted to warm up, it must be isolated from the vacuum chamber so that condensed material does not migrate into the chamber. Also the trap should be vented so that the evaporating condensate does not build up a dangerously high pressure.

An optically dense baffle of the type shown in Figure 3.23 is usually placed over the inlet of a diffusion pump. These baffles may be air-cooled, water-cooled, or cooled by a small refrigerator. When placed between a diffusion pump and a liquid-nitrogen trap or molecular-sieve trap, a baffle will significantly reduce the rate of contamination of the trap. When a baffled pump is used without a trap, the partial pressure of oil vapor in the vacuum system is reduced to the vapor pressure of oil at the temperature of the baffle.

3.5.6 Molecular Beams and Gas Jets

Very often it is necessary to introduce a gaseous sample into a vacuum system. The sample may be contained in a chamber with holes suitably located to admit probes such as light beams or electron beams. However, the walls of such a chamber may prove to be a hindrance to

COOLANT

Figure 3.23 Cooled baffle.

the proposed experiment. In this case the sample may be introduced as an uncontained, but directed beam of atoms or molecules. The absence of walls is only one of several advantages to using a gas beam. Because of the directed velocities of the particles in the beam, it is possible to maintain the sample in a collisionless environment while still obtaining useful densities. The beam can be crossed with another beam or directed at a surface to obtain collisions of a specified orientation. In other cases, the expansion of a jet of gas results in extreme cooling to give a sample of gas with only a small number of quantum states populated. Lucas succinctly stated the case for atomic or molecular beams when he observed that "beams are employed in experiments where collisions . . . are either to be studied, or to be avoided."[9]

A gas beam is created by permitting the gas to flow into a vacuum through a tube. The result depends upon whether the flow exiting the tube is in the molecular or viscous flow regime. Of course the beam shape can be determined by appropriate apertures downstream of the channel through which the gas flows into the vacuum.

We begin with the molecular-flow case, where the mean free path λ of the gas at the outlet of the gas channel is greater than the diameter d of the channel. For the case of flow from a region of relatively high pressure into a vacuum through a round aperture (that is, a tube of length $l = 0$), the flux in a direction θ to the normal to the aperture surface is

$$I(\theta) = I(\theta = 0)\cos\theta \quad (\text{atoms sec}^{-1}\text{sr}^{-1}),$$

and the beam width $H = 120°$ (full width at half maximum). The beam can be narrowed by increasing the length l of the tube or channel through which the gas flows into the vacuum. A theoretical description of the gas beam issuing from such a channel depends upon the molecular diameter σ and the nature of the scattering of the molecules from the walls. A number of both theoretical and experimental investigations have been carried out in an effort to describe a gas beam in terms of easily measured parameters. In comparing experiment and theory it has turned out that the observed flux and the sharpness of the beam fall short of theoretical expectation by a factor of 2 to 5.

In general, both experimental and theoretical results

imply that to obtain useful fluxes and reasonable collimation, a beam source consisting of a tube or channel must be at least ten times greater in length than its diameter, and the gas pressure behind this channel should be such that $d < \lambda < l$.

Lucas has devised a neat description of gas beams in the molecular-flow regime in terms of a set of reduced parameters.[9] The gas pressure P behind the channel (in torr), the beam width H (degrees), the gas flux I (atoms sec^{-1} sr^{-1}), and the throughput Q (atoms sec^{-1}) are related to the corresponding reduced parameters by

$$P = \frac{P_R}{l\sigma^2},$$

$$H = \frac{H_R d}{l},$$

$$I = \left(\frac{T}{295\,M}\right)^{1/2}\frac{d^2 I_R}{l\sigma^2},$$

$$Q = \left(\frac{T}{295\,M}\right)^{1/2}\frac{d^3 Q_R}{l^2\sigma^2},$$

where T is the absolute temperature, M is the molecular weight, d and l are in cm, and σ in Å (10^{-8} cm). For pressures roughly in the range where $d < \lambda < l$, the reduced half angle, flux, and throughput are related to the reduced pressure by

$$H_R = 2.48 \times 10^2 \sqrt{P_R},$$

$$I_R = 1.69 \times 10^{20} \sqrt{P_R},$$

$$Q_R = 2.16 \times 10^{21} P_R.$$

From his model calculations, Lucas has discovered that to obtain maximum intensity for a given half angle H, the reduced pressure P_R should not be less than unity.

For design work, a useful "optimum" equation can be derived by combining the above and setting $P_R = 1$ to give

$$I = 6.7 \times 10^{17}\left(\frac{T}{295\,M}\right)^{1/2}\frac{dH}{\sigma^2} \quad (\text{optimum}).$$

Then, for any gas and a given channel diameter and beam half angle, the axial intensity can be determined.

The tube length is fixed by the equations for H and $H_R(P_R = 1)$ above, and the input pressure by the equation for P. Note, however, that operating at somewhat higher pressures gives some control over the flux and throughput without significantly departing from the optimum condition.

The construction of low-pressure, single-channel gas-beam sources is fairly straightforward. An excellent source can be made of a hypodermic needle cut to the appropriate length. These needles are made of stainless steel, they are available in a wide range of lengths and diameters, and they come with a mounting fixture that is reasonably gastight.

The goal of high intensity in a sharp beam is incompatible with a single-channel source since for a fixed half angle, the gas load (throughput) increases more rapidly than the flux as the channel diameter is increased. The solution to this problem is to use an array of many tubes, each having a small aspect ratio (i.e., $d/l \ll 1$).[10] Tubes with diameters as small as 2×10^{-4} cm and lengths of 1×10^{-1} cm arranged in an array a centimeter or more across are commercially available in glass (e.g. from Galileo Electro-Optics Corp.).

When the gas pressure behind an aperture or nozzle leading to a vacuum is increased to the extent that the mean free path is much smaller than the dimensions of the aperture, a whole new situation ensues. Not only is the density of the resultant gas jet much greater than in the molecular-flow case, but the shape of the jet changes and the gas becomes remarkably cold. This is a direct result of collisions between molecules in the gas as it expands from the orifice into the vacuum. Because of collisions the velocities of individual molecules tend toward that of the bulk gas flow, just as an individual in a crowd tends to be dragged along with the crowd. The translational temperature of the gas, defined by the width of its velocity distribution, decreases, while the bulk-flow velocity increases. This conversion of random molecular motion into directed motion continues until the gas becomes too greatly rarefied by expansion, at which point the final temperature is frozen in. Because the mass-flow velocity increases while the local speed of sound, proportional to the square root of the translational temperature, decreases, the Mach number rises and the flow becomes supersonic. Inelastic molecular collisions in the expanding gas also cause internal molec-ular energy to flow into the kinetic energy of bulk flow, with the result that there may be substantial rotational and vibrational relaxation. Rotational temperatures less than 1 K and vibrational temperatures less than 50 K have been obtained. At these low temperatures only a small number of quantum states are occupied, a situation that is ideal for a variety of spectroscopic studies.

The low temperatures obtained in a supersonic jet can present some problems. Chief among these is that of condensation. Collisions in the high-density region of the jet cause dimer or even polymer formation. With high-boiling samples, bulk condensation may occur. Of course, if one wishes to study dimers or clusters this condensation is desirable. To avoid dimer formation, the sample gas is mixed at low concentration with helium. This *seeded gas* sample is then expanded in a jet. The degree of clustering of the seed-gas molecules is controlled by adjusting its concentration in the helium carrier.

The cooling effect as well as the directionality of the jet is lost if the expanding gas encounters a significant pressure of background gas in the vacuum chamber. Unfortunately, optimum operating conditions require a large throughput of gas into the chamber. The extent of cooling depends upon the probability of binary collisions, which is proportional to the product $P_0 d$ of the pressure behind the nozzle and the diameter of the nozzle. Furthermore, to minimize condensation the ratio d/P_0 should be as large as possible. The result, in practice, is that a large-capacity vacuum pumping system is needed. In typical continuously operating supersonic jet apparatus, source pressures of 10 to 100 atm have been used with nozzle diameters of 0.01 cm down to 0.0025 cm. The conductance of an aperture under these conditions is roughly

$$C = 15d^2 \text{ liter sec}^{-1},$$

when the diameter is in centimeters. This implies a throughput on the order of 10 torr liter sec^{-1}. To obtain a mean free path of several tens of centimeters, a base pressure of about 10^{-4} torr is required. Thus, a pump speed approaching 10,000 liter sec^{-1} may be necessary. Typically several large diffusion pumps are used. When possible the jet is aimed straight down the throat of a pump.

In many experiments the cold molecules in a super-sonic jet are only probed periodically—as, for example, when doing laser spectroscopy with a pulsed laser. In this event the gas load imposed by the jet can be greatly reduced by operating the jet in a synchronously pulsed mode. A number of fast valves for this purpose have been devised.[11] The simplest are based upon inexpensive automobile fuel injector valves.[12] Jet sources providing gas pulses of duration less than a millisecond at a rate of 10 Hz are easily fabricated and are compatible with very modest vacuum systems.

Another efficient means of dealing with the background-gas problem has been demonstrated.[13] This relies upon the fact that interaction of the expanding gas with the background gas gives rise to a shock wave surrounding the gas jet. If the pressure P_0 behind the nozzle is increased, it is possible to achieve a mode of operation where the expanding gas behind the shock wave is unaffected by the background gas. In the region upstream of the shock front the gas behaves like a free jet expanding into a perfect vacuum. The distance from the nozzle to the shock front is

$$l = 0.67d \left(\frac{P_0}{P} \right)^{1/2},$$

where P_0 is the pressure in the nozzle, P the background pressure, and d the nozzle diameter.

When the nozzle pressure is sufficiently high to achieve a free jet length of usable dimensions, the background pressure will increase greatly. This state of affairs has a distinct advantage, since at a high background pressure, a large throughput can be achieved with a pump of moderate speed. For example, to obtain a free length $l = 1$ cm with a 0.01-cm nozzle and a source pressure of 10 atm, the background pressure should be about 400 mtorr. The throughput in this case is about 10 torr liter sec^{-1}, and the required pumping speed, 25 liter sec^{-1}, could be achieved by a large rotary mechanical pump or a Roots pump of modest capacity.

The fabrication of a nozzle is straightforward, although some difficulty may be encountered in making the necessary small hole. A skillful mechanical techni-

cian can drill a hole as small as 0.01-cm diameter. Smaller holes can be made by spark or electrolytic erosion, or by swaging a hole closed on a piece of hard wire and then withdrawing the wire. One of the first small, high-pressure nozzles[13] was made by drilling down the axis of a stainless-steel rod to within 0.1 mm of the end with a drill bit having a sharp conical point. A 0.0025-cm-diameter hole was then made through the remaining metal by spark erosion.

For many experiments, the gas flowing into a vacuum system through an aperture or a channel produces a beam that is too broad or too divergent for the intended application. In this case the beam shape can be defined by one or more apertures placed downstream from the source. Often it is advantageous to build these apertures into partitions that separate the vacuum housing into a succession of chambers. Each chamber can be evacuated with a separate pump. The pressure in the first will be highest, so that the large throughput obtained here will significantly reduce the gas load on the next pump. This scheme is called *differential pumping*.

An aperture downstream of a supersonic jet but close enough to be in the viscous flow region is called a *skimmer*. The design of these skimmers is critical, since they tend to produce turbulence, which will destroy the directionality of the flowing gas. A skimmer is usually a cone with its tip cut off and the truncated edge ground to knife sharpness. The details of the design, location, and fabrication of a skimmer are beyond the scope of this book, and the interested reader should refer to a specialized text on this subject.[14]

3.5.7 Electronics and Electricity *in Vacuo*

Electrical insulation inside a vacuum system is not generally a problem. At pressures below 10^{-4} torr a gap of 1 mm is adequate insulation up to at least 5000 volts. Initially, sparks may occur between closely spaced parts because of high field gradients around whiskers of metal. These whiskers quickly evaporate and sparks do not recur. High-voltage discharges can also occur along surfaces in a vacuum system. This is particularly a problem if the surfaces are dirty or hygroscopic. Inorganic salts on the surface of an insulator tend to absorb moisture and become conducting. After cleaning, insulators should be rinsed with distilled water followed by ethanol and then dried with hot air.

If the spacing between wires in a vacuum system cannot be reliably maintained, then insulation is required. Teflon-insulated hookup wire may be used down to 10^{-7} torr, although air bleeding out from under the insulation will slow pumpdown. At very low pressures and high temperatures, wires can be insulated by stringing ceramic beads or pieces of Pyrex tubing over them.

Solder should not be used for electrical connections in a vacuum system, because the lead in solder and soldering flux contaminate the vacuum. Mechanical connections are preferable. Connections to electrodes can be made by wrapping a wire under the head of a screw and tightening the screw. Wires may be jointed by slipping the ends into a piece of tubing and crimping the tubing onto the wire. Tungsten, molybdenum, nichrome, or stainless-steel wires may be spot-welded together. In welding refractory metals, a more secure weld is obtained if a piece of nickel foil is interposed between the wires.

Electronic devices tend to overheat in a vacuum, since the only cooling is by radiation. Electronic components such as power transistors and integrated circuits that must dissipate more than about $\frac{1}{2}$ watt are particularly unreliable. Such devices should be placed in vacuum-tight, air-filled boxes that are thermally connected to the vacuum wall.

A number of electrical feedthroughs are illustrated in Figure 3.24. The tungsten-wire feedthrough (a) is made by sealing the wire directly into Pyrex as described in Section 2.3.10. A simple electrical feedthrough for glass systems may be made by soldering or brazing an electrode into a Kovar-to-Pyrex graded seal as in (b). Ceramic-insulated terminal end bushings of the type shown in (c) are available commercially (e.g. from Ceramaseal). After an electrode wire is welded or brazed into the center hole, these bushings are easily joined to a metal vacuum wall by welding or brazing. There are, in addition to the feedthroughs shown in Figure 3.24, a number of different types sold by manufacturers of vacuum equipment.

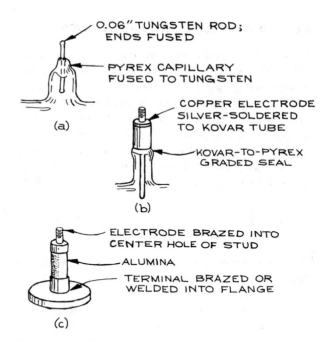

Figure 3.24   Electrical feedthroughs employing (a) a tungsten-to-glass seal; (b) a Kovar-to-glass seal; (c) a ceramic-to-metal terminal bushing.

3.6   VACUUM-SYSTEM DESIGN AND CONSTRUCTION

Before beginning the design of a vacuum system, the size and shape of the vacuum chamber must be determined. The desired ultimate pressure and the composition of the residual gas in the chamber must be specified. In order to choose pumps for the system, a number of parameters must be determined in at least a semiquantitative manner. One must estimate the gas load on the pumps from outgassing as well as from gas introduced into the vacuum. The maximum tolerable pumpdown time should be specified.

The amount of money and time available are important considerations when designing a vacuum system. It is instructive to spend a few evenings leafing through vacuum-equipment manufacturers' catalogs in order to become familiar with the specifications and cost of commercial apparatus.

3.6.1   Some Typical Vacuum Systems

A diffusion-pumped vacuum system is shown schematically in Figure 3.25. For illustration, suppose the vacuum chamber has a volume of 10 liters and a surface area of 3000 cm², and is constructed of stainless steel with an initial outgassing rate of 10^{-7} torr liter sec^{-1} cm^{-2}. If the pump stack consists of a 2-in. oil diffusion pump with a speed of 150 liter sec^{-1} and a trap and baffle with a conductance of 300 liter sec^{-1}, then the net speed of the pump station will be

$$S = \left(\frac{1}{150 \text{ liter sec}^{-1}} + \frac{1}{300 \text{ liter sec}^{-1}} \right)^{-1} = 100 \text{ liter sec}^{-1}.$$

The ultimate pressure attainable against the outgassing load will be

$$P = \frac{Q}{S} \frac{(10^{-7} \text{ torr liter sec}^{-1} \text{ cm}^{-2}) \times (3000 \text{ cm}^2)}{100 \text{ liter sec}^{-1}}$$

$$= 3 \times 10^{-6} \text{ torr}.$$

After a day of pumping and perhaps a light bake to 100°C, the outgassing rate should fall below 10^{-8}

torr liter sec^{-1} cm^{-2} and the ultimate pressure will decrease to 3×10^{-7} torr or less.

To determine the speed required of the backing pump, first estimate the maximum throughput of the diffusion pump. If the diffusion pump is operated at pressures up to its stalling pressure of about 5×10^{-2} torr, the maximum throughput will be

$$Q_{max} = P_{max} S = (5 \times 10^{-2} \text{ torr}) \times (100 \text{ liter sec}^{-1})$$

$$= 5 \text{ torr liter sec}^{-1}.$$

To keep the foreline pressure below 300 mtorr at this throughput, the speed of the backing pump should be

$$S_p = \frac{Q_{max}}{P} = \frac{5 \text{ torr liter sec}^{-1}}{3 \times 10^{-1} \text{ torr}}$$

$$= 16 \text{ liter sec}^{-1}.$$

This requirement can be met by the smallest single-stage or double-stage mechanical pump.

The conductance of the foreline should be at least twice the speed of the forepump, so that the forepump is not strangled by the foreline. To achieve a conduc-

Figure 3.25   Schema of a diffusion-pumped vacuum system.

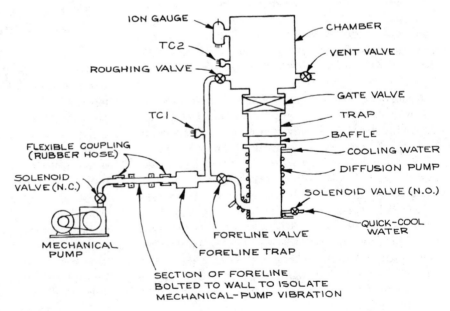

tance of 30 litersec^{-1} at a pressure of 300 mtorr in a foreline 1 m long, the diameter (from Section 3.2.4) must be

$$D = \left(\frac{CL}{180P} \right)^{1/4} \text{cm} \qquad (L \text{ in cm})$$

$$= \left(\frac{30 \times 100}{180 \times 3 \times 10^{-1}} \right)^{1/4} = 2.7 \text{ cm}.$$

In this case a 1-in. copper water pipe would make an excellent foreline.

A number of inexpensive safeguards are incorporated in the design in Figure 3.25 so that the system will be failsafe in the event of a loss of electrical power. A normally open (N.O.) solenoid-operated water valve wired in parallel with the pump heater admits water to the diffusion pump quick-cooling coils if the power fails. Also, the diffusion pump is isolated from the mechanical pump by a normally closed (N.C.) solenoid valve to prevent air or oil from being sucked through the mechanical pump into the foreline. It is helpful to place a ballast volume in the foreline to maintain a low foreline pressure while the diffusion pump cools after a power outage. A useful failsafe mechanism, not shown in Figure 3.25, would be a water-pressure sensor in the water line that interrupts the electrical power if the water fails. Another useful, but expensive, accessory would be an electrically controlled, pneumatically activated gate valve that can isolate the vacuum chamber from the pump.

To activate a system of the type shown in Figure 3.25, starting with all valves closed and the pumps off, proceed as follows:

1. Turn on the mechanical pump.

2. When the pressure indicated by thermocouple gauge 1 (TC1) is below 200 mtorr, open the foreline valve.

3. When TC1 again indicates a pressure below 200 mtorr, turn on the cooling water and activate the diffusion pump. The pump will require about 20 minutes to reach operating temperature. When operating it makes a crackling sound.

4. Before pumping on the chamber with the diffusion pump, the pressure in the chamber must be reduced to a rough vacuum. Close the foreline valve and open the roughing valve. When the pressure indicated by TC2 is below 200 mtorr, close the roughing valve and open the foreline valve. The diffusion pump should not be operated for more than 2 minutes with the foreline valve closed. If necessary, interrupt the roughing procedure, close the roughing valve, and reopen the foreline valve for a moment to ensure that the diffusion-pump outlet pressure does not rise above about 200 mtorr.

5. Open the gate valve that isolates the diffusion pump from the chamber. Within about a minute the pressure indicated by TC2 should fall to a few millitorr and the ionization gauge can be turned on.

A diffusion-pumped system of the type shown in Figure 3.25 is one of the most convenient and least expensive systems for obtaining high or even ultrahigh vacuum. The system can be fabricated of metal or glass using either an oil or a mercury diffusion pump. Without a trap, a mercury pump will yield an ultimate pressure of about 10^{-3} torr. A baffled, but untrapped, oil diffusion pump on this system will result in an ultimate pressure slightly below 10^{-6} torr. Using a mercury diffusion pump with a liquid-nitrogen-cooled trap or an oil diffusion pump with either a molecular-sieve trap or a cold trap, an ultimate pressure of 10^{-8} torr can be achieved. An ultimate pressure of 10^{-10} torr is possible if the vacuum chamber is baked and only very low-vapor-pressure materials are exposed to the vacuum.

A small, inexpensive system, capable of producing an oil-free vacuum of about 10^{-10} torr in a 1-liter volume, is illustrated in Figure 3.26. The main residual gases in such a system are primarily water vapor and rare gases. The system must be heated to drive water vapor from the walls, and since none of the pumps are particularly efficient for rare gases, the system is purged with nitrogen before being evacuated in order to displace the rare gases.[15]

The pumpdown procedure for this vacuum system is as follows:

Figure 3.26 A small, clean, inexpensive ultrahigh-vacuum system.

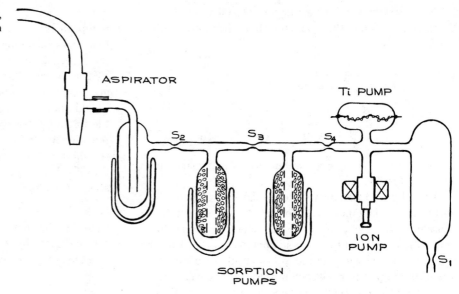

Figure 3.27 An oil-free ultrahigh-vacuum system.

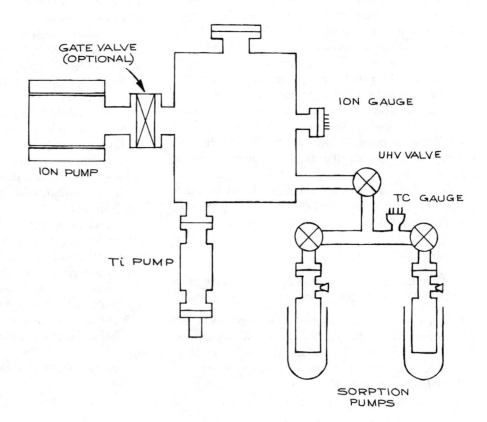

1. Flow pure, dry nitrogen through the system from S_1 to the aspirator. Do not activate the aspirator. The nitrogen gas can be obtained from a Dewar filled with liquid nitrogen.

2. As nitrogen flows through the system, the molecular-sieve sorption pumps are heated to 300°C and the trap is cooled with liquid nitrogen. The remainder of the system should be heated to about 100°C with heating tape or by brushing the glass from S_1 toward V_1 with a cold flame or with hot air from a heat gun.

3. After about an hour, close pinchoff S_1 by heating the glass with a torch, and turn on the aspirator. The pressure will quickly fall to about 20 torr.

4. Seal off constriction S_2 and refrigerate the first molecular-sieve pump. The pressure will fall to about 10^{-4} torr. Continue to heat the system.

5. Seal off S_3 and cool the second sorption pump. The pressure will fall to about 10^{-7} torr. Continue heating the system for a short time to drive off the remaining adsorbed gases.

6. Pass a current through the filament of the titanium pump to evaporate a layer of titanium onto the glass envelope, and activate the ion pump.

7. Seal off S_4 and deposit a fresh layer of titanium. The pressure should fall well below 10^{-9} torr.

When sealing off the constriction it is wise to proceed slowly so that the pumps have time to take up any gas that is liberated.

A similar system can be made with the pinchoffs replaced by all-metal high-vacuum valves. In this case it would also be possible to construct the entire system of stainless steel.

A large, oil-free ultrahigh-vacuum system is illustrated in Figure 3.27. The system is roughed down with sorption pumps, and the ion pump is depended upon for the removal of rare gases. The system must be baked. A number of variations on this system are possible. The rough-pumping operation can be accelerated and the gas load on the sorption pumps reduced if the system is first roughed down with one of the compressed-air aspirators sold for this purpose. The pumping at high

vacuum can be carried out with a cryopump rather than a getter pump. The choice of the high-vacuum pump or combination of pumps depends upon the composition of the residual gas as well as the composition of gases to be admitted to the system at high vacuum.

3.6.2 The Construction of Metal Vacuum Apparatus

Metal vacuum chambers are usually made up of one or more cylindrical sections, because a thin cylindrical shape has the strength to withstand external pressure and because metal tube stock is readily available. Even when tube stock is unavailable, a cylindrical section is easily fabricated by rolling sheet metal. The ends of a cylindrical chamber are most conveniently closed with flat plates as shown in Figure 3.28. A flat end plate must be quite thick to withstand atmospheric pressure, as discussed below.

The ability of a cylinder to resist collapsing under external pressure depends upon the length of the cylinder between supporting flanges, the diameter, the wall thickness, and the strength of the material. The A.S.M.E. has published standards for the wall thickness of cylindrical vessels under external pressure.[16] The data in

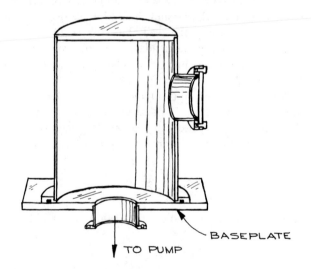

Figure 3.28   Typical metal vacuum chamber.

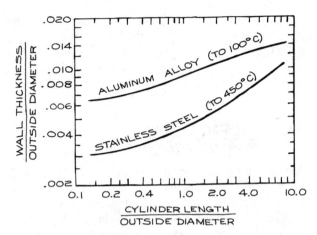

Figure 3.29 Minimum wall thickness for aluminum alloy and stainless-steel tubes under atmospheric pressure as a function of length between supporting flanges.

Figure 3.29 have been abstracted from the A.S.M.E. specifications for type 3003-T0 aluminum alloy and type 304 stainless steel respectively. These recommendations should be more than adequate for any of the 4000-, 6000-, or 7000-series tempered aluminum alloys (such as 2024-T4, 6061-T6 or 7075-T6) and adequate for most stainless steels. Soft aluminum such as the 1000-series alloys should not be used for vacuum chambers. Holes cut for ports in the side of a cylinder reduce the strength of the cylinder and should be avoided. If side ports are necessary the wall thickness should be increased over the recommendations of the code. Dents and out-of-roundness in a cylinder weaken it and should be avoided.

A flat end plate will bend a surprising amount under atmospheric pressure. The deflection at the center of a flat circular end plate clamped at its edges (e.g. by welding or otherwise firmly affixing it to a cylinder) is

$$\delta = \frac{3PR^4(1-\mu^2)}{16Ed^3},$$

and the maximum tensile stress, which occurs at the edge, is

$$s_{max}(\text{edge}) = \frac{3}{4}\left(\frac{R}{d}\right)^2 P,$$

where R is the radius of the plate, d the thickness, P the external pressure, μ Poisson's ratio, and E the modulus of elasticity or Young's modulus of the material.[17] For metals Poisson's ratio is about 0.3. For steels with a Young's modulus of 3×10^7 psi, an acceptable value of the ratio of radius to thickness, R/d, is 30. In this case the relative deflection is

$$\frac{\delta}{R} = 0.002 \qquad (\text{steel}, R/d = 30).$$

A slightly thicker aluminum plate is required to achieve the same result:

$$\frac{\delta}{R} = 0.002 \qquad (\text{aluminum}, R/d = 20).$$

The deflection of an end plate is greater when the edges are not securely clamped, as is the case for almost all removable end plates. For a circular plate with unclamped edges,

$$\delta = \frac{3PR^4(5+\mu)(1-\mu)}{16Ed^3},$$

and the maximum tensile stress, which occurs at the center, is

$$s_{max}(\text{center}) = \frac{3}{8}\left(\frac{R}{d}\right)^2 (3+\mu)P.$$

A deflection of about $0.003R$ is usually acceptable for a demountable end plate:

$$\frac{\delta}{R} = 0.003 \qquad (\text{steel}, R/d = 20),$$

and

$$\frac{\delta}{R} = 0.003 \qquad (\text{aluminum}, R/d = 15).$$

Steel is about three times heavier than aluminum; so for a given maximum deflection, an aluminum plate is less than half as heavy as a steel plate. A vacuum-chamber design that incorporates a stainless-steel cylinder with a demountable aluminum end plate has much to say in its

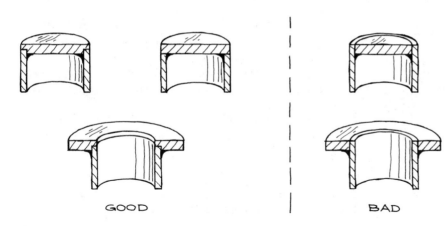

GOOD

BAD

Figure 3.30   The design of joints to be soldered or brazed.

favor. Welds in the stainless cylinder are much stronger and more reliable than would be the case if the cylinder were made of aluminum. A removable aluminum end plate is lighter and much less expensive to machine than a steel one.

Metal parts may be joined by soldering, welding, or brazing. Heliarc welding produces the cleanest, strongest joint, but soldering and brazing result in less distortion of the joined parts and are more convenient operations in the laboratory. As shown in Figure 3.30, a joint that is to be soldered or brazed should be designed so that the metal parts take the thrust of the atmosphere and the soldering or brazing material serves only as a sealant. If possible, the joint should be designed to provide positive location of mating parts. The surfaces to be joined should be cleaned before assembly by sandblasting or by polishing with sandpaper so that the solder or brazing metal will flow into and completely fill the joint. This is necessary to achieve maximum strength and to eliminate the narrow gap, which is difficult to pump out and is liable to collect flux and other material that will contaminate the vacuum system. Lead-tin solder should only be used for rough vacuum applications such as a diffusion-pump foreline. Brazed joints are acceptable in high vacuum, but usually the joints should not be heated over about 100°C *in vacuo*, since most brazing alloys contain high-vapor-pressure constituents such as zinc or cadmium. Brazing is an excellent way of attaching thin metal parts, such as bellows, which can easily be burned through in a welding operation.

Parts to be joined by welding are designed so that the rate of heat dissipation while welding is the same to each part. This is necessary to prevent destructive thermal stresses from being frozen into the finished joint. When joining a heavy flange to a thin tube as shown in Figure 3.31, a notch or a groove is cut in the thick piece of metal near the joint to control the rate of dissipation of heat from the joint. Whenever possible, welds in the wall of a vacuum chamber should be on the inside. If an outside weld is necessary, it should be a full-penetration weld. If neither an inside nor a full-penetration outside weld is possible, the mating parts should fit together loosely so that the gap between is wide open.

Metal parts inside a vacuum system may be joined by welding or brazing, but in most cases nut-and-bolt assembly is more convenient. Special care must be taken to prevent air from being trapped under a screw in a

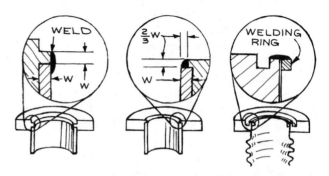

Figure 3.31   The design of joints to be heliarc-welded.

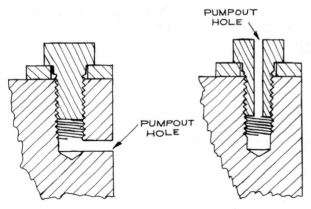

Figure 3.32   Pumpout holes for blind screw holes.

blind hole. The placement of pumpout holes for blind screw holes is illustrated in Figure 3.32.

3.6.3 Surface Preparation

Surfaces that are to be exposed to a vacuum should be free of substances that have a significant vapor pressure, and they should be as smooth as possible to minimize the microscopic surface area and thus minimize the amount of adsorbed gas. The substances that must be removed from the surfaces of vacuum apparatus are mainly hydrocarbon oils and greases, and inorganic salts that are hygroscopic and outgas water vapor.

The two preferred treatments for metal vacuum-system surfaces are electropolishing and bead blasting. Conventional wheel polishing and buffing is unsatisfactory, since these processes tend to flatten surface burrs and trap gas underneath. Electropolishing is conveniently carried out in the laboratory, although most machine shops are prepared to do it routinely. An electropolishing solution for stainless steel recommended by Armco Steel consists of 50 parts by volume of citric acid and 15 parts by volume of sulfuric acid plus enough water to make 100 parts of solution. The solution should be used at a temperature of 90°C. A current density of about 0.1 A cm^{-2} at 6 to 12 volts is required. A copper cathode is used with the piece to be polished serving as the anode.

Blasting with 20–30-μm glass beads effectively reduces the adsorbing area of a metal surface and thus reduces the rate of outgassing from the surface. Glass beading can be carried out in a conventional sandblasting apparatus. The machine should be carefully cleaned of coarse grits before use.

Glass may be effectively cleaned with a 10% solution of HF. The routine cleaning procedure for metal parts is as follows:

1. Scrub with a strong solution of detergent (Alconox or liquid dishwashing detergent is fine).

2. Rinse with very hot water.

3. Rinse with distilled water.

4. Rinse with pure methanol.

Do not touch clean parts with bare hands. Disposable plastic gloves (free of talc) are convenient for handling vacuum apparatus after cleaning.

An ultrasonic cleaner is particularly effective for preparing vacuum parts. The cleaner should be filled with a detergent solution. If two cleaners are available, fill one with detergent solution and the other with distilled water, and use them in sequence with a hot-water rinse between.

A vapor-phase degreaser is one of the most effective tools for cleaning metal parts after fabrication. The part to be cleaned is suspended in a cloud of hot solvent vapor. Vapor condenses on the part, heating it and rinsing it with pure hot solvent. As illustrated in Figure 3.33, a vapor degreaser is easily and inexpensively constructed. Any steel container, from a coffee can to a 55-gallon drum, can be used, depending on the size of the objects to be cleaned. The cooling coil around the top of the can is necessary to condense solvent vapors before they escape the container. Trichloroethylene is the preferred solvent. A vapor degreaser should always be operated under a fume hood to prevent contamination of the laboratory with toxic solvent vapors. Objects to be cleaned in the degreaser should first be washed with detergent and rinsed so as not to load the degreaser with volatile oils.

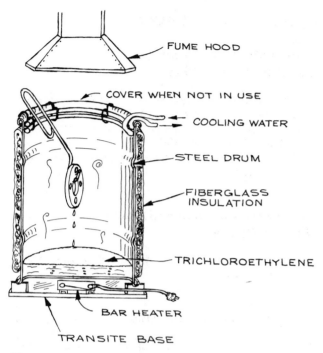

FUME HOOD

COVER WHEN NOT IN USE

COOLING WATER

STEEL DRUM

FIBERGLASS INSULATION

TRICHLOROETHYLENE

BAR HEATER

TRANSITE BASE

Figure 3.33   Vapor-phase degreaser.

3.6.4   Leak Detection

A leak that raises the base pressure of a system above about 10^{-6} torr can usually be found by probing suspected locations on the outside of the vacuum chamber with a liquid or vapor for which the gauge sensitivity or pump speed is very different from that for air. A squeeze bottle of acetone or a spray can of liquid Freon cleaner is a useful tool. These liquids will usually cause a very abrupt increase in indicated pressure as they flow through a leak, but sometimes rapid evaporation of a liquid through a leak will cause the liquid to freeze and temporarily plug the leak, causing the pressure to fall. A disadvantage of this method is that the solvent may contaminate O-rings. A small jet of helium is also a useful leak probe, since an ionization gauge is very insensitive to helium. The indicated pressure will fall when helium is introduced into a leak.

In a glass system a leak that raises the pressure into the range from 10 mtorr to several torr can be located with a Tesla coil. The surface of the glass is brushed with the discharge from the Tesla coil. The discharge is preferentially directed toward the leak, and a bright white spot will reveal the location as the discharge passes through. Avoid very thin glass walls and glass-to-metal seals, as the Tesla discharge can hole the glass in these fragile areas.

For very small leaks in high-vacuum systems a mass-spectrometer leak detector is needed.

3.6.5   Ultrahigh Vacuum

To achieve pressures much below 10^{-7} torr, baking is required in order to remove water and hydrocarbons from vacuum-system walls. Heating to 50–100°C for several hours will improve the ultimate pressure of most systems by an order of magnitude. A true ultrahigh-vacuum system must be baked to at least 250°C for several hours, and a system contaminated with hydrocarbons should be baked at 400°C. After a rigorous preliminary bake *in vacuo* to remove deeply adsorbed material, the system may be baked more gently after subsequent exposures to air.

A small system can be wrapped with heating tape and then covered with fiberglass or asbestos insulation. For larger systems an oven constructed of Transite or sheet metal insulated with fiberglass can be erected around the vacuum chamber and heated with bar heaters. Batts of fiberglass insulation intended for use in automobile engine compartments are readily available. Do not use home-insulating fiberglass materials. Metal seals must be used in a metal vacuum system that is to be baked. It is, however, often necessary to join the vacuum chamber to its pumping station with an O-ring-sealed flange. In this case, a cooling coil should be installed around the flange (as in Figure 3.21) to prevent melting the O-ring while the chamber is being baked.

CITED REFERENCES

1. N. Milleron, Res. Dev., 21, No. 9, 40, 1970.
2. The derivation of the conductance equations is given by

C.M. VanAtta, *Vacuum Science and Engineering*, McGraw-Hill, New York, 1965, pp. 44–62.

3. M.A. Biondi, Rev. Sci. Instr., 24, 989, 1953.

4. A.T.J. Hayward, J. Sci. Instr., 40, 173, 1963; C. Veillon, Rev. Sci. Instr., 41, 489, 1970.

5. H. Ishii and K. Nakayama, *Transactions of the 2nd International Congress on Vacuum Science and Technology*, Vol. 1, Pergamon Press, Elmsford, N.Y., 1961, p. 519; R.J. Tunnicliffe and J.A. Rees, Vacuum, 17, 457, 1967; T. Edmonds and J.P. Hobson, J. Vac. Sci. Tech., 2, 182, 1965.

6. G.R. Brown, P.J. Sowinski, and R. Pertel, Rev. Sci. Instr., 43, 334, 1972.

7. An application of this process is described by J.H. Moore and C.B. Opal, Space Sci. Instr., 1, 377, 1975.

8. The trap shown in Figure 3.20(*b*) is described by W.W. Roepke and K.G. Pung, Vacuum, 18, 457, 1968.

9. C.B. Lucas, Vacuum, 23, 395, 1973.

10. R.H. Jones, D.R. Olander, and V.R. Kruger, J. Appl. Phys., 40, 4641, 1969; D.R. Olander, J. Appl. Phys., 40, 4650, 1969; D.R. Olander, J. Appl. Phys., 41, 2769, 1970; D.R. Olander, R.H. Jones, and W.J. Siekhaus, J. Appl. Phys., 41, 4388, 1970; W.J. Siekhaus, R.H. Jones, J. Appl. Phys., 41, 4392, 1970.

11. M.G. Liverman, S.M. Beck, D.L. Monts, and R.E. Smalley, J. Chem. Phys., 70, 192, 1979; W.R. Gentry and C.F. Giese, Rev. Sci. Instr., 49, 595, 1978.

12. D. Bassi, S. Iannotta, and S. Niccolini, Rev. Sci. Instr., 52, 8, 1981; F.M. Behlen, Chem. Phys. Lett., 60, 364, 1979.

13. R.E. Smalley, D.H. Levy, and L. Wharton, J. Chem. Phys., 64, 3266, 1976.

14. J.B. Anderson, R.P. Andres, and J.B. Fenn, in *Advances in Chemical Physics*, Vol. 10, *Molecular Beams*, J. Ross, Ed., Wiley, New York, 1966, Chapter 8; H. Pauly and J.P. Toennies, in *Methods of Experimental Physics*, Vol. 7, Part A, B. Bederson and W.L. Fite, Eds., Academic Press, New York, 1968, Chapter 3.1.

15. This system is an adaptation of a design described by N.W. Robinson, *Ultra-high Vacuum*, Chapman and Hall, London, 1968, pp. 72–73.

16. *A.S.M.E. Boiler and Pressure Vessel Code*, Section VIII, Division 1, Appendix V, 1974.

17. J.F. Harvey, *Pressure Vessel Design*, Van Nostrand, Princeton, N.J., 1963, pp. 89–91.

GENERAL REFERENCES

Comprehensive Texts on Vacuum Technology

S. Dushman, *Scientific Foundations of Vacuum Technique*, 2nd edition, J.M. Lafferty, Ed., Wiley, New York, 1961.

G. Lewin, *Fundamentals of Vacuum Science and Technology*, McGraw-Hill, New York, 1965.

C.M. Van Atta, *Vacuum Science and Engineering*, McGraw-Hill, New York, 1965.

Criteria for Selection and Sizing Vacuum Pumps

J.F. O'Hanlon, *A User's Guide to Vacuum Technology*, Wiley, New York, 1980.

Design of Vacuum Systems

N.T.M. Dennis and T.A. Heppell, *Vacuum System Design*, Chapman and Hall, London, 1968.

G.W. Green, *The Design and Construction of Small Vacuum Systems*, Chapman and Hall, London, 1968.

R.P. LaPelle, *Practical Vacuum Systems*, McGraw-Hill, New York, 1972.

Outgassing Data

W.A. Campbell, Jr., R.S. Marriott, and J.J. Park, *A Compilation of Outgassing Data for Spacecraft Materials*, NASA Technical Note TND-7362, NASA, Washington, D.C., 1973.

Properties of Materials Used in Vacuum Systems

W. Espe, *Materials of High Vacuum Technology*: Vol. 1, *Metals and Metaloids*; Vol. 2, *Silicates*; Vol. 3, *Auxiliary Materials*, Pergamon Press, Oxford, 1968 (a translation of the original German published in 1960).

Sealing Ceramics and Glass to Metal, Heat-Treating, Cleaning, Building Joints, and Feedthroughs

F. Rosebury, *Handbook of Electron Tube and Vacuum Techniques*, Addison-Wesley, Reading, Mass., 1969.

Ultrahigh Vacuum

P. A. Redhead, J. P. Hobson, and E. V. Kornelsen, *The Physical Basis of Ultrahigh Vacuum*, Chapman and Hall, London, 1968.

R. W. Roberts and T. A. Vanderslice, *Ultrahigh Vacuum and Its Applications*, Prentice-Hall, Englewood Cliffs, N. J., 1963.

W. Robinson, *The Physical Principles of Ultrahigh Vacuum Systems and Equipment*, Chapman and Hall, London, 1968.

MANUFACTURERS AND SUPPLIERS

Ace Glass, Inc.
P.O. Box 688
1430 Northwest Blvd.
Vineland, NJ 08360
(Glass vacuum accessories and glass process pipe)

Alloy Products Company
1045 Perkins Ave.
Waukesha, WI 53186

Balston, Inc.
703 Massachusetts Ave.
Lexington, Mass. 02173

Ceramaseal, Inc.
New Lebanon Center
New York, NY 12126

Eutectic Welding Alloys Co.
40-42 172nd St.
Flushing, NY 11358

Galileo Electro-Optics Corp.
Galileo Park
Sturbridge, MA 01518

Glas-Col Apparatus Co.
709 Hulman St.
Terre Haute, IN 47802

Industrial Tectonics, Inc.
P.O. Box 1128
Ann Arbor, MI 48106

Ladish Co.
Cudahy, WI 53110

Lebold-Heraeus,
200 Seco Rd.,
Monroeville, PA 15146

MKS Instruments
22 3rd Ave.
Burlington, MA 01803

Neslab Instruments, Inc.
871 Islington St.,
Portsmouth, NH 03801
(Refrigerators and immersion coolers)

Texwipe Company
Hillsdale, NJ 07642

Tra-Con, Inc.
55-T North St.
Medford, MA 02155

Wallace & Tiernan
25 Main St.
Belleville, NJ 07109

CHAPTER 4

OPTICS

This chapter discusses techniques, devices, and systems that are used in the production, controlled transmission, spectral analysis and control, modulation, and detection of light. These discussions will be appropriate to experiments where light is used to stimulate some physical, chemical, or biological phenomenon; experiments where a phenomenon under study leads directly or indirectly to the production of light (where a study of the characteristics of this light gives information about the phenomenon); and experiments where the characteristics of light are modified by interaction with a system under study.

4.1 OPTICAL TERMINOLOGY

Light is electromagnetic radiation with a wavelength between 0.1 nm and 100 μm. Of course, this selection of wavelength range is somewhat arbitrary; short-wavelength vacuum-ultraviolet light between 0.1 and 10 nm (so called because these wavelengths—and a range above up to about 200 nm—are absorbed by air and most gases) might as easily be called soft X-radiation. At the other end of the wavelength scale, between 100 and 1000 μm (the *submillimeter* wave region of the spectrum), lies a spectral region where conventional optical methods become difficult, as does the extension of

microwave techniques for the centimeter- and millimeter-wave regions. The use of optical techniques in this region, and even in the millimeter region, is often called quasi-optics.[1,2] Only a few experimental techniques and devices that are noteworthy in this region will be mentioned here.

Table 4.1 summarizes the important parameters that are used to characterize light and the media through which it passes. A few comments on the table are appropriate. Although the velocity of light in a medium depends both on the relative magnetic permeability μ_r and dielectric constant ϵ_r of the medium, for all practical optical materials $\mu_r = 1$, so the refractive index and dielectric constant are related by

$$n = \sqrt{\epsilon_r} . \tag{1}$$

When light propagates in an anisotropic medium, such as a crystal of lower than cubic symmetry, n and ϵ_r will, in general, depend on the direction of propagation of the wave and its polarization state. The velocity of light *in vacuo* is currently the most precisely known of all the physical constants—its value is known[3] within 40 cm sec^{-1}. A redefinition of the meter has been proposed[4] based on the cesium-atomic-clock frequency standard[5] and a velocity of light of exactly 2.99792458×10^8 m sec^{-1}. The velocity, wavelength, and wavenumber of light traveling in the air have slightly different values

Table 4.1   FUNDAMENTAL PARAMETERS OF ELECTROMAGNETIC RADIATION
AND OPTICAL MEDIA

Parameter	Symbol	Value	Units
Velocity of light *in vacuo*	$c_0 = (\mu_0 \varepsilon_0)^{-1/2}$	2.99792458×10^8	$\mathrm{m\,sec}^{-1}$
Permeability of free space	μ_0	$4\pi \times 10^{-7}$	$\mathrm{henry\,m}^{-1}$
Permittivity of free space	ε_0	8.85416×10^{-12}	$\mathrm{farad\,m}^{-1}$
Velocity of light in a medium	$c = (\mu_r \mu_0 \varepsilon_r \varepsilon_0)^{-1/2} = c_0/n$		$\mathrm{m\,sec}^{-1}$
Refractive index	$n = (\mu_r \varepsilon_r)^{1/2}$		(dimensionless)
Relative permeability of a medium	μ_r	Usually 1	(dimensionless)
Dielectric constant of a medium	ε_r		(dimensionless)
Frequency	$\nu = c/\lambda$		Hz 10^9 Hz = 1 GHz (gigahertz) 10^{12} Hz = 1 THz (terahertz)
Wavelength *in vacuo*	$\lambda_0 = c_0/\nu$		m
Wavelength in a medium	$\lambda = c/\nu = \lambda_0/n$		m 10^{-3} m = 1 mm (millimeter) 10^{-6} m = 1 μm (micrometer) = 1 μ (micron) 10^{-9} m = 1 nm (nanometer) = 1 mμ (millimicron) 10^{-12} m = 1 pm (picometer) 10^{-10} m = 1 Å (angstrom)
Wavenumber	$\bar{\nu} = 1/\lambda$		cm^{-1} (Kayser)
Wavevector	$\mathbf{k}, |\mathbf{k}| = 2\pi/\lambda$		m^{-1}
Photon energy	$E = h\nu$		J 1.60202×10^{-19} J = 1 eV (electron volt)
Electric field of wave	$\mathbf{E}$		$\mathrm{V\,m}^{-1}$
Magnetic field of wave	$\mathbf{H}$		$\mathrm{A\,m}^{-1}$
Poynting vector	$\mathbf{P} = \mathbf{E} \times \mathbf{H}$		$\mathrm{W\,m}^{-2}$
Intensity	$I = \langle |\mathbf{P}| \rangle_{\mathrm{av}} = |E|^2/2Z$		$\mathrm{W\,m}^{-2}$
Impedance of medium	$Z = \dfrac{E_x}{H_y} = -\dfrac{E_y}{H_x} = \left(\dfrac{\mu_r \mu_0}{\varepsilon_r \varepsilon_0}\right)^{1/2}$		Ω
Impedance of free space	$Z_0 = (\mu_0/\varepsilon_0)^{1/2}$	376.7	Ω

than they have *in vacuo*. Tables that give corresponding values of $\bar{\nu}$ *in vacuo* and in standard air are available.[6]

A plane electromagnetic wave traveling in the *z*-direction can, in general, be decomposed into two independent, linearly polarized components. The electric and magnetic fields associated with each of these components are themselves mutually orthogonal and transverse to the direction of propagation. They can be written as (E_x, H_y) and (E_y, H_x). The ratio of the mutually orthogonal $\mathbf{E}$ and $\mathbf{H}$ components is called the *impedance* Z of the medium:

$$\frac{E_x}{H_y} = \frac{-E_y}{H_x} = Z = \sqrt{\frac{\mu_r \mu_0}{\epsilon_r \epsilon_0}}. \tag{2}$$

The negative sign in Equation (2) arises because (y, x, z) is not a right-handed coordinate system.

The Poynting vector

$$\mathbf{P} = \mathbf{E} \times \mathbf{H} \qquad (3)$$

is a vector that points in the direction of energy propagation of the wave, as shown in Figure 4.1. The average magnitude of the Poynting vector is called the *intensity* and is given by

$$I = \langle |\mathbf{P}| \rangle_{av} = \frac{|\mathbf{E}|^2}{2Z}. \qquad (4)$$

The factor of 2 comes from time-averaging the square of the sinusoidally varying electric field.

The wavevector $\mathbf{k}$ points in the direction perpendicular to the phase front of the wave (the surface of constant phase). In an isotropic medium $\mathbf{k}$ and $\mathbf{P}$ are always parallel.

The photon flux corresponding to an electromagnetic wave of average intensity I and frequency ν is

$$N = \frac{I}{h\nu} = \frac{I\lambda}{hc}, \qquad (5)$$

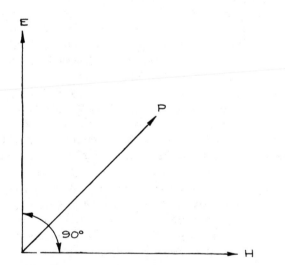

Figure 4.1 Orientation of the mutually orthogonal electric-field vector $\mathbf{E}$, magnetic-field vector $\mathbf{H}$, and Poynting vector $\mathbf{P}$.

where h is Planck's constant, 6.6×10^{-34} J sec. For a wave of intensity 1 W m^{-2} and wavelength 1 μm *in vacuo*, $N = 5.04 \times 10^{18}$ photons sec^{-1} m^{-2}. Photon energy is sometimes measured in electron volts (eV): 1 eV $= 1.60202 \times 10^{-19}$ J. A photon of wavelength 1 μm has an energy of 1.24 eV. It is often important, particularly in the infrared, to know the correspondence between photon and thermal energies. The characteristic thermal energy at absolute temperature T is kT. At 300 K, $kT = 4.14 \times 10^{-21}$ J $= 208.6$ cm$^{-1} = 0.026$ eV.

4.2 ANALYSIS OF OPTICAL SYSTEMS

Before embarking on a detailed discussion of the properties and uses of passive optical components and systems containing them, it is worthwhile reviewing some of the methods that are useful in analyzing the way in which light passes through an optical system.

Optical materials have refractive indices that vary with wavelength. This phenomenon is called *dispersion*. It causes a wavelength dependence of the properties of an optical system containing transmissive components. However, the change of index with wavelength is very gradual, and often negligible, unless the wavelength approaches a region where the material is not transparent. The discussions that follow will therefore usually be simplified by assuming that the light is *monochromatic* (that is, it contains only a small spread of wavelength components).

4.2.1 Simple Reflection and Refraction Analysis

The phenomena of reflection and refraction are most easily understood in terms of plane electromagnetic waves—those in which the direction of energy flow (the ray direction) is unique. Other types of wave, such as spherical waves and Gaussian beams, are also important in optical science; however, the part of their wavefront that strikes an optical component can frequently be

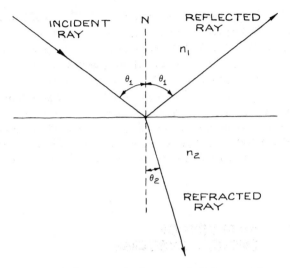

Figure 4.2 Reflection and refraction of a light ray at the boundary between two different isotropic media of refractive indices n_1 and n_2, respectively. The case shown is for $n_2 > n_1$. The incident, reflected, and transmitted rays and the surface normal N are coplanar.

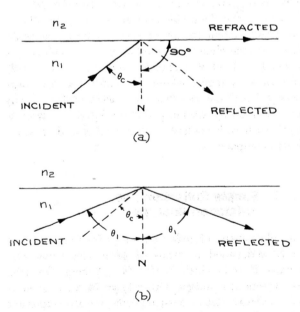

Figure 4.3 (a) Critical angle ($n_1 > n_2$); (b) total internal reflection ($n_1 > n_2$). N is the surface normal.

approximated as a plane wave, so plane-wave considerations of reflection and refraction still hold true.

When light is reflected from a plane mirror, or the planar boundary between two media of different refractive index, the *angle of incidence* is always equal to the *angle of reflection*, as shown in Figure 4.2. This is the fundamental *law of reflection*.

When a light ray crosses the boundary between two media of different refractive index, the *angle of refraction* θ_2, shown in Figure 4.2, is related to the angle of incidence θ_1 by *Snell's law*:

$$\frac{\sin \theta_1}{\sin \theta_2} = \frac{n_2}{n_1}. \tag{6}$$

This result does not hold true in general unless both media are isotropic (gas, liquid, or a crystal of cubic symmetry).

Since $\sin \theta_2$ cannot be greater than unity, if $n_2 < n_1$ there is a maximum angle of incidence for which there can be a refracted wave. It is called the *critical* angle θ_c and is given by

$$\sin \theta_c = n_2 / n_1, \tag{7}$$

as illustrated in Figure 4.3(a).

If θ_1 exceeds θ_c, the boundary acts as a very good mirror, as illustrated in Figure 4.3(b). This phenomenon is called *total internal reflection*. Several types of reflecting prisms discussed in Section 4.3.4 operate this way. When total internal reflection occurs, there is no transmission of energy through the boundary. However, the fields of the wave do not go abruptly to zero at the boundary. There is an *evanescent wave* on the other side of the boundary, the field amplitudes of which decay exponentially with distance. For this reason, other optical components should not be brought too close to a totally reflecting surface, or energy will be coupled to them via the evanescent wave and the efficiency of total internal reflection will be reduced. With extreme care this effect can be used to produce a variable-reflectivity, totally internally reflecting surface, as can be seen in Figure 4.4.

One or both of the media in Figure 4.2 may be anisotropic, like calcite, crystalline quartz, ammonium

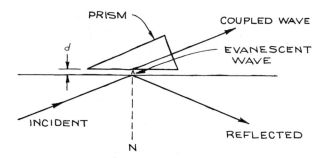

Figure 4.4 Schema of an evanescent-wave coupler. The amount of intensity reflected or coupled is varied by adjusting the spacing d; shown greatly exaggerated in the figure, it is typically on the order of the wavelength. The surface boundary and adjacent prism face must be accurately flat and parallel.

dihydrogen phosphate (ADP), potassium dihydrogen phosphate (KDP), or tellurium. Then the incident wave in general will split into two components, of which one obeys Snell's law (the ordinary wave) and one does not (the extraordinary wave). This phenomenon is called *double refraction*.[7-9]

If an optical system contains only planar interfaces, the path of a ray of light through the system can be easily calculated using only the law of reflection and Snell's law. This simple approach neglects diffraction effects, which become significant unless the lateral dimensions (*apertures*) of the system are *all* much larger than the wavelength (say 10^4 times larger). The behavior of light rays in more complex systems containing nonplanar components, but where diffraction effects are negligible, is better described with the aid of *paraxial-ray* analysis.[10] Transmitted and reflected intensities and polarization states cannot be determined by the above methods and are most easily determined by the *method of impedances*.

4.2.2 Paraxial-Ray Analysis

A plane wave is characterized by a unique propagation direction given by the wave vector **k**. All fields associated with the wave are, at a given time, equal at all points in infinite planes orthogonal to the propagation direction. In real optical systems such plane waves do not exist, as the finite size of the elements of the system restricts the lateral extent of the waves. Nonplanar optical components will cause further deviations of the wave from planarity. Consequently, the wave acquires a ray direction that varies from point to point on the phase front. The behavior of the optical system must be characterized in terms of the deviations its elements cause to the bundle of rays that constitute the propagating, laterally restricted wave. This is most easily done in terms of paraxial rays. In a cylindrically symmetric optical system (for example, a coaxial system of spherical lenses or mirrors), *paraxial rays* are those rays whose directions of propagation occur at sufficiently small angles θ to the symmetry axis of the system that it is possible to replace $\sin\theta$ or $\tan\theta$ by θ—in other words, paraxial rays obey the small-angle approximation.

(a) Matrix Formulation. In an optical system whose symmetry axis is in the z-direction, a paraxial ray in a given cross section ($z = $ constant) is characterized by its distance r from the z-axis and the angle r' it makes with that axis. Suppose the values of these parameters at two planes of the system (an *input* and an *output* plane) are $r_1 r_1'$ and $r_2 r_2'$, respectively, as shown in Figure 4.5(a). Then in the paraxial-ray approximation, there is a linear relation between them of the form

$$r_2 = Ar_1 + Br_1',$$
$$r_2' = Cr_1 + Dr_1', \tag{8}$$

or, in matrix notation,

$$\begin{pmatrix} r_2 \\ r_2' \end{pmatrix} = \begin{pmatrix} A & B \\ C & D \end{pmatrix}\begin{pmatrix} r_1 \\ r_1' \end{pmatrix}. \tag{9}$$

Here

$$\mathbf{M} = \begin{pmatrix} A & B \\ C & D \end{pmatrix}$$

is called the *ray transfer matrix*. Its determinant is usually unity, i.e., $AD - BC = 1$.

Optical systems made of isotropic material are generally reversible—a ray that travels from right to left with

Figure 4.5 (*a*) Generalized schematic diagram of an optical system, showing a typical ray and its paraxial-ray parameters at the input and output planes; (*b*) focal points F_1 and F_2 and principal rays of a generalized optical system.

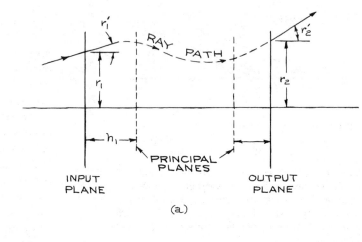

(a)

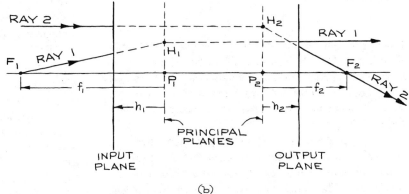

(b)

input parameters r_2, r_2' will leave the system with parameters r_1, r'. Thus

$$\begin{pmatrix} r_1 \\ r_1' \end{pmatrix} = \begin{pmatrix} A' & B' \\ C' & D' \end{pmatrix} \begin{pmatrix} r_2 \\ r_2' \end{pmatrix}, \qquad (10)$$

where the reverse ray transfer matrix is

$$\begin{pmatrix} A' & B' \\ C' & D' \end{pmatrix} = \begin{pmatrix} A & B \\ C & D \end{pmatrix}^{-1}.$$

The ray transfer matrix allows the properties of an optical system to be described in general terms by the location of its *focal points* and *principal planes*, whose location is determined from the elements of the matrix. The significance of these features of the system can be illustrated with the aid of Figure 4.5(*b*). An input ray

that passes through the *first focal point* F_1 (or would pass through this point if it did not first enter the system) emerges traveling parallel to the axis. The intersection point of the extended input and output rays, point H_1 in Figure 4.5(*b*), defines the location of the *first principal plane*. Conversely, an input ray traveling parallel to the axis will emerge at the output plane and pass through the second focal point F_2 (or appear to have come from this point). The intersection of the extension of these rays, point H_2, defines the location of the *second principal plane*. Rays 1 and 2 in Figure 4.5(*b*) are called the *principal rays* of the system. The location of the principal planes allows the corresponding emergent ray paths to be determined as shown in Figure 4.5(*b*). The dashed lines in this figure, which permit the geometric construction of the location of output rays 1 and 2, are called *virtual ray paths*. Both

F_1 and F_2 lie on the axis of the system. The axis of the system intersects the principal planes at the *principal points*, P_1 and P_2, in Figure 4.5(*b*). The distance, f_1, from the first principal plane to the first focal point is called the *first focal length*; f_2 is called the *second focal length*.

In most practical situations, the refractive indices of the media to the left of the input plane (the *object space*) and to the right of the output plane (the *image space*) are equal. In this case, several simplifications arise:

$$f_1 = f_2 \equiv f = -\frac{1}{C},$$
$$h_1 = \frac{D-1}{C}, \qquad (11)$$
$$h_2 = \frac{A-1}{C}.$$

h_1 and h_2 are the distances of the input and output planes from the principal planes, measured in the sense shown in Figure 4.5(*b*).

Thus, if the elements of the transfer matrix are known, the location of the focal points and principal planes is determined. Graphical construction of ray paths through the system using the methods of *ray tracing* is then straightforward [see Section 4.2.2(b)].

In using the matrix method for optical analysis, a consistent sign convention must be employed. In the present discussion, a ray is assumed to travel in the positive z-direction from left to right through the system. The distance from the first principal plane to an object is measured positive from right to left—in the negative z-direction. The distance from the second principal plane to an image is measured positive from left to right—in the positive z-direction. The lateral distance of the ray from the axis is positive in the upward direction, negative in the downward direction. The acute angle between the system-axis direction and the ray, say r_1' in Figure 4.5(*a*), is positive if a counterclockwise motion is necessary to go from the positive z-direction to the ray direction. When the ray crosses a spherical interface, the radius of curvature is positive if the interface is convex to the input ray. The use of ray transfer matrices in optical-system analysis can be illustrated with some specific examples.

(i) *Uniform optical medium.* In a uniform optical medium of length d, no change in ray angle occurs, as illustrated in Figure 4.6(*a*); so

$$r_2' = r_1',$$
$$r_2 = r_1 + dr_1'.$$

Therefore,

$$\mathbf{M} = \begin{pmatrix} 1 & d \\ 0 & 1 \end{pmatrix}. \qquad (12)$$

The focal length of this system is infinite and it has no specific principal planes.

(ii) *Planar interface between two different media.* At the interface, as shown in Figure 4.6(*b*), we have $r_1 = r_2$, and from Snell's law, using the approximation $\sin\theta \simeq \theta$,

$$r_2' = \frac{n_1}{n_2} r_1'.$$

Therefore,

$$\mathbf{M} = \begin{pmatrix} 1 & 0 \\ 0 & n_1/n_2 \end{pmatrix}. \qquad (13)$$

(iii) *A parallel-sided slab of refractive index n bounded on both sides with media of refractive index 1* [*Figure 4.6(c)*]. In this case,

$$\mathbf{M} = \begin{pmatrix} 1 & d/n \\ 0 & 1 \end{pmatrix}. \qquad (14)$$

The principal planes of this system are the boundary faces of the optically dense slab.

(iv) *Thick lens.* The ray transfer matrix of the thick lens shown in Figure 4.7(*a*) is the product of the three transfer matrices:

$$\mathbf{M}_3\,\mathbf{M}_2\,\mathbf{M}_1 = \text{(matrix for second spherical interface)}$$
$$\times\,\text{(matrix for medium of length } d\text{)}$$
$$\times\,\text{(matrix for first spherical interface)}.$$

Note the order of these three matrices; $\mathbf{M}_1$ comes on the

Figure 4.6 Simple optical systems for illustrating the application of ray transfer matrices, with input and output planes marked 1 and 2: (*a*) uniform optical medium; (*b*) planar interface between two different media; (*c*) a parallel-sided slab of refractive index *n* bounded on both sides with media of refractive index 1; (*d*) thin lens (r_2' is a negative angle; for a thin lens, $r_2 = r_1$); (*e*) a length of uniform medium plus a thin lens; (*f*) two thin lenses.

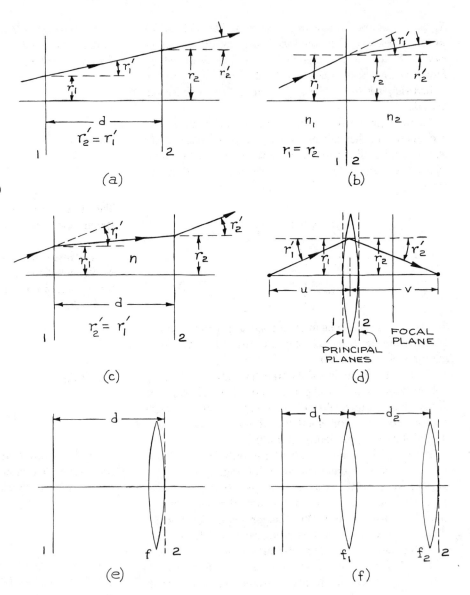

right because it operates first on the column vector that describes the input ray.

At the first spherical surface,

$$n'(r_1' + \phi_1) = nr = n(r_2'' + \phi_1),$$

and, since $\phi_1 = r_1/R_1$, this equation can be rewritten as

$$r_2'' = \frac{n'r_1'}{n} + \frac{(n'-n)r_1}{nR_1}.$$

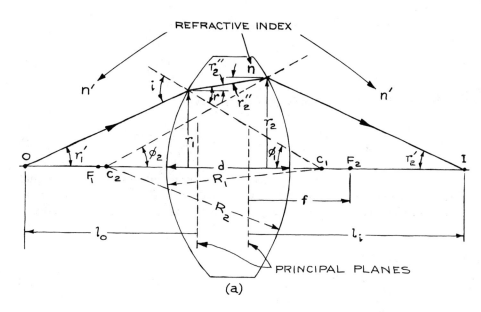

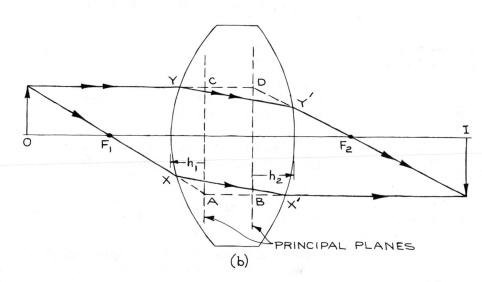

Figure 4.7 (*a*) Diagram illustrating ray propagation through a thick lens; (*b*) diagram showing the use of the principal planes to determine the principal ray paths through the lens. The dashed paths *XABX'* and *YCDY'* are the virtual-ray paths; the solid lines *XX'* and *YY'* are the real-ray paths.

The transfer matrix at the first spherical surface is

$$\mathbf{M}_1 = \begin{pmatrix} 1 & 0 \\ \dfrac{n'-n}{nR_1} & \dfrac{n'}{n} \end{pmatrix} = \begin{pmatrix} 1 & 0 \\ -\dfrac{D_1}{n} & \dfrac{n'}{n} \end{pmatrix}, \quad (15)$$

where $D_1 = (n - n')/R_1$ is called the *power* of the surface. If R_1 is measured in meters, the units of D_1 are *diopters*.

In the paraxial approximation all the rays passing through the lens travel the same distance d in the lens.

Thus,

$$\mathbf{M}_2 = \begin{pmatrix} 1 & d \\ 0 & 1 \end{pmatrix}. \qquad (16)$$

The ray transfer matrix at the second interface is

$$\mathbf{M}_3 = \begin{pmatrix} 1 & 0 \\ \dfrac{n-n'}{n'R_2} & \dfrac{n}{n'} \end{pmatrix} = \begin{pmatrix} 1 & 0 \\ -\dfrac{D_2}{n'} & \dfrac{n}{n'} \end{pmatrix}, \quad (17)$$

which is identical in form to $\mathbf{M}_1$. Note that in this case both r_2' and R_2 are negative. The overall transfer matrix of the thick lens is

$$\mathbf{M} = \mathbf{M}_3\mathbf{M}_2\mathbf{M}_1 = \begin{pmatrix} 1 - \dfrac{dD_1}{n} & \dfrac{dn'}{n} \\ \dfrac{dD_1D_2}{nn'} - \dfrac{D_1}{n'} - \dfrac{D_2}{n'} & 1 - \dfrac{dD_2}{n} \end{pmatrix}.$$

$$(18)$$

If Equations (18) and (11) are compared, it is clear that the locations of the principal planes of the thick lens are

$$h_1 = \frac{d}{\dfrac{n}{n'}\left(1 + \dfrac{D_1}{D_2} - \dfrac{dD_1}{n}\right)}, \qquad (19)$$

$$h_2 = \frac{d}{\dfrac{n}{n'}\left(1 + \dfrac{D_2}{D_1} - \dfrac{dD_2}{n}\right)}. \qquad (20)$$

A numerical example will best illustrate the location of the principal planes for a biconvex thick lens. Suppose that

$$n' = 1 \quad \text{(air)}, \qquad n = 1.5 \quad \text{(glass)},$$
$$R_1 = -R_2 = 50 \text{ mm},$$
$$d = 10 \text{ mm}.$$

In this case, $D_1 = D_2 = 0.01$ and, from Equations (19) and (20), $h_1 = h_2 = 3.226$ mm. These principal planes are symmetrically placed inside the lens. Figure 4.7(b) shows how the principal planes can be used to trace the principal-ray paths through a thick lens.

From Equations (18) and (11),

$$r_2' = -\frac{r_1}{f} + \left(1 - \frac{dD_2}{n}\right)r_1', \qquad (21)$$

where the focal length is

$$f = \left(\frac{D_1 + D_2}{n'} - \frac{dD_1D_2}{nn'}\right)^{-1}. \qquad (22)$$

If l_o is the distance from the object O to the first principal plane, and l_i the distance from the second principal plane to the image I in Figure 4.7(a), then

$$r_1' = \frac{r_1}{l_o - h_1},$$
$$r_2' = -\frac{r_2}{l_i - h_2}. \qquad (23)$$

Using Equations (19)–(23), it can be shown that

$$\frac{1}{l_o} + \frac{1}{l_i} = \frac{1}{f}. \qquad (24)$$

This is the fundamental imaging equation. It should be noted that it holds true only when the medium on both sides of the lens is the same.

(v) **Thin lens.** If a lens is sufficiently thin that to a good approximation $d = 0$, the transfer matrix is

$$\mathbf{M} = \begin{pmatrix} 1 & 0 \\ -\dfrac{D_1 + D_2}{n'} & 1 \end{pmatrix}. \qquad (25)$$

As shown in Figure 4.6(d), the principal planes of such a thin lens are at the lens. The focal length of the thin lens is f, where

$$\frac{1}{f} = \frac{D_1 + D_2}{n'} = \left(\frac{n}{n'} - 1\right)\left(\frac{1}{R_1} - \frac{1}{R_2}\right), \qquad (26)$$

so the transfer matrix can be written very simply as

$$\mathbf{M} = \begin{pmatrix} 1 & 0 \\ -1/f & 1 \end{pmatrix}. \qquad (27)$$

The focal length of the lens depends on the refractive indices of the lens material and of the medium within which it is immersed. In air,

$$\frac{1}{f} = (n-1)\left(\frac{1}{R_1} - \frac{1}{R_2}\right). \tag{28}$$

For a biconvex lens, R_2 is negative and

$$\frac{1}{f} = (n-1)\left(\frac{1}{|R_1|} + \frac{1}{|R_2|}\right). \tag{29}$$

For a biconcave lens,

$$\frac{1}{f} = -(n-1)\left(\frac{1}{|R_1|} + \frac{1}{|R_2|}\right). \tag{30}$$

The focal length of any diverging lens is negative. For a thin lens the object and image distances are measured to a common point. It is common practice to rename the distances l_o and l_i in this case, so that the imaging Equation (24) reduces to its familiar form

$$\frac{1}{v} + \frac{1}{u} = \frac{1}{f}. \tag{31}$$

Then u is called the *object distance* and v the *image distance*.

(vi) *A length of uniform medium plus a thin lens [Figure 4.6(e)].* This is a combination of systems in Sections 4.2.2(a) (i and v); its overall transfer matrix is found from Equations (12) and (27) as

$$M = \begin{pmatrix} 1 & 0 \\ -1/f & 1 \end{pmatrix}\begin{pmatrix} 1 & d \\ 0 & 1 \end{pmatrix}$$

$$= \begin{pmatrix} 1 & d \\ -1/f & 1 - d/f \end{pmatrix}. \tag{32}$$

(vii) *Two thin lenses.* As a final example of the use of ray transfer matrices consider the combination of two thin lenses shown in Figure 4.6(f). The transfer matrix of this combination is

$$M = (\text{matrix of second lens})$$
$$\times (\text{matrix of uniform medium})$$
$$\times (\text{matrix of first lens}),$$

which can be shown to be

$$M = \begin{pmatrix} 1 - \dfrac{d_2}{f_1} & d_1 + d_2 - \dfrac{d_1 d_2}{f_1} \\[3mm] -\dfrac{1}{f_1} - \dfrac{1}{f_2} + \dfrac{d_2}{f_1 f_2} & 1 - \dfrac{d_1}{f_1} - \dfrac{d_2}{f_2} + \dfrac{d_1 d_2}{f_1 f_2} \end{pmatrix}. \tag{33}$$

The focal length of the combination is

$$f = \frac{f_1 f_2}{(f_1 + f_2) - d_2}. \tag{34}$$

The optical system consisting of two thin lenses is the standard system used in analyses of the stability of lens waveguides and optical resonators.[10, 11]

(viii) *Spherical mirrors.* The object and image distances from a spherical mirror also obey Equation (31), where the focal length of the mirror is $R/2$; f is positive for a concave mirror, negative for a convex mirror. Positive object and image distances for a mirror are measured positive in the normal direction away from its surface. If a negative image distance results from Equation (31), this implies a *virtual image* (behind the mirror).

(b) **Ray Tracing.** Practical implementation of paraxial-ray analysis in optical-system design can be very conveniently carried out graphically by *ray tracing*. In ray tracing a few simple rules allow geometrical construction of the principal-ray paths from an object point. (These constructions do not take into account the nonideal behavior, or *aberrations*, of real lenses, which will be discussed later.) The first principal ray from a point on the object passes through (or its projection passes through) the first focal point. From the point where this ray, or its projection, intersects the first principal plane, the output ray is drawn parallel to the axis. The actual

Figure 4.8 Ray-tracing techniques for locating image and ray paths for (*a*) a converging thin lens; (*b*) a diverging thin lens. *F* = focal point; *C* = center of lens. The principal rays and a general ray pencil are shown in each case.

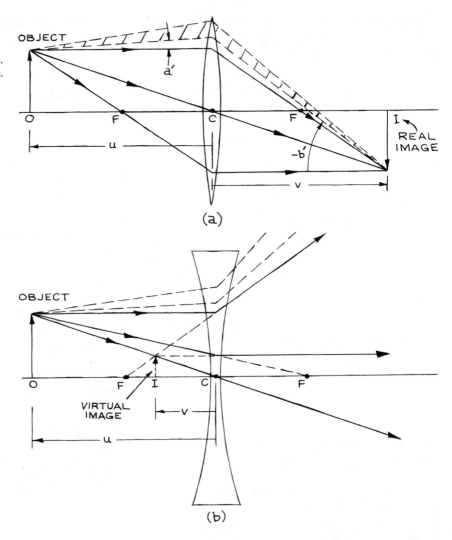

ray path between input and output planes can be found in simple cases—for example, the path *XX'* in the thick lens shown in Figure 4.7(*b*). The second principal ray is directed parallel to the axis; from the intersection of this ray, or its projection, with the second principal plane, the output ray passes through (or appears to have come from) the second focal point. The actual ray path between input and output planes can again be found in simple cases—for example, the path *YY'* in the thick lens shown in Figure 4.7(*b*). The intersection of the two principal rays in the image space produces the image point that corresponds to the original point on the object. If only the back projections of the output principal rays appear to intersect, this intersection point lies on a *virtual* image.

In the majority of applications of ray tracing a quick analysis of the system is desired. In this case, if all lenses in the system are treated as thin lenses, the position of the principal planes need not be calculated beforehand and ray tracing becomes particularly easy. For a thin lens a third principal ray is useful for determining the image location. The ray from a point on the object that passes through the center of the lens is not deviated by the lens.

The use of ray tracing to determine the size and position of the real image produced by a convex lens and the virtual image produced by a concave lens are shown in Figure 4.8. Figure 4.8(a) shows a converging lens; the input principal ray parallel to the axis actually passes through the focal point. Figure 4.8(b) shows a diverging lens; the above ray now emerges from the lens so as to appear to have come from the focal point. More complex systems of lenses can be analyzed the same way. Once the image location has been determined by the use of the principal rays, the path of any group of rays, a ray *pencil*, can be found. See, for example, the cross-hatched ray pencils shown in Figure 4.8.

The use of ray-tracing rules to analyze spherical-mirror systems is similar to those described above, except that the ray striking the center of the mirror in this case reflects so that the angle of incidence equals the angle of reflection.

(c) Imaging and Magnification. In Figure 4.8(a) the ratio of the height of the image to the height of the object is called the *magnification M*. In the case of a thin lens,

$$M = \frac{b}{a} = \frac{v}{u}. \tag{35}$$

For a more general system the magnification can be obtained from the ray-transfer matrix equation

$$\begin{pmatrix} b \\ b' \end{pmatrix} = \begin{pmatrix} A & B \\ C & D \end{pmatrix} \begin{pmatrix} a \\ a' \end{pmatrix} = \mathbf{M} \begin{pmatrix} a \\ a' \end{pmatrix}, \tag{36}$$

where a' is the angle of a ray through a point on the object and b' through the corresponding point on the image. The matrix $\mathbf{M}$ in this case includes the entire system from object O to image I. So in Figure 4.8(a),

$\mathbf{M} =$ (matrix for uniform medium of length u)

$\times$ (matrix for lens)

$\times$ (matrix for uniform medium of length v). (37)

For imaging, b must be independent of a', so $B = 0$. The magnification is

$$M = \frac{b}{a} = A, \tag{38}$$

and so the ray transfer matrix of the imaging system can be written as

$$\mathbf{M} = \begin{pmatrix} M & 0 \\ -\dfrac{1}{f} & \dfrac{1}{M} \end{pmatrix}, \tag{39}$$

where it should be noted that the result $\det(\mathbf{M}) = 1$ has been used.

The *angular magnification* of the system is defined as

$$m = \left(\frac{b'}{a'} \right)_{a=0},$$

which gives $m = 1/M$, or $Mm = 1$, a useful general result.

4.2.3   The Use of Impedances in Optics

The method of impedances is the easiest way to calculate the fraction of incident intensity transmitted and reflected in an optical system. It is also the easiest way to follow the changes in polarization state that result when light passes through an optical system.

As we have seen in Section 4.1, the impedance of a plane wave traveling in a medium of relative permeability μ_r and dielectric constant ϵ_r is

$$Z = \sqrt{\frac{\mu_r \mu_0}{\epsilon_r \epsilon_0}} = Z_0 \sqrt{\frac{\mu_r}{\epsilon_r}}. \tag{40}$$

If $\mu_r = 1$, as is usually the case for optical media, the impedance can be written as

$$Z = Z_0 / n. \tag{41}$$

This impedance relates the transverse $\mathbf{E}$ and $\mathbf{H}$ fields of the wave:

$$Z = \frac{E_{\text{tr}}}{H_{\text{tr}}}. \tag{42}$$

When a plane wave crosses a planar boundary between two different media, the components of both $\mathbf{E}$ and $\mathbf{H}$ parallel to the boundary have to be continuous across

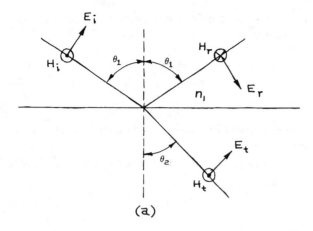

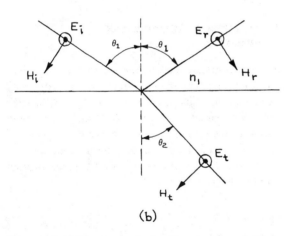

Figure 4.9 Reflection and refraction at a planar boundary between two different dielectric media: (*a*) wave polarized in the plane of incidence (*P*-polarization); (*b*) wave polarized perpendicular to the plane of incidence (*S*-polarization).

that boundary. Figure 4.9(*a*) illustrates a plane wave polarized in the plane of incidence striking a planar boundary between two media of refractive indices n_1 and n_2. In terms of the *magnitudes* of the vectors involved,

$$E_i \cos \theta_1 + E_r \cos \theta_1 = E_t \cos \theta_2, \tag{43}$$

$$H_i - H_r = H_t. \tag{44}$$

Equation (44) can be written as

$$\frac{E_i}{Z_1} - \frac{E_r}{Z_1} = \frac{E_t}{Z_2}. \tag{45}$$

It is easy to eliminate E_t between Equations (43) and (45) to give

$$\rho = \frac{E_r}{E_i} = \frac{Z_2 \cos \theta_2 - Z_1 \cos \theta_1}{Z_2 \cos \theta_2 + Z_1 \cos \theta_1}, \tag{46}$$

where ρ is the *reflection coefficient* of the surface. The fraction of the incident energy reflected from the surface is called the reflectance, $R = \rho^2$. Similarly, the *transmission coefficient* of the boundary is

$$\tau = \frac{E_t \cos \theta_2}{E_i \cos \theta_1} = \frac{2 Z_2 \cos \theta_2}{Z_2 \cos \theta_2 + Z_1 \cos \theta_1}. \tag{47}$$

By a similar treatment applied to the geometry shown in Figure 4.9(*b*) it can be shown that for a plane wave polarized perpendicular to the plane of incidence,

$$\rho = \frac{Z_2 \sec \theta_2 - Z_1 \sec \theta_1}{Z_2 \sec \theta_2 + Z_1 \sec \theta_1}, \tag{48}$$

$$\tau = \frac{2 Z_2 \sec \theta_2}{Z_2 \sec \theta_2 + Z_1 \sec \theta_1}. \tag{49}$$

If the effective impedance for a plane wave polarized in the plane of incidence (*P-polarization*) and incident on a boundary at angle θ is defined as

$$Z' = Z \cos \theta, \tag{50}$$

and for a wave polarized perpendicular to the plane of incidence (*S-polarization*) as

$$Z' = Z \sec \theta, \tag{51}$$

then a universal pair of formulae for ρ and τ results:

$$\rho = \frac{Z_2' - Z_1'}{Z_1' + Z_2'}, \tag{52}$$

$$\tau = \frac{2 Z_2'}{Z_1' + Z_2'}. \tag{53}$$

It will be apparent from an inspection of Figure 4.9 that Z' is just the ratio of the electric-field component parallel to the boundary and the magnetic-field component parallel to the boundary. For reflection from an ideal mirror, $Z_2' = 0$.

In normal incidence Equations (46) and (48) become identical and can be written as

$$\rho = \frac{Z_2 - Z_1}{Z_2 + Z_1} = \frac{n_1 - n_2}{n_1 + n_2}. \tag{54}$$

Note that there is a change of phase of π in the reflected field relative to the incident field when $n_2 > n_1$.

Since intensity $\propto$ (electric field)^2, the fraction of the incident energy that is reflected is

$$R = \rho^2 = \left(\frac{n_1 - n_2}{n_1 + n_2}\right)^2. \tag{55}$$

R increases with the index mismatch between the two media, as shown in Figure 4.10. If there is no absorption of energy at the boundary, the fraction of energy transmitted, called the *transmittance*, is

$$T = 1 - R = \frac{4 n_1 n_2}{(n_1 + n_2)^2}. \tag{56}$$

(a) **Reflectance for Waves Incident on an Interface at Oblique Angles.** If the wave is not incident normally, it must be decomposed into two linearly polarized components, one polarized in the plane of incidence, and the other polarized perpendicular to the plane of incidence.

For example, consider a plane-polarized wave incident on an air glass ($n = 1.5$) interface at an angle of incidence of $30°$, with a polarization state exactly intermediate between the S-polarization and the P-polarization. The angle of refraction at the boundary is found from Snell's law:

$$\sin \theta_2 = \frac{\sin 30°}{1.5},$$

and so $\theta_2 = 19.47°$.

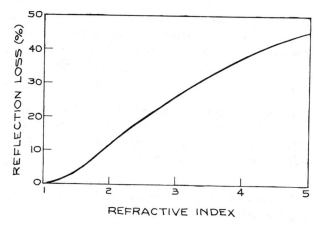

Figure 4.10 Reflection loss per surface for normal incidence as a function of refractive index.

The effective impedance of the P-component in the air is

$$Z_{P1}' = 376.7 \cos \theta_1 = 326.23 \text{ ohm},$$

and in the glass,

$$Z'_{P2} = \frac{376.7}{1.5} \cos \theta_2 = 236.77 \text{ ohm}.$$

Thus, from Equation (52), the reflection coefficient for the P-component is

$$\rho_P = \frac{236.77 - 326.33}{236.77 + 326.23} = -0.159.$$

The fraction of the intensity associated with the P-component that is reflected is $\rho_P^2 = 0.0253$.

For the S-component of the input wave,

$$Z_{S1}' = \frac{376.7}{\cos \theta_1} = 434.98 \text{ ohm},$$

$$Z_{S2}' = \frac{376.7}{1.5 \cos \theta_2} = 266.37 \text{ ohm},$$

$$\rho_S = \frac{266.37 - 434.98}{266.37 + 434.98} = -0.240.$$

The fraction of the intensity associated with the S-polarization component that is reflected is $\rho_S^2 = 0.0578$. Since the input wave contains equal amounts of S- and P-polarization, the overall reflectance in this case is

$$R = \langle \rho^2 \rangle_{av} = 0.0416 \simeq 4\%.$$

Note that the reflected wave now contains more S-polarization than P, so the polarization state of the reflected wave has been changed—a phenomenon that will be discussed further in Section 4.3.6(b).

(b) Brewster's Angle.   Returning to Equation (46), it might be asked whether the reflectance is ever zero. It is clear that ρ will be zero if

$$n_1 \cos \theta_2 = n_2 \cos \theta_1, \qquad (57)$$

which, from Snell's law [Equation (7)], gives

$$\cos \theta_1 = \frac{n_1}{n_2} \sqrt{1 - \left(\frac{n_1}{n_2}\right)^2 \sin^2 \theta_1}, \qquad (58)$$

giving the solution

$$\theta_1 = \theta_B = \arcsin \sqrt{\frac{n_2^2}{n_1^2 + n_2^2}} = \arctan \frac{n_2}{n_1}. \qquad (59)$$

The angle θ_B is called *Brewster's angle*. A wave polarized in the plane of incidence and incident on a boundary at this angle is totally transmitted. This fact is put to good use in the design of low-reflection-loss windows in laser systems, as will be seen in Section 4.6.2. If Equation (48) is inspected carefully, it will be seen that there is no angle of incidence that yields zero reflection for a wave polarized perpendicular to the plane of incidence.

(c) Transformation of Impedance through Multilayer Optical Systems.   The impedance concept allows the reflection and transmission characteristics of multilayer optical systems to be evaluated very simply. If the incident light is incoherent (a concept discussed in more detail later in Section 4.5.1), then the overall transmission of a multilayer structure is just the product of the

transmittances of its various interfaces. For example, an air-glass interface transmits about 96% of the light in normal incidence. The transmittance of a parallel-sided slab is $0.96 \times 0.96 = 92\%$. This simple result ignores the possibility of interference effects between reflected and transmitted waves at the two faces of the slab. If the faces of the slab are very flat and parallel, and if the light is coherent, such effects cannot be ignored. In this case, the method of transformed impedances is useful.

Consider the three-layer structure shown in Figure 4.11(a). The path of a ray of light through the structure is shown. The angles $\theta_1, \theta_2, \theta_3$ can be calculated from Snell's law. As an example consider a wave polarized in the plane of incidence. The effective impedances of media 1, 2, and 3 are

$$Z_1' = Z_1 \cos \theta_1 = \frac{Z_0 \cos \theta_1}{n_1},$$

$$Z_2' = Z_2 \cos \theta_2 = \frac{Z_0 \cos \theta_2}{n_2}, \qquad (60)$$

$$Z_3' = Z_3 \cos \theta_3 = \frac{Z_0 \cos \theta_3}{n_3}.$$

It can be shown that the reflection and transmission coefficients of the structure are exactly the same as the equivalent structure in Figure 4.11(b) for normal incidence, where the effective thickness of layer 2 is now

$$d' = d \cos \theta_2. \qquad (61)$$

The reflection and transmission coefficients of the structure can be calculated from its equivalent structure using the transformed-impedance concept.[11]

The transformed impedance of medium 3 at the boundary between media 1 and 2 is

$$Z_3'' = Z_2' \left(\frac{Z_3' \cos k_2 d' + i Z_2' \sin k_2 d'}{Z_2' \cos k_2 d' + i Z_3' \sin k_2 d'} \right), \qquad (62)$$

where $k_2 = 2\pi/\lambda_2 = 2\pi n_2/\lambda_0$. The reflection coefficient of the whole structure is now just

$$\rho = \frac{Z_3'' - Z_1'}{Z_3'' + Z_1'}, \qquad (63)$$

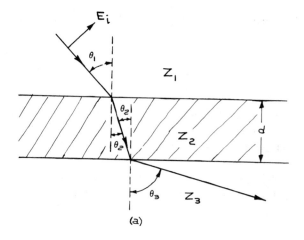

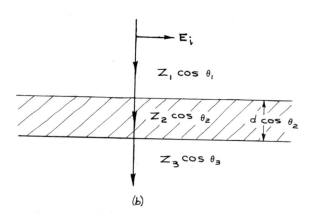

Figure 4.11 (*a*) Wave polarized in the plane of incidence passing through a dielectric slab of thickness d and impedance Z_2 separating two semiinfinite media of impedances Z_1 and Z_3, respectively; (*b*) equivalent structure for normal incidence.

and its transmission coefficient

$$\tau = \frac{2Z_3''}{Z_3'' + Z_1'}. \tag{64}$$

In a structure with more layers, the transformed impedance formula (62) can be used sequentially, starting at the last optical surface and working back to the first. More examples of the use of the technique of transfor-

mation of impedance will be given in Sections 4.3.1(e) and 4.3.7(b).

4.2.4 Gaussian Beams

Gaussian beams are propagating-wave solutions of Maxwell's equations that are restricted in lateral extent even in free space. They do not need any beam-confining reflective planes, as do the confined electromagnetic-field modes of waveguides.[12] Whereas the field components of a plane transverse electromagnetic wave of angular frequency ω propagating in the z-direction are of the form

$$V = V_0 e^{i(\omega t - kz)}, \tag{65}$$

where V_0 is a constant, a Gaussian beam is of the form

$$V = \Psi(x, y, z) e^{i(\omega t - kz)} = U(x, y, z) e^{i\omega t}, \tag{66}$$

where, for example, for a particular value of z, $\Psi(x, y, z)$ gives the spatial variation of the fields in the xy plane. $\Psi^*\Psi$ gives the relative intensity distribution in the plane; for a Gaussian beam this gives a localized intensity pattern. The various Gaussian-beam solutions of Maxwell's equations are denoted as TEM_{mn} modes; a detailed discussion of their properties is given by Kogelnik and Li.[11] The output beam from a laser is generally a TEM mode or a combination of TEM modes. Many laser systems operate in the fundamental Gaussian mode, denoted TEM_{00}. This is the only mode that will be considered in detail here.

For the TEM_{00} mode, $\Psi(x, y, z)$ has the form

$$\Psi(x, y, z) = \exp\left\{-i\left[P(z) + \frac{kr^2}{2q(z)}\right]\right\}, \tag{67}$$

where $P(z)$ is a *phase factor*, $q(z)$ is called the *beam parameter*, k equals $2\pi/\lambda$, and r^2 equals $x^2 + y^2$. The beam parameter $q(z)$ is usually written in terms of the *phase-front curvature* $R(z)$ of the beam, and its *spot size* $w(z)$, as

$$\frac{1}{q} = \frac{1}{R} - \frac{i\lambda}{\pi w^2}. \tag{68}$$

q and P obey the following relations:

$$\frac{dq}{dz} = 1, \qquad (69)$$

$$\frac{dP}{dz} = -\frac{i}{q}. \qquad (70)$$

For a particular value of z, the intensity variation in the xy plane (the *mode pattern*) is, from Equation (67),

$$\Psi^*\Psi = \exp\left[\frac{-ikr^2}{2}\left(\frac{1}{q(z)} - \frac{1}{q^*(z)}\right)\right], \qquad (71)$$

which from Equation (68) gives

$$\Psi\Psi^* = e^{-2r^2/w^2}. \qquad (72)$$

Thus, $w(z)$ is the distance from the axis of the beam $(x = y = 0)$ where the intensity has fallen to $1/e^2$ of its axial value and the fields to $1/e$ of their axial magnitude. The radial distribution of both intensity and field strength of the TEM_{00} mode is Gaussian, as shown in Figure 4.12. Clearly, from Equation (69)

$$q = q_0 + z, \qquad (73)$$

where q_0 is the value of the beam parameter at $z = 0$. From Equation (70),

$$P(z) = -i\ln q + \text{constant}$$
$$= -i\ln(q_0 + z) + \text{constant}. \qquad (73a)$$

Writing the constant in Equation (71) as $\theta + i\ln q_0$, the full spatial variation of the Gaussian beam is

$$U = \exp\left\{-i\left[kz - i\ln\left(1 + \frac{z}{q_0}\right) + \theta \right.\right.$$
$$\left.\left. + \frac{kr^2}{2}\left(\frac{1}{R(z)} - \frac{i\lambda}{\pi w^2}\right)\right]\right\}. \qquad (73b)$$

The phase angle θ is usually set equal to zero. In the plane $z = 0$,

$$U = \exp\left\{-i\frac{kr^2}{2}\left(\frac{1}{R(0)} - \frac{i\lambda}{\pi w(0)^2}\right)\right\}. \qquad (73c)$$

The surface of constant phase at this point is defined by the equation

$$\frac{kr^2}{2R(0)} = \text{constant}. \qquad (74)$$

If the constant is taken to be zero and $R(0)$ infinite, then r becomes indeterminate and the surface of constant phase is the plane $z = 0$. In this case, $w(0)$ can have its minimum value anywhere. This minimum value is called the *minimum spot size* w_0, and the plane $z = 0$ is called the *beam waist*. At the beam waist,

$$\frac{1}{q_0} = \frac{-i\lambda}{\pi w_0^2}. \qquad (75)$$

Using Equations (68) and (73), for any arbitrary value of z,

$$w^2(z) = w_0^2\left[1 + \left(\frac{\lambda z}{\pi w_0^2}\right)^2\right]. \qquad (76)$$

The radius of curvature of the phase front at this point is

$$R(z) = z\left[1 + \left(\frac{\pi w_0^2}{\lambda z}\right)^2\right]. \qquad (77)$$

From Equation (76) it can be seen that the Gaussian beam expands in both the positive and negative z-directions from its beam waist along a hyperbola that has asymptotes inclined to the axis at an angle

$$\theta_{\text{beam}} = \arctan\frac{\lambda}{\pi w_0}, \qquad (78)$$

as illustrated in Figure 4.12(b). The surfaces of constant phase of the Gaussian beam are in reality parabolic. However, for $r^2 \ll z^2$ (which is generally true, except close to the beam waist), they are spherical surfaces with the radius of curvature $R(z)$. Although they will not be discussed further here, the higher-order Gaussian beams, denoted TEM_{mn}, are also characterized by radii of curvature and spot sizes that are identical to those of the TEM_{00} mode and obey Equations (76) and (77).

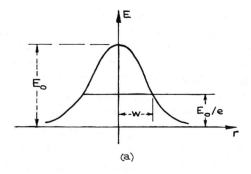

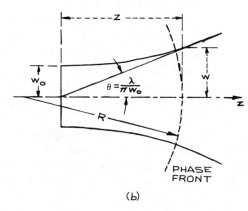

Figure 4.12 (a) Radial amplitude variation of the TEM$_{00}$ Gaussian beam; (b) contour of a Gaussian beam.

A lens can be used to focus a laser beam to a small spot, or systems of lenses may be used to expand the beam and recollimate it (i.e., minimize the beam divergence). In such an application a thin lens will not alter the transverse intensity pattern of the beam at the lens, but it will alter its radius of curvature. Far enough from the beam waist, the radius of curvature of a Gaussian beam behaves exactly as a true spherical wave, since for $z \gg \pi w_0^2/\lambda$, Equation (77) becomes

$$R(z) = z. \qquad (79)$$

Now, when a spherical wave of radius R_1 strikes a thin lens, the object distance is clearly also R_1—the distance to the point of origin of the wave. Therefore, the radius of curvature R_2 immediately after passage

through the lens must obey

$$\frac{1}{R_2} = \frac{1}{R_1} - \frac{1}{f}, \qquad (80)$$

as shown in Figure 4.13. Thus, if w is unchanged at the lens, the beam parameter after passage through the lens obeys

$$\frac{1}{q_2} = \frac{1}{q_1} - \frac{1}{f}. \qquad (81)$$

It is straightforward to use this result in conjunction with Equations (68) and (73) to give the minimum spot size w_f of a TEM$_{00}$ Gaussian beam focused by a lens as

$$w_f = f\left[\left(\frac{\lambda}{\pi w_1}\right)^2 + \frac{w_1^2}{R_1^2}\right]^{1/2}, \qquad (82)$$

where w_1 and R_1 are the laser-beam spot size and radius of curvature at the input face of the lens. If the lens is placed a great distance from, or very close to, the beam waist, Equation (82) reduces to

$$w_f = f\theta_B, \qquad (83)$$

where θ_B is the beam divergence at the input face of the lens.

If θ_B is a small angle, then w_f is located almost at the focal point of the lens—in reality, very slightly closer to the lens. It is worthwhile comparing the result given by Equation (83) with the result obtained for a plane wave

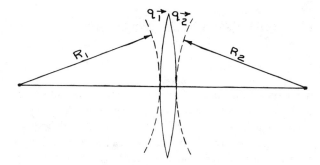

Figure 4.13 Transformation of a Gaussian beam by a thin lens.

being focused by a lens. In the latter case, the lens diameter D is the factor that limits the lateral extent of the wave being focused. Diffraction theory[7,13] shows that, in this case, 84% of the energy is focused into a region of diameter

$$S = 2.44\lambda f/D. \qquad (84)$$

It should be noted that the spot size to which a laser beam can be focused cannot be reduced without limit merely by reducing the focal length. In the limit the lens becomes a sphere and, of course, is then no longer a thin lens. In practice, to focus a laser beam to a small spot, the value of θ_B should be first reduced by expanding the beam and then recollimating it—generally with a Galilean telescope, as will be seen in Section 4.3.3(a). To prevent diffraction effects the lens aperture should be larger than the spot size at the lens: $D = 2.8w$ is a common size used. In this case, if the focusing lens is placed at the beam waist of the collimated beam, the focal-spot diameter is

$$2w_f = \frac{5.6\lambda f}{\pi D} = \frac{1.78\lambda f}{D}. \qquad (85)$$

In practice, it is difficult to manufacture spherical lenses with very small values of f/D (called the $f/number$) that achieve the diffraction-limited performance predicted by this equation; commercially available spherical lenses[14] achieve values of $2w_f$ of about 10λ. However, smaller spot sizes are possible with aspheric lenses.

4.3 OPTICAL COMPONENTS

4.3.1 Mirrors

Whenever light passes from one medium to another of different refractive index there is some reflection, so the interface acts as a partially reflecting mirror. By applying an appropriate single-layer or multilayer coating to the interface between the two media, the reflection can be controlled so that the reflectance has any desired value between 0 and 1. Both flat and spherical mirrors made in this way are available—there are numerous suppliers.[15] If no transmitted light is required, high-reflectance mirrors can be made from metal-coated substrates or from metals themselves. Mirrors that reflect and transmit roughly equal amounts of incident light are often referred to as *beamsplitters*.

(a) **Flat Mirrors.** Flat mirrors are used to deviate the path of light rays without any focusing. These mirrors can have their reflective surface on the front face of any suitable substrate, or on the rear face of a transparent substrate. Front-surface, totally reflecting mirrors have the advantage of producing no unwanted additional or *ghost* reflections; however, their reflective surface is exposed. Rear-surface mirrors produce ghost reflections, as illustrated in Figure 4.14, unless their front surface is antireflection-coated; but the reflective surface is protected. Most household mirrors are made this way. The cost of flat mirrors depends on their size and on the degree of flatness required.

Mirrors for high-precision applications—for example, in visible lasers—are normally specified to be flat between $\lambda/10$ and $\lambda/20$ for visible light. This degree of flatness is not required when the mirror is merely for light collection and redirection, for example to reflect light onto the surface of a detector. Mirrors for this sort

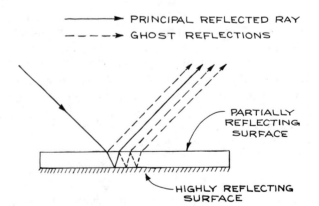

Figure 4.14 Production of ghost reflections by a rear-surface mirror whose front surface is not antireflection-coated.

of application are routinely flat to within a few wavelengths of visible light. Excellent mirrors for this purpose can be made from "float" plate glass,[16] which is flat to 1 or 2 wavelengths per inch.

(b) Spherical Mirrors. Spherical mirrors are widely used in laser construction, where the radii of curvature are typically rather long—frequently 20 m radius or more. They can be used whenever light must be collected and focused. However, spherical mirrors are only good for focusing nearly parallel beams of light that strike the mirror close to normal. In common with spherical lenses, spherical mirrors suffer from various imaging defects called aberrations, which are discussed in Section 4.3.3(c). A parallel beam of light that strikes a spherical mirror far from normal is not focused to a small spot. When used this way the mirror is *astigmatic*, that is, off-axis rays are focused at different distances in the horizontal and vertical planes. This leads to blurring of the image, or at best the focusing of a point source into a line image.

(c) Paraboloidal and Ellipsoidal Mirrors. Paraboloidal mirrors will produce a parallel beam of light when a point source is placed at the focus of the paraboloidal surface. Thus, these mirrors are very useful in projection systems. They are usually made of polished metal, although versions using Pyrex substrates are also available. An important application of parabolic mirrors is in the off-axis focusing of laser beams, using off-axis mirrors in which the axis of the paraboloid does not pass through the mirror. When they are used in this way there is complete access to the focal region without any shadowing, as shown in Figure 4.15. Spherical mirrors, on the other hand, do not focus well if they are used in this way. Off-axis paraboloidal mirrors, as well as metal axial paraboloidal mirrors, are available from Melles Griot, Space Optics Research Lab, and Optics Plus. (Here, and wherever suppliers are mentioned in the text, the list is intended to be representative but not necessarily exhaustive.)

Ellipsoidal mirrors are also used for light collecting. Light that passes through one focus of the ellipsoid will,

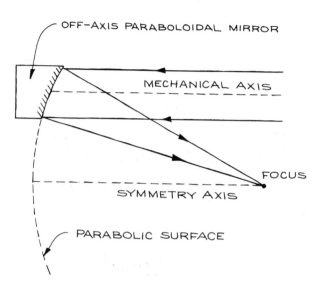

Figure 4.15 Off-axis paraboloidal mirror.

after reflection, also pass through the other. These mirrors, like the all-metal paraboloidal mirrors, are generally made by electroforming. The surface finish obtained in this way is quite good, although not so good as can be obtained by optical polishing. Rhodium-plated electroformed ellipsoidal mirrors are available from Melles Griot.

(d) Beamsplitters. Beamsplitters are semitransparent mirrors that both reflect and transmit light over a range of wavelengths. A good beamsplitter has a multilayer dielectric coating on a substrate that is slightly wedge-shaped to eliminate interference effects, and antireflection-coated on its back surface to minimize ghost images. The ratio of reflectance to transmittance of a beamsplitter depends on the polarization state of the light. The performance is usually specified for light linearly polarized in the plane of incidence (*P*-polarization) or orthogonal to the plane of incidence (*S*-polarization). Cube beamsplitters are pairs of identical right-angle prisms cemented together on their hypotenuse faces. Before cementing, a metal or dielectric semireflecting layer is placed on one of the hypotenuse faces. Antireflection-coated cube prisms have virtually

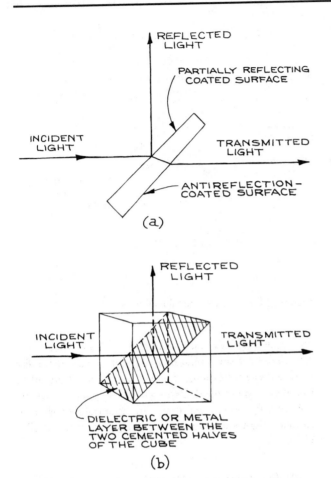

Figure 4.16 (a) Conventional planar beamsplitter (wedge angle greatly exaggerated); (b) cube beamsplitter.

no ghost image problems and are more rigid than plate type beamsplitters. The operation of both types of beamsplitter is illustrated in Figure 4.16.

(e) Dielectric Coatings. Several different kinds of single-layer and multilayer dielectric coatings are available commercially on the surface of optical components. These coatings can either reduce the reflectance or enhance it in some spectral region.

(i) Single-layer antireflection coatings. The mode of operation of a single-layer antireflection (AR) coating can be conveniently illustrated by the method of transformed impedances. Suppose medium 3 in Figure 4.11 is the surface to be AR-coated. Apply to the surface a coating with an effective thickness

$$d' = \lambda_2/4, \qquad (86)$$

where $\lambda_2 = c_0/\nu n_2$. The actual thickness of layer 2 is

$$d = d'/\cos\theta_2. \qquad (87)$$

The transformed impedance of medium 3 at the first interface is found by substitution in Equation (62):

$$Z_3'' = Z_2'^2/Z_3'. \qquad (88)$$

To reduce the reflection coefficient to zero, we need $Z_3'' = Z_1'$; so the effective impedance of the antireflection layer must be

$$Z_2' = \sqrt{Z_1' Z_3'}. \qquad (89)$$

Thus, to eliminate reflection of waves linearly polarized in the plane of incidence and incident at angle θ_1, n_2 must be chosen to satisfy

$$\frac{\cos\theta_2}{n_2} = \sqrt{\frac{\cos\theta_1\cos\theta_3}{n_1 n_3}}. \qquad (90)$$

For use in normal incidence, $n_2 = \sqrt{n_1 n_3}$ and $d = \lambda_2/4$. In normal incidence, an AR coating works for any incident polarization. To minimize reflection at an air–flint-glass ($n = 1.7$) interface, we would need

$$n_2 = \sqrt{1.7} = 1.3.$$

Magnesium fluoride with a refractive index of 1.38 and cryolite (sodium aluminum fluoride) with a refractive index of 1.36 come closest to meeting this requirement in the visible region of the spectrum. Optical components such as camera lenses are usually coated with one of these materials to minimize reflection at 550 nm. The slightly greater Fresnel reflection that results in the blue and red gives rise to the characteristic purple color of these components in reflected light.

(ii) *Multilayer antireflection coatings.* The minimum reflectance of a single-layer antireflection-coated substrate is $(n_2^2 - n_1 n_3)/(n_2^2 + n_1 n_3)^2$. However, available robust optical coating materials such as MgF_2 do not have a sufficiently high refractive index to reduce the reflectance to zero with a single layer. Two-layer coatings, often called V-coatings because of the shape of their transmission characteristic, reduce reflection better than a single layer, as shown in Figure 4.17(*b*). Multilayer coatings can be used to reduce reflectance over a broader wavelength region than a V-coating, as shown in Figure 4.18. Such broad-band coatings are usually of proprietary design: they are available commercially from many sources, as are coatings of other types.[15] Some companies offer both optical components and coatings, while others specialize in coatings and will coat customers' own materials.

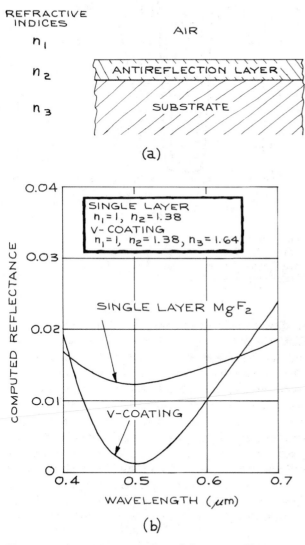

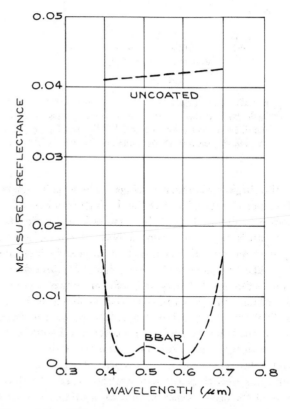

Figure 4.17   (*a*) Geometry of single-layer antireflection coating; (*b*) reflectances of single- and double-layer antireflection coatings on a substrate of refractive index 1.64. (From *Handbook of Lasers*, R. J. Pressley, Ed., CRC Press, Cleveland, 1971; by permission of CRC Press, Inc.)

Figure 4.18   Reflectance of a typical broad-band antireflection coating (BBAR) showing considerable reduction in reflectance below the uncoated substrate. (From *Handbook of Lasers*, R. J. Pressley, Ed., CRC Press, Cleveland, 1971; by permission of CRC Press, Inc.)

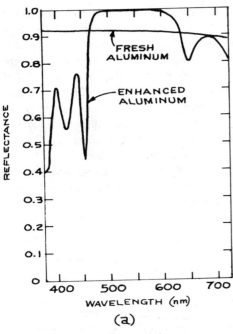

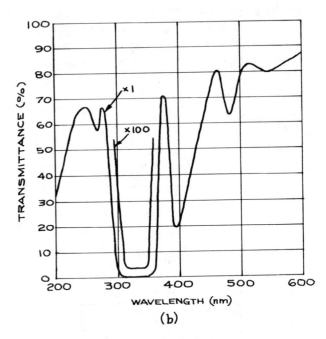

(a)

(b)

Figure 4.19 (*a*) Reflectance of a plain aluminum mirror and an aluminum mirror with four dielectric overlayers (from *Handbook of Lasers*, R. J. Pressley, Ed., CRC Press, Cleveland, 1971; by permission of CRC Press); (*b*) typical transmittance characteristic of a narrow-band, maximum-reflectance, multilayer dielectric coating on a dielectric substrate.

(iii) *High-reflectance coatings*. The usually high reflectance of a metal surface can be further enhanced, as shown in Figure 4.19(*a*), by a multiple dielectric coating consisting of an even number of layers, each a quarter wavelength thick, and of alternately high and low refractive index, with a low-refractive-index material next to the metal. A very high reflectance over a narrow wavelength range can be achieved, as shown in Figure 4.19(*b*), by using a dielectric substrate, such as fused quartz, and an odd number of quarter-wavelength layers of alternately high and low refractive index with a high-index layer on the substrate. Broad-band high-reflectance coatings are also available where the thickness of the layers varies around an average value of a quarter wavelength.

(f) Pellicles. Pellicles are beam-splitting mirrors made of a high-tensile-strength polymer stretched over a flat metal frame. The polymer film can be coated to modify the reflection-transmission characteristics of the film. The polymer generally used, nitrocellulose, transmits in the visible and near infrared to about 2 μm. These devices have some advantages over conventional coated-glass or quartz beamsplitters; the thinness of the polymer film virtually eliminates spherical and chromatic aberrations [see Section 4.3.3(c)] when diverging or converging light passes through them, and ghost-image problems are virtually eliminated. However, they will produce some wavefront distortion, typically about 2 waves per inch, and are not suitable for precision applications. Pellicles are available from several suppliers, such as Edmund Scientific and Melles Griot.

4.3.2 Windows

An optical window serves as a barrier between two media, for example, as an observation window on a vacuum system or on a liquid or gas cell, or as a

Brewster window on a laser. When light passes through a window which separates two media, there is in general some reflection from the window and a change in the state of polarization of both the reflected and the transmitted light. If the window material is not perfectly transparent at the wavelength of interest, there is also absorption of light in the window, which at high light intensities will cause the window to heat. This will cause optical distortion of the transmitted wave, and at worst—in high-power laser applications—damage the window on its surface, internally, or both. Additionally, if the two surfaces of a window are both very flat (roughly speaking, one wave per centimeter or better) and close to parallel, the window will act as an *etalon* [see Sections 4.3.7(b) and 4.7.4] and exhibit distinct variations in transmission with wavelength. To circumvent this difficulty, which is a particular nuisance in experiments using lasers, most precision flat optical windows are constructed as a slight wedge, with an angle usually about 30′.

The details of the reflection, refraction, and change of polarization state that occur when light strikes a window surface are dealt with in detail in Sections 4.2.3 and 4.3.6(b). These considerations also apply when light strikes a lens or prism. The reflection at such surfaces can be reduced by a single-layer or multilayer dielectric antireflection coating. In optical systems, such as multielement camera lenses, where light crosses many such surfaces, antireflection coatings are very desirable. Otherwise a severe reduction in transmitted light intensity will result, as illustrated in Figure 4.20.

4.3.3 Lenses and Lens Systems

(a) **Simple and Galilean Telescopes.** The operation of a simple telescope is illustrated in Figure 4.21(a). The objective is usually an *achromat* (a lens in which chromatic aberration—the variation of focal length with wavelength—has been minimized). It has a long focal length f_1 and produces a real image of a distant object, which can then be examined by the eyepiece lens of focal length f_2. This design, which produces an inverted image, is often called an *astronomical* telescope. If desired, the final image can be erected with a third lens,

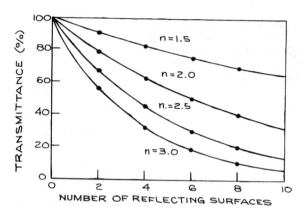

Figure 4.20 Transmittance of a multielement optical system as a function of the refractive index of the elements.

in which case the device is called a *terrestrial* telescope. The eyepiece of the telescope can be a single lens, but composite eyepiece designs such as the Ramsden and the Kellner[7] are also common. Most large astronomical telescopes use spherical mirrors as objectives, and various configurations are used such as the Newtonian, Cassegrain, and Schmidt.[7,17-19] Since most telescope development has been for astronomical applications, it is outside the scope of this book to give a detailed discussion; for further details the reader should consult Born and Wolf,[7] Levi,[17] Meinel,[18] or Brouwer and Walther.[19]

The Galilean telescope illustrated in Figure 4.21(b), first constructed by Galileo in 1609, is the earliest telescope of which there exists definite knowledge. It produces no real intermediate image, but the final image is erect. In *normal* adjustment the focal point of the diverging eyepiece is outside the telescope and coincident with that of the objective. In the simple telescope the two focal points also coincide, but are between the two. The magnification M produced by either type of telescope can be written as

$$M = -f_1/f_2. \tag{91}$$

Since the simple telescope has two positive (converging) lenses, its overall magnification is negative, which indicates that the final image is inverted.

Simple and Galilean telescopes have practical uses in the laboratory that are distinct from their traditional use for observing distant objects.

Figure 4.21 Ray paths through telescopes in *normal* adjustment (object and final image at infinity): (*a*) astronomical telescope; (*b*) Galilean telescope.

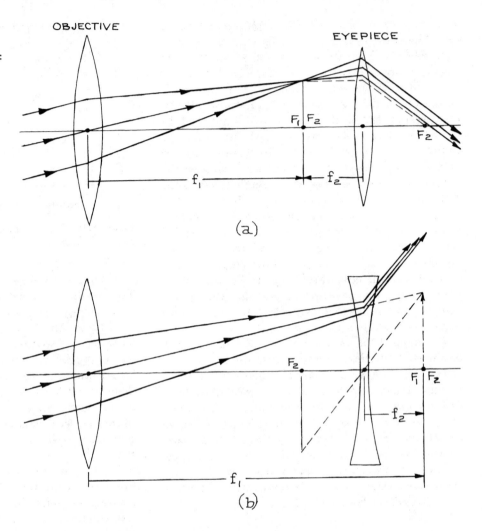

(b) **Laser-Beam Expanders and Spatial Filters.** Laser beams can be expanded and recollimated (or focused and recollimated) with simple or Galilean telescope arrangements, as illustrated in Figure 4.22. For optimum recollimation the spacing of the two lenses should be adjustable, as fine adjustment about the spacing $f_2 + f_1$ will be necessary for recollimating a Gaussian beam. In this application the Galilean telescope has the advantage that the laser beam is not brought to an intermediate focus inside the beam expander. Gas breakdown at such an internal focus can occur with high-power laser beams, although this problem can be solved in simple-telescope beam expanders by evacuat-

ing the telescope. Since laser beams are highly monochromatic, beam expanders need not be constructed from achromatic lenses. However, attempts should be made to minimize spherical aberration and beam distortion. It is best to use precision antireflection-coated lenses if possible. Laser-beam-expanding telescopes that are very well corrected for spherical aberration are available from Janos, Jodon, Klinger, Melles Griot, Oriel, and Special Optics, among others.[15] Infrared laser-beam expanders usually have ZnSe or germanium lenses. The use of biconvex and biconcave lenses distributes the focusing power of the lenses over their surfaces and minimizes spherical aberration. Better cancellation of

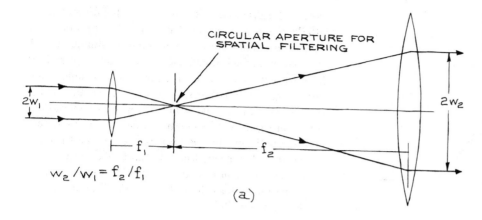

CIRCULAR APERTURE FOR
SPATIAL FILTERING

$2w_1$

$2w_2$

f_1 f_2

$w_2/w_1 = f_2/f_1$

(a)

Figure 4.22 Laser-beam expanders; (a) focusing type with spatial filter; (b) Galilean type.

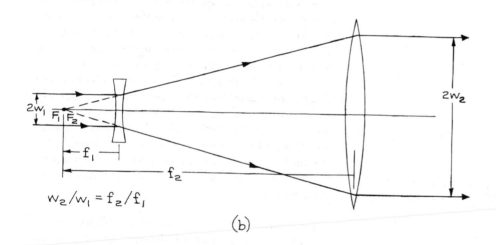

$2w_1$ $F_1 F_2$

$2w_2$

f_1

f_2

$w_2/w_1 = f_2/f_1$

(b)

spherical aberration is possible in Galilean beam expanders than with simple telescopes.

If a small circular aperture is placed at the common intermediate focus of a simple-telescope beam expander, the device becomes a spatial filter as well. Although ideally the output beam from a laser emitting a TEM_{00} mode has a Gaussian radial intensity profile, in practice the radial profile may have some irregular structure. Such beam irregularities may be produced in the laser or by passage of the beam through some medium. If such a beam is focused through a small enough aperture and then recollimated, the irregular structure on the radial intensity profile can be removed and a smooth profile restored, as illustrated in Figure 4.23. This is called *spatial filtering*. The minimum aperture diameter that should be used is

$$D_{\min} = \frac{2f_1\lambda}{\pi w_1},\tag{92}$$

where f_1 is the focal length of the focusing lens and w_1 is the spot size at that lens. Typical aperture sizes used in commercial spatial filters for visible lasers are on the order of 10 μm. This aperture must be accurately and symmetrically positioned at the focal point, so it is usually mounted on an adjustable XY translation stage.

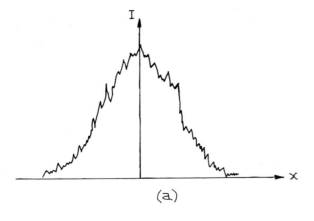

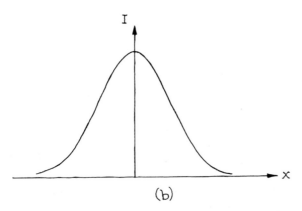

Figure 4.23   Intensity variation $I(x)$ along a diameter of a TEM_{00} laser beam possessing spatial noise structure: (a) before and (b) after spatial filtering.

(c) Lens Aberrations.   Real lenses suffer from various forms of aberration, which lead the image of an object to be an imperfect reconstruction of it. We give a brief description of the effect of the most important of these aberrations in optical images so that the experimentalist can assess whether a given form of aberration is detrimental in a particular experimental situation. If so, a special lens combination can often be obtained that minimizes that aberration—occasionally at the expense of worsening others.

(i) *Chromatic aberrations.* Because the refractive index of the material of a lens varies with wavelength, so does its focal length. Therefore rays of different wavelengths from an object form images in different locations. Chromatic aberration is almost eliminated with the use of an achromatic doublet. This is a pair of lenses, usually consisting of a positive crown-glass lens and a negative flint-glass lens, cemented together. These two lenses cancel each other's chromatic aberrations exactly at specific wavelengths in the blue and red, and almost exactly in the region in between. Good camera lenses are well corrected for chromatic aberrations.

Achromatic doublets are available from numerous suppliers, including Edmund Scientific, Klinger, Melles Griot, Oriel, and Rolyn.

(ii) *Spherical aberrations.* In the paraxial-ray approximation, monochromatic light originating from an axial point again passes through a single point after transversing a lens or system of lenses. This is no longer true for rays with large angles of incidence: different zones of the lens aperture have different focal lengths, depending on their distance from the axis. This effect is called *spherical aberration*. It can be minimized in a single lens by distributing the curvature between its two surfaces without altering the focal length. So, for example, a biconvex lens will produce less spherical aberration than a plano-convex one of the same focal length. Spherical aberration can be eliminated in an achromatic doublet with appropriate choice of curvatures for the constituent lens elements.

(iii) *Coma.* When an object point is off the axis of a lens, its image is produced in different lateral positions by different zones of the lens. Called *coma*, this can be controlled by an appropriate choice of lens curvatures.

(iv) *Astigmatism and field curvature.* All the aberrations discussed so far are defects in the imaging of *meridional rays*, that is, rays in the plane containing the axis of the lens and the line from the object point through the center of the lens. The imaging is different for *sagittal rays*, that is, rays in the sagittal plane, which is perpendicular to the meridional plane, as illustrated in Figure 4.24. The resulting aberration is called *astigmatism*. It can be controlled by lens curvature and refractive-index variations and by the use of apertures to

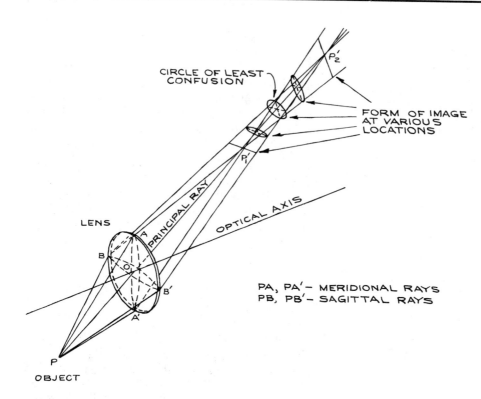

CIRCLE OF LEAST CONFUSION

FORM OF IMAGE AT VARIOUS LOCATIONS

LENS

PRINCIPAL RAY

OPTICAL AXIS

PA, PA' — MERIDIONAL RAYS
PB, PB' — SAGITTAL RAYS

OBJECT

Figure 4.24 Illustrating astigmatism. Meridional rays PA and PA' are imaged at P_1', sagittal rays PB and PB' are imaged at P_2'. (After A. C. Hardy and F. H. Perrin, *Principles of Optics*, McGraw-Hill, New York, 1932; by permission of McGraw-Hill Book Company, Inc.)

restrict the range of angles and off-axis distances at which rays can traverse the lens.

Other forms of aberration also occur, such as *field curvature*, in which an object plane orthogonal to the lens axis is imaged as a curved surface.

(v) *Distortion*. When the aforementioned aberrations have been eliminated, each object point appears as a point in the image, and the images of object points on a plane orthogonal to the axis are also on such a plane. However, the magnification of an object line segment may vary with its distance from the axis. This is called *distortion*. It takes two common forms: *pincushion* and *barrel* distortion, as illustrated in Figure 4.25. Distortion is sensitive to the lens shape and spacing, and to the size and position of apertures in the system—called *stops*—that restrict the regions over which rays can traverse the optical system. The improper positioning of stops in an imaging system is called *vignetting*; see for example Figure 5.4.

For further details of aberrations and how to deal with them, we refer the reader to standard texts on optics, such as those by Born and Wolf,[7] Ditchburn,[13] and Levi.[17]

(d) Aspheric Lenses. An aspheric lens has one non-spherical surface and a second surface that is either concave, convex, or plano, as shown in Figure 4.26. With such a lens it is possible to obtain a much smaller focal length than with a conventional spherical lens of the same diameter, without increasing the spherical aberration of the lens. An ideal aspheric lens exactly cancels the spherical aberration that would otherwise be present in the optical system. Such lenses can collect and focus light rays over a much larger solid angle than conventional lenses and can be used at *f*/numbers as low as 0.6. (The *f/number*, a frequently used lens parameter, is the ratio of the focal length to the available aperture diameter.) Aspheric lenses save space and energy. They can be used as collection lenses very close

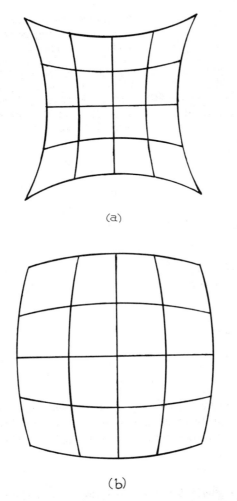

(a)

(b)

Figure 4.25 (*a*) "Pincushion" distortion; (*b*) "barrel" distortion.

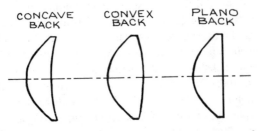

CONCAVE BACK   CONVEX BACK   PLANO BACK

Figure 4.26 Cross sections of three common types of aspheric lens.

to small optical sources such as high-pressure mercury of xenon arc lamps, and for collecting radiation over a large solid angle and focusing it onto a detector element. Because aspheric lenses are generally designed only to minimize spherical aberrations, they will contribute to chromatic aberration, coma, distortion, astigmatism, and curvature of field. Consequently, an aspheric lens should not be used under circumstances where these aberrations will be compounded by transmission through additional components of the optical system. Glass aspheric lenses are available in a range of diameters and focal lengths from several suppliers;[15] aspheric lenses in more exotic materials can be obtained as custom items.

(e) **Fresnel Lenses.** A *Fresnel lens* is an aspheric lens whose surface is broken up into many concentric annular rings. Each ring refracts incident rays to a common focus, so that a very large-aperture, small-f/number, thin aspheric lens results. Figure 4.27 illustrates the construction and focusing characteristics of such a lens. These lenses are generally manufactured from precision molded Acrylic. Because the refractive index of Acrylic varies little with wavelength, from 1.51 at 410 nm to 1.488 at 700 nm, these lenses can be very free of spherical aberration throughout the visible spectrum. Fresnel lenses should be used so that they focus to the plane side of the lens, and their surfaces should not be handled. They are inexpensive and available from very many suppliers, including Edmund Scientific, Melles Griot, and Oriel. However, they should not be considered for use in applications where a precision image or diffraction-limited operation is required.

(f) **Cylindrical Lenses.** Cylindrical lenses are planar on one side and cylindrical on the other. In planes perpendicular to the cylinder axis they have focusing properties identical to a spherical lens, but they do not focus at all in planes containing the cylinder axis. Cylindrical lenses focus an extended source into a line, and so are very useful for imaging sources onto monochromator slits, although perfect matching of f/numbers is not possible in this way. These lenses are also widely used for focusing the output of solid-state and nitrogen lasers

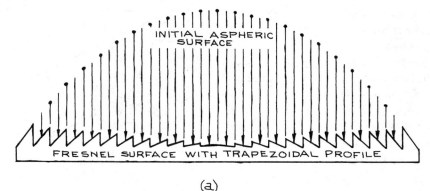

INITIAL ASPHERIC SURFACE

FRESNEL SURFACE WITH TRAPEZOIDAL PROFILE

(a)

Figure 4.27 (*a*) A Fresnel lens, made up of trapezoidal concentric sections (it replaces a bulky aspheric lens); (*b*) illustration of how the focusing characteristics of a Fresnel lens result from refraction at the individual surface grooves. (Courtesy of Melles Griot, Inc.)

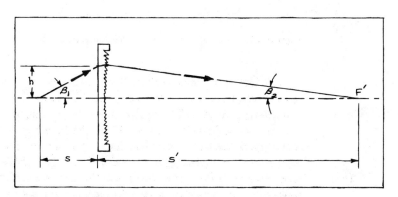

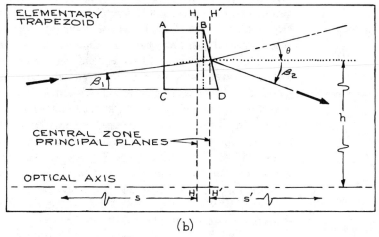

(b)

into a line image in the dye cell of dye lasers [see Section 4.6.3(j)]. Cylindrical lenses are available from Esco, Bond Optics, Melles Griot, Rolyn, Optics for Research, Optics Plus, Oriel, and Infrared Optics, among various other companies.

4.3.4 Prisms

Prisms are used for the dispersion of light into its various wavelength components and for altering the direction of beams of light. Prisms are generally available in most materials that transmit ultraviolet, visible, and infrared light. However, high-refractive-index semi-conductor materials, such as silicon, germanium, and gallium arsenide, are rarely used for prisms. Consult Section 4.4 for a list of suppliers of optical materials.

The deviation and dispersion of a ray of light passing through a simple prism can be described with the aid of Figure 4.28. The various angles $\alpha, \beta, \gamma, \delta$ satisfy Snell's law irrespective of the polarization state of the input beam, provided the prism is made of an isotropic material. It is easy to show that

$$\sin \delta = \sin A \left(n^2 - \sin^2 \alpha \right)^{1/2} - \cos A \sin \alpha \tag{93}$$

and

$$D = \alpha + \delta - A. \tag{94}$$

The exit ray will not take the path shown if γ is greater than the critical angle.

The *dispersion* of the prism is defined as

$$\frac{d\delta}{d\lambda} = \frac{d\delta}{dn} \frac{dn}{d\lambda}. \tag{95}$$

Substituting from Equation (93), this becomes

$$\frac{d\delta}{d\lambda} = \frac{\sin A}{\cos \beta \cos \delta} \frac{dn}{d\lambda}. \tag{96}$$

When the prism is used in the position of *minimum deviation*,

$$\beta = \gamma = \tfrac{1}{2} A \tag{97}$$

and

$$\alpha = \delta = \tfrac{1}{2} \left(D_{\min} + A \right). \tag{98}$$

By the use of the sine rule, Equation (96) can be written in the form

$$\frac{d\delta}{d\lambda} = \frac{t}{W} \frac{dn}{d\lambda}, \tag{99}$$

where t is the distance traveled by the ray through the prism and W is the dimension shown.

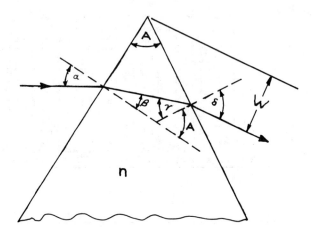

Figure 4.28   Passage of a ray of light through a prism.

A simple, yet important, application of a prism as a dispersing element is in the separation of a laser beam containing several well-spaced wavelengths into its constituent parts. Situations where this might be necessary include isolation of the 488.8-nm line in the output of an argon ion laser oscillating simultaneously at 476.5, 488.8, and 5145 nm and other lines, or isolation of a particular Stokes frequency in the output of a high-pressure-hydrogen stimulated Raman cell pumped by a frequency-doubled Nd:YAG laser.[20] If the spatial width of the laser beam involved is known, then an equation such as (96) will determine at what distance from the prism the different wavelengths are spatially separated from each other. The advantages of prisms in this application are that they can introduce very little scattered light, produce no troublesome ghost images, and can be cut so that they operate with $\alpha = \delta = \theta_B$, thereby eliminating reflection losses at their surfaces. Brewster-angle prisms are used in this way inside the cavity of a laser to select oscillation at a particular frequency. A modified prism called a reflecting *Littrow prism* is frequently used in this way, as shown in Figure 4.29. A Littrow prism is designed so that for a particular wavelength the refracted ray on entering the prism travels normally to the exit face. Thus, if the exit face is reflectively coated, for the specified wavelength the incident light is returned along its original path. Generally, a dispersing optical element used *in Littrow* has this retroreflective characteristic at a particular wavelength.

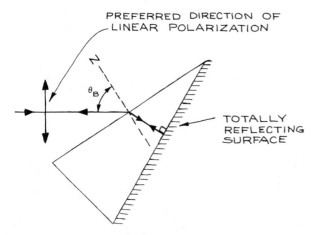

PREFERRED DIRECTION OF LINEAR POLARIZATION

θ_B

TOTALLY REFLECTING SURFACE

Figure 4.29 Reflecting prism used in Littrow at Brewster's angle. Only light of a specific wavelength will be refracted at the entrance face of the prism and then reflected back on itself at the coated reflecting face.

If a beam of parallel monochromatic light passes through a prism, unless the prism is used in the symmetrical minimum-deviation position, the width of the beam will be increased or decreased in one dimension. This effect is illustrated in Figure 4.30. By the use of two identical prisms in an inverted configuration, the compression or expansion can be accomplished without an angular change in beam direction. Prism beam expanders employing this principle are used in some commercial dye lasers for expanding a small-diameter, intracavity laser beam onto a diffraction grating to achieve greater laser line narrowing.

Triangular prisms with a wide range of shapes and vertex angles are available. The most common types are the equilateral prism, the right-angle prism, and isosceles and Littrow prisms designed for Brewster-angle

operation. Such prisms are readily available in glass and fused silica from Oriel and Melles Griot, among others.[15] Triangular and other prisms also find wide use as reflectors, utilizing total internal reflection inside the prism or a reflective coating on appropriate faces. Some specific reflective applications of prisms are worthy of special note.

(a) Right-Angle Prisms. Right-angle prisms can be used to deviate a beam of light through 90° as shown in Figure 4.31(a). Antireflection coatings on the faces shown are desirable in exacting applications. A right-angle prism can also be used to reflect a beam back parallel to its original path, as shown in Figure 4.31(b). The retroreflected beam is shifted laterally. The prism will operate in this way provided the incident beam is in the plane of the prism cross section. Thus, the retroreflection effect is independent of rotation about the axis A in Figure 4.31(b). If the angle of incidence differs significantly from zero, the roof faces of the prism need to be coated; otherwise total internal reflection at these surfaces may not result. If such a prism is rotated about the axis B, only in a single orientation does retroreflection result. This fact is put to good use in spinning-roof-prism laser Q-switches, where rotation about the axis B of a roof prism laser cavity reflector leads to an alignment orientation only once per revolution, independent of instability about A. Roof prisms are available from numerous suppliers, including Edmund Scientific, Rolyn, Oriel, Melles Griot, Karl Lambrecht (KLC), and Ealing.

Porro prisms are roof prisms with rounded corners on the hypotenuse face and beveled edges. They are widely used in pairs for image erection in telescopes and binoculars, as shown in Figure 4.32(a). Amici prisms are right-angle prisms where the hypotenuse face has been

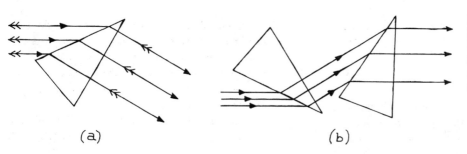

(a)   (b)

Figure 4.30 (a) One-dimensional beam expansion (or compression) with a prism; (b) one-dimensional beam expansion without beam deflection using two oppositely oriented, similar prisms.

Figure 4.31 (*a*) Right-angle prism used for 90° beam deflection; (*b*) right-angle prism used for retroreflection. Incident and reflected ray are parallel only if the incident beam is in the plane of prism cross section; however, in this orientation, retroreflection is independent of orientation about the axis *A* within a large angular range.

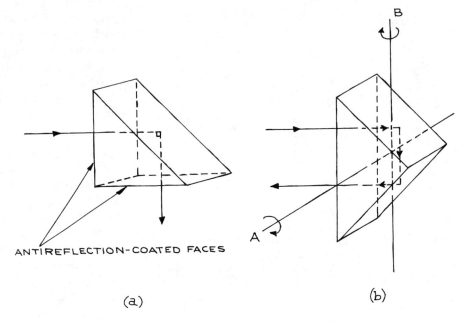

ANTIREFLECTION-COATED FACES

(a)

(b)

Figure 4.32 (*a*) Two Porro prisms used as an image-erecting element; (*b*) Amici prism. (Courtesy of Melles Griot, Inc.)

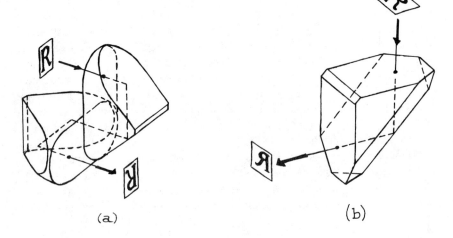

(a)

(b)

replaced by a 90° roof. An image viewed through an Amici prism is left-to-right reverted and top-to-bottom inverted as shown in Figure 4.32(*b*).

(b) Dove, Penta, Polygon, Rhomboid, and Wedge Prisms. These prisms are illustrated in Figure 4.33. Dove prisms are used to rotate the image in optical systems; rotation of the prism at a given angular rate causes the image to rotate at twice this rate. Dove prisms have a length-to-aperture ratio of about 5, so they must be used with reasonably well-collimated light. They are available from Esco, Melles Griot, Rolyn, KLC, and Edmund Scientific.

Penta prisms deviate a ray of light by 90° without inversion (turning upside down) or reversion (right-left

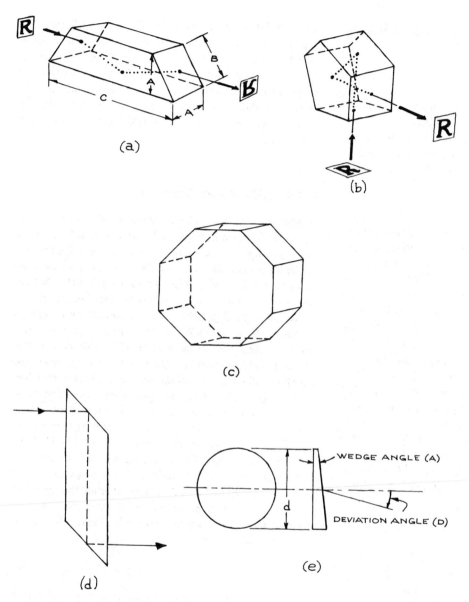

Figure 4.33 (a) Dove prism; (b) penta prism; (c) octagonal prism; (d) rhomboid prism; (e) wedge prism. [(a), (b), and (e) courtesy of Melles Griot, Inc.]

reversal) of the image. This 90° deviation applies to all rays incident on the useful aperture of the prism, irrespective of their angle of incidence. Penta prisms do not operate by total internal reflection, so their reflective surfaces are coated. These prisms are useful for 90° deviation when vibrations or other effects prevent their alignment being well controlled. They are available from Melles Griot, Rolyn, and KLC.

Polygon prisms, usually octagonal, are available from Rolyn and are used for high-speed light-ray deviation, for example, in high-speed rotary prism cameras.

Rhomboid prisms are used for lateral deviation of a light ray. When used in imaging applications, there is no change of orientation of the image. Rhomboid prisms are available from Rolyn and Edmund Scientific.

Wedge prisms are used in beam steering. The minimum angular deviation D of a ray passing through a thin wedge prism of apex angle A is

$$D = \arcsin(n \sin A) - A \simeq (n-1)A, \qquad (100)$$

where n is the refractive index of the prism material. A combination of two identical wedge prisms can be used to steer a light ray in any direction lying within a cone of semivertical angle $2D$ about the original ray direction. Wedge prisms are available from Melles Griot and Rolyn.

(c) Corner-Cube Prisms (Retroreflectors). Corner-cube prisms are exactly what their name implies, prisms with the shape of a corner of a cube, cut off orthogonal to one of its triad (body-diagonal) axes. The front face of the resultant prism is usually polished into a circle as shown in Figure 4.34. As a result of three total internal reflections, these prisms reflect an incident light ray back parallel to its original direction, no matter what the angle of incidence is. The reflected ray is shifted laterally by an amount that depends on the angle of incidence and the point of entry of the incident ray on the front surface of the prisms. These prisms are invaluable in experiments where a light beam must be reflected back to its point of origin from some (usually distant) point—for example, in long-path absorption measurements through the atmosphere with a laser beam. They are available commercially from Edmund Scientific,

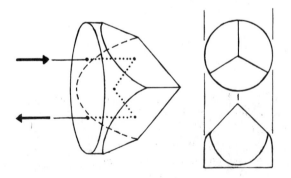

Figure 4.34 Corner-cube prism. (Courtesy of Melles Griot, Inc.)

KLC, Melles Griot, Oriel, Perkin-Elmer, and Rolyn. The angular deviation of the reflected ray from its original direction can be held to less than a half second of arc in the best prisms. For infrared applications beyond the transmission of fused silica, corner-cube prisms are not readily available, but retroreflectors can be made by using three square mirrors butted together to form a hollow cube corner.

4.3.5 Diffraction Gratings

A diffraction grating is a planar or curved optical surface covered with straight, parallel, equally spaced grooves. Such gratings can be made for use in transmission, but are more commonly used in reflection. Light incident on the grooved face of a reflection grating is diffracted by the grooves; the angular intensity distribution of the diffracted light depends on the wavelength and angle of incidence of the incident light and on the spacing of the grooves. This can be illustrated with reference to the plane grating shown in Figure 4.35. The grooves are usually produced in an aluminum or gold layer that has been evaporated onto a flat optical substrate. For high-intensity laser applications the substrate should have high thermal conductivity. Master gratings ruled on metal are available for this purpose from PTR, Diffraction Products, Jobin-Yvon, and Bausch and Lomb.[15] If the light diffracted at angle β from a given groove differs in phase from light diffracted from the adjacent groove by an integral multiple m of 2π, a maximum in diffracted intensity will be observed. This condition can be expressed as

$$m\lambda = d(\sin\alpha \pm \sin\beta). \qquad (101)$$

The plus sign applies if the incident and the diffracted ray lie on the same side of the normal to the grating surface; m is called the *order* of the diffraction. For $m = 0$, $\alpha = \beta$ and the grating acts like a mirror. In any other order, the diffraction maxima of different wavelengths lie at different angles. The actual distribution of diffracted intensity among the various orders depends on the profile of the grating grooves. If the grooves are planar, cut at an angle θ to the plane of the grating,

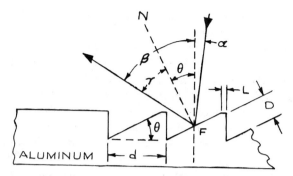

Figure 4.35 Blazed groove profile. The thickness of the aluminum is generally equal to the blaze wavelength, and the depth of the groove is about half the thickness. The dimensions shown are approximately those for a grating ruled with 1200 grooves per millimeter, blazed at 750 nm. D = depth of ruling; d = groove spacing; L = unruled land; α = angle of incidence; β = angle of diffraction; γ = angle of reflection; θ = blaze angle. The line N is normal to the groove face F. (From D. Richardson, "Diffraction Gratings," in *Applied Optics and Optical Engineering*, Vol. 5, R. Kingslake, Ed., Academic Press, New York, 1969; by permission of Academic Press.)

maximum diffracted intensity into a particular order results if the angle of diffraction and the angle of reflection are the same. In this case,

$$\beta - \theta = \gamma. \tag{102}$$

θ is called the *blaze* angle. The wavelength for which the angle of reflection from the groove face and the diffraction angle are the same is called the *blaze* wavelength, λ_B. It is common to use a grating in a Littrow configuration so that the diffracted ray lies in the same direction as the incident ray. In this case,

$$m\lambda = 2d\sin\beta. \tag{103}$$

The wavelength at which the light reflects normally from the grating groove satisfies

$$m\lambda_B = 2d\sin\theta. \tag{104}$$

Thus, for example, a grating could be specified to be blazed at 600 nm in the first order; it would, of course, be simultaneously blazed in the second order for 300

nm, the third order for 200 nm, and so on. There are many suppliers of both plane and concave gratings, particularly Bausch and Lomb, Jarrell-Ash, Jobin-Yvon, and PTR.

(a) Resolving Power. The resolving power of a grating is a measure of its ability to separate two closely spaced wavelengths. The resolving power depends both on the dispersion and the size of the grating. For a fixed angle of incidence, the *dispersion* is defined as

$$\left(\frac{d\beta}{d\lambda}\right)_\alpha = \frac{m}{d\cos\beta}. \tag{105}$$

Thus, the dispersion can be increased by increasing the number of lines per millimeter ($1/d$), by operating in a high order, and by using a large angle of incidence (grazing incidence). The incidence, however, must not be so flat as to make the projection of the groove width perpendicular to the incoming light smaller than the wavelength of the light.

The resolving power of a grating is $\Delta\lambda/\lambda$, where λ and $\lambda + \Delta\lambda$ are two closely spaced wavelengths that are just resolved by the grating. The limiting resolution depends on the projected width of the grating perpendicular to the diffracted beam. This width is $Nd\cos\beta$, where N is the total number of lines in the grating. Diffraction theory predicts that the angular resolution of an aperture of this size is $\Delta\phi \simeq \lambda/(Nd\cos\beta)$. The angle between the two closely spaced wavelengths, from the dispersion relation, is

$$\Delta\beta = \frac{m\Delta\lambda}{d\cos\beta}. \tag{106}$$

In the limit of resolution, $\Delta\phi = \Delta\beta$, which gives

$$\lambda/\Delta\lambda = mN. \tag{107}$$

Thus, from Equation (101),

$$\frac{\lambda}{\Delta\lambda} = \frac{Nd(\sin\alpha \pm \sin\beta)}{\lambda}, \tag{108}$$

and so it is clear that large gratings used at high angles

give the highest resolving power. The resolving powers available from 1-cm-wide gratings range up to about 5×10^5.

(b) Efficiency. The efficiency of a grating is a measure of its ability to diffract a given wavelength into a particular order of the diffraction pattern. Blazing of the grating is the main means for obtaining efficiency in a particular order; without it, the diffracted energy is distributed over many orders. Gratings are available with efficiencies of 95% or more at their blaze wavelength, so they are virtually as efficient as a mirror, yet retain wavelength selectivity.

(c) Defects in Diffraction Gratings. In principle, if an ideal grating is illuminated with a plane monochromatic beam of light, diffracted maxima occur only at angles that satisfy the diffraction equation (101). In practice, however, this is not so. Gratings manufactured by ruling a metal-coated substrate with a ruling engine exhibit undesirable additional maxima.

Periodic errors in the spacing of the ruled grooves produce *Rowland ghosts*. These are spurious intensity maxima, usually symmetrically placed with respect to an expected maximum and usually lying close to it. The strongest Rowland ghosts from a modern ruled grating will be less than 0.1% of the expected diffraction peaks. Lyman ghosts occur at large angular separations from their parent maximum, usually at positions corresponding to a simple fraction of the wavelength of the parent maximum—$\frac{4}{9}$ or $\frac{5}{9}$, for example. They are also associated with slow periodic errors in the ruling process. Lyman-ghost intensities from modern grating are exceedingly weak (0.001% of the parent or less). Ghosts can be a problem when used for the detection of weak emissions in the presence of a strong laser signal, the ghosts of which can (and have been) mistakenly identified as other real spectral lines. If there is any suspicion of this, the weak signal should be checked for its degree of correlation with the laser signal—a linear correlation would be strong evidence for a ghost. The development of unruled holographic gratings, which are essentially perfect, plus the improvement of ruled gratings made by interferometrically controlled ruling engines, has considerably reduced the problem of ghosts.

There are other diffraction-grating defects. *Satellites* are misplaced spectral lines, which can be numerous, occurring very close to the parent, usually so close that they can only be discerned under conditions of high resolution: they arise from small local variations in groove spacing. *Scattering* gives rise to an apparent weak continuum over all diffraction angles when a grating is illuminated with an intense monochromatic source such as a laser. Scattering can arise from microscopic dust particles or nongroovelike, random defects on the diffraction-grating surface. In practice it does not follow that a diffraction grating that shows apparent blemishes, such as broad, shaded bands, will perform defectively. One should never attempt to remove such apparent visual blemishes by cleaning or polishing the grating.

(d) Specialized Diffraction Gratings. Concave gratings are frequently used in vacuum-ultraviolet spectrometers and spectrographs, as they combine the functions of both dispersing element and focusing optics. Thus, fewer reflective surfaces are required—a highly desirable feature of an instrument used at short wavelengths, where reflectances of all materials decrease markedly.

Gratings for use at long wavelengths, above 50 μm, use relatively few grooves per millimeter. Such gratings are generally ruled directly on metal and are frequently called *echelettes* (little ladders) because of the shape of their grooves.

Echelle gratings are special gratings designed to give very high dispersion, up to 10^6 for ultraviolet wavelengths, when operated in a very high order. They have the surface profile illustrated in Figure 4.36, where the dimension D is much larger than in a conventional grating, and can range up to several micrometers. Echelles are widely used as the wavelength tuning element in pulsed dye lasers because of their high dispersion and efficient operation. Other uses of diffraction gratings, such as in the production of moiré fringes and in interferometers, will not be discussed here. These subjects have been dealt with by Girard and Jacquinot.[21]

(e) Practical Considerations in Using Diffraction Gratings. The main considerations in specifying a diffraction grating are the wavelength region where maximum efficiency is required (which will specify the blaze

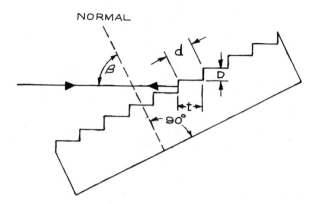

Figure 4.36 Echelle grating used in Littrow.

angle) and the number of grooves per millimeter and the size of the grating (which will determine the resolving power). Most gratings are rectangular or circular, the latter being primarily for laser-cavity wavelength selection. Diffraction gratings should be mounted with the care due all precision components. Their surfaces should *never* be touched and should be protected from dust. It is important to note that if a grating is illuminated with a given wavelength, say 300 nm, then diffraction maxima will occur in the same positions as would be found for 600 and 900 nm. This difficulty can be avoided by using appropriate filters to prevent unwanted wavelengths from passing through the system.

4.3.6 Polarizers

(a) **Polarized Light.** If the electric vector of an electromagnetic wave always points in the same direction as the wave propagates through a medium, then the wave is said to be *linearly polarized*. The *direction of linear polarization* is defined as the direction of the electric displacement vector **D**, where

$$\mathbf{D} = \epsilon_r \epsilon_0 \mathbf{E}. \tag{109}$$

Except in anisotropic media, where ϵ_r is a tensor, **D** and **E** are parallel and the direction of linear polarization can be taken as the direction of **E**. If a combination of two linearly polarized plane waves of the same frequency

but having different phases, magnitudes, and polarization directions is propagating in the z-direction, the resultant light is said to be *elliptically polarized*. Such a pair of waves, in general, have resultant electric fields in the x- and y-directions that can be written as

$$E_x = E_1 \cos \omega t, \tag{110}$$

$$E_y = E_2 \cos(\omega t + \phi), \tag{111}$$

where ϕ is the phase difference between these two resultant field components. These are the parametric equations of an ellipse. If $\phi = \pm \pi/2$ and $E_1 = E_2 = E_0$, then

$$E_x^2 + E_y^2 = E_0^2, \tag{112}$$

which is the equation of a circle. This represents *circularly polarized* light. Then the instantaneous angle that the total electric field vector makes with the x-axis, as illustrated in Figure 4.37, is

$$\alpha = \arctan \frac{E_y}{E_x} = \arctan(\mp \tan \omega t) = \mp \omega t. \tag{113}$$

For $\phi = \pi/2$ the resultant electric vector rotates counterclockwise viewed in the direction of propaga-

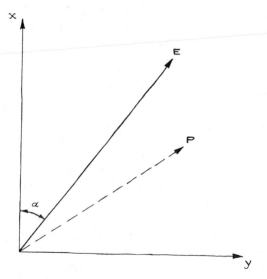

Figure 4.37 Instantaneous direction of the electric vector of an electromagnetic wave.

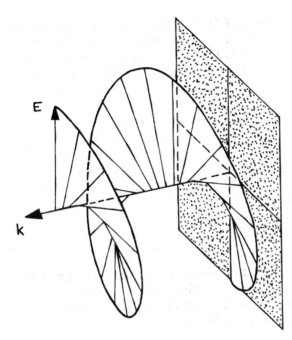

Figure 4.38   Left-hand circularly polarized light. (From E. Wahlstrom, *Optical Crystallography*, 3rd edition, Wiley, New York, 1960; by permission of John Wiley and Sons, Inc.)

tion—this is *right-hand circularly polarized light*. If $\phi = -\pi/2$ the rotation is clockwise—this is *left-hand circularly polarized light*. The motion of the total electric vector as it propagates is illustrated in Figure 4.38. If $\phi = 0$ or π we have *linearly polarized light*. Just as circularly polarized light can be viewed as a superposition of two linearly polarized waves with orthogonal polarizations, a linearly polarized wave can be regarded as a superposition of left- and right-hand circularly polarized waves. If an electromagnetic wave consists of a superposition of many independent linearly polarized waves of independent phase, amplitude, and polarization direction, it is said to be *unpolarized*.

In anisotropic media—media with lower than cubic symmetry—there is at least one direction, and at most two directions, along which light can propagate with no change in its state of polarization, independent of its state of polarization. This direction is called the *optic axis*. *Uniaxial* crystals have one such axis; *biaxial* crystals have two. When a wave does not propagate along the

optic axis in such crystals, it is split into two polarized components with orthogonal linear polarizations. These two components are called the *ordinary* and *extraordinary* waves. They travel with different phase velocities, characterized by two different refractive indices, n_o and $n_e(\theta)$, respectively. This phenomenon is referred to as *birefringence*.

D and **E** are not necessarily parallel in an anisotropic medium; however, **D** and **H** are orthogonal to the wave vector of any propagating wave. Consequently the Poynting vector of the wave does not, in general, lie in the same direction as the wave vector. The direction of the Poynting vector is the direction of energy flow—the ray direction. When a plane wave crosses the boundary between an isotropic and an anisotropic medium, the path of the ray will not, in general, satisfy Snell's law. The angles of refraction of the ordinary and extraordinary rays will be different. This phenomenon is called *double refraction*. In uniaxial crystals, however, the ordinary ray direction at a boundary does satisfy Snell's law. For further details of these and other optical characteristics of anisotropic media, the reader should consult Born and Wolf[7] and Wahlstrom.[8]

If linearly polarized light propagates into a birefringent material, unless its polarization direction matches the allowed direction of the ordinary or extraordinary wave, it will be split into two components polarized in these two allowed directions. These two components propagate at different velocities and experience different phase changes on passing through the crystal. For light of free-space wavelength λ_0 passing through a uniaxial crystal of length L, the birefringent phase shift is

$$\Delta\phi = \frac{2\pi L}{\lambda_0}\left[n_e(\theta) - n_o\right]. \qquad (114)$$

In uniaxial crystals the allowed polarization directions are perpendicular to the optic axis (for the ordinary wave) and in the plane containing the propagation direction and the optic axis (for the extraordinary wave), as shown in Figure 4.39. If the input wave is polarized at an angle β to the ordinary polarization direction and $\Delta\phi = (2n+1)\pi$, the output wave will remain linearly polarized, but its direction of polarization will have been

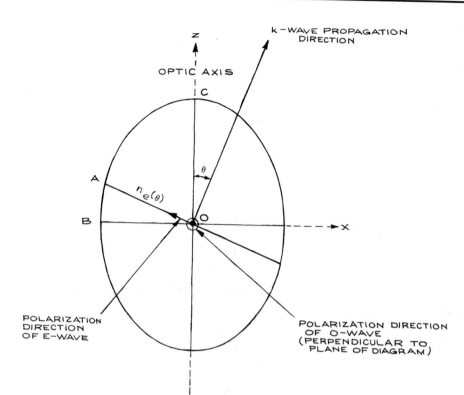

Figure 4.39 Cross section of an ellipsoidal figure called the indicatrix, which allows determination of the refractive indices and permitted polarization directions for a wave traveling in a uniaxial crystal. The equation of the ellipse shown is $(x^2/n_o^2)+(z^2/n_e^2)=1$, where $OB = n_o$ and $OC = n_e$. $OA = n_e(\theta)$ is the effective extraordinary refractive index for a wave traveling at angle θ to the optic axis. The ordinary refractive index for this wave is still n_o and is independent of θ.

rotated by an angle of 2β. It is usual to make $\beta = 45°$ in which case the crystal rotates the plane of polarization by 90°. Such a device is called a $(2n+1)$th-order *half-wave plate*. On the other hand, if $\beta = 45°$ and $\Delta\phi = (2n+1)\pi/2$, the ordinary and extraordinary waves recombine to form circularly polarized light. Such a device is called a $(2n+1)$th-order *quarter-wave plate*. If β is not 45°, a quarter-wave plate will convert linearly polarized light into elliptically polarized light.

(b) Polarization Changes on Reflection. As shown in Section 4.2.3, if a linearly polarized wave reflects from a dielectric surface (or mirror), its state of linear polarization can be changed. For example, if a vertically polarized wave traveling horizontally striking a mirror at 45° is polarized in the plane of incidence, then after reflection it will be traveling vertically and will be horizontally polarized. A reflection off a second mirror at 45°, so oriented that the plane of incidence is perpendicular to the polarization, will produce a horizontally traveling, horizontally polarized wave. This exemplifies how successive reflections can be used to rotate the plane of polarization of a linearly polarized wave and is illustrated in Figure 4.40.

If circular or elliptically polarized light reflects from a dielectric surface or mirror, its polarization state will in general be changed. As an example consider reflection at a metal mirror. From Equations (46) and (48), for the *in*-plane and perpendicular polarized components of the incident wave, respectively, we have

$$\rho_{\parallel} = -1,$$
$$\rho_{\perp} = -1. \tag{115}$$

These reflection coefficients ensure that the tangential component of electric field goes to zero at the conducting surface of the mirror. As illustrated in Fig-

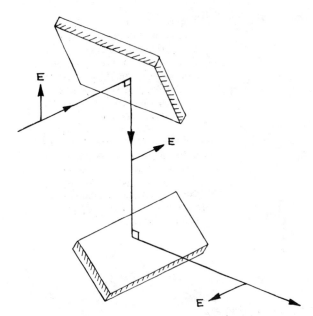

Figure 4.40 Rotation of the plane of polarization of a linearly polarized beam by successive reflection at two mirrors.

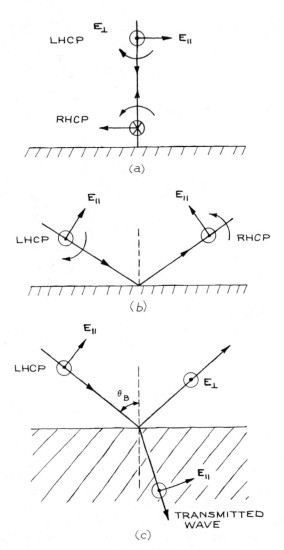

Figure 4.41 Changes of polarization state on reflection: (*a*) and (*b*) conversion of left-hand circularly polarized light (LHCP) to right-hand circularly polarized light (RHCP) on reflection at a perfect metal mirror; (*c*) conversion of LHCP to linear polarization on reflection at Brewster's angle.

ure 4.41(*a*) and (*b*), this leads to a conversion of left- to right-hand circularly polarized light on reflection. The change of polarization state on reflection from a dielectric interface depends on the angle of incidence and whether $n_2 > n_1$. A special case illustrated in Figure 4.41(*c*) shows left-hand circularly polarized light being converted to linearly polarized light on reflection from an interface placed at Brewster's angle. Such changes of polarization state must be taken into account when designing an optical system to work with polarized light. The change in polarization of a beam of light as it passes through a series of optical components can be calculated very conveniently by the use of Jones or Mueller calculus, which uses matrix methods to describe the state of polarization of the beam and its interaction with each component. Details of these techniques are given by Shurcliff.[9]

(c) Linear Polarizers. A linear polarizer changes unpolarized light to linearly polarized light or changes polarized light to a desired linear polarization.

The simplest linear polarizers are made from dichroic materials—materials that transmit one polarization, either ordinary or extraordinary, and strongly absorb the other. Modern dichroic linear polarizers, the commonest of which is Polaroid, are made of polymer films in which long-chain molecules with appropriate absorbing side groups are oriented by stretching. The stretched film is then sandwiched between glass or plastic sheets. These polarizers are inexpensive, but cannot be used to transmit high intensities because they absorb all polarizations to some degree. They cannot be fabricated to very high optical quality, and generally only transmit about 50% of light already linearly polarized for maximum transmission. However, they do work when the incident light strikes them at any angle up to grazing incidence. The extinction coefficient K that can be achieved with crossed polarizers is defined as

$$K = \log_{10} \frac{T_0}{T_{90}}, \qquad (116)$$

where T_0 and T_{90} are the transmittances of parallel and crossed polarizers, respectively. K is a useful quantity for specifying a linear polarizer—the larger its value, the better the polarizer. For dichroic polarizers, values of K up to about 10^4 can be obtained.

Higher quality, but more expensive, linear polarizers can be made by using the phenomenon of double refraction in transparent birefringent materials such as calcite, crystalline quartz, or magnesium fluoride. Figure 4.42 shows schematically the way in which various polarizing prisms of this type operate. Some of these polarizers generate two orthogonally polarized output beams separated by an angle, while others, for example the Glan-Taylor, reject one polarization state by total internal reflection at a boundary. Extinction coefficients as high as 10^6 and good optical quality can be obtained from such polarizers. However, the acceptance angle of these polarizers is generally quite small, although it can range up to about 38° with an Ahrens polarizer. Suppliers of polarizing prisms include Karl Lambrecht Corporation, Inrad, and Broomer Corporation.

Infrared polarizers (beyond about 7 μm) are frequently made in the form of very many fine, parallel, closely spaced metal wires. Waves with their electric vector perpendicular to the wires are transmitted; the parallel polarization is reflected. Such polarizers are available from Coherent Inc.

Linear polarizers for any wavelength where a transparent window material is available can be made using a stack of plates placed at Brewster's angle. Unpolarized light passing through one such plate becomes slightly polarized, since the component of incident light polarized in the plane of incidence is completely transmitted, whereas only about 90% of the energy associated with the orthogonal polarization is transmitted. A pile of 25 plates gives a high degree of polarization. Stacked-plate polarizers are rather cumbersome, but they are most useful for polarizing high-energy laser beams where prism polarizers would suffer optical damage. They are available commercially from Inrad.

(d) Retardation Plates. Quarter-wave plates are generally used for converting linear to circularly polarized light. They should be used with the optic-axis direction at 45° to the incident linear polarization direction. They are available commercially in two forms: multiple- and single-order. Multiple-order plates produce a phase change (retardation) between the ordinary and extraordinary waves of $(2n+1)\pi/2$. However, this phase difference is very temperature-sensitive—a thickness change of one wavelength will produce a substantial change in retardation. Single order plates are very thin, their thickness being

$$L = \frac{\lambda_0}{4(n_e - n_o)}. \qquad (117)$$

For example, with calcite, which has $n_o = 1.658$, $n_e = 1.486$, and with $\lambda_0 = 500$ nm this thickness is only 726.7 nm. Consequently, most single-order plates are made by stacking a $(2n+1)$-order quarter-wave plate on top of a $2n$-order plate whose optic axis is orthogonal to that of the first plate. The net retardation is just $\pi/2$ and is very much less temperature-sensitive.

Direct single-order quarter-wave plates for low-intensity applications can be made from mica, which cleaves naturally in thin slices. The production of a plate for a particular wavelength is a trial-and-error procedure.[9, 22] Quarter-wave plates are invaluable, in combination with a linear polarizer, for reducing reflected glare and for constructing optical isolators, as shown in Figure 4.43.

Half-wave plates are generally used for rotating the plane of polarization of linearly polarized radiation.

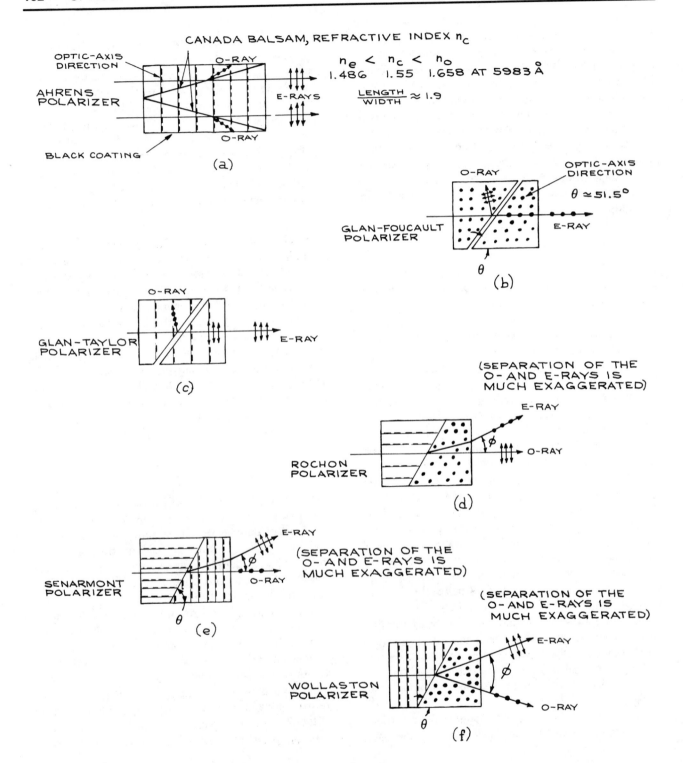

CANADA BALSAM, REFRACTIVE INDEX n_c

$n_e < n_c < n_o$
1.486   1.55   1.658 AT 5983 Å

$\dfrac{\text{LENGTH}}{\text{WIDTH}} \approx 1.9$

OPTIC-AXIS DIRECTION

O-RAY

AHRENS POLARIZER

E-RAYS

BLACK COATING

(a)

O-RAY

OPTIC-AXIS DIRECTION

$\theta \approx 51.5°$

GLAN-FOUCAULT POLARIZER

E-RAY

θ

(b)

O-RAY

GLAN-TAYLOR POLARIZER

E-RAY

(c)

(SEPARATION OF THE O- AND E-RAYS IS MUCH EXAGGERATED)

E-RAY

ϕ

O-RAY

ROCHON POLARIZER

(d)

E-RAY

SENARMONT POLARIZER

ϕ

O-RAY

(SEPARATION OF THE O- AND E-RAYS IS MUCH EXAGGERATED)

θ (e)

(SEPARATION OF THE O- AND E-RAYS IS MUCH EXAGGERATED)

E-RAY

ϕ

WOLLASTON POLARIZER

O-RAY

θ (f)

Figure 4.42 Construction of various polarizers using birefringent crystals: (a) Ahrens polarizer—three calcite prisms cemented together with Canada balsam, whose refractive index is intermediate between n_o and n_e for calcite. The O-ray suffers total internal reflection at the cement and is absorbed in the black coating on the sides. (b) Glan-Foucault polarizer—two calcite prisms separated with an air gap. The O-ray is totally internally reflected and either absorbed in the sides or transmitted if the sides are polished. A Glan-Thompson polarizer is similar. (c) Glan-Taylor polarizer—similar to a Glan-Foucault polarizer except for the optic-axis orientation, which ensures that the transmitted E-ray passes through the air gap nearly at Brewster's angle; consequently, transmission losses are much reduced from those of a Glan-Foucault. For high-power laser applications, the side faces can be polished at Brewster's angle. (d) Rochon polarizer—two calcite prisms with orthogonal optic axes cemented together with Canada balsam. ϕ depends on the interface angle. The intensities of the transmitted O- and E-rays are different. Other birefringent materials can be used; in the case of quartz, for example, the E-ray would exit below the O-ray in the diagram shown here. (e) Senarmont polarizer—two calcite prisms cemented together with Canada balsam. The angle ϕ depends on θ. This type of polarizer is less commonly used than the Rochon polarizer. (f) Wollaston polarizer—two calcite prisms cemented together with Canada balsam. The O- and E-ray transmitted intensities are different. The beam separation angle ϕ depends on the interface angle θ.

Multiple- and zero-order plates are available; as in the case of quarter-wave plates, the operating wavelength must be specified in buying one. This operating wavelength can be tuned to some extent by tilting the retardation plate.

Retardation plates are available from several suppliers, including Karl Lambrecht Corporation, Optics for Research, Inrad, II–VI Inc., Airtron, Melles Griot, and Oriel. Continuously adjustable retardation plates, called Babinet-Soleil compensators, are available from Karl Lambrecht Corporation, Melles Griot, II–VI Inc.,

Continental Optical, and Optics for Research, among others.[15]

(e) Retardation Rhombs. Because, in general, there is a different phase shift on total internal reflection for waves polarized in and perpendicular to the plane of incidence, a retarder that is virtually achromatic can be made by the use of two internal reflections in a rhomboidal prism of appropriate apex angle and refractive index. Two examples are shown in Figure 4.44. The Fresnel rhomb, for example, uses glass of index 1.51,

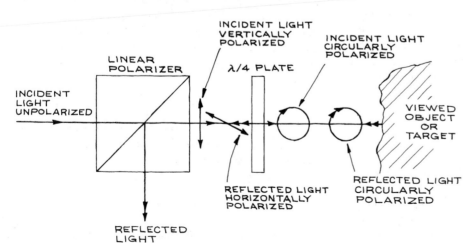

Figure 4.43 An optical isolator constructed from a linear polarizer and a quarter-wave plate.

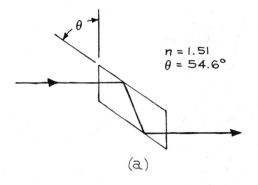

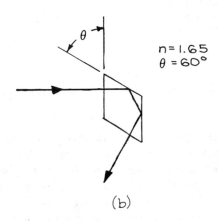

Figure 4.44   Rhomb retarders: (*a*) Fresnel; (*b*) Mooney.

and with an apex angle of approximately 54.6° produces a retardation of exactly 90°. However, the optimum angle must be determined by trial and error, and the desired retardation will only be obtained for a specific angle of incidence. Fresnel rhombs are available from KLC and II–VI.

4.3.7 Filters

Filters are used to select a particular wavelength region from light containing a broader range of wavelengths than is desired. Thus, for example, they allow red light to be obtained from a white light source, or they allow the isolation of a particular sharp line from a lamp or laser in the presence of other sharp lines or continuum emission. Although filters do not have high resolving power (see Section 4.3.5), they have much higher optical efficiency at their operating wavelength than higher-resolving-power wavelength-selective instruments, such as prism or grating monochromators. That is, they have a high ratio of transmitted flux to input flux in the wavelength region desired. Filters of several types are available for different applications.

(a) Color Filters.   A color filter selectively transmits a particular spectral region while absorbing or reflecting others. These simple filters are glasses or plastics containing absorbing materials such as metal ions or dyes, which have characteristic transmission spectra. Such color filters are available from Corning, Schott, Chance-Pilkington, Rolyn, Oriel, and Kodak (Wratten filters). Transmission curves for these filters are available from the manufacturers and in tabulations of physical and chemical data.[5,23]

(b) Band-Pass Filters.   A band-pass filter is usually a *Fabry-Perot etalon* of small thickness. Although etalons will be discussed in further detail in Section 4.7.4, a brief discussion is in order here.

In its simplest form the Fabry-Perot etalon consists of a plane parallel-sided slab of optical material of refractive index n and thickness d with air on both sides. In normal incidence the device has maximum transmission for wavelengths for which the thickness of the device is an integral number of half wavelengths. In this case, from Equation (62),

$$Z_3'' = Z_3' = Z_1',$$

and there is zero reflection. For incidence at angle θ, regardless of the polarization state of the light, there is maximum transmission for wavelengths (*in vacuo*) that satisfy

$$m\lambda_0 = 2d\cos\theta_2, \qquad (118)$$

where θ_2 is the angle of refraction in the etalon.

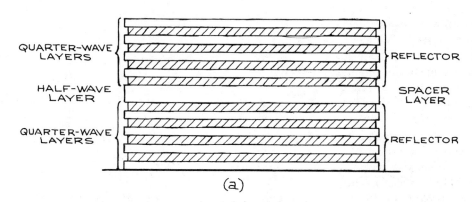

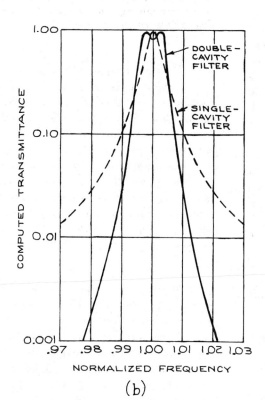

Figure 4.45 (a) Construction of an all-dielectric single-cavity Fabry-Perot interference filter; (b) theoretical transmittance characteristics of single- and double-cavity filters. (From *Handbook of Lasers*, R. J. Pressley, Ed., CRC Press, Cleveland, 1971; by permission of CRC Press, Inc.)

Thus a Fabry-Perot etalon is a "comb" filter. (Figure 4.108 shows an example of a transmission characteristic of such a device.) If the thickness of the etalon is small, the transmission peaks are broad. In a typical band-pass filter operated as an etalon, all the transmission peaks but the desired one can be suppressed by additional absorbing or multilayer reflective layers. The spectral width of the transmission peak of the filter can be reduced by stacking individual etalon filters in series. A single etalon filter might consist, for example, of two multilayer reflective stacks separated by a half-wavelength-thick layer as shown in Figure 4.45(a), while a two-etalon filter would have two half-wavelength-thick layers bounded by multilayer reflective stacks. Commercially available band-pass filters for use in the visible usually have transmission peaks in a range from 1 nm to 50 nm (full width at half maximum transmission). The narrowest band-pass filters are frequently called spike filters. Filters to transmit particular wavelengths such as 632.8 nm (He-Ne laser), 514.5 and 488 nm (argon ion laser), 404.7, 435.8, 546.1, 577, and 671.6 nm (mercury lamp), 589.3 nm (sodium lamp), and other wavelengths are available as standard items.[15] Filters at nonstandard wavelengths can be fabricated as custom items. Band-pass filters for use in the ultraviolet usually have lower transmittance, typically $\simeq 20\%$, than filters in the visible, whose transmittance usually ranges from 40% to 70%, being lowest for the narrowest passband. Infrared band-pass filters are also readily obtained, at least out to about 10 μm, and have good peak transmittance. The narrowest available band pass for a filter whose peak transmission is λ_{peak} can usually be estimated as $\lambda_{peak}/50$.

Band-pass filters are usually designed for use in normal incidence. Their peak of transmission can be moved to shorter wavelengths by tilting the filter, although the sharpness of the transmission peak will be degraded by this procedure. If a filter designed for peak wavelength λ_0 is tilted by an angle θ, its transmission maximum will shift to a wavelength

$$\lambda_s = \lambda_0 \sqrt{1 - \frac{\sin^2\theta}{n^2}}, \qquad (119)$$

where n is the refractive index of the half-wavelength-thick layer of the filter. For small tilt angles,

$$\lambda_0 - \lambda_s = \frac{\lambda_0 \sin^2\theta}{2n^2}. \qquad (120)$$

Thus, a band-pass filter whose peak-transmission wavelength is somewhat longer than desired can be optimized by tilting.

(c) Long- and Short-Wavelength-Pass Filters.

A *long-wavelength-pass* filter is one that transmits a broad

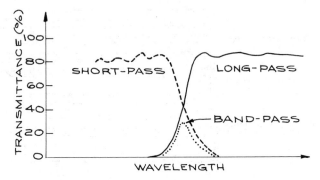

Figure 4.46   Schematic transmission of long- and short-pass optical filters showing the band-pass transmission characteristics of the two in combination.

spectral region beyond a particular cutoff wavelength, λ_{min}. A *short-wavelength-pass* filter transmits in a broad spectral region below its cutoff wavelength λ_{max}. A combination of a long-wavelength-pass and a short-wavelength-pass filter can be used to produce a band-pass filter, as shown in Fig. 4.46. Although long and short wavelength-pass filters usually involve multilayer dielectric stacks, the inherent absorption and transmission characteristics of materials can be utilized. Semiconductors exhibit fairly sharp long-wavelength-pass behavior beginning at the band gap energy. For example, germanium is a long-pass filter with $\lambda_{min} = 1.8\ \mu$m.

(d) Reststrahlen Filters.

When ionic crystals are irradiated in the infrared, they reflect strongly when their absorption coefficient is high and their refractive index changes sharply. This high reflection results from resonance between the applied infrared frequency and the natural vibrational frequency of ions in the crystal lattice. The characteristic reflected light from different crystals, termed *Reststrahlen*, allows specific broad spectral regions to be isolated by reflection from the appropriate crystal. Some examples of *Reststrahlen* filters are given in Figure 4.47.

(e) Christiansen Filters.

In the far infrared, the alkali halides exhibit *anomalous dispersion*: at certain wavelengths their refractive indices pass through unity.

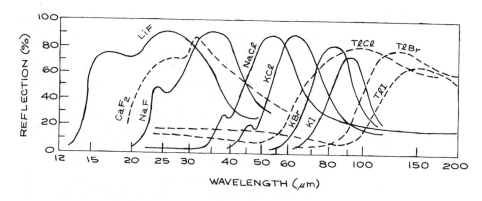

Figure 4.47 *Reststrahlen* filters. (Courtesy of Harshaw Chemical Co.)

(Normal dispersion involves an increase of refractive index with wavelength.) Thus a powdered alkali halide, which would generally attenuate a transmitted wave severely because of scattering, will transmit well at wavelengths where its refractive index is the same as that of air. Filters using such an alkali halide powder held between parallel plates are called *Christiansen* filters and provide sharp transmission peaks at certain wavelengths listed in Table 4.2.

(f) Neutral-Density Filters. Neutral-density filters are designed to attenuate light uniformly over some broad spectral region. The ideal neutral-density filter should have a transmittance that is independent of wavelength. These filters are usually made by depositing a thin metal layer on a transparent substrate. The layer is kept sufficiently thin that some light is transmitted through it. The *optical density D* of a neutral-density filter is defined as

$$D = \log_{10}(1/T), \qquad (121)$$

where T is the transmittance of the filter. T is controlled both by reflection from and by absorption in the metal film. If such filters are stacked in series, the optical density of the combination is the sum of the optical densities of the individual filters, provided the filters are positioned so that multiple reflection effects do not occur between them in the direction of interest.

Neutral-density filters are used for calibrating optical detectors and for attenuating strong light signals falling on detectors to ensure that they respond linearly. Because neutral-density filters usually reflect and transmit light, they can be used as beam splitters and beam combiners for any desired intensity ratio.

(g) Light Traps. If a beam of light must be totally absorbed (for example, in an application where any reflected light from a surface could interfere with some underlying weak emission), a light trap can be constructed. The two types illustrated in Figure 4.48 work well: the Wood's horn, in which a beam of light entering the trap is gradually attenuated by a series of reflections at an absorbing surface made in the form of a

Table 4.2 CHRISTIANSEN FILTERS

Crystal	Wavelength of Maximum Transmission (μm)
LiF	11.2
NaCl	32
NaBr	37
KCl	37
RbCl	45
TlCl	45
NaI	49
CsCl	50
KBr	52
CsBr	60
TlBr	64
KI	64
RbBr	65
RbI	73
TlI	90

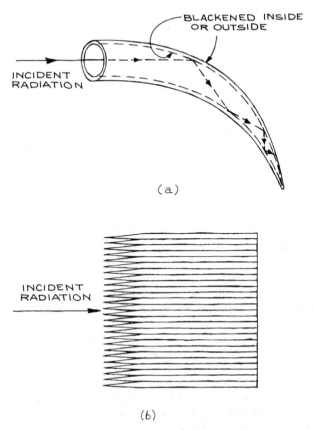

Figure 4.48   Light traps: (*a*) Wood's horn; (*b*) stacked razor blades.

curved cone (usually made of glass); and a stack of razor blades, which absorb incident light very efficiently. Light traps are available commercially from Klinger.

4.3.8 Devices for Positional and Orientational Adjustment of Optical Components

Provision frequently has to be made for translation, rotation, or tilt adjustment of optical components such as mirrors, lenses, prisms, crystals, and diffraction gratings. For a given component, the maximum number of possible adjustments is six: translation along three mutually orthogonal axis and rotation about these (or other)

axes. Mounts can be made that provide all six of these adjustments simultaneously. Usually, however, they are restricted and have no more degrees of freedom than are necessary. Thus, for example, a translation stage is usually designed for motion in only one dimension; independent motion along other axes can then be obtained by stacking one-dimensional translation stages. Orientation stages are usually designed for rotation about one axis (polar rotation) or two orthogonal axes (mirror mounts or tilt tables). Once again, such mounts can be stacked to provide additional degrees of freedom. It is often unnecessary for angular adjustment to take place about axes that correspond to translation directions.

The ideal optical mount provides independent adjustment in each of its degrees of freedom without any interaction with other potential degrees of freedom. It should be sturdy, insensitive to extraneous vibrational disturbances, (such as air or structure-borne acoustic vibrations), and free of backlash during adjustment. Generally it should be designed on kinematic principles (see Section 1.6.1). The effect of temperature variations on an optical mount can also be reduced, if desired, by constructing the mount from low-coefficient-of-expansion material, such as Invar, or by using thermal-expansion compensation techniques. Very high rigidity in an optical mount generally implies massive construction. When weight is a limitation, honeycombed material or material appropriately machined to reduce excess weight can be used. Most commercially available mounts are not excessively massive but represent a compromise between size, weight, and rigidity. Generally, mounts do not need to be made any more rigid than mechanical considerations involving Young's modulus and likely applied forces on the mount would dictate.

The design of optical mounts on kinematic principles involves constraining the mount just enough so that, once adjusted, it has no degrees of freedom. Some examples will illustrate how kinematic design is applied to optical mounts.

(a) **Two-Axis Rotators.** Within this category the commonest devices are adjustable mirror or grating mounts and tilt tables used for orienting components such as prisms. Devices that adjust orientation near the horizontal are usually called tilt adjusters. Mirror holders, on the other hand, involve adjustment about a vertical or near-vertical axis. A basic device of either kind involves a fixed stage equipped with three spherical or hemispherical ball bearings, which locate, respectively, in a V-groove, in a conical hole, and on a plane surface machined on a movable stage. A typical device is shown in Figure 4.49. The movable stage makes contact with the three balls on the fixed stage at just six points—sufficient to prevent motion in its six degrees of freedom—and is held in place with tension springs between the fixed and movable stages. Adjustment is provided by making two of the locating balls the ends of fine-thread screws or micrometer heads. Suitable micrometer heads for this purpose are manufactured by Starrett, Brown and Sharpe, Moore and Wright, Shardlow, and Mitutoyo and are available from most wholesale machine-tool suppliers. The ball of the micrometer should be very accurately centered on the spindle. The adjustment of the mount shown in Figure 4.49 is about two almost orthogonal axes. There is some translation of the center of the movable stage during

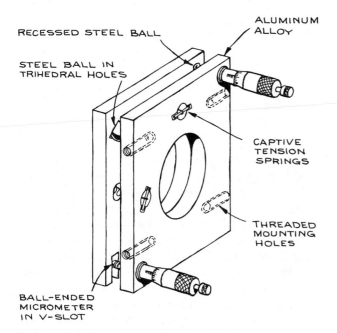

RECESSED STEEL BALL

STEEL BALL IN
TRIHEDRAL HOLES

ALUMINUM
ALLOY

CAPTIVE
TENSION
SPRINGS

THREADED
MOUNTING
HOLES

BALL-ENDED
MICROMETER
IN V-SLOT

Figure 4.49   Kinematic double-hinge mirror mount.

angular adjustment, which can be avoided by the use of a true gimbal amount.

A few of the constructional details of the mount in Figure 4.49 are worthy of note. Stainless steel, brass, and aluminum are all suitable materials of construction; an attempt to reduce the thermal expansion of the mount by machining it from a low-expansion material such as Invar may not meet with success. Extensive machining of low-expansion-coefficient alloys generally increases the coefficient unless they are carefully annealed afterward. Putting the component in boiling water and letting it gradually cool will lead to partial annealing. True kinematic design requires ball location in a trihedral hole, in a V-slot, and on a plane surface (or in three V-slots, but this no longer permits angular adjustment about orthogonal axes). If the mount is made of aluminum, the locating hole, slot, and plane on the movable stage should be made of stainless steel, or some other hard material, shrunk-fit into the main aluminum body. A true trihedral slot can be made with a punch as shown in Figure 4.50(a), or it can be machined with a 45° milling cutter as shown in Figure 4.50(b). A conical hole is not truly kinematic; a compromise design is shown in Figure 4.51. Although motion of the mount in Figure 4.49 results from direct micrometer adjustment, various commercial mounts incorporate a reduction mechanism to increase the sensitivity of the mount. Typical schemes that are used involve adjustment of a wedge-shaped surface by a moving ball, as shown in Figure 4.52, or the use of a differential screw drive. The two or more movable stages of an adjustable mount can be held together with captive tension springs as in Figure 4.49. However, designs involving only small angular adjustment can utilize the bending of a thin section of material.

The basic principle of a gimbal mount for two-axis adjustment is shown in Figure 4.53. True orthogonal motion about two axes can be obtained in such a design without translation of the center of the adjusted component. However, a kinematically designed gimbal mount is much more complicated in construction than the mount shown in Figure 4.53. Precision gimbal mounts should be purchased rather than built. Very many companies manufacture adequate devices with a variety of designs, and their prices are lower than the cost of

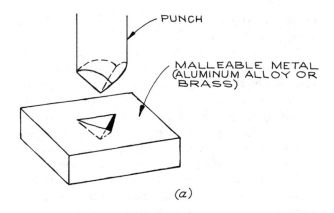

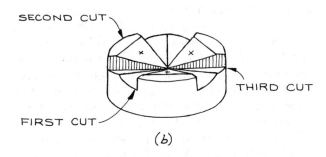

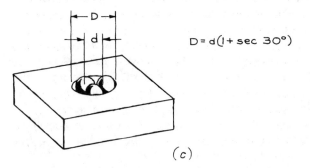

$$D = d(1 + \sec 30°)$$

Figure 4.50 Methods for making trihedral locating holes for kinematic design: (*a*) trihedral steel punch; (*b*) trihedral hollow machined with 45° milling tool; (*c*) steel balls fitting tightly in a hole.

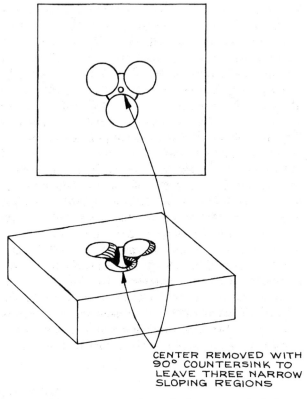

Figure 4.51 Compromise design for kinematic trihedral hole.

J. A. Noll, and Oriel. Most of these manufacturers also manufacture good non-gimbal mounts similar to the one shown in Figure 4.49.

(b) Translation Stages.   Translation stages are generally designed to provide linear motion in a single dimension and can be stacked to provide additional degrees of freedom. Mounts for centering optical components, particularly lenses, incorporate motion in two orthogonal (or near-orthogonal) axes in a single mount. One-dimensional kinematic translation stages involve precision screw motion of a movable stage along linear roller bearings on a fixed stage (see also Section 1.6.4). A simple design is shown in Figure 4.54, although variants are possible. Suppliers of precision, roller bearing translators include Aerotech, Ardel, Ealing, Klinger,

duplicating such devices in the laboratory. Suppliers of gimbal mounts include Aerotech, Ardel Kinematic, Burleigh, Ealing Optics, Jodon, Klinger Scientific, Lansing Research, Newport Research Corporation,

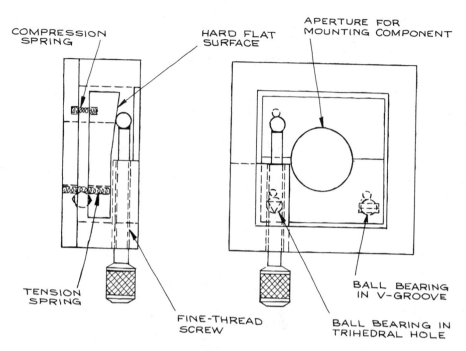

COMPRESSION SPRING

HARD FLAT SURFACE

APERTURE FOR MOUNTING COMPONENT

TENSION SPRING

FINE-THREAD SCREW

BALL BEARING IN V-GROOVE

BALL BEARING IN TRIHEDRAL HOLE

Figure 4.52 Cutaway views of a kinematic, single-axis rotator incorporating a reduction mechanism involving a ball bearing on an angled surface. Two rotators of this kind could be incorporated in a single mount to give two-axis gimbal adjustment.

Lansing Research, Newport Research and Oriel. Vertical translators using a fine-thread screw drive are available from Oriel. Less precise, but generally adequate, non-ball-bearing translation stages are available from Newport Research Corporation and Velmex. Mounts for centering optical components such as lenses are available from Aerotech, Ardel, Ealing, Klinger Scientific, and Newport Research, among others. A simple kinematic design for a centering mount is illustrated in Figure 4.55.

(c) Rotators and Other Mounts. Rotators are used, for example, for orienting polarizers, retardation plates, prisms, and crystals. They provide rotation about a single axis. Precision devices using ball bearings are available from Aerotech, Ardel, Ealing, Klinger, and Newport Research Corporation, among various other suppliers.

There are numerous more complicated optical mounts than those described in the previous sections, which provide in a single mount any combination of degrees of freedom desired. Motor-driven versions of all these

mounts are generally available. For details, the catalogs of the manufacturers previously mentioned should be consulted.

(d) Optical Benches and Components. For optical experiments that involve the use of one or more optical components in conjunction with a source and/or a detector in some sort of linear optical arrangement, construction using an optical bench and components is very convenient, particularly in breadboarding applications. The commonest such bench design is based on the equilateral triangular bar developed by Zeiss and typically machined from lengths of stabilized cast iron. Aluminum benches of this type are also available, but are less rigid. Components are mounted on the bench in rod-mounted holders that fit into carriers that locate on the triangular bench. A cross-sectional diagram of a typical bench and carrier is shown in Figure 4.56.

Coarse adjustment along the length of the bench is provided by moving the carrier manually. On many benches this is accomplished with a rack-and-pinion arrangement. A wide range of mounts and accessories

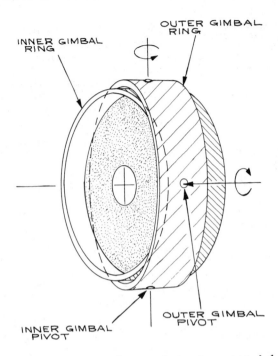

Figure 4.53 Schematic diagram of a true two-axis gimbal mount. The center of the mounted component does not translate during rotation.

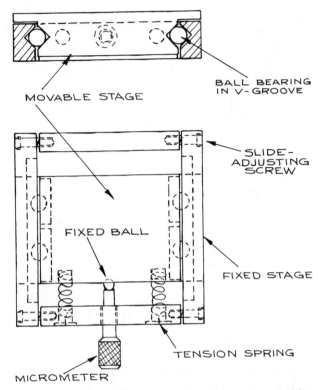

Figure 4.54 Section drawings of a simple one-dimensional translation stage.

are available, providing for vertical, lateral, and longitudinal adjustment of mounted-component positions. Holders for filters, for centering lenses, and for holding and rotating prisms and polarizers are readily obtained. Precision gimbal mounts, translators, and rotators can be readily incorporated into the structure. Examples of devices that lend themselves well to optical-bench construction are telescopes, collimators, laser-beam expanders and spatial filters, optical isolators, and electro-optic light modulators. Triangular-cross-section optical benches and accessories are available from Ealing, Klinger Scientific, Precision Tool, and Rolyn, among others. Other designs of optical benches are also available, but the widespread use of the Zeiss equilateral design offers the convenience that mounts made by different manufacturers can be used interchangeably. For very high-precision optical-bench arrangements, precision-machined lathe-bed optical benches are recommended.

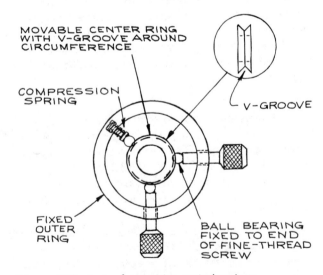

Figure 4.55 A simple precision centering stage.

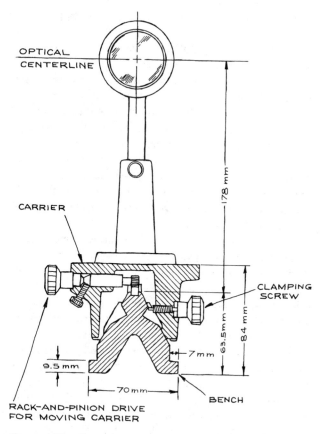

OPTICAL CENTERLINE

CARRIER

CLAMPING SCREW

178 mm

84 mm

63.5 mm

7 mm

9.5 mm

70 mm

BENCH

RACK-AND-PINION DRIVE FOR MOVING CARRIER

Figure 4.56 Typical dimensions of a Zeiss triangular optical bench with carrier and mounted optical component. (Courtesy of Ealing Corporation.)

Ealing guarantees the straightness of its best-quality bench to 12.5 μm or better over a 2.4-m length.

4.3.9 Optical Tables and Vibration Isolation

A commercially made optical table provides the perfect base for construction of an optical system. Such tables have tops made of granite, steel, invar, superinvar (very expensive), or ferromagnetic or nonmagnetic stainless steel. When vibrational isolation of the table top is required, the top is mounted on pneumatic legs. Such an arrangement effectively isolates the table top from

floor vibrations above a few hertz. To minimize acoustically driven vibrations of the table top itself, the better metal ones employ a laminated construction with a corrugated or honeycomb metal cellular core bounded with epoxy to the top and bottom table surfaces. Granite table tops are very much less intrinsically resonant than metal tops and represent the ultimate in dimensional stability and freedom from undesirable vibrational modes. Complete pneumatically isolated optical table systems, nonisolated tables, and table tops alone are available from Ealing, Lansing, Newport Research Corporation, and Modern Optics. In selecting a table, points of comparison to look for include the following:

1. The overall flatness of the table top (flatness within 50 μm is good, although greater flatness can be obtained).

2. The low-frequency vibration isolation characteristics of the pneumatic legs.

3. The rigidity of the table top (how much it deforms with an applied load at the center or end).

4. The isolation from both vertical and horizontal vibrations.

5. The frequencies and damping of the intrinsic vibrational modes.

Other criteria such as cost and table weight may be important. Typical sizes of readily available tops range up to 5×12 feet (larger tables are also available) and in thickness from 8 inches to 2 feet. The thicker the table, the greater its rigidity and stability. Most commercial tables are available with $\frac{1}{4}$-20-threaded holes spaced on 1-in. centers over the surface of the table. Ferromagnetic table tops permit the use of magnetic hold-down bases for mounting optical components. Such mounts are available from numerous optical suppliers. However, it is cheaper, and just as satisfactory, to buy magnetic bases designed for machine-tool use from a machine-tool supply company.

Steel and granite surface plates that can be used as small, nonisolated optical tables are available from machine-tool supply companies. If it is desired to construct a vibration-isolated table in the laboratory, a 2-in.-thick aluminum plate resting on underinflated automobile-tire inner tubes works rather well, although the plate will have resonant frequencies. Very flat aluminum plate, called *jig* plate, can be obtained for this application at a cost not much greater than ordinary aluminum plate. Just mounting a heavy plate on dense polyurethane foam also works surprisingly well. In fact, the vibrational isolation of a small optical arrangement built on an aluminum plate and mounted on a commercial optical table is increased by having a layer of thick foam between plate and table top.

Even the best-constructed precision optical arrangement on a pneumatic isolation table is subject to airborne acoustic disturbances. A very satisfactory solution to this problem is to surround the system with a wooden particle-board box lined with acoustic absorber. Excellent absorber material, a foam-lead-foam-lead-foam sandwich called Hushcloth DS, is available from American Acoustical Products.

4.3.10 Alignment of Optical Systems

Helium-neon lasers are now so widely available and inexpensive that they must be regarded as the universal tool for the alignment of optical systems. Complex systems of lenses, mirrors, beamsplitters, and so on can be readily aligned with the aid of such a laser: even visible-opaque infrared components can be aligned in this way by the use of reflections from their surfaces. Fabry-Perot interferometers, perhaps the optical instruments most difficult to align, are easily aligned with the aid of a laser. The laser beam is shone orthogonally through the first mirror of the interferometer (easily accomplished by observing part of the laser beam reflect back on itself), and the second interferometer mirror is adjusted until the resultant multiple-spot pattern between the mirrors coalesces into one spot. The virtual-pivot-beam laser developed by Lansing Research Corporation is a particularly powerful alignment tool. The beam from this laser can be directed to a point on the first optical component in a sequence and then pivoted about this same point (external to the laser) to define an appropriate alignment direction.

4.3.11 Mounting Optical Components

A very wide range of optical devices and systems can be constructed using commercially available mounts designed to hold prisms, lenses, windows, mirrors, light sources, and detectors. Some commercially made optical mounts merely hold the component without allowing precision adjustment of its position. Many of the rod-mounted lens, mirror, and prism holders designed for use with standard optical benches fall into this category, although they generally have some degree of coarse adjustability. However, a wide range of commercially available optical mounts allow precision adjustment of the mounted component.

The construction of a complete optical system from commercial mounts can be very expensive, while providing in many cases unnecessary, and consequently undesirable, degrees of freedom in the adjustment and orientation of the various components of the system. Although extra adjustability is often desirable when an optical system is in the "breadboard" stage, in the final version of the system it reduces overall stability. An adjustable mount can never provide the stability of a fixed mount of comparable size and mass. Stability in the mounting of optical components is particularly important in the construction of precision optical systems for use, for example, in interferometry and holography.

A custom-made optical mount must allow the insertion and stable retention of the optical component without damage to the component, which is usually made of some fragile (glassy or single-crystal) material. If possible, the precision surface or surfaces of the component should not contact any part of the mount. The component should be clamped securely against a rigid surface of the mount. It should not be clamped between two rigid metal parts: thermal expansion or contraction of the metal may crack the component or loosen it. The clamping force should be applied by an intermediary rubber pad or ring. Some examples of how circular mirrors or windows can be mounted in this way are shown in Figure 4.57. If a precision polished surface of the component must contact part of the mount (for example, when a precision lens, window, mirror, or prism is forming a vacuum-tight seal), the metal surface

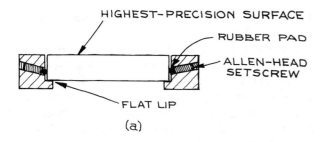

(a)

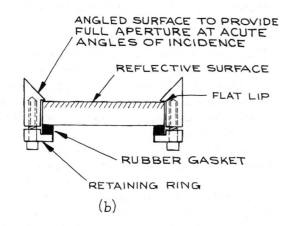

(b)

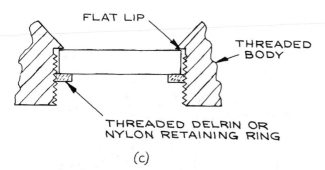

(c)

Figure 4.57 Ways for mounting delicate optical components: (*a*) no contact between precision surface of component and mount; (*b*) precision surface in minimal contact with mount and held in place with retaining ring; (*c*) precision surface in minimal contact with bottom of recess in mount and held in place with threaded plastic ring.

should be finely machined: any hole over which the component is mounted *must* be free of burrs. Ideally, the metal with which the component is in contact should have a lower hardness than the component. A thin piece of paper placed between the component and metal will protect the surface of the component without significantly detracting from the rigidity of the mounting.

Generally speaking, unless an optical component has one of a few standard shapes and sizes, it will not fit directly in a commercial mount and a custom adaptor will need to be made.

It is often necessary to mount a precision optical component (usually a flat window or mirror, but occasionally a lens or spherical mirror) so that it provides a vacuum- or pressure-tight seal. There are two important questions to be asked when this is done: can the component, of a certain clear aperture and thickness, withstand the pressure differential to which it will be subjected, and will it be distorted optically either by this pressure differential or the clamping forces holding it in place?

There are various ways of mounting a circular window so that it makes a vacuum- or pressure-tight seal. Demountable seals are generally made by the use of rubber O-rings (see Section 3.5.2) for vacuum, or flat rubber gaskets for pressure. More or less permanent seals are made with the use of epoxy resin, by running a molten bead of silver chloride around the circumference of the window where it sits on a metal flange, or in certain cases by directly soldering or fusing the window in place. Some crystalline materials, such as germanium, can be soldered to metal; many others can be soldered with indium if they, and the part to which they are to be attached, are appropriately treated with gold paint beforehand.[24] If a window is fused in place, its coefficient of expansion must be well matched to that of the part to which it is fixed. Suitably matched glass, fused silica, and sapphire windows on metal are available. Either the glass and metal are themselves matched, or an intermediate graded seal is used (see Section 2.2.4). To decide what thickness-to-unsupported-diameter ratio is appropriate for a given pressure differential, it must be determined whether the method of mounting corre-

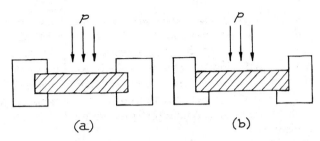

Figure 4.58 Window mounting arrangements: (*a*) clamped or fixed edge (maximum stress at edge); (*b*) unclamped or freely supported edge (maximum stress at center).

sponds to Figure 4.58(*a*) or (*b*). In either case, the maximum stress in the window must be kept below the modulus of rupture *Y* of the window by an appropriate safety factor S_f. A safety factor of 4 is generally adequate for determining a safe window thickness, but a larger factor may be necessary to avoid deformation of the window that would lead to distortion of transmitted wavefronts. The desired thickness-to-free-diameter ratio for a pressure differential *p* is found from the formula

$$\frac{t}{D} = \sqrt{\frac{KpS_f}{4Y}}, \qquad (122)$$

where *K* is a constant whose value can be taken as 0.75

for a clamped edge and 1.125 for a free edge (see also Section 3.6.2).

A window mounted with an O-ring in a groove, as shown in Figure 4.59(*a*), may not provide an adequate seal at moderately high pressures, say about 1000 psi, unless the clamping force of the window on the ring is great enough to prevent the pressure differential deforming the ring in its grooves and creating a leak. A better design for high-pressure use, up to about 7000 psi, is shown in Figure 4.59(*b*): the clamping force and internal pressure act in concert to compress the ring into position. Clamping an optical-quality window or an O-ring seal of the sort shown in Figure 4.59(*a*) may cause slight bending of the window.

If an optical window has been fixed "permanently" in place with epoxy resin, it may be possible to recover the window, if it is made of a robust material, by heating the seal to a few hundred degrees or by soaking it in methylene chloride or a similar commercial solvent such as "Vis-Strip".[25]

If optical windows are to be used in pressure cells where there is a pressure differential of more than 500 bars (≃ 7000 psi), O-rings are no longer adequate and special window-mounting techniques for very high pressures must be used.[26-28] Optical cells for use at pressures up to 100,000 psi are available from American Instrument Company (Aminco).

Figure 4.59 Mounting arrangements for optical windows in vacuum and pressure applications: (*a*) simple vacuum window; (*b*) high-pressure window.

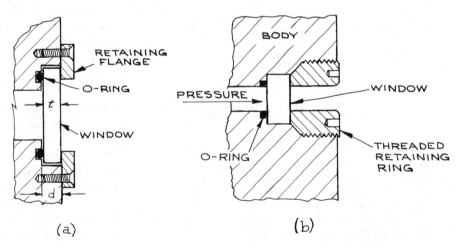

4.3.12 Cleaning Optical Components

It is always desirable for the components of an optical system to be clean. At the very least, particles of dust on mirrors, lenses, and windows increase scattering; organic films lead to unnecessary absorption, possible distortion of transmitted wavefronts, and—in experiments with ultraviolet light sources—possible unwanted fluorescence. In optical systems employing lasers, dust and films will frequently lead to visible, undesirable interference and diffraction effects. At worst, dirty optical components exposed to high-intensity laser beams can be permanently damaged, as absorbing dust and films can be burnt into the surface of the component.

The method used for cleaning optical components, such as mirrors, lenses, windows, prisms, and crystals, will depend on the nature of both the optical component and the type of contaminant to be removed. For components made of hard materials such as borosilicate glass, sapphire, quartz, or fused silica, gross amounts of contamination can be removed by cleaning in an ultrasonic tank. The tank should be filled with water and the components to be cleaned placed in a suitable pure solvent, such as trichloroethylene or acetone, contained in a clean glass or plastic beaker. The ultrasonic cleaning is accomplished by placing the beaker in the tank of water. Dust can be removed from the surface of hard materials, including most modern hard dielectric reflecting and antireflecting coatings, by brushing with a soft camel's-hair brush. Suitable brushes are available from most photographic stores. Lens tissue and solvent can be used to clean contaminants from both hard and medium-hard surfaces, provided the correct method is used. The category of medium-hard materials includes materials such as arsenic trisulphide, silicon, germanium, gallium arsenide, KRS-5, and tellurium. Information about the relative hardness of these and other materials is contained in Table 4.3. Dust should first be removed from the component by wiping, *very gently*, in one direction only, with a folded piece of dry lens tissue. The tissue should be folded so that the fingers do not come near any part of the tissue that will contact the surface. Better still, if a cylinder of dry nitrogen is available, dust can be blown from the surface of the

component. However, because the gas can itself contain particles, it should be passed through a suitable filter, such as fine sintered glass, before use. Clean gas can be used in this way to remove dust from the surface of even very delicate components, such as aluminum and gold-coated mirrors, and diffraction gratings. Commercial "dust-off" sprays are available that use a liquified propellant gas that vaporizes on leaving the spray-can nozzle. These sprays are not recommended for cleaning delicate or expensive components. They can deposit unvaporized propellant on the component, or condense water vapor into the surface of a hygroscopic material.

There are two good methods for cleaning hard and medium-hard components with lens tissue and solvent. The lens tissue should be folded as indicated previously, and a few drops of solvent placed on the tissue in the area to be used for the cleaning. The tissue should be wiped just once in a single direction across the component and then discarded. Fingers should be kept away from the solvent-impregnated part of the tissue, and it may be desirable to hold the component to be cleaned with a piece of dry tissue to avoid the transmission of finger grease to its surface. Hemostats, available from surgical supply stores, are very convenient for holding tissue during cleaning operations. To clean a single, fairly flat optical surface, such as a dielectric coated mirror or window, place a single lens tissue flat on the surface. Put one drop of clean solvent (spectroscopically pure grades are best) on the tissue over the component. Then draw the tissue across the surface of the component so that the dry portion of the tissue follows the solvent-soaked part and dries off the surface.

Several suitable solvents exist for these cleaning operations, such as acetone, trichloroethylene, and methyl, ethyl, and isopropyl alcohols. Diethyl ether is a very good solvent for cleaning laser mirrors, but its extreme flammability makes its use undesirable.

If there is any doubt as to the ability of an optical component to withstand a particular solvent, the manufacturer should be consulted. This applies particularly to specially coated components used in ultraviolet and some infrared systems. Some delicate or soft components cannot be cleaned at all without running the risk of damaging them permanently, although it is usually

Table 4.3   CHARACTERISTICS OF OPTICAL MATERIALS

Material	Useful Transmission Range ($\geq$ 10% transmission) in 2-mm Thickness	Index of Refraction [wavelength (μm) in parentheses]	Knoop Hardness	Melting Point (°C)
LiF	0.104–7	1.60(0.125), 1.34(4.3)	100	870
MgF$_2$	0.1216–9.7	$n_o = 1.3777$, $n_e = 1.38950(0.589)$[f]	415	1396
CaF$_2$	0.125–12	1.47635(0.2288), 1.30756(9.724)	158	1360
BaF$_2$	0.1345–15	1.51217(0.3652), 1.39636(10.346)	65	1280
Sapphire (Al$_2$O$_3$)	0.15–6.3	$n_o = 1.8336(0.26520)$, $n_o = 1.5864(5.577)$,[f] n_e slightly less than n_o	1525–2000[c]	2040 $\pm$ 10
Fused silica (SiO$_2$)	0.165–4[d]	1.54715(0.20254), 1.40601(3.5)	615	1600
Pyrex 7740	0.3–2.7	1.474(0.589), $\simeq$ 1.5(2.2)	$\simeq$ 600	820[g]
Vycor 7913	0.26–2.7	1.458(0.589)	—	1200
As$_2$S$_3$	0.6–0.13	2.84(1.0), 2.4(8)	109	300
RIR 2	$\simeq$ 0.4–4.7	1.75(2.2)	$\simeq$ 600	$\simeq$ 900
RIR 20	$\simeq$ 0.4–5.5	1.82(2.2)	542	760
NaF	0.13–12	1.393(0.185), 0.24(24)	60	980
RIR 12	$\simeq$ 0.4–5.7	1.62(2.2)	594	$\simeq$ 900
MgO	0.25–8.5	1.71(2.0)	692	2800
Acrylic	0.340–1.6	1.5066(0.4101), 1.4892(0.6563)	—	Distorts at 72
Silver chloride (AgCl)	0.4–32	2.134(0.43), 1.90149(20.5)	9.5	455
Silver bromide (AgBr)	0.45–42	2.313(0.496), 2.2318(0.671)	$\geq$ 9.5	432
Kel-F	0.34–3.8	—	—	—
Diamond (type IIA)	0.23–200 +	2.7151(0.2265), 2.4237(0.5461)	5700–10,400[c]	—
NaCl	0.21–25	1.89332(0.185), 1.3403(22.3)	18	803
KBr	0.205–25	1.55995(0.538), 1.46324(25.14)	7	730
KCl	0.18–30	1.78373(0.19), 1.3632(23)	—	776
CsCl	0.19– $\simeq$ 30	1.8226(0.226), 1.6440(0.538)	—	646
CsBr	0.21–50	1.75118(0.365), 2.55990(39.22)	19.5	636
KI	0.25–40	2.0548(0.248), 1.6381(1.083)	5	723
CsI	0.235–60	1.98704(0.297), 1.61925(53.12)	—	621
SrTiO$_3$	0.4–7.4	2.23(2.2), 2.19(4.3)	620	2080
SrF$_2$	0.13–14	1.438(0.538)	130	1450
Rutile (TiO$_2$)	0.4–7	$n_o = 2.5(1.0)$, $n_e = 2.7(1.0)$[f]	880	1825
Thallium bromide (TlBr)	0.45–45	2.652(0.436), 2.3(0.75)	12	460
Thallium bromoiodide (KRS-5)	0.56–60	2.62758(0.577), 2.21721(39.38)	40	414.5
Thallium chlorobromide (KRS-6)	0.4–32	2.3367(0.589), 2.0752(24)	39	423.5
ZnSe	0.5–22	2.40(10.6)	150	—
Irtran 2 (ZnS)	0.6–15.6	2.26(2.2), 2.25(4.3)	354	800
Si	1.1–15[e]	3.42(5.0)	1150	1420
Ge	1.85–30[e]	4.025(4.0), 4.002(12.0)	692	936
GaAs	1–15	3.5(1.0), 3.135(10.6)	750	1238
CdTe	0.9–16	2.83(1.0), 2.67(10.6)	45	1045
Te	3.8–8 +	$n_o = 6.37(4.3)$, $n_e = 4.93(4.3)$[f]	—	450
CaCO$_3$	0.25–3	$n_o = 1.90284(0.200)$, $n_e = 1.57796(0.198)$[f] $n_o = 1.62099(2.172)$, $n_e = 1.47392(3.324)$	135	894.4[h]

[a] Parallel to c-axis.

[b] Perpendicular to c-axis.

[c] Depends on crystal orientation.

[d] Depends on grade.

Table 4.3 (*continued*)

Material	Thermal-Expansion Coefficient $(10^{-6}/°C)$	Solubility in Water [g/(100 g), 20°C]	Soluble in	Comments
LiF	9	0.27	HF	Scratches easily
MgF_2	16	7.6×10^{-3}	HNO_3	—
CaF_2	25	1.1×10^{-3}	NH_4 salts	Not resistant to thermal or mechanical shock
BaF_2	26	0.12	NH_4Cl	Slightly hygroscopic, sensitive to thermal shock
Al_2O_3	6.66,[a] 5.0[b]	9.8×10^{-5}	NH_4 salts	Very resistant to chemical attack, excellent material
SiO_2	0.55	0.00	HF	Excellent material
Pyrex	3.25	0.00	HF, hot H_2PO_4	Excellent mechanical, optical properties
Vycor	0.8	0.00	HF, hot H_2PO_4	Excellent mechanical, optical properties
As_2S_3	26	5×10^{-5}	alcohol	Nonhygroscopic
RIR 2	8.3	0.00	1% HNO_3	Good mechanical, optical properties
RIR 20	9.6	—	—	Good mechanical, optical properties
NaF	36	4.2	HF	Lowest ref. index of all known crystals
RIR 12	8.3	—	—	Good mechanical and optical properties
MgO	43	6.2×10^{-4}	NH_4 salts	Nonhygroscopic; surface scum forms if stored in air
Acrylic	110–140	0.00	Methylene chloride	Easily scratched, available in large sheets
AgCl	30	1.5×10^{-4}	NH_4OH, KCN	Corrosive, nonhygroscopic, cold-flows
AgBr	—	12×10^{-6}	KCN	Cold-flows
Kel-F	—	—	—	Soft, easily scratched
Diamond	0.8	0.00	—	Hardest material known; thermal conductivity $6 \times$ Cu at room temp., chemically inert
NaCl	44	36	H_2O, glycerine	Corrosive, hygroscopic
KBr	—	65.2	H_2O, glycerine	Hygroscopic
KCl	—	34.35	H_2O, glycerine	Hygroscopic
CsCl	—	186	Alcohol	Very hygroscopic
CsBr	48	124	H_2O	Soft, easily scratched, hygroscopic
KI	—	144.5	Alcohol	Soft, easily scratched, hygroscopic, sensitive to thermal shock
CsI	50	160	Alcohol	Soft, easily scratched, very hygroscopic
$SrTiO_3$	9.4	—	—	Refractive index $\simeq \sqrt{5}$
SrF_2	—	1.17×10^{-2}	hot HCl	Slightly sensitive to thermal shock
TiO_2	9	0.00	H_2SO_4	Nonhygroscopic, nontoxic,
TlBr	—	0.0476	Alcohol	Flows under pressure, toxic
KRS-5	51	<0.0476	HNO_3, aqua regia	Cold-flows, nonhygroscopic, toxic
KRS-6	60	0.32	HNO_2, aqua regia	Cold-flows, nonhygroscopic, toxic
ZnSe	8.5	0.001	—	Very good infrared material, also transparent to some visible light
ZnS	—	6.5×10^{-5}	HNO_3, H_2SO_4	
Si	4.2	0.00	HF + HNO_3	Resistant to corrosion, must be highly polished to reduce scattering losses at surface
Ge	5.5	0.00	Hot H_2SO_4, aqua regia	
GaAs	5.7	0.00	—	Very good high-power infrared-laser window material
CdTe	4.5	—	—	—
Te	16.8	0.00	H_2SO_4, HNO_3	Poisonous, soft, easily scratched
$CaCO_3$	—	1.4×10^{-3}	Acids, NH_4Cl	—
	—	—	—	—

[e]Long-wavelength limit depends on purity of material.
[f]Birefringent.
[g]Softening temperature.
[h]Decomposition temperature.

safe to remove dust from them by blowing with clean, dust-free gas. It is worth noting that diffraction gratings frequently look as though they have surface blemishes. However, these blemishes frequently look much worse than they actually are and do not noticeably affect performance. *Never* try to clean diffraction gratings. It is also very difficult to clean aluminum- or gold-coated mirrors that are not overcoated with a hard protective layer such as silicon monoxide. Copper mirrors are soft and should be cleaned with extreme care.

Most optical components used in the visible and near ultraviolet regions of the spectrum are hard; soft materials most frequently find use in infrared systems. In this region several soft crystalline materials such as sodium chloride, potassium chloride, cesium bromide and cesium iodide are often used. These materials can be cleaned, essentially by repolishing their surfaces. For sodium and potassium chloride windows this is particularly straightforward. Fold a piece of soft, clean muslin several times to form a large pad. Place the pad on a flat surface—a piece of glass is best. Stretch the pad tight on the surface and fix it securely at the edges with adhesive tape. Dampen an area of the pad with a suitable solvent such as trichloroethylene or alcohol. Place the window to be polished on the solvent-soaked area, and work it back and forth in a figure-eight motion. Gradually work the disc onto the dry portion of the pad. This operation can be repeated as many times as necessary to restore the surface of the window. Do not attempt to clean windows of this kind, which are hygroscopic, by wiping them with solvent-soaked tissue. The evaporation of solvent from the surface will condense water vapor onto it and cause damage. The above recommended cleaning procedure is quite satisfactory for cleaning even laser windows. Small residual surface blemishes and slightly foggy areas do not detract significantly from the transmission of such windows in the middle infrared and beyond—which in any case is the only spectral region where they should be used. For removing slightly larger blemishes and scratches from such soft windows, electronic-grade aluminum oxide powder (available from Linde) can be mixed with the cleaning solvent on the pad in the first stages of polishing.

Perhaps the best way of all for cleaning hard and medium-hard optical components is to vapor-degrease them (see Figure 3.32). This is a very general procedure for cleaning precision components. Place some trichloroethylene or isopropyl alcohol in a large beaker (500 ml or larger) to a depth of about 1 cm. Suspend the components to be cleaned in the top of the beaker so that their critical surfaces face downward. An improvised wire rack is useful for accomplishing this. Cover the top of the beaker tightly with aluminum foil. Place the beaker on an electric hot plate, and heat the solvent until it boils. The solvent vapor thus produced is very pure. It condenses on, and drips off, the suspended item being cleaned, effectively removing grease and dirt. If large items are being cleaned, it may be necessary to blow air on the aluminum foil with a small fan. At the end of several minutes turn off the heat and remove the clean items while they are still warm. This cleaning procedure is recommended for components that must be extremely clean, such as laser Brewster windows.

4.4 OPTICAL MATERIALS

The choice of materials for the various components of an optical system such as windows, prisms, lenses, mirrors, and filters is governed by several factors:

1. Wavelength range to be covered.
2. Environment and handling that components must withstand.
3. Refractive-index considerations.
4. Intensity of radiation to be transmitted or reflected.
5. Cost.

4.4.1 Materials for Windows, Lenses, and Prisms

For the purposes of classification, the characteristics of various materials for use in transmissive applications will be considered in three spectral regions:

1. Ultraviolet, 100–400 nm.

2. Visible and near infrared, 400 nm—2 μm.

3. Middle and far infrared, 2–1000 μm.

There are, of course, many materials that can be used in part of two or even three of these regions. Table 4.3 summarizes the essential characteristics of all the common, and most of the rarely used, materials in these three regions. The useful transmission range given for each material is only a guide; the transmission at the ends of this range can be increased and the useful wavelength range of the component extended by using thinner pieces of material—if this is possible. The refractive index of all these materials varies with wavelength, as is illustrated for several materials in Figure 4.60. The Knoop hardness[29] is a static measure of material hardness based on the size of impression made in the material with a pyramidal diamond indenter under specific conditions of loading, time, and tempera-

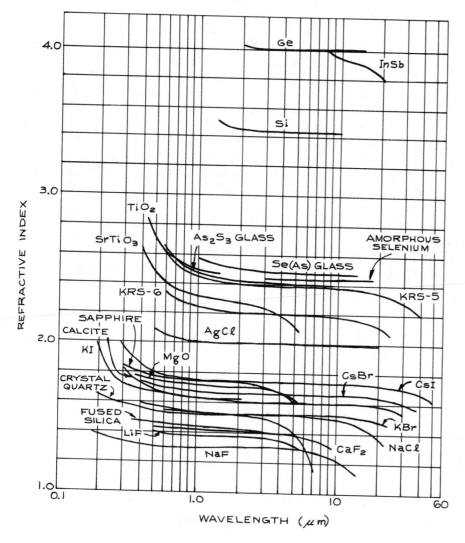

Figure 4.60 The refractive indices of various optical materials. [Adapted, with permission, from W. C. Wolfe and S. S. Ballard, "Optical Materials, Films and Filters for Infrared Instrumentation," Proceedings of the IRE, Vol. 47, pp. 1540–1546, 1959. © 1959 IRE (now IEEE).]

ture. Roughly speaking, a material with a Knoop hardness above about 60 is hard enough to withstand the cleaning procedures described in Section 4.3.12.

Transmission curves for many of the useful optical materials summarized in Table 4.3 are collected together in Figures 4.61–4.70.[15] Most of these materials are available from several suppliers, a partial list of whom is given at the end of this chapter. Some of the materials listed in Table 4.3 are worthy of brief extra comment.

(a) Ultraviolet Transmissive Materials. The following are suitable for use in the UV region.

(i) *Lithium fluoride.* Lithium fluoride (LiF, Figure 4.61) has useful transmittance further into the vacuum ultraviolet than any other common crystal. The transmission of vacuum-ultraviolet quality crystals is more than 50% at 121.6 nm for 2-mm thickness, and thinner crystals have useful transmission down to 104 nm. However, the short-wavelength transmission of the material deteriorates on exposure to atmospheric moisture or high-energy radiation. Moisture does not affect the infrared transmission, which extends to 7 μm. Several grades of LiF are available. Vacuum-ultraviolet-grade material is soft, but visible-grade material is hard. LiF can be used as a vacuum-tight window material by sealing with either silver chloride or a suitable epoxy.

(ii) *Magnesium fluoride.* Vacuum-ultraviolet-grade magnesium fluoride (MgF$_2$, Figure 4.61) transmits farther into the ultraviolet than any common material except LiF. The transmission at 121.6 nm is 35% or more for 2-mm thickness. MgF$_2$ is recommended for use as an ultraviolet transmissive component in space work, as it is only slightly affected by hard radiation. MgF$_2$ is birefringent and is used for making polarizing components in the ultraviolet. Irtran 1 is a polycrystalline form of magnesium fluoride manufactured by Kodak, which is not suitable for ultraviolet or visible applications. Its useful transmission extends from 500 nm to 9 μm.

(iii) *Calcium fluoride.* Calcium fluoride (CaF$_2$, Figure 4.61) is an excellent, hard material that can be used for optical components from 125 nm to beyond 10 μm. Vacuum-ultraviolet-grade material transmits more than 50% at 125.7 nm for a 2-mm path length. Calcium fluoride is not significantly affected by atmospheric moisture at ambient temperature. Calcium fluoride lenses are available from Oriel and Unique Optical. Irtran 3 is a polycrystalline form of CaF$_2$, which is not suitable for ultraviolet or visible applications but transmits in the infrared to 11.5 μm.

(iv) *Barium fluoride.* Although barium fluoride (BaF$_2$, Figure 4.62) is slightly more water-soluble than calcium or magnesium fluoride, it is more resistant than either of these to hard radiation. It is a good general-

Figure 4.61 Transmission curves of lithium fluoride, magnesium fluoride, and calcium fluoride windows of specified thickness.

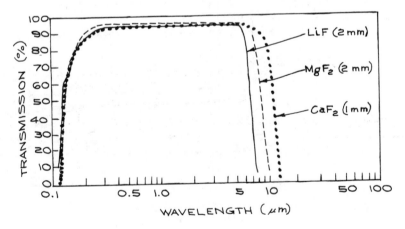

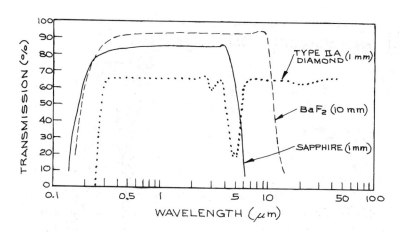

Figure 4.62 Transmission curves of sapphire, barium fluoride, and type-IIA diamond windows of specified thickness.

purpose optical material from ultraviolet to infrared. It has excellent transmission to beyond 10 μm, and windows that are not too thick (< 2 mm) can be used in CO_2-laser applications except at very high energy densities. Barium fluoride lenses are available from Unique Optical.

(v) *Synthetic sapphire.* Synthetic sapphire (corundum, Al_2O_3; Figure 4.62) is probably the finest optical material available in applications from 300 nm to 4 μm. It is very hard, strong, and resistant to moisture and chemical attack (even HF below 300°C). The useful transmissive range extends from 150 nm to 6 μm, the transmission of a 1-mm thickness being 21% and 34%, respectively, at these two wavelengths.

Sapphire is also resistant to hard radiation, has high thermal conductivity, and can be very accurately fabricated in a variety of forms. It has low dispersion, however, and is not very useful as a prism material. Chromium-doped Al_2O_3 (ruby) is widely used in laser applications.

(vi) *Fused silica.* Fused silica (SiO_2, Figure 4.63) is the amorphous form of crystalline quartz. It is almost as good an optical material as sapphire and is much cheaper. Its useful transmission range extends from about 170 nm to about 4.5 μm. Special ultraviolet-transmitting grades are manufactured (such as Spectrosil and Suprasil), as well as infrared grades (such as Infrasil) from which undesirable absorption features at 1.38,

2.22, and 2.72 μm due to residual OH radicals have been removed. Fused silica has a very low coefficient of thermal expansion, 5×10^{-7} K^{-1} in the temperature range from 20 to 900°C, and is very useful as a spacer material in applications such as Fabry-Perot etalons and laser cavities. Fused-silica optical components such as windows, lenses, prisms, and etalons are readily available. Crystalline quartz is birefringent and is widely used in the manufacture of polarizing optics, particularly quarter-wave and half-wave plates.

(vii) *Diamond.* Windows made of type-IIA diamond (Figure 4.62) have a transmittance that extends from 230 nm to beyond 200 μm, although there is some absorption between 2.5 and 6.0 μm. The high refractive index of diamond, $n \simeq 2.4$, over a very wide band, leads to a reflection loss of about 34% for two surfaces. Diamond windows are extremely expensive, ($10,000–20,000 for a 1-cm-diameter, 1-mm-thick window). However, some properties of diamond are unique: its thermal conductivity is 6 times that of pure copper at 20°C, and it is the hardest material known, very resistant to chemical attack, with a low thermal-expansion coefficient and high resistance to radiation damage.

(b) Visible- and Near-Infrared-Transmissive Materials. All the materials so far considered as ultraviolet-transmitting are also excellent for use in the visible and

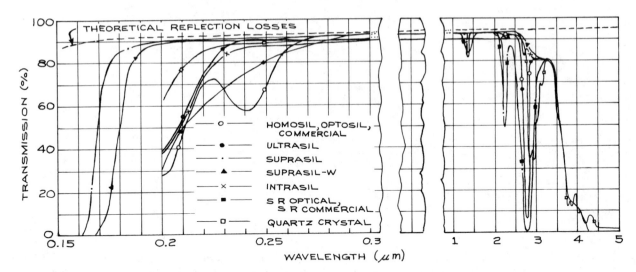

Figure 4.63 Transmission characteristics of various grades of fused silica; all measurements made on 10-mm-thick materials. (Courtesy of Heraeus-Amersil, Inc.)

near infrared. However, there are several materials that transmit well in the visible and near infrared but are not suitable for ultraviolet use.

(i) Glasses. There are an extremely large number of types of glass used in the manufacture of optical components. They are available from many manufacturers, such as Bausch and Lomb, Chance, Corning, and Schott. A few of these glasses are widely used in the manufacture of lenses, windows, prisms, and other components.

These include the borosilicate glasses Pyrex, BK7/A, and crown (BSC), as well as Vycor, which is 96% fused quartz. Transmission curves for Pyrex No. 7740 and Vycor No. 7913 are shown in Figure 4.64. These glasses are not very useful above 2.5 μm, their transmission at 2.7 μm is down to 20% for a 10-mm thickness. However, in the spectral region between 350 nm and 2.5 μm they are excellent transmissive materials. They are inexpensive, hard, and chemically resistant, and can be fabricated and polished to high precision. Special glasses for

Figure 4.64 Transmission curves for Pyrex 7740 and Vycor 7913.

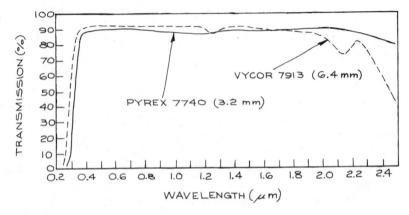

use further into the infrared are available from Bausch and Lomb and from Corning. However, these glasses do not appear to offer any particular advantages over more desirable materials such as infrared-grade fused silica or sapphire.

For special applications, very many colored filter glasses can be obtained. Glasses that transmit visible but not infrared, or vice versa, or that transmit some near ultraviolet but no visible, and many other combinations are available. The reader should consult the catalogs published by the various manufacturers and suppliers such as Corning, Schott, Chance-Pilkington, Kodak, Oriel, or Rolyn.

Because the composition of glass is continuously variable, special glasses whose expansion coefficients match selected metals can be made. These glasses are useful for sealing to these metals in order to make windows on vacuum chambers. Corning glass No. 7056 is often used in this way, as it can be sealed to the alloy Kovar. Vacuum-chamber windows of quartz or sapphire are also available,[30] but their construction is complex, and consequently they are costly. Various grades of flint glass are available that have high refractive indices and are therefore useful for prisms. For a tabulation of the types available the reader is referred to Kaye and Laby.[5] The refractive indices available range up to 1.93 for Chance-Pilkington Double Extra Dense Flint glass No. 927210, but this and other high-refractive-index glasses suffer long-term damage such as darkening when exposed to the atmosphere.

(ii) *Plastics.* In certain noncritical applications, such as observation windows, transparent plastics such as polymethylmethacrylate (Acrylic, Plexiglas, Lucite, Perspex) or polychlorotrifluoroethylene (Kel-F) are cheap and satisfactory. However, they cannot be easily obtained in high optical quality (with very good surface flatness, for example) and are easily scratched. The useful transmission range of Acrylic windows 1 cm thick runs from about 340 nm to 1.6 μm. A transmission-versus-wavelength curve for this material is shown in Figure 4.65. One important application of Acrylic is in the manufacture of Fresnel lenses [see Section 4.3.3(e)]. Kel-F is useful up to about 3.8 μm, although it shows some decrease in transmission near 3 μm. It is similar to

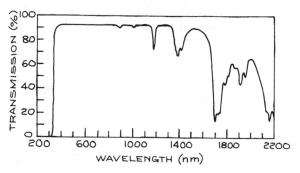

Figure 4.65 Transmission versus wavelength of 3.175-mm-thick Acrylic. (Courtesy of Melles Griot, Inc.)

polytetrafluoroethylene (Teflon) and is very resistant to a wide range of chemicals—even gaseous fluorine.

(iii) *Arsenic trisulfide.* Arsenic trisulfide (As_2S_3, Figure 4.66) is a red glass that transmits well from 800 nm to 10 μm. Its usefulness stems from its relatively low price, ease of fabrication, nontoxicity, and resistance to moisture. Although primarily an infrared-transmissive material, it is also transmissive in the red, which facilitates the alignment of optical systems containing it. However, As_2S_3 is relatively soft and will cold-flow over a long period of time. A wide range of As_2S_3 lenses are available from Unique Optical Company and from Servo Corporation.

(c) **Middle- and Far-Infrared-Transmissive Materials.** The following are useful well into the IR region.

(i) *Sodium chloride.* Sodium chloride (NaCl, Figure 4.67) is one of the most widely used materials for infrared-transmissive windows. Although it is soft and hygroscopic, it can be repolished by simple techniques and maintained exposed to the atmosphere for long periods without damage simply by maintaining its temperature higher than ambient. Two simple ways of doing this are to mount a small tungsten-filament bulb near the window, or to run a heated wire around the periphery of small windows. High-precision sodium chloride windows are in widespread use as Brewster windows in high-energy, pulsed CO_2 laser systems.

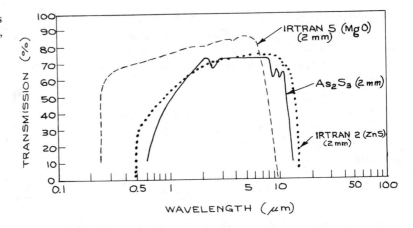

Figure 4.66 Transmission curves for Irtran 2 (ZnS), Irtran 5 (MgO), and As_2S_3 windows of specified thickness.

(ii) *Potassium chloride.* Potassium chloride (KCl, Figure 4.67) is very similar to sodium chloride. It is also an excellent material for use as a Brewster window in pulsed CO_2 lasers, as its transmission at 10.6 μm is slightly greater than that of sodium chloride.

(iii) *Cesium bromide.* Cesium bromide (CsBr, Figure 4.67) is soft and extremely soluble in water, but has useful transmission to beyond 40 μm. It will suffer surface damage if the relative humidity exceeds 35%.

(iv) *Cesium iodide.* Cesium iodide (CsI, Figure 4.68) is used for both infrared prisms and windows; it is similar to CsBr, but its useful transmission extends beyond 60 μm.

(v) *Thallium bromoiodide (KRS-5).* Thallium bromoiodide (TlBrI, Figure 4.69), widely known as KRS-5, is an important material because of its wide transmission range in the infrared—from 600 nm to beyond 40 μm for a 5-mm thickness—and its small solubility in water. KRS-5 will survive atmospheric exposure for long periods of time and can even be used in liquid cells in direct contact with aqueous solutions. Its refractive index is high: reflection losses at 30 μm are 30% for two surfaces. It is toxic and has a tendency to cold-flow. KRS-5 lenses are available from Unique Optical.

(vi) *Zinc selenide.* Chemical-vapor-deposited zinc selenide (ZnSe, Figure 4.69), a fairly recently developed material, is finding widespread use in infrared-laser ap-

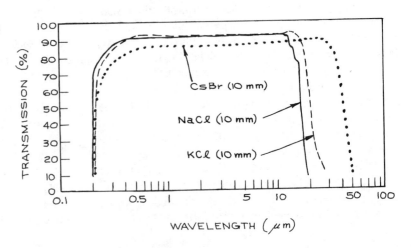

Figure 4.67 Transmission curves for cesium bromide, potassium chloride, and sodium chloride windows of specified thickness.

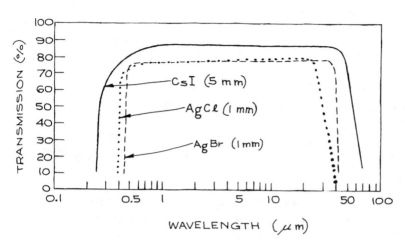

Figure 4.68 Transmission curves for cesium iodide, silver bromide, and silver chloride windows of specified thickness.

plications. It is transmissive from 500 nm to 22 μm. Zinc selenide is hard enough that it can be cleaned easily. It is resistant to atmospheric moisture and can be fabricated into precision windows and lenses. Its transmissive qualities in the visible (it is orange-yellow in color) makes the alignment of infrared systems using it much more convenient than in systems using germanium, silicon, or gallium arsenide. Irtran 4 is a polycrystalline form of zinc selenide manufactured by Kodak.

(vii) *Plastic films.* Several polymer materials are sold commercially in film form, including polyethylene (Polythene, Polyphane, Poly-Fresh, Dinethene, etc.), polyvinylidene chloride copolymer (Saran Wrap), poly-

thylene terephthalate (Mylar, Melinex, Scotchpar), and polycarbonate (Lexan). These plastic materials can be used to wrap hygroscopic crystalline optical components to protect them from atmospheric moisture. If the film is wrapped tightly, it will not substantially affect the optical quality of the wrapped components in the infrared. However, care should be taken to use a polymer film that transmits the wavelength for which the protected component is to be used. For example, several commercial plastic food wraps can be used to wrap NaCl or KCl windows for use in CO_2-laser applications. Not all will prove satisfactory, however, and a particular plastic film should be tested for transmission before use. A note of caution: plastic films contain plasticizers, the vapor from which may be a problem.

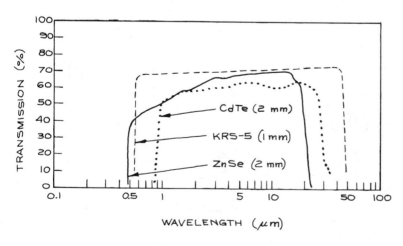

Figure 4.69 Transmission curves for zinc selenide, cadmium telluride, and KRS-5 windows of specified thickness.

(viii) *Plastics and other materials for far-infrared use.* Plastics are widely used for windows, beam-splitters, and light guides in the far infrared. A Mylar film is widely used as the beam-splitting element in far-infrared Michelson interferometers, and polyethylene, polystyrene, nylon, and Teflon are used as windows in far-infrared spectrometers and lasers. Black polyethylene excludes visible and near-infrared radiation but transmits far-infrared. Lenses and stacked-sheet polarizers for far-infrared use can be made from polyethylene. Several crystalline and semiconducting materials, although they absorb in the middle infrared, begin to transmit again in the far infrared. A 1-mm-thick quartz window has 70% transmission at 100 μm and more than 90% transmission from 300 to 1000 μm. The alkali and alkaline-earth halides such as NaF, LiF, KBr, KCl, NaCl, CaF_2, SrF_2, and BaF_2 show increasing transmission for wavelengths above a critical value that varies from one crystal to another but is in the 200–400-μm range. Further details of the specialized area of far-infrared materials and instrumentation are available in the literature.[31–33]

(d) Semiconductor Materials. Several of the most valuable infrared optical materials are semiconductors. In very pure form these materials should be transmissive to all infrared radiation with wavelength longer than that corresponding to the band gap. However, in practice the long-wavelength transmission will be governed by the presence of impurities—a fact that is put to good use in the construction of infrared detectors. Semiconductors make very convenient "cold mirrors," as they reflect visible radiation quite well but do not transmit it.

(i) *Silicon.* Silicon (Si, Figure 4.70) is a hard, chemically resistant material that transmits wavelengths beyond about 1.1 μm. It does, however, show some absorption near 9 μm, so it is not suitable for CO_2-laser windows. It has a high refractive index (the reflection loss for two surfaces is 46% at 10 μm), but can be antireflection-coated. Because of its high refractive index it should always be highly polished to reduce surface scattering. Silicon has a high thermal conductivity and a low thermal expansion coefficient, and it is resistant to mechanical and thermal shock. Silicon mirrors can therefore handle substantial infrared-laser power densities. A wide range of silicon lenses are available from Unique Optical.

(ii) *Germanium.* Germanium (Ge, Figure 4.70) transmits beyond about 1.85 μm. Very pure samples can be transparent into the microwave region. It is somewhat brittle, but is still one of the most widely used window and lens materials in infrared-laser applications. It is chemically inert and can be fabricated to high precision. Germanium has good thermal conductivity and a low coefficient of thermal expansion; it can be soldered to metal. In use its temperature should be kept

Figure 4.70 Transmission curves for the semiconductor materials germanium, gallium arsenide, and silicon.

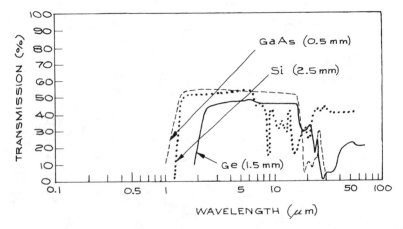

below 40°C, as it exhibits *thermal runaway*—its absorption increases with increasing temperature. A wide range of germanium lenses are available from Unique Optical.

(iii) *Gallium arsenide.* Gallium arsenide (GaAs, Figure 4.70) transmits from 1 to 15 μm. It is a better, although more expensive, material than germanium, particularly in CO- and CO_2-laser applications. It is hard and chemically inert, and maintains a very good surface finish. It can handle large infrared power densities because it does not exhibit thermal runaway until it reaches 250°C. It is also used for manufacturing infrared-laser electro-optic modulators.

(iv) *Cadmium telluride.* Cadmium telluride (CdTe, Figure 4.69) is another excellent material for use between 1 and 15 μm. It is quite hard, chemically inert, and takes a good surface finish. It is also used for manufacturing infrared-laser electro-optic modulators. Irtran 6 is a polycrystalline form of cadmium telluride available from Kodak; it transmits to 31 μm.

(v) *Tellurium.* Tellurium is a soft material that is transparent from 3.3 μm to beyond 11 μm. It is not affected by water. Its properties are highly anisotropic, and it finds use in CO_2-laser nonlinear-optical applications such as second-harmonic generation. Tellurium should not be handled, as it is toxic and can be absorbed through the skin.

4.4.2 Materials for Mirrors and Diffraction Gratings

(a) Metal Mirrors. The best mirrors for general broad-band use have pure metallic layers, vacuum-deposited or electrolytically deposited on glass, fused-silica, or metal substrates. The best metals for use in such reflective coatings are aluminum, silver, gold, and rhodium. Solid metal mirrors with highly polished surfaces made of metals such as stainless steel, copper, zirconium-copper, and molybdenum are also excellent, particularly in the infrared. The reflectance of a good

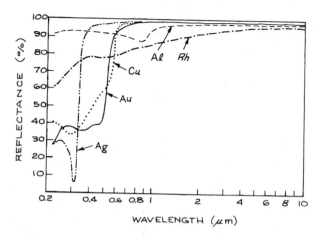

Figure 4.71 Reflectance of freshly deposited films of aluminum, copper, gold, rhodium, and silver as a function of wavelength from 0.2 to 10 μm. (After G. Hass, "Filmed Surfaces for Reflecting Optics," J. Opt. Soc. Am., 45, 945–952, 1955; by permission of the Optical Society of America.)

evaporated metal coating exceeds that of the bulk metal. Flat solid-metal mirrors can be made fairly cheaply by machining the mirror blank, surface-grinding it, and finally polishing the surface. If precision polishing facilities are not available, there are numerous optical component suppliers who will undertake the polishing of such surface-ground blanks to flatnesses as good as $\lambda/10$ in the visible. Finished flat and spherical solid-metal mirrors are available from Spawr. Solid-metal mirrors are most valuable for handling high-power infrared laser beams when metal-coated mirrors cannot withstand the power dissipation in the slightly absorbing mirror surface.

The spectral reflectances in normal incidence of several common metal coatings are compared in Figure 4.71. Several points are worthy of note. Gold and copper are not good for use as reflectors in the visible region. Unprotected gold, silver, copper, and aluminum are very soft. Aluminum rapidly acquires a protective layer of oxide after deposition; this significantly reduces the reflectance in the vacuum ultraviolet and contributes to increased scattering throughout the spectrum. Aluminum and gold are frequently supplied commercially with a thin protective dielectric layer, which increases their resistance to abrasion without significantly affecting their reflectance, and also protects the aluminum from oxidation. The protective layer is usually a $\lambda/2$ layer (at 550 nm) of SiO for aluminum mirrors used in the visible region of the spectrum. Aluminum coated with a thin layer of MgF$_2$ can be used as a reflector in the vacuum ultraviolet, although the reflectance is substantially reduced below about 100 nm, as shown in Figure 4.72. Many optical instruments operating in the vacuum ultraviolet, including some spectrographs, have at least one reflective component. Above 100 nm, coated aluminum is almost always used for this reflector. Below 100 nm, platinum and indium have superior reflectance to aluminum. However, at very short wavelengths, below about 40 nm, the reflectance of platinum in normal incidence is down to only 10%, as shown in Figure 4.72. Therefore, reflective components in this wavelength region are generally used at grazing incidence, as the reflectance under these conditions can remain high. For example, at 0.832 nm the reflectance of an aluminum film at a grazing angle of $\frac{1}{2}°$ is above 90%.

Freshly evaporated silver has the highest reflectance of any metal in the visible. However, it tarnishes so rapidly that it is rarely used except on internal reflection surfaces. In this case the external surface is protected with a layer of Inconel or copper and a coat of paint.

(b) Multilayer Dielectric Coatings. Extremely high-reflectance mirrors (up to 99.99%) can be made by using multilayer dielectric films involving alternate high- and low-refractive-index layers, ranging around $\lambda/4$ thick, deposited on glass, metal, or semiconductor substrates. The layers are made from materials that are transparent in the wavelength region where high reflectance is required. Both narrow-band and broad-band reflective-coated mirrors are available commercially. Figure 4.19 shows an example of each. Multilayer or single-layer dielectric-coated mirrors having almost any desired reflectance at wavelengths from 150 nm to 20 μm are also commercially available.[15]

(c) Substrates for Mirrors. The main factors influencing the choice of substrate for a mirror are (1) the dimensional stability required, (2) thermal dissipation, (3) mechanical considerations such as size and weight, and (4) cost. Glass is an excellent substrate for most totally reflective mirror applications throughout the spectrum, except where high thermal dissipation is necessary. Glass is inexpensive and strong, and takes a surface finish as good as $\lambda/200$ in the visible. Fused silica and certain ceramic materials such as Zerodur or Cervit (both available from Corning Glass Works, Optical Products Department) are superior, but more expensive, alternatives to glass when slightly greater dimensional stability and a lower coefficient of thermal expansion are required. Fused silica has the lowest coefficient of thermal expansion of any readily available material. Some ceramics and the alloy Superinvar have lower values, but generally over a restricted temperature range; also, machining or polishing of these materials can produce stresses in them and increase their expansion coefficients. Metal mirrors are frequently used when a very high light flux must be reflected and even the small absorption loss in the reflecting surface necessitates cooling of the substrate. However, metals are not so dimensionally stable as glasses or ceramics, and their use as mirrors should be avoided in precision applications, particularly in the visible or ultraviolet.

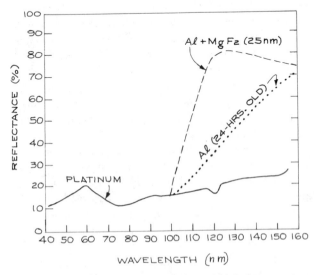

Figure 4.72   Normal-incidence reflectance of platinum, aluminum with a 25-nm-thick overcoat of MgF$_2$, and unprotected aluminum 24 hours after film deposition.

Partially reflecting, partially transmitting mirrors are used in many optical systems, such as interferometers and lasers. In this application the substrate for the mirror has to transmit in the spectral region being handled. The substrate material should possess the usually desirable properties of hardness and dimensional stability, plus the ability to be coated. For example, germanium, silicon, and gallium arsenide are frequently used as partially transmitting mirror substrates in the near infrared. Only the least expensive partially reflecting mirrors use thin metal coatings. Such coatings are absorbing, and multilayer, dielectric-coated reflective surfaces are much to be preferred.

There are numerous suppliers of totally and partially reflecting mirrors.[15] The experimentalist only has to be aware of his needs and specify the mirror appropriately.

Mirrors that reflect visible radiation and transmit infrared (*cold mirrors*) or transmit visible radiation and reflect infrared (*hot mirrors*) are available as standard items from several suppliers.[15]

(d) Diffraction Gratings. Diffraction gratings are generally ruled on glass, which may then be coated with aluminium or gold for visible or infrared reflective use, respectively. Most commercial gratings are replicas made from a ruled master, although high-quality holographic gratings are now available from Jobin Yvon. For high-power infrared-laser applications, gold-coated replica gratings are unsatisfactory and master gratings ruled on copper should be used. Such gratings are available from PTR, Bausch & Lomb, Jobin-Yvon, and Diffraction Products.

4.5 OPTICAL SOURCES

Optical sources fall into two categories: incoherent sources such as mercury or xenon arc, tungsten filament, and sodium lamps; and coherent sources (lasers). There are many different types of laser, of which some exhibit a high degree of coherence and others are not significantly more coherent than a line source such as a low-pressure mercury lamp.

Incoherent sources fall into two broad categories: line

sources and continuum sources. *Line sources* emit most of their radiation at discrete wavelengths, which correspond to strong spectral emission features of the excited atom or ion that is the emitting species in the source. *Continuum sources* emit over some broad spectral region, although their radiant intensity varies with wavelength.

4.5.1 Coherence

There are two basic coherence properties of an optical source. One is a measure of its relative spectral purity. The other depends on the degree to which the wavefronts coming from the source are uniphase or of fixed spatial phase variation. A wavefront is said to be *uniphase* if it has the same phase at all points on the wavefront. The Gaussian TEM_{00} mode emitted by many lasers has this property.

The two types of coherence are illustrated by Figure 4.73. If the phases of the electromagnetic field at two points, A and B, longitudinally separated in the direction of propagation, have a fixed relationship, then the wave is said to be *temporally coherent* for times corresponding to the distance from A to B. The existence of such a fixed phase relationship could be demonstrated by combining waves extracted at points separated like A and B and producing an interference pattern. The Michelson interferometer [see Section 4.7.6(a)] demonstrates the existence of such a fixed phase relationship. The maximum separation of A and B for which a fixed phase relationship exists is called the *coherence*

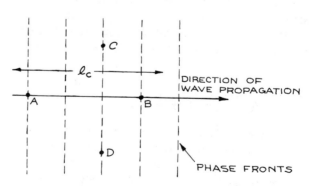

Figure 4.73 Temporal and spatial coherence.

length of the source l_c. The coherence length of a source is related to its *coherence time* τ_c by

$$l_c = c\tau_c. \qquad (123)$$

Conceptually l_c is the average length of the uninterrupted wave trains coming from the source between random phase interruptions. Fourier theory demonstrates that the longer in time a sinusoidal wave train is observed, the narrower is its frequency spread. Thus the spectral width of a source can be related to its coherence time by writing

$$\Delta \nu \simeq 1/\tau_c. \qquad (124)$$

The most coherent conventional lamp sources are stabilized low-pressure mercury lamps, which can have τ_c ranging up to about 10 ns. Lasers, however, can have τ_c ranging up to at least 1 ms.

Spatial coherence is a measure of phase relationships in the wavefront transverse to the direction of wave propagation. In Figure 4.73, if a fixed phase relationship exists between points C and D in the wavefront, the wave is said to be spatially coherent over this region. The *area of coherence* is a region of the wavefront within which all points have fixed phase relationships. Spatial coherence can be demonstrated by placing pinholes at different locations in the wavefront and observing interference fringes. Temporally incoherent sources can exhibit spatial coherence. Small sources (so-called point sources) fall into this category. The light from a star can be spatially coherent.

Extended, temporally incoherent sources have a low degree of spatial coherence because light coming from different parts of the source has different phases. Lasers emitting a TEM$_{00}$ (fundamental) mode have very good spatial coherence over their whole beam diameter.

4.5.2 Radiometry: Units and Definitions

Radiometry deals with the measurement of amounts of light. In radiometric terms the characteristics of a light source can be specified in several ways.

Radiant power, W, measured in watts, is the total amount of energy emitted by a light source per second. The spectral variation of radiant power can be specified in terms of the radiant power density per unit wavelength interval, W_λ. Clearly,

$$W = \int_0^\infty W_\lambda \, d\lambda. \qquad (125)$$

If a light source emits radiation only for some specific duration—which may be quite short in the case of a flashlamp—it is more useful to specify the source in terms of its *radiant energy* output, Q_e, measured in joules. If the source emits radiation for a time T, we can write

$$Q_e = \int_0^T W(t) \, dt. \qquad (126)$$

The amount of power emitted by a source in a particular direction per unit solid angle is called *radiant intensity* I_e, and is measured in watts per steradian. In general,

$$W = \oint I_e(\omega) \, d\omega, \qquad (127)$$

where the integral is taken over a closed surface surrounding the source. If I_e is the same in all directions, the source is said to be an *isotropic radiator*. At a distance r from such a source, if r is much greater than the dimensions of the source, the *radiant flux* crossing a small area ΔS is

$$\phi_e = \frac{I_e \Delta S}{r^2}. \qquad (128)$$

The *irradiance* at this point, measured in $\mathrm{W\,cm^{-2}}$, is

$$E_e = \frac{I_e}{r^2}, \qquad (129)$$

which is equal to the average value of the Poynting vector measured at the point. The radiant flux emitted per unit area of a surface (whether this be emitting or merely reflecting and scattering radiation) is called the *radiant emittance* M_e, and it is measured in $\mathrm{W\,m^{-2}}$.

For an extended source, the radiant flux emitted per

unit solid angle per unit area of the source is called its *radiance* L_e:

$$L_e = \frac{\delta I_e}{\delta S_n}, \qquad (130)$$

where the area δS_n is the projection of the surface element of the source in the direction being considered. When the light emitted from a source or scattered from a surface has a radiance that is independent of viewing angle, the source or scatterer is called a perfectly diffuse or *Lambertian* radiator. Clearly, for such a source, the radiant intensity at an angle θ to the normal to the surface is

$$I_e(\theta) = I_e(0)\cos\theta. \qquad (131)$$

The total flux emitted per unit area of such a surface is its radiant emittance, which in this case is

$$M_e = \pi I_e(0). \qquad (132)$$

Illuminated diffusing surfaces made of finely ground glass or finely powered magnesium oxide will behave as Lambertian radiators.

For plane waves, since all the energy in the wave is transported in the same direction, the concepts of radiant intensity and emittance are not useful. It is customary to specify the radiant flux crossing unit area normal to the direction of propagation, and call this the *intensity I* of the plane wave. Because lasers emit radiation into an extremely small solid angle, they have very high radiant intensity, and it is once again more usual to refer simply to the *intensity* of the laser beam at a point as the energy flux per second per unit area. The total power output of a laser is

$$W = \int_{\text{beam}} I\, dS. \qquad (133)$$

4.5.3 Photometry

The response of the human eye gives rise to a nonlinear and wavelength-dependent subjective impression of radiometric quantities. The response function of the hu-

man eye extends roughly from 400 to 700 nm, with a peak at 555 nm for the *photopic* (light-adapted) eye. Measures in photometry take into account this so-called *relative spectral luminous efficiency* $V(\lambda)$. Thus, for example, in physiological photometry the *luminous flux* F is related to radiant flux $\phi_e(\lambda)$ by

$$F = K\int_0^\infty V(\lambda)\phi_e(\lambda)\, d\lambda, \qquad (134)$$

where K is a constant. When F is measured in lumens and $\phi_e(\lambda)$ in watts, $K \equiv 679.6\ \text{lumen W}^{-1}$. Other photometric quantities that may be encountered in specifications of light sources are as follows:

1. The *luminous intensity* measured in candela, where

$$1\ \text{candela} = 1\ \text{lumen sr}^{-1}.$$

2. The *illumination*, measured in candela cm^{-2}, which is a measure of the amount of light reaching an area.

3. The *luminance*, also measured in candela cm^{-2}, which is a measure of the amount of light leaving a surface in a given direction.

For a Lambertian source the luminance is independent of the observation direction. Photometric description of the characteristics of light sources should be avoided in strict scientific work, but some catalogs of light sources use photometric units to describe lamp performance. For further details of photometry, and other concepts such as color in physiological optics, the reader should consult Levi[17] or Fry.[34]

4.5.4 Line Sources

Line sources are used as wavelength standards for calibrating spectrometers and interferometers; as sources in atomic absorption spectrometers; in interferometric arrangements for testing optical components, such as Twyman-Green interferometers [Section 4.7.6(a)]; in a few special cases for optically pumping solid-state and gas lasers; and for illumination.

The emission lines from a line source are not infinitely sharp. Their shape is governed by the actual conditions and physical processes occuring in the source. The variation of the radiant intensity with frequency across a line whose center frequency is ν_0 is described by its *lineshape function* $g(\nu, \nu_0)$, where

$$\int_{-\infty}^{\infty} g(\nu, \nu_0)\, d\nu = 1. \qquad (135)$$

The extension of the lower limit of this integral to negative frequencies is done for formal theoretical reasons connected with Fourier theory and need not cause any practical problems, since for a sharp line the major contribution to the integral in Equation (135) comes from frequencies close to the center frequency ν_0. There are three main types of lineshape function:[35] Lorentzian, Gaussian, and Voigt.

The *Lorentzian* lineshape function is

$$g_L(\nu, \nu_0) = \frac{2}{\pi \Delta \nu}\, \frac{1}{1 + \left[(\nu - \nu_0)/\Delta \nu\right]^2}, \qquad (136)$$

where $\Delta \nu$ is the frequency spacing between the half-intensity points of the line (the full width at half maximum height, or FWHM). Spectral lines at long wavelengths (in the middle and far infrared) and lines emitted by heavy atoms at high pressures and/or low temperatures frequently show this type of lineshape.

The *Gaussian* lineshape function is

$$g_D(\nu, \nu_0) = \frac{2}{\Delta \nu_D}\left(\frac{\ln 2}{\pi}\right)^{1/2} \exp\left\{-\left[2\frac{\nu - \nu_0}{\Delta \nu_D}\right]^2 \ln 2\right\}, \qquad (137)$$

where $\Delta \nu_D$ is the FWHM. Gaussian lineshapes are usually associated with visible and near-infrared lines emitted by light atoms in discharge-tube sources at moderate pressures. In this case, the broadening comes from the varying Doppler shifts of emitting species, whose velocity distribution in the gas is Maxwellian. However, emitting ions in real crystals sometimes have this type of lineshape because of the random variations of ion environment within a real crystal produced by dislocations, impurities, and other lattice defects. A Lorentzian and a Gaussian lineshape are compared in Figure 4.74.

Frequently, the broadening processes responsible for Lorentzian and Gaussian broadening are simultaneously operative, in which case the resultant lineshape is a convolution of the two and is called a *Voigt profile*.

The low-pressure mercury lamp is the most commonly used narrow-line source. These lamps actually operate with a mercury-argon or mercury-neon mixture. The principal lines from a mercury-argon lamp are listed in Table 4.4. Numerous other line sources are also available. In particular, hollow-cathode lamps emit the strongest spectral line of any desired element for use in atomic absorption spectrometry. Such lamps are available from Oriel and Baird Atomic, among others.

4.5.5 Continuum Sources

A continuum source in conjunction with a monochromator can be used to obtain radiation whose wavelength is tunable throughout the emission range of the source.

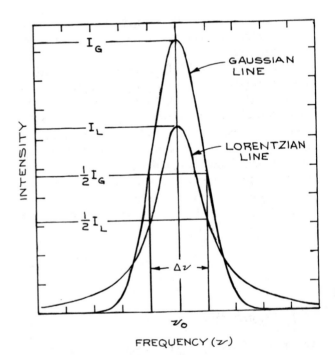

Figure 4.74 Gaussian and Lorentzian lineshapes of the same FWHM, $\Delta \nu$.

Table 4.4 CHARACTERISTIC LINES FROM A
MERCURY LAMP

Wavelength[a] (μm)
0.253652
0.313156
0.313184
0.365015
0.365483
0.366328
0.404656
0.435835
0.546074
0.576960
0.579066
0.69075
0.70820
0.77292
1.0140
1.1287
3.9425

Note: Extensive listings of calibration lines from other sources can be found in G.R. Harrison, *M.I.T. Wavelength Tables*, M.I.T. Press, Cambridge, Mass, 1969; and in A.R. Striganov and N.S. Sventitskii, *Tables of Spectral Lines of Neutral and Ionized Atoms*, IFT/Plenum Press, New York, 1968.
[a] *In vacuo.*

However, if the wavelength region transmitted by the monochromator is made very small, not very much energy will be available in the wavelength region selected. Even so, continuum sources find extensive use in this way in absorption and fluorescence spectrometers. Certain continuum sources, called blackbody sources, have very well characterized radiance as a function of wavelength and are used for calibrating both the absolute sensitivity of detectors and the absolute radiance of other sources.

(a) Blackbody Sources. All objects are continuously emitting and absorbing radiation. When an object is in thermal equilibrium with its surroundings, it emits and absorbs radiation in any spectral interval at equal rates. An object that absorbs all radiation incident on is called a *blackbody*—its *absorbivity* α is equal to unity. Such a body is also the most efficient of all emitters—its *emissivity* ϵ is also unity. In general, for any object emitting

and absorbing radiation at wavelength λ, $\epsilon_\lambda = \alpha_\lambda$. Highly reflecting, opaque objects, such as polished metal surfaces, do not absorb radiation efficiently; nor, when heated, do they emit radiation efficiently.

The simplest model of a blackbody source is a heated hollow object with a small hole in it. Any radiation entering the hole has minimal chance of reemerging. Consequently, the radiation leaving the hole will be characteristic of the interior temperature of the object. The energy density distribution of this *blackbody radiation* in frequency is

$$\rho(\nu) = \frac{8\pi h\nu^3}{c^3}\frac{1}{e^{h\nu/kT}-1}, \quad (138)$$

where $\rho(\nu)\,d\nu$ is the energy stored ($\mathrm{Jm^{-3}}$) in a small frequency band $d\nu$ at ν. The energy density distribution in wavelength is

$$\rho(\lambda) = \frac{8\pi hc}{\lambda^5}\frac{1}{e^{hc/\lambda kT}-1}. \quad (139)$$

This translates into a spectral emittance (the total power emitted per unit wavelength interval into a solid angle 2π by unit area of the blackbody) given by

$$M_{e\lambda} = \frac{C_1}{\lambda^5(e^{C_2/\lambda T}-1)}, \quad (140)$$

where $C_1 = 2\pi hc^2$, called the *first radiation constant*, has the value $3.7405\times10^{16}\,\mathrm{W m^2}$, and $C_2 = ch/k$, called the *second radiation constant*, has the value $1.43878\times10^{-2}\,\mathrm{mK}$.

A true blackbody is also a diffuse (Lambertian) radiator. Its radiance is independent of the viewing angle. For such a source,

$$M_{e\lambda} = \pi L_{e\lambda}. \quad (141)$$

The variation of $L_{e\lambda}$ with wavelength for various values of the temperature is shown in Figure 4.75. The wavelength of maximum emittance, λ_m, at temperature T obeys Wien's displacement law,

$$\lambda_m T = 2.8978\times10^{-3}\,\mathrm{mK}. \quad (142)$$

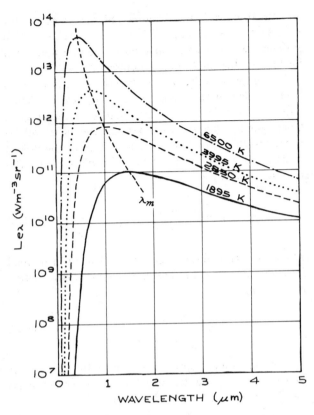

Figure 4.75 Spectral radiance $L_{e\lambda}$ of a blackbody source at various temperatures. (From A. A. Kruithof, "Modern Light Sources," in *Advanced Optical Techniques*, A.C.S. Van Heel, Ed., North-Holland, Amsterdam, 1967; by permission of North-Holland Publishing Company.)

The total radiant emittance of a blackbody at temperature T is

$$M_e = \int_0^\infty M_{e\lambda}\,d\lambda = \frac{2\pi^5 k^4}{15 c^2 h^3} T^4 = \sigma T^4. \quad (143)$$

This is a statement of the Stefan-Boltzmann law. The coefficient σ, called the *Stefan-Boltzmann constant*, has a value of 5.6697×10^{-8} $\mathrm{W\,m^{-2}\,K^{-4}}$. The known parameters $M_{e\lambda}$ and M_e of a blackbody allow it to be used as an absolute calibration source in radiometry. If a detector responds to photons, the spectral emittance

in terms of photons, N_λ, may be useful:

$$N_\lambda = \frac{M_{e\lambda}}{hc/\lambda}. \quad (144)$$

Curves of N_λ are given by Kruse, McGlauchlin, and McQuistan.[36]

A source whose spectral emittance is identical to that of a blackbody apart from a constant multiplicative factor is called a graybody. The constant of proportionality, ϵ, is called the *emissivity*. Several continuum sources, such as tungsten-filament lamps, carbon arcs, and flashlamps, are approximately graybody emitters within certain wavelength regions.

(b) Practical Blackbody Sources. The radiant emittance of a blackbody increases at all wavelengths as the temperature of the blackbody is raised, so a practical blackbody should, ideally, be a heated body with a small emitting aperture that is kept as hot as possible. Kruse, McGlauchlin, and McQuistan[36] describe such a source, illustrated in Figure 4.76, that can be operated at temperatures as high as 3000 K. A 25-μm-thick tungsten ribbon 2 cm wide is rolled on a 3-mm-diameter copper mandrel and seamed with a series of overlapping spot welds. A hole about 0.75 mm in diameter is made in the foil, and the copper dissolved out with nitric acid under a fume hood. The resulting cylinder is mounted on 1-mm-diameter Kovar or tungsten rod feedthroughs in a glass envelope and heated from a high-current, low-voltage power supply. The glass envelope should be fitted with a window that is transmissive to the wavelength region desired from the source.

Another design of blackbody source is shown in Figure 4.77. This design is based on a heated copper cylinder, containing a conical cavity of 15° semivertical angle, that is allowed to oxidize during operation (so that it becomes nonreflective and consequently of high emissivity). The cylinder is heated by an insulated heater wire wrapped around its circumference. If nichrome wire is used, the cylinder can be heated to about 1400 K. This assembly is mounted in a ceramic tube (alumina is quite satisfactory) or potted in high-temperature ceramic cement. For high-temperature operation the

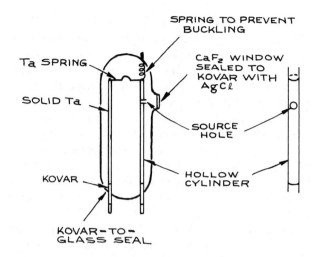

Figure 4.76 Construction details of a simple high-temperature blackbody source. (From P. W. Kruse, L. D. McGlauchlin, and R. B. McQuistan, *Elements of Infrared Technology: Generation, Transmission, and Detection*, Wiley, New York, 1962; by permission of John Wiley and Sons, Inc.)

whole assembly can be mounted inside a water-cooled block.

The most popular blackbody source available commercially is the Globar, a rod of bonded silicon carbide available from Carborundum, Perkin-Elmer, and Oriel. For further details of the advantages and disadvantages of this and various other blackbody sources, the reader is referred to Hudson.[37]

(c) Tungsten-Filament Lamps. Tungsten-filament lamps are approximately graybodies in the visible with an emittance between 0.45 and 0.5. Such lamps are frequently described in terms of their *color temperature* T_c, which is the temperature at which a blackbody would have a spectral emittance closest in shape to the lamp's. The color temperature will depend on the operating conditions of the lamp. Lamps with calibrations of color temperature versus operating current are available from EG&G and G.E.

Figure 4.77 Construction details of an NBS-type blackbody source.

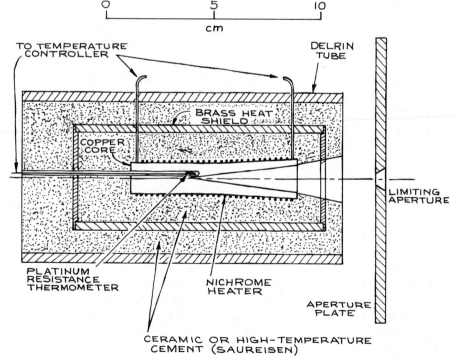

Tungsten-filament lamps can most conveniently be operated in the laboratory with a variable transformer. However, for best stability and freedom from ripple on their output, they should be operated from a stabilized d.c. supply. Typical supply requirements range up to a few hundred volts. Lamps with wattage ratings up to 1 kW are readily available.

Small tungsten-filament lamps that can be used as point sources are available from Oriel. Very-long-life, constant-efficiency tungsten-halogen lamps are available, in which the lamp envelope usually contains a small amount of iodine. In operation, the iodine vaporizes and recombines with tungsten that has evaporated from the filament and deposited on the inside of the lamp envelope. The tungsten iodate thus formed diffuses to the hot filament, where it decomposes, redepositing tungsten on the filament. The constant replacement of the filament in this way allows it to be operated at very high temperature and radiant emittance. Because the lamp envelope must withstand the chemical action of hot iodine vapor and high temperatures, it is made of quartz. Hence such lamps are frequently called quartz-iodine lamps. Such lamps can be quite compact: a 1-kW lamp will have a filament length of about 1 cm. The NBS standard of spectral irradiance consists of a quartz-iodine lamp with a coiled-coil tungsten filament operating at about 3000 K and calibrated from 250 nm to 2.6 μm against a blackbody source. Such calibrated lamps are available from EG&G. Because they are intense sources of radiant energy, these lamps can also be used for heating. In particular, they are often placed inside complex vacuum systems to bake out internal components that are well insulated thermally from the chamber walls.

(d) Continuous Arc Lamps.

High-current electrical discharges in gases, with currents that typically range from 1 to 100 A, can be intense sources of continuum or line emission, and sometimes both at the same time. For substantial continuum emission the most popular such lamps are the high-pressure xenon, high-pressure mercury, and high-pressure mercury-xenon lamps. The arc in these lamps typically ranges up to about 5 cm

long and 6.2 mm in diameter (for a 10-kW lamp—100-V, 100-A input). Because of their small size, arc lamps have much higher spectral radiance (brightness) than quartz-iodine lamps of comparable wattage. In the visible region at 500 nm a typical xenon arc lamp shows 1.9 times the output of a quartz-iodine lamp; at 350 nm, 14 times; and at 250 nm, 200 times. In addition, because of their small size, high-pressure arc lamps lend themselves well to the illumination of monochromator slits in spectroscopic applications. Lower-wattage arc lamps come close to being point sources and are ideal for use in projection systems and for obtaining well-collimated beams.

There are two different kinds of high-pressure arc lamps: those where the discharge is confined to a narrow quartz capillary (which must be water-cooled), and those where the discharge is not confined (which usually operate with forced-air cooling). The former are available from Illumination Industries/PEK and are used for pumping CW solid-state lasers. (Krypton arc lamps are better than xenon arc lamps for pumping Nd^{3+} lasers, as their emission is better matched to the absorption spectrum of Nd^{3+} ions.)

Because high-pressure arc lamps operate at very high pressures when hot (up to 200 bars), they must be housed in a rugged metal enclosure to contain a possible lamp explosion. The mounting must be such as to allow stress-free expansion during warmup. Generally speaking, commercial lamp assemblies should be used. The power-supplies requirements are somewhat unusual. An initial high-voltage pulse is necessary to strike the arc, and then a lower voltage, typically in the range 70–120 V, to establish the arc. When the arc is fully established, the operating voltage will drop to perhaps as low as 10 V. Arcs containing mercury need a further increase in operating voltage as they warm up and their internal mercury pressure increases. Complete lamp assemblies and power supplies are available from several companies, among them, Oriel and Conrad-Hanovia.

High-pressure arc lamps give substantial continuum emission with superimposed line structure, as can be seen in Figure 4.78. These lamps are not efficient sources of infrared radiation. However, they give substantial UV emission, and care should be taken in their

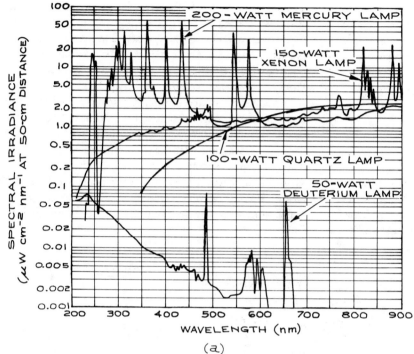

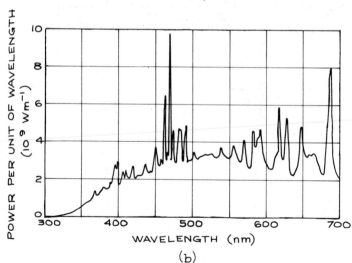

Figure 4.78 (*a*) Spectral irradiance of various moderate-power gas-discharge lamps (courtesy of Oriel Corporation, Stamford, Conn.); (*b*) spectral energy distribution for a high-power (10-kW) xenon arc lamp with a long discharge column (from A. A. Kruithof, "Modern Light Sources," in *Advanced Optical Techniques*, A. C. S. Van Heel, Ed., North-Holland, Amsterdam, 1967; by permission of North-Holland Publishing Company).

use to avoid eye or skin exposure. Their UV output will also generate ozone, and provision should be made for venting this safely from the lamp housing.

Deuterium lamps are efficient sources of ultraviolet emission with very little emission at longer wavelengths, as shown in Figure 4.78(*a*). They are available from Oriel.

Discharges in high-pressure noble gases can also be

used as intense continuum sources of vacuum-ultraviolet radiation. This radiation arises from noble-gas excimer emission, which in the case of helium, for example, arises from a series of processes that can be represented as

$$He + e \rightarrow He^{*},$$

$$He^{*} + He \rightarrow He_2^{*},$$

$$He_2^{*} \rightarrow He + He + h\nu.$$

The spectral regions covered by the excimer continua are He, 105–400 nm; Ar, 107–165 nm; Kr, 124–185 nm; and Xe, 147–225 nm. Because there are no transmissive materials available for wavelengths below about 110 nm, sources below this wavelength are used without windows. Radiation leaves the lamp through a small slit, or through a multicapillary array. The latter is a close-packed array of many capillary tubes, which presents considerable resistance to the passage of gas but is highly transparent to light. Multicapillary arrays can be obtained from Galileo Electro-Optics. To maintain the lamp pressure, gas is continuously admitted. Gas that passes from the lamp into the rest of the experiment (for example a vacuum-ultraviolet monochromator) is continuously pumped away by a high-speed vacuum pump, as shown in Figure 4.79. This technique is called *differential pumping*. For further details of vacuum-ultraviolet sources and technology, the reader is referred to Samson.[38]

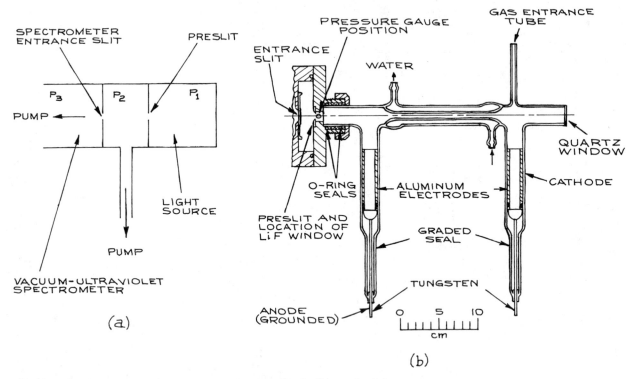

Figure 4.79 (*a*) Differential pumping arrangement (P_1, P_2, P_3 are the light source, differential pumping unit, and spectrometer operating pressures, respectively); (*b*) light source for the production of the noble-gas continua—in a differentially pumped mode, no LiF window would be used. (From R. E. Huffman, Y. Tanaka, and J. C. Larrabee, "Helium Continuum Light Source for Photoelectric Scanning in the 600–1100 Å Region," Appl. Opt., 2, 617–623, 1963; by permission of the Optical Society of America.)

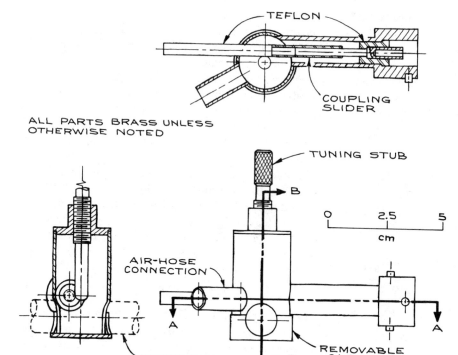

TEFLON

COUPLING
SLIDER

ALL PARTS BRASS UNLESS
OTHERWISE NOTED

TUNING STUB

B

AIR-HOSE
CONNECTION

A

A

B

DISCHARGE
TUBE

REMOVABLE
CAP

0   2.5   5
cm

Figure 4.80 Microwave cavity for exciting a gas discharge in a cylindrical quartz tube (design used by Opthos). (After F. C. Fehsenfeld, K. M. Evenson, and H. P. Broida, "Microwave Discharge Cavities Operating at 2450 MHz," Rev. Sci. Instr., 36, 294–298, 1965; by permission of the American Institute of Physics.)

(e) Microwave Lamps. Small microwave-driven discharge lamps are very useful where a low-power atomic-emission line source is desired, particularly if the atomic emission is desired from some reactive species such as atomic chlorine or iodine. A small cylindrical quartz cell containing the material to be excited, usually with the addition of a buffer of helium or argon, is excited by a microwave source inside a small tunable microwave cavity as shown in Figure 4.80. Suitable cavities for this purpose, designed for operation at 2450 MHz, are available from Opthos Instruments; power supplies for these lamps are available from Opthos and Baird Atomic.

(f) Flashlamps. The highest-brightness incoherent radiation is obtained from short-pulse flashlamps. By discharging a capacitor through a gas-discharge tube, much higher discharge currents and input powers are possible than in d.c. operation. As the current density or pressure of a flashlamp is increased, the emission from the lamp shifts from radiation that is characteristic of the fill gas, with lines superimposed on a continuum, to an increasingly close approximation to blackbody emission corresponding to the temperature of the discharge gas. Commercial flashlamps can be roughly divided into two categories. In long-pulse lamps, fairly large capacitors (100–10,000 μF), charged to moderately high voltages (typically up to about 5 kV), are discharged slowly (on time scales from 100 μs to 1 ms) through high-pressure discharge tubes. In short-pulse, high-peak-power lamps, smaller, low-inductance capacitors (typically 0.1–10 μF), charged to high voltages (10–30 kV), are discharged rapidly (on time scales down to 1 μs) through lower-pressure discharge tubes.

(i) Long-pulse lamps. Long-pulse lamps are generally filled with xenon,[39] but krypton lamps (for pumping

Nd³⁺ lasers) and alkali-metal lamps are also available.[40] Fill pressures are typically on the order of 0.1–1 bar. Although the discharge current in such lamps can run to tens of thousands of A cm⁻², at low repetition frequencies (≤ 0.1 Hz) ambient cooling is all that is necessary. Three of the most commonly used circuits for operating such lamps are shown in Figure 4.81. In all three of

these circuits, the capacitor, or pulse forming network shown in Figure 4.81(c), is charged through a resistor R. The capacitor is discharged through the lamp by triggering the flashlamp with one of the trigger circuits shown in Figure 4.82. The lamp itself behaves both resistively and inductively when it is fired. If the inductance and resistance of the lamp are not sufficiently large, the capacitor may discharge too rapidly, which can lead to damage of both lamp and capacitor. Generally speaking, an additional series inductor will be desirable to control the discharge. The problem of selecting the appropriate inductor for a particular capacitor size and lamp has been dealt with in detail by Markiewicz and Emmett.[41] Slow flashlamps have a non-linear $V - I$ characteristic, which can be approximated by

$$V = K_0\sqrt{|I|} , \qquad (145)$$

where the sign of V is taken to be the same as the sign of I. The value of K_0, measured in $\Omega\,A^{1/2}$, is a parameter specified by a manufacturer for a given lamp. Given this value and the capacitor size to be used, the calculations of Markiewicz and Emmett allow a suitable value of series inductor to be chosen. Other factors must be taken into account in designing the flashlamp circuit: the maximum power loading, usually specified as the explosion energy of the lamp (which will depend on the discharge-pulse duration), and the maximum repetition frequency of the lamp. Usually, the larger the capacitor stored energy, the lower the permitted repetition frequency will be. As lamps and discharge energies get smaller, repetition frequencies can be extended, to several kilohertz for the smallest low-energy lamps such as those used in stroboscopes. The explosion energy of the lamp is the minimum input required to cause lamp failure in one shot. To obtain long flashlamp life, a lamp should be operated only at a fraction of its explosion energy. For example, at 50% of the explosion energy the expected lifetime is from 100 to 1000 flashes, while at 20% it is from 10^5 to 10^6 flashes. The lamp should be mounted freely and not clamped rigidly at its ends when it is operated.

The spectral output of a flashlamp varies slightly during the flash. For moderate- and high-power lamps (≥ 100 J in 1 ms), the spectral output approaches that

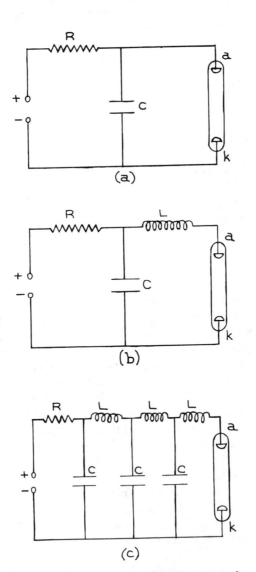

Figure 4.81 Operating circuits for flashlamps (a = lamp anode; k = lamp cathode): (a) RC discharge; (b) RLC critically damped discharge; (c) pulse forming network.

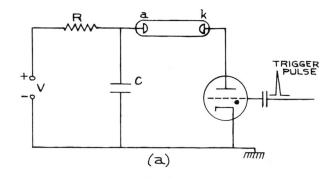

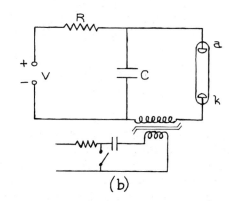

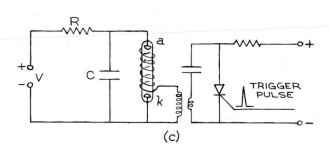

of a blackbody as shown in Figure 4.83.[42] Small-power lamps exhibit the spectral features of their fill gas, as shown in Figure 4.84.

A word about the triggering schemes shown in Figure 4.82. External triggering, accomplished by switching a high-voltage pulse from a transformer to a trigger wire wrapped around the lamp, is perhaps the simplest scheme; it is used only with long-pulse lamps and does not give quite so good time synchronization as series-spark-gap or thyratron-switched operation. The other methods are more complex but can be also used with short-pulse, high-power lamps. Series triggering allows the incorporation of triggering and lamp series inductance in a single unit. Trigger transformers of various types are available from EG&G and ILC. EG&G also supplies a range of excellent thyratrons.

(ii) *Short-pulse high-power lamps.* Flashlamps that can handle the rapid discharge of hundreds of joules on time scales down to a few microseconds are available commercially from ILC, EG&G, Candela, and Xenon Corporation, among others. To achieve such rapid dis-

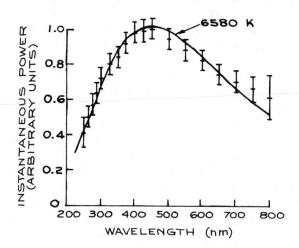

Figure 4.82 Flashlamp triggering schemes (*a* = lamp anode; *k* = lamp cathode): (*a*) overvoltage triggering (*V* > self-flash voltage of flashlamp; switching is accomplished in this case with a thyraton, but a triggered spark gap or ignitron can also be used); (*b*) series triggering (a saturable transformer is used with a fast-risetime 10–20-kV pulse with sufficient energy to trigger the lamp and saturate the core); (*c*) external triggering (the flashlamp is ionized by a trigger wire wrapped around the outside of the lamp and connected to the 5–15-kV secondary of a high-voltage pulse transformer).

Figure 4.83 Spectral distribution of power radiated by an EG&G FX42 xenon-filled flashlamp (76 mm long by 7 mm bore) operated in a critically damped mode with 500 J discharged in 1 ms. The spectral distribution was observed 0.7 ms from flash initiation. The line is the blackbody radiation curve of best fit. (After J. G. Edwards, "Some Factors Affecting the Pumping Efficiency of Optically Pumped Lasers," Appl. Opt., 6, 837–843, 1967; by permission of the Optical Society of America.)

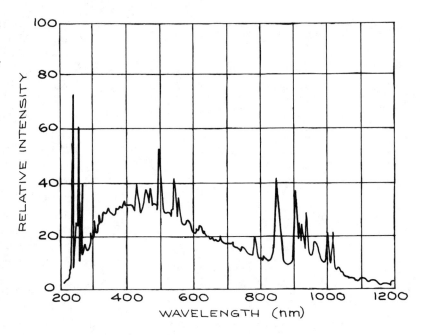

Figure 4.84   Spectral distribution of intensity from an ILC 4L2 xenon flashlamp (51 mm long by 4 mm bore) operated in a critically damped mode with 10 J discharged in 115 μs. (Courtesy of ILC.)

charges, these lamps are generally operated in a low-inductance discharge circuit. When these lamps are fired at high peak power inputs, a severe shock wave is generated in the lamp. To withstand the shock wave, the lamps are designed with special reentrant electrode seals. If lamps designed for long-pulse operation are discharged too rapidly, their electrodes are quite likely to pop off because the electrode seals are not shock-resistant.

The spectral emission of short-pulse, high-energy lamps is generally quite close to that of a high-temperature blackbody, perhaps as hot as 30,000 K. The fill pressure in these lamps is often lower than in long-pulse lamps, typically from 0.1 to a few tens of millibars. The fill gas is usually xenon, but other noble gases such as krypton or argon can be used. If a short-pulse lamp has a narrow discharge-tube cross section, at high power inputs material can ablate from the wall and a substantial part of the emission from the lamp can come from this material. Such lamps can be made with heavy-walled discharge-tube bodies of silica, glass, or even Plexiglas. These lamps can be made fairly simply. A good design described by Baker and King[43] is illustrated in Figure 4.85. It can be operated in a nonablating or ablating

mode (at high or low pressure, respectively). When operated in the ablating regime, such lamps can be filled with air as the nature of the fill gas is unimportant. Figure 4.85 also shows the use of a field-distortion-triggered spark gap for firing the lamp—a spark-gap design that offers quiet, efficient switching of rapid discharge capacitors. Rapid-discharge (low-inductance) capacitors suitable for fast flashlamp and other applications are available from Maxwell Laboratories, Hi Voltage Components, CSI Technologies, and KGM Electronics, among others.

4.6   LASERS

Lasers are now so widely used in physics, chemistry, and engineering that they must be regarded as the experimentalist's most important type of optical source. Generally speaking, anyone who wants a laser for an experiment should buy one. This is certainly true in the case of helium-neon, argon-ion, and helium-cadmium gas lasers and all solid-state crystalline, glass, or semiconductor lasers. True, these lasers can be built in the

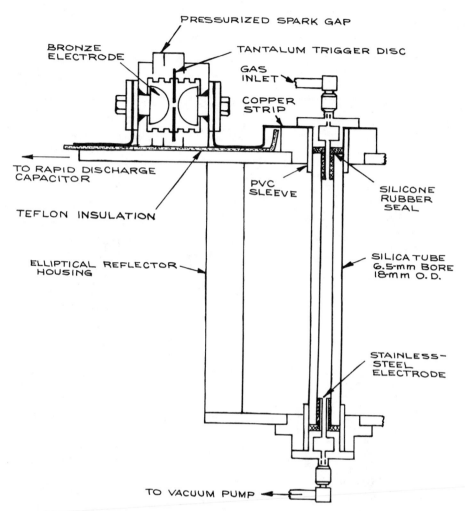

PRESSURIZED SPARK GAP

BRONZE ELECTRODE

TANTALUM TRIGGER DISC

GAS INLET

COPPER STRIP

TO RAPID DISCHARGE CAPACITOR

TEFLON INSULATION

PVC SLEEVE

SILICONE RUBBER SEAL

ELLIPTICAL REFLECTOR HOUSING

SILICA TUBE 6.5-mm BORE 18-mm O.D.

STAINLESS-STEEL ELECTRODE

TO VACUUM PUMP

Figure 4.85 High-pulse-energy, fast flashlamp design for conventional or ablation-mode operation. (From H. J. Baker and T. A. King, "Optimization of Pulsed UV Radiation from Linear Flashtubes," J. Phys. E.: Sci. Instr., 8, 219–223, 1975. Copyright 1975 by The Institute of Physics; used with permission.)

laboratory, but this is the province of the laser specialist and will be found time-consuming and unproductive for scientists in other specialties. On the other hand, there are some lasers, such as nitrogen, exciplex, CO_2, and CO gas lasers and dye lasers, that it may sometimes pay to build for oneself even though models are available commercially. A detailed discussion of how to construct all these lasers will not be given here, but, to illustrate how it is accomplished, design features of some specific systems will be discussed.

Table 4.5 lists some of the characteristic features of the more commonly used lasers presently available com-

mercially. Before briefly describing some of these laser systems, some background material will be presented that is pertinent to a discussion of lasers in general. More detailed information regarding the physics of laser operation is given in the books by Yariv,[10] Lengyel,[44] and Siegman.[45]

Some comments on Table 4.5 are in order. It is not meant to be an exhaustive compilation of all the laser types, wavelengths, and power outputs that can be obtained. In particular, far-infrared, DF, and fluorine lasers are not listed. The lasers listed in the table fall into two categories: pulsed and CW (continuous wave).

Table 4.5 CHARACTERISTICS OF IMPORTANT LASERS

CW Gas Lasers

Type	Operating Wavelengths (μm)	Output Power (W)		Suppliers[a]
		TEM$_{00}$	Multimode	
Ar ion	0.3336 0.3344 0.3358 0.3511[c] 0.3638[c]	0.04–3[b]		Coherent, Spectra-Physics, Lexel, Control Laser
	0.4545 0.4579 0.4658 0.4765 0.4880 0.4965 0.5017 0.5145 0.5287 1.0923	0.005–40[b]	0.03–24[b]	Coherent, Spectra-Physics, Lexel, Control Laser, American Laser
Kr ion	0.3375 0.3507 0.3564	2[b]		Coherent
	0.4067 0.4131 0.4619 0.4680 0.4762 0.4825 0.5309 0.5681 0.6470 0.6764 0.7993	0.05–6[b]	0.3–6[b]	Coherent, Spectra-Physics[e], Lexel, Control Laser, American Laser
He-Ne	0.6328[d]	5×10^{-4}–0.1		Spectra-Physics, Coherent, Hughes, Metrologic, Tropel[e]
	1.15	2×10^{-3}		Spectra-Physics, Jodon

[a] For a more extensive listing, see *Laser Focus Buyers Guide* (Reference 15).
[b] Available power outputs vary from line to line.
[c] Strongest lines.
[d] Most readily available wavelength.
[e] Only current suppliers of single-frequency He-Ne laser.

Table 4.5 (*continued*)

CW Gas Lasers (*continued*)				

Type	Operating Wavelengths (μm)	Output Power (W) TEM$_{00}$	Multimode	Suppliers[a]
	3.39	4×10^{-3}		Spectra-Physics, Jodon
He-Cd	0.3250	$(1-8) \times 10^{-3}$	$1.5 - 10 \times 10^{-3}$	Liconix, RCA
	0.4416	$(1-4) \times 10^{-2}$		Liconix, RCA
CO	5–6.5[f]	2		Edinburgh Inst.
CO$_2$	9–11[f] 10.6[d]	1–15,000		Apollo Lasers, GTE Sylvania, Coherent, Photon Sources, Edinburgh Inst.

Pulsed Gas Lasers					

Type	Operating Wavelengths (μm)	Output Energy (J/pulse) TEM$_{00}$	Multimode	Pulses per Second	Pulse Length (μs)	Suppliers
Ar ion	0.4545 0.4579 0.4658 0.4765 0.4880 0.4965 0.5017 0.5145 0.5287	$2 \times 10^{-8} - 7 \times 10^{-6}$		$\leqslant 4 \times 10^{6}$[g]	$0.2-15 \times 10^{-3}$	Spectra Physics
Ar F	0.193	0.08	0.0006–10.5	0–1000	0.01–0.02	Tachisto, Lumonics, Oxford Lasers, Lambda Physik,[h] Math Sciences NW

[f]Molecular lasers that offer discrete tunability over several lines.
[g]If a mode-locked CW laser is used, pulse repetition rates beyond 10^8 pps and pulse lengths of 200 ps are possible.
[h]Lambda Physik Lasers are available from Coherent, Inc.

Table 4.5   *(continued)*

		Pulsed Gas Lasers *(continued)*				
Type	*Operating Wavelengths* (μm)	*Output Energy* (J/pulse) TEM_{00}	*Multimode*	*Pulses per Second*	*Pulse Length* (μs)	*Suppliers*
CO_2	9–11	3×10^{-5}	0.251–60	0–10^5	0.04–250	Edinburgh Inst., Lumonics, Tachisto, Apollo Lasers, Laakmann
	10.6	0.035–8000	0.05–10^4	0.1–2500	0.04–250	Lumonics, Tachisto, Gen-Tec Systems, Science Software, Maxwell Labs
CO	5–6	0.008	0.04	1–5	1–4	Lumonics
Cu Vapor	$\begin{Bmatrix} 0.5105 \\ 0.5782 \end{Bmatrix}$	0.0002	0.0025	2000–6000	0.02	Laser Consultants, GE Space Sciences
HF	2.6–3.0	0.3–0.5	1	1–3	0.5	Lumonics
Kr ion	0.4067 0.4131 0.4619 0.4680 0.4762 0.4825 0.5681 0.6470 0.6764	$10^{-8} - 3 \times 10^{-6}$		$4 \times 10^{6\,g}$	$\leqslant 0.015$	Spectra Physics
Kr Cl	0.222		0.01–0.075	1–100	0.005–0.03	Lumonics, Oxford Lasers
Kr F	0.248	10	0.03–100	1–100	0.007–1	Lumonics, Tachisto, Maxwell Labs, Sopra, Oxford Lasers, Lambda Physik, Math Sciences NW

Table 4.5   (*continued*)

		Pulsed Gas Lasers (*continued*)				

Type	Operating Wavelengths (μm)	Output Energy (J/pulse) TEM$_{00}$	Multimode	Pulses per Second	Pulse Length (μs)	Suppliers
N$_2$	0.3371	$2 \times 10^{-5} - 10^{-3}$	5×10^{-6} –0.01	1–1000	3×10^{-4}–0.01	Lambda Physik, Molectron, NRG Inc., Lumonics, Math Sciences NW
Xe ion	0.4954 0.5159 0.5353 0.5395 0.5419		2×10^{-4}	150	0.5	Holotron
Xe Br	0.282		1.4–17×10^{-3}	1–100	0.008–0.03	Lambda Physik, Tachisto
Xe Cl	0.308	0.1–5	0.0006–80	1–20	0.004–0.01	Lumonics, Tachisto, Oxford Lasers, Lambda Physik, Maxwell Labs
Xe F	0.351		0.0007–50	1–1000	0.006–1	Lumonics, Tachisto, Maxwell Labs, Oxford Lasers, Lambda Physik, Sopra

		CW Solid-State Lasers		

Type	Operating Wavelengths (μm)	Output Energy (J) TEM$_{00}$	Multimode	Suppliers
F-center	2.3–3.3 (tunable)	1–20×10^{-3}	1–15×10^{-3}	Burleigh
Ho:YLF	2.06		5	Sanders Associates

Table 4.5 (*continued*)

CW Solid-State Lasers (*continued*)				
Type	*Operating Wavelengths* (μm)	*Output Energy* (J)		*Suppliers*
		TEM$_{00}$	*Multimode*	
Nd:YAG	0.265[i]		0.025	Control Laser
	0.530[j]	1–2		Holobeam, Quantronix
	1.06	0.1–18	25–150	Holobeam, Quantronix, Korad/Hadron, Control Laser, GTE Sylvania, General Photonics, JEC Associates
	1.318	0.1–0.2	0.025–0.075	General Photonics, JEC Associates

Pulsed Solid-State Lasers						
Type	*Operating Wavelengths* (μm)	*Output Energy* (J)		*Pulses per Second*	*Pulse Length*[k] (μs)	*Suppliers*
		TEM$_{00}$	*Multimode*			
Alexandrite	0.73–0.78		0.1–7	⩽ 20	0.1–200	Allied Chemical
Er:YLF	0.85		0.05	10	0.15	Sanders
	1.73		0.005	10	0.15	Sanders
Ho:YLF	2.06		0.001–0.075	10–5000	0.07–100	Sanders
Nd:glass	0.266[l]	0.05–0.1		0.02–0.1	$(3–15)\times10^{-3}$	Quantel
	0.355[m]	0.1–0.3		0.02–0.1	$(3–15)\times10^{-3}$	Quantel
	0.532[n]	0.25–1		0.02–0.1	$(3–15)\times10^{-3}$	Quantel
	1.06	10^{-3}–3[o]	0.3–1000	0.02–1	0.003–8000	Quantel, Apollo Lasers, Holobeam, Korad/Hadron, Lasag, CILAS, Laser, Inc.

[i]Frequency quadrupled from the 1.06-μm Nd:YAG output wavelength.
[j]Frequency doubled from the 1.06-μm Nd:YAG output wavelength.
[k]Shorter pulses can be obtained by mode locking.
[l]Frequency quadrupled from 1.06-μm output.
[m]Frequency tripled from 1.06-μm output.
[n]Frequency doubled from 1.06-μm output.
[o]Much higher energies can be obtained from oscillator amplifier configurations.

Table 4.5 (*continued*)

Pulsed Solid-State Lasers (*continued*)

Type	Operating Wavelengths (μm)	Output Energy (J)		Pulses per Second	Pulse Length[k] (μs)	Suppliers
		TEM$_{00}$	Multimode			
Nd:YAG	0.213[p]	0.025		2–22	0.004	Quanta Ray
	0.265[l]	0.0006–1.1	0.003–0.1	0.1–50	0.004–0.018	Quanta Ray, ILS, Quantel, JK Lasers
	0.355[m]	0.001–0.2	0.01–0.06	0.5–50	$(5$–$17)\times10^{-3}$	Quantel, ILS, Quanta Ray
	0.53	0.003–0.6	0.003–0.5	0.5–50	0.002–0.017	Holobeam, ILS, Quanta Ray, Quantel, General Photonics, Apollo Lasers, Korad/Hadron, JK Lasers
	1.06	0.01–10	0.1–400	0–300	0.003–20,000	Apollo Lasers, Quantel, Quanta Ray, Molectron, ILS, General Photonics, Lasag, Holobeam, CILAS, Raytheon, WEC Engineering, Laser Nucleonics
Ruby	0.6943	0.02–10	0.3–400	0–10	0.015–3000	Holobeam Laser, Apollo Lasers, Quantel, Laser Nucleonics, Korad/Hadron, WEC Engineering

[p] Fifth harmonic from 1.06-μm output.

Table 4.5   (*continued*)

CW Tunable Dye Lasers (*pumped with ion lasers*)			
Tuning Range (μm)	*Power*	*Linewidth*	*Supplier*
0.4–1	1	20–30 GHz	Spectra Physics
0.41–0.83	Varies	120 MHz–60 GHz	Coherent

Pulsed Tunable Dye Lasers						
Tuning Range (μm)	*Peak Power* (W)	*Pulses per Second*	*Pulse Length* (ns)	*Linewidth* (nm)	*Pumping Method*	*Supplier*
0.19–3	$(2–55) \times 10^6$	0–40	3–8	0.001–0.002	Nd:YAG or excimer laser	Quantel Quantaray
0.217–1.175	$10^4–10^6$	0–100	2–16	$10^{-4}–0.5$	N_2 laser, excimer, Nd:YAG	Molectron Sopra Lambda Physik NRG Instruments SA
0.22–0.96	$10^5–10^8$	0.1–25	200–600	0.001–0.4	Coaxial flashlamp	Phase-R
0.34–0.8	$2 \times 10^4–2 \times 10^6$	0.1–200	160–400	0.1	Coaxial flashlamp or dual liner lamp	Candela
0.265–0.730[q]	$10^3–10^4$	0–30	1000	0.0025–0.2	Linear flashlamp	Chromatix

[q]Tunable operation out to 2.60 μm is possible with an available optical parametric oscillator.

The energy outputs listed for the pulsed lasers, together with the available pulse repetition rates and pulse lengths, represent the ranges that are readily available commercially. They are not intended to imply that all possible combinations can be obtained simultaneously. Indeed, in the case of pulsed solid-state lasers, the highest available energy outputs are generally available only at the lowest pulse repetition rates. In the case of the CW lasers, available power outputs vary from line to line and from manufacturer to manufacturer. To the intended purchaser of a laser system, we recommend the following questions:

1. What fixed wavelength or wavelengths must the laser supply?

2. Is tunability required?

3. What frequency and amplitude stability are required?

4. What laser linewidth is required?

5. What power or pulse energy is required?

6. Are high operating reliability and long operating lifetime required?

7. For time-resolved experiments, what pulse length is needed?

The desirable attributes in each of these areas are unlikely to occur simultaneously. For example, frequency tunability is not readily available throughout the spectrum without resorting to the specialized techniques of nonlinear optics.[46-48] High-energy and high-power lasers are generally less stable both in amplitude and in frequency, and will generally be less reliable and need more "hands on" attention than their low-energy or low-power counterparts. When a decision is made to purchase a laser system, compare the specifications of lasers supplied by different manufacturers; discuss the advantages and disadvantages of different systems with others who have purchased them previously. Reputable laser manufacturers are generally very willing to supply the names of previous purchasers.

To some experimental scientists a laser is merely a rather monochromatic directional lightbulb; for others, detailed knowledge of its operating principles and characteristics is essential. Certain aspects of laser designs and operating characteristics, however, are very general and are worthy of some discussion.

4.6.1 General Principles of Laser Operation

A laser is an optical-frequency oscillator; in common with electronic circuit oscillators, it consists of an amplifier with feedback. The optical-frequency amplifying part of a laser can be a gas, a crystalline or glassy solid, a liquid, or a semiconductor. This medium is maintained in an amplifying state, either continuously or on a pulsed basis, by pumping energy into it appropriately. In a gas laser the input energy comes from electrons (in a gas discharge or an electron beam), or from an optical pump (which may be a lamp or a laser). Solid-state crystalline or glassy lasers receive their pumping energy from continuous or pulsed lamps.

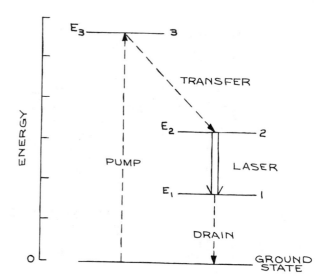

Figure 4.86   Schematic partial energy-level diagram of a laser system.

Liquid lasers can be pumped with a flashlamp or, on a continuous or pulsed basis, by another laser. Semiconductor lasers are $p - n$ junction devices and are operated by passing pulsed or continuous electrical currents though them.

The energy *sublevels* of a system are single states with their own characteristic energy. A set of sublevels having the same energy is called an energy *level*. The amplifying state in a laser medium results if a population inversion can be achieved between two sublevels of the medium. This can be seen with reference to Figure 4.86, which shows a schematic partial energy-level diagram of a typical laser system. Input energy excites ground-state particles (atoms, molecules, or ions) of the medium into the state (or states) indicated as 3. These particles then transfer themselves or excite other particles preferentially to sublevel 2. In an ideal system negligible excitation of particles into sublevel 1 should occur. If the populations of the sublevels 2 and 1 are n_2 and n_1, respectively, and $n_2 > n_1$, this is called a *population inversion*. The medium then becomes capable of amplifying radiation of frequency

$$\nu = \frac{E_2 - E_1}{h},$$

(146)

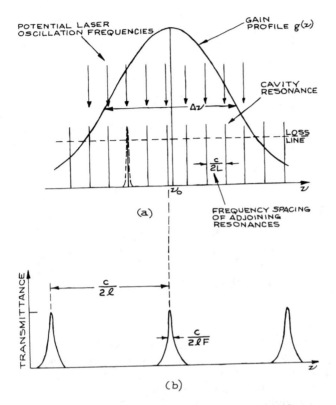

POTENTIAL LASER OSCILLATION FREQUENCIES

GAIN PROFILE $g(\nu)$

CAVITY RESONANCE

LOSS LINE

$\Delta\nu$

$\dfrac{c}{2L}$

ν_0

(a)

FREQUENCY SPACING OF ADJOINING RESONANCES

TRANSMITTANCE

$\dfrac{c}{2\ell}$

$\dfrac{c}{2\ell F}$

(b)

Figure 4.87 (*a*) Laser gain profile showing position of cavity resonances and potential laser oscillation frequencies near cavity resonances where gain lies above loss (*L* is length of laser cavity); (*b*) transmission characteristic of intracavity etalon, on same frequency scale as (*a*), for single-mode operation (*l* is the etalon thickness and *F* its finesse).

where *h* is Planck's constant. In practice, energy levels of real systems are of finite width, so the medium will amplify radiation over a finite bandwidth. The gain of the amplifier varies with frequency in this band and is specified by a gain profile $g(\nu)$ as shown in Figure 4.87. Actual laser oscillation occurs when the amplifying medium is placed in an optical resonator, which provides the necessary positive feedback to turn the amplifier into an oscillator. Most optical resonators consist of a pair of concave mirrors, or one concave and one flat mirror, placed at opposite ends of the amplifying medium and aligned parallel. A laser resonator is in essence a Fabry-Perot (see Section 4.7.4), generally of large spacing. The radii and spacing of the mirrors will

determine to a large degree the type of Gaussian beam that the laser will emit. At least one of the mirrors is made partially transmitting, so that useful output power can be extracted. The choice of optimum mirror reflectance depends on the type of laser and its gain.

Low-gain lasers such as the helium-neon have high-reflectance mirrors, (98–99%), while higher-gain lasers such as pulsed nitrogen, CO_2, or Nd^{3+}: YAG can operate with much lower reflectance values. Under certain circumstances when their gain is very high, lasers will emit laser radiation without any deliberately applied feedback. Lasers that operate in this fashion are frequently called *superradiant*, although this is a misnomer: such lasers are generally just amplifying their own spontaneous emission very greatly in a single pass.

4.6.2 General Features of Laser Design

Even though the majority of laser users do not build their own, an awareness of some general design features of laser systems will help the user to understand what can and cannot be done with a laser. It will also assist the user who wishes to make modifications to a commercial laser system to increase its usefulness or convenience in a particular experimental situation. Figure 4.88 shows a schematic diagram of a typical gas laser, which incorporates most of the desirable design features of a precision system. The amplifying medium of a gas laser is generally a high- or low-pressure gas excited in a discharge tube. The excitation may be pulsed (usually by capacitor discharge through the tube), d.c., or occasionally a.c. Some gas lasers are also excited by electron beams, or by optical pumping with either a lamp or another laser. Far infrared lasers in particular are frequently pumped with CO_2 lasers. Depending on the currents that must be passed through the discharge tube of the laser, it can be made of Pyrex, quartz, beryllium oxide, graphite, or segmented metal. These last three are used in high-current-density discharge tubes for CW ion lasers. Several types of pulsed high-pressure or chemical lasers, such as CO_2, HF, DF, and HCl, can be excited in structures built of Plexiglas or Kel-F.

Because most gas lasers are electrically inefficient, a

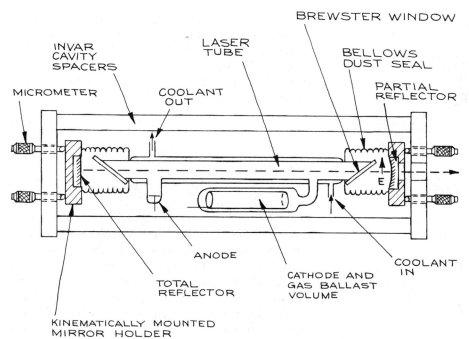

INVAR
CAVITY
SPACERS

MICROMETER

LASER
TUBE

COOLANT
OUT

BREWSTER WINDOW

BELLOWS
DUST SEAL

PARTIAL
REFLECTOR

E

ANODE

TOTAL
REFLECTOR

CATHODE AND
GAS BALLAST
VOLUME

COOLANT
IN

KINEMATICALLY MOUNTED
MIRROR HOLDER

Figure 4.88 Schematic diagram of a typical gas laser showing some of the desirable features of a well-engineered research system. In high-discharge-current systems, some form of internal or external gas-return path from cathode to anode must be incorporated into the structure.

large portion of their discharge power is dissipated as heat. If ambient or forced-air cooling cannot keep the discharge tube cool enough, water cooling is used. This can be done in a closed cycle, using a heat exchanger. Some lasers need the discharge tube to run at temperatures below ambient. In such cases a refrigerated coolent such as ethylene glycol can be used. Many gas lasers use discharge tubes fitted with windows placed at Brewster's angle. This permits linearly polarized laser oscillation to take place in the direction indicated in Figure 4.89; light bouncing back and forth between the laser mirrors passes through the windows without reflection loss. Mirrors fixed directly on the discharge tube can also be used—this is common in commercial He-Ne lasers. In principle such lasers should be unpolarized, although in practice various slight anisotropies of the structure often lead to at least some polarization of the output beam.

The two mirrors that constitute the laser resonator, unless these are fixed directly to the discharge tube, should be mounted in kinematically designed mounts and held at fixed spacing L by a thermally stable

resonator structure made of Invar or quartz. Because the laser generates one or more output frequencies which are close to integral multiples of $c/2L$, any drift in mirror spacing L causes changes in output frequency ν. This frequency change $\Delta\nu$ satisfies

$$\frac{\Delta\nu}{\nu} = \frac{\Delta L}{L}. \qquad (147)$$

For a laser 1 m long, even with an Invar-spaced resonator structure, the temperature of the structure would need to be held constant within 10 mK to achieve a frequency stability of only 1 part in 10^8. The discharge tube should be thermally isolated from the resonator structure unless the whole forms part of a temperature-stabilized arrangement.

If single-frequency operation is desired, the laser should be made very short—so that $c/2L$ becomes larger than $\Delta\nu$ in Figure 4.87—or operated with an intracavity etalon (see Sections 4.6.4 and 4.7.4). Visible lasers that have the capability of oscillating at several different wavelengths are generally tuned from line to

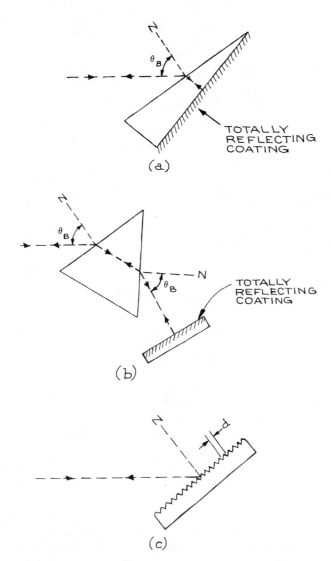

(a)

TOTALLY
REFLECTING
COATING

(b)

TOTALLY
REFLECTING
COATING

(c)

Figure 4.89 Methods for wavelength-selective reflection in laser systems: (*a*) Brewster's-angle Littrow prism; (*b*) intracavity Brewster's-angle prism; (*c*) reflective diffraction grating. N = surface normal.

faces at Brewster's angle. Infrared gas lasers that can oscillate at several different wavelengths are tuned by replacing one resonator mirror with a gold-coated or solid metal diffraction grating mounted in Littrow, as shown in Figure 4.89(*c*). The laser oscillation wavelength then satisfies

$$m\lambda = 2d\sin\theta, \qquad (148)$$

where d is the spacing of the grating grooves, θ is the angle of incidence, and m is an integer.

A laser will oscillate only if its gain exceeds all the losses in the system, which include mirror transmission losses, absorption and scattering at mirrors and windows, and any inherent losses of the amplifying medium itself—the last generally being significant only in solid-state, liquid, and semiconductor lasers. Consequently, it is particularly important to keep the mirrors and windows of a laser system clean and free of dust. Most commercial systems incorporate a flexible, sealed enclosure between Brewster windows and mirrors to exclude contaminants.

Figure 4.90 shows some design features of a simple solid-state laser oscillator. This design incorporates a linear pulsed or continuous lamp and a crystalline or glass laser rod with Brewster windows mounted inside a metal elliptical reflector, which serves to reflect pumping light efficiently into the laser rod. Although solid-state lasers are rarely used in experiments where extreme frequency stability and narrow linewidth operation are required, it is still good practice to incorporate features such as stable, kinematic resonator design and thermal isolation of the hot lamp from the resonator structure. Elliptical reflector housings for solid-state lasers (and flashlamp-pumped dye lasers) can be made in two halves by horizontally milling a rectangular aluminum block with the axis of rotation of the milling cutter set at an angle of arccos(a/b), where a and b are the semiminor and semimajor axes of the ellipse. This process is repeated for a second aluminum block. The two halves are then given a high polish, or plated, and joined together with locating dowel pins. For further details of solid-state laser design and particulars of other arrangements for optical pumping the reader should consult Röss.[49] Flashlamp-pumped dye lasers share some design features with solid-state lasers. The main difference is

line by replacing one laser mirror with a Littrow prism whose front face is set at Brewster's angle and whose back face is given a high-reflectance coating as shown in Figure 4.89(*a*). Alternatively, as shown in Figure 4.89(*b*), one can use a separate intracavity prism, designed so the intracavity beam passes through both its

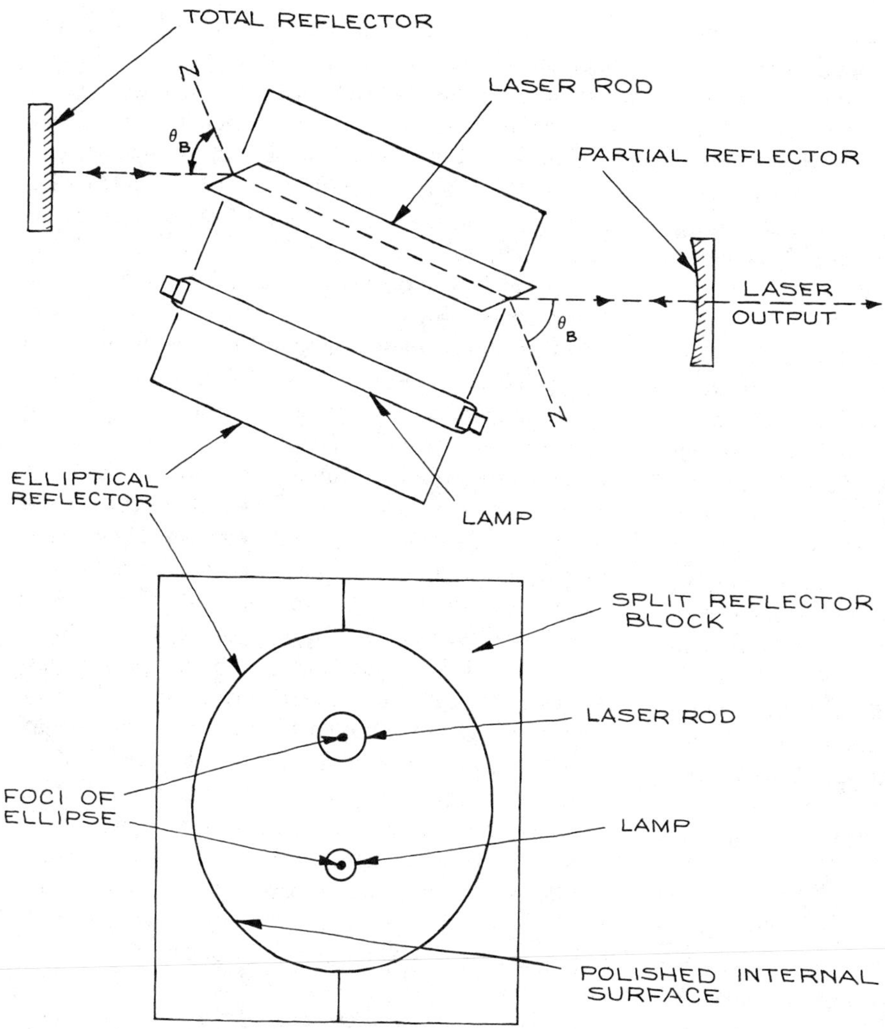

Figure 4.90 Schematic diagram of a simple solid-state laser system.

that the former require excitation of short pulse duration ($\lesssim 1 \ \mu s$) with special flashlamps in low-inductance discharge circuitry.[50-53]

4.6.3 Specific Laser Systems

(a) **Gaseous Ion Lasers.** Gaseous ion lasers, of which the argon, krypton, and helium-cadmium are the most important, generate narrow-linewidth, highly coherent visible and ultraviolet radiation with powers in the range from milliwatts to a few tens of watts CW. Helium-cadmium lasers use low-current-density (a few $A\,cm^{-2}$) air-cooled discharge structures, which in many respects are similar to those of helium-neon lasers. They are limited to a few tens of milliwatts in conveniently available output power. Argon and krypton ion lasers use very high-current-density discharges (100–2000 $A\,cm^{-2}$) in special refractory discharge structures made of beryllia, segmented graphite, or segmented metal. Commercially available powers range up to a few tens of watts, but outputs up to 1 kW in the visible have been reported. Comprehensive information about ion lasers can be found elsewhere.[54,55] Argon and krypton ion lasers are widely used for pumping CW dye lasers. Helium-cadmium lasers can be built in the laboratory if long-term reliability and convenience are not a prime consideration. Home construction of CW argon or krypton ion lasers is not recommended. Pulsed ion lasers operate at high current densities but have low duty cycles, so that ambient cooling is adequate. These lasers are not in widespread experimental use.

(b) **Helium-Neon Lasers.** Of all types of laser, helium-neon lasers come closest to being the ideal classical monochromatic source. They can have very good amplitude (= 0.1%) and frequency stability (1 part in 10^8 without servo frequency control). Servo-frequency-controlled versions are already in use as secondary frequency standards, and have excellent temporal and spatial coherence.[56] Typical available power outputs range up to 50 mW. Single-frequency versions with power outputs of 1 mW, which are ideal for interferometry and optical heterodyne experiments, are available from Spectra-Physics and Tropel. Although 632.8 nm is the routinely available wavelength from He-Ne lasers, versions with 1.15- or 3.39-μm output are also available. Helium-neon lasers of low power are very inexpensive, costing from about $200, and are ideal for alignment purposes.

(c) **CO_2 Lasers.** Both pulsed and CW CO_2 lasers are easy to construct in the laboratory. A typical CW version would incorporate most of the features shown in Figure 4.89. A water-cooled Pyrex discharge tube of internal diameter 10–15 mm, 1–2 m long, and operating at a current of about 50 mA, is capable of generating tens of watts output at 10.6 μm. The best operating gas mixture is CO_2-N_2-He in the ratio $1:1:8$ or $1:2:8$ at a total pressure of about 25 mbar in a 10-mm-diameter tube. ZnSe is the best material to use for the Brewster windows, as it is transparent to red light and permits easy mirror alignment. The optimum output mirror reflectance depends on the ratio of the tube length to diameter, L/d, roughly according to[57]

$$R = 1 - L/500d. \qquad (149)$$

The output mirror is usually a dielectric-coated germanium, gallium arsenide, or zinc selenide substrate. Such mirrors are available from Janos, Laser Optics, Exotic Materials, Inc., Broomer Research Corp., Coherent, Infrared Industries, Unique Optical, and Valpey, among others. The total reflector can be gold-coated in lasers with outputs of a few tens of watts. For tunable operation, gratings are available from PTR, Jobin-Yvon, Bausch and Lomb, Diffraction Products Inc., and Perkin-Elmer. When a grating is used to tune a laser with Brewster windows, the grating should be mounted so that the E vector of the laser beam is orthogonal to the grating grooves. An excellent review of CW CO_2 gas lasers has been given by Tyte.[57]

Transversely-excited atmospheric-pressure (TEA) CO_2 lasers are now widely used when high-energy, pulsed infrared energy near 10 μm is required. These lasers operate in discharge structures where the current flow is transverse to the resonator axis through a high-pressure (≥ 0.5 bar) mixture of CO_2, N_2, and He, typically in the ratio $1:1:5$. To achieve a uniform, pulsed glow discharge through a high-pressure gas, special electrode structures and preionization techniques are used to inhibit the formation of localized spark discharges. Many very-high-energy types utilize electron-beam excitation. Discharge-excited TEA lasers are relatively easy to construct in the laboratory. A device with a discharge volume 60 cm long$\times 5$ cm wide with a 3-cm electrode spacing excited with a 0.2-μF capacitor charged to 30 kV will generate several joules in a pulse about 150 ns long. There are very many different designs for CO_2 TEA lasers; particulars of various types can be found in the *IEEE Journal of Quantum Electronics* and the *Journal of Applied Physics*. A particularly good design that is easy to construct is derived from a vacuum-ultraviolet photopreionized design described by Seguin and Tulip.[58] A diagram of the discharge structure is shown in Figure 4.91, together with its associated capacitor-discharge circuitry. The cathode is an aluminum Rogowski $2\pi/3$ profile,[59] the anode is a brass mesh, beneath which is a section of double-sided copper-clad printed circuit board machined into a matrix of separate copper sections on its top side. This printed circuit board provides a source of very many surface sparks for

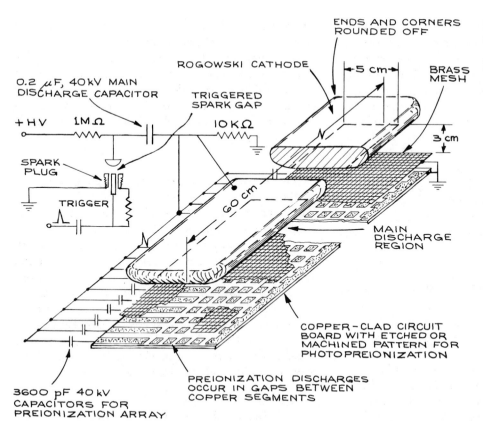

Figure 4.91 Cutaway view of vacuum-ultraviolet photopreionized Rogowski TEA laser structure.

photopreionization of the main discharge. The laser is fitted with two separate capacitor banks: each circuit-board spark channel is energized by a 3600-pF Sprague or similar "doorknob" capacitor, while the main discharge is energized with a 0.2-μF low-inductance capacitor. Such capacitors are available from Hi Voltage Components, Maxwell Labs, and KGM Electronics among others. Both capacitor banks are switched with a single spark gap, which incorporates a Champion marine spark plug for triggering. The end of the spark plug, which has an annular gap, is ground down to its ceramic insulator and is triggered by an EG&G Model TM-11A Trigger generator. A good double Rogowski TEA laser design has been given by Seguin, Manes, and Tulip.[60] TEA lasers based on the multiple-pin-discharge design first described by Beaulieu[61] are also easy to construct, but they do not give output energies comparable to

those of volume-excited designs. They do, however, lend themselves well to laser oscillation in many different gases—for example, HF formed by discharge excitation of H_2-F_2 mixtures.

(d) CO Lasers. CW CO lasers are essentially similar in construction to CO_2 lasers except that the discharge tube must be maintained at low temperatures—certainly below 0°C. They operate in a complex mixture of He, CO, N_2, O_2, and Xe with (for example) 21 mbar He, 0.67 mbar N_2, 0.013 mbar O_2, and 0.4 mbar Xe.

(e) Exciplex Lasers. Exciplex lasers operate in high-pressure mixtures such as Xe-F_2, Xe-Cl_2, Kr-Cl_2, Kr-F_2, and Ar-F_2. They are commonly also called "excimer" lasers, although this is not strictly correct scientific terminology. Both discharge TEA and E-beam-pumped configurations are used. Under electron bombardment a series of reactions occur that lead to the formation of an excited complex (exciplex) such as XeF^*, which is unstable in its ground state and so dissociates immediately on emitting light. Exciplex lasers available commercially are sources of intense pulsed ultraviolet radiation. The radiation from these lasers is not inherently of narrow linewidth, because the laser transition takes place from a bound to a repulsive state of the exciplex, and therefore is not of well-defined energy. However, exciplex lasers are attractive sources for pumping dye lasers. Development of new and existing exciplex lasers is an active field, and improvements in operating performance and reliability are sure to be forthcoming.

(f) Nitrogen (N_2) Lasers. Molecular-nitrogen lasers are sources of pulsed ultraviolet radiation at 337.1 nm. Commercially available devices operate on a flowing nitrogen fill at pressures of several tens of mbar, and generate output powers up to about 1 MW in pulses up to about 10 ns long. These lasers are reliable and easy to operate, and are widely used for pumping pulsed dye lasers. Nitrogen lasers, by virtue of their internal kinetics, only operate in a pulsed mode. They use very rapid transversely excited discharges, usually between two parallel, rounded electrodes energized with small, low-inductance capacitors charged to about 20 kV. The whole discharge arrangement must be constructed to have very low inductance. Special care must be taken with insulation in these devices; the very rapidly changing electric fields present can punch through an insulating material to ground under circumstances where the same applied d.c. voltage would be very adequately insulated. There are several good N_2-laser designs in the literature.[62-67] These designs fall into two general categories: Blumlein-type distributed-capacitance discharge structures, and designs based on discrete capacitors. The latter, although they do not give such high-energy outputs from a given laser as the very rapid-discharge Blumlein-type structures, are much easier to construct in the laboratory and are likely to be much more reliable. The capacitors used in these lasers need to be of low inductance and capable of withstanding the high voltage reversal and rapid discharge to which they will be subjected. Suitable capacitors for this application can be obtained from Murata or Sprague.

The output beam from a transversely excited N_2 laser is by no means Gaussian. It generally takes the form of a rectangular beam, typically $\simeq 0.5$ cm $\times 3$ cm, with poor spatial coherence. Such a beam can be focused into a line image with a cylindrical lens for pumping a dye cell in a pulsed dye laser.

(g) CW Solid-State Lasers. The most important laser in this category is the Nd:YAG laser. Its principal output wavelength is 1.06 μm, but other infrared wavelengths are available. The 1.06 μm can be efficiently doubled to yield a CW source of green radiation, although such a source is not competitive with an argon ion laser in this application. F-center lasers[68] are doped-crystal lasers that are pumped with argon, krypton, or dye lasers. They provide tunable operation between 2.2 and 3.3 μm, but are somewhat inconvenient because the crystal must be cooled with liquid nitrogen. Recent developments with other tunable crystalline lasers[69,70] may eventually render F-center lasers obsolete.

(h) Pulsed Solid-State Lasers. Three lasers are important in this category Nd:YAG, Nd:glass and ruby. Nd:YAG and Nd:glass lasers oscillate at the same wavelength, 1.06 μm. Nd:YAG is generally used in high-pulse-repetition-rate systems and/or where good beam quality is desired. High-energy Nd laser systems frequently incorporate an Nd:YAG oscillator and a series of Nd:glass amplifiers. Ruby lasers are still widely used in situations in which a high-energy visible output (694.3 nm) is required.

All these pulsed solid state lasers use fairly long flash excitation ($\simeq 1$ ms). Unless the laser pulse output is controlled, they generate a very untidy optical output, consisting of several hundred microseconds of random optical pulses about 1 μs wide, spaced a few microseconds apart. This mode of operation, called *spiking*, is rarely used. Usually the laser is operated in a Q-switched mode. In this mode the laser cavity is blocked with an optical shutter such as an electro-optic modulator or Kerr cell, as illustrated in Figure 4.92. An appropriate time after flashlamp ignition, after the population inversion

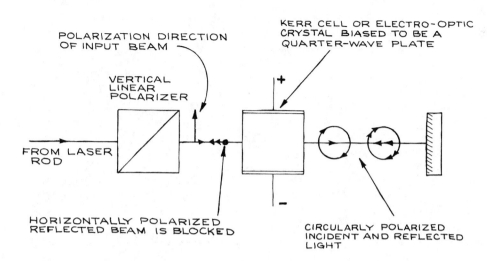

POLARIZATION DIRECTION OF INPUT BEAM

VERTICAL LINEAR POLARIZER

KERR CELL OR ELECTRO-OPTIC CRYSTAL BIASED TO BE A QUARTER-WAVE PLATE

FROM LASER ROD

HORIZONTALLY POLARIZED REFLECTED BEAM IS BLOCKED

CIRCULARLY POLARIZED INCIDENT AND REFLECTED LIGHT

Figure 4.92 Schematic arrangement for Q-switching. Laser oscillation occurs when the voltage bias on the Kerr cell or electro-optic crystal is removed, thereby allowing the reflected beam to remain linearly polarized in the vertical direction.

in the laser rod has had time to build to a high value, the shutter is opened. This operation, which changes the Q of the laser cavity from a low to a high value, causes the emission of a very large short pulse of laser energy. The Q-switched pulse can contain nearly as much energy as would be emitted in the "spiking" mode and is typically 10–20 ns long. Passive Q-switching can also be accomplished with a "bleachable" dye solution placed in an optical cell inside the laser cavity. The dye is opaque at low incident light intensities and inhibits the buildup of laser oscillation. However, when the laser acquires enough gain to overcome the intracavity dye absorption loss, laser oscillation begins, the dye bleaches, and a Q-switched pulse results. A bleachable dye cell with a short relaxation time, placed close to one of the laser mirrors, will frequently lead to mode-locked operation[71,72] within the Q-switched pulse (see Section 4.6.4).

(i) Semiconductor Lasers. Two important types of semiconductor lasers are available commercially: tunable diode lasers, which utilize lead-salt semiconductors such as PbSSe, PbSnSe, and PbSnTe and can provide tunable operation over limited regions anywhere between 2 and 30 μm; and GaAs or GaAlAs lasers, which operate at fixed wavelengths in the region between 0.8 and 0.9 μm.[73] The latter find most practical use in communication and information-processing systems, or whenever a low-power source of fairly coherent radia-

tion in the near infrared is required. Semiconductor lasers operate using a population inversion generated near a p-n junction by the passage of current. These lasers are very small: the active length is usually under 1 mm, and the active region of the junction where oscillation occurs is very narrow, $\simeq 2$ μm. Consequently the output radiation from a semiconductor laser diverges into a large solid angle. For an 800-nm device, for example, $\theta_{\text{beam}} = \lambda/\pi w_0 \simeq 0.25$ rad $\simeq 15°$. As a result, these lasers are used with external collimating optics. The tunable lead-salt diode lasers are now widely used in high-resolution spectroscopy.[74] They generally emit several modes, which, because of the very short length of the laser cavity, are spaced far enough in wavelength for a single one to be isolated by a monochromator. The output laser linewidths that are obtainable in this way are about 10^{-4} cm^{-1} (3 MHz). Scanning is accomplished by changing the temperature of the cryogenically cooled semiconductor or the operating current. Complete tunable-semiconductor-laser spectroscopic systems are available from Laser Analytics or New England Research Center, Inc. Tunable diode lasers have opened up a very large area of high-resolution spectroscopy.

(j) CW Dye Lasers. CW dye lasers usually take the form of a planar jet of dye solution continuously sprayed at Brewster's angle from a slit nozzle, collected, and

Figure 4.93   Hänsch-type pulsed dye laser.

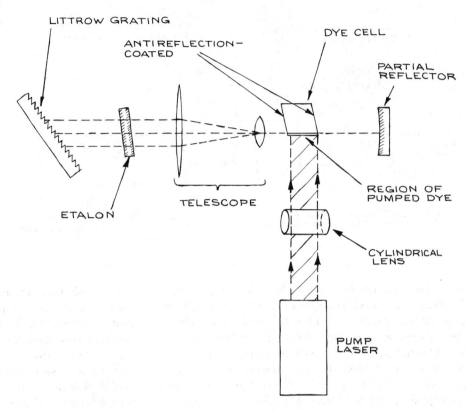

recirculated. The jet stream is in a spherical mirror laser cavity, where it is pumped by the focused radiation of a CW ion laser. These lasers can generally be regarded as wavelength converters—they have essentially the linewidth, frequency, and amplitude stability of the pumping source. For example, narrow-linewidth operation requires an etalon-controlled single-frequency ion laser. CW dye lasers are quite efficient at converting the wavelength of the pump laser; at pump powers far enough above threshold, the conversion efficiency ranges from 10 to 45%. For further details of CW-dye-laser operation consult Schäfer.[50]

(k) Pulsed Dye Lasers.   CW dye lasers are not as frequently constructed in the laboratory as are pulsed dye lasers. In its simplest form, a pulsed dye laser consists of a small rectangular glass or quartz cell placed on the axis of an optical resonator and excited with a focused line image from a pump laser. The most com-

monly used pump lasers are N_2 and frequency-doubled Nd:YAG, although ruby, exciplex, xenon-ion, and copper-vapor lasers are also used. In its simplest form this type of pulsed dye laser will generate a broad-band laser output ($\simeq 5$–10 nm). Various additional components are added to the basic design to give tunable, narrow-linewidth operation. A very popular design that incorporates all the desirable features of a precision-tunable visible source has been described by Hänsch.[75] The essential components of a Hänsch-type pulsed dye laser are shown in Figure 4.93. The pump laser is focused, with a cylindrical and/or a spherical lens, to a line image in the front of a dye cell. The line of excited dye solution (for example a 5×10^{-3} molar solution of Rhodamine 6G in ethanol) is the gain region of the laser. The laser beam from the gain region is expanded with a telescope in order to illuminate a large area of a high dispersion, Littrow-mounted echelle grating. Rotation of the grating tunes the center wavelength of the laser. Without the telescope, only a small portion of the

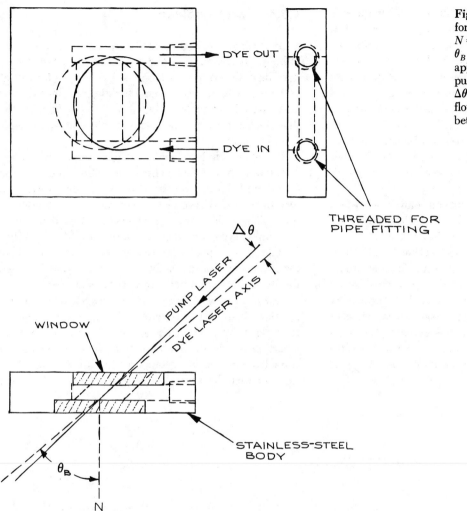

Figure 4.94 Flowing dye cell for end-pumped operation. N = normal to windows; θ_B = Brewster's angle, approximately the same for both pump and dye laser wavelengths; $\Delta\theta$ = a small angle. Dye solution flows laminarly in the channel between the windows.

grating is illuminated, high resolution is not achieved, and the laser linewidth will not be very narrow. With telescope and grating alone, the laser linewidth will be of the order of 0.005 nm at 500 nm ($\simeq 0.2$ cm^{-1}). Even narrower linewidth can be achieved by including a tilted Fabry-Perot etalon in the cavity. A suitable etalon will have a finesse of 20 and a free spectral range (see Section 4.7.4) below about 1 cm^{-1} in this case. Linewidths as narrow as 4×10^{-4} nm can be achieved. A few experimental points are worthy of note: the laser is tuned by adjusting grating and/or etalon. Alignment of the telescope to give a parallel beam at the grating is

quite critical. The dye should be contained in a cell with slightly skewed faces to prevent spurious oscillation. The dye solution can remain static when low-power (10–100 kW) pump lasers are used, magnetically stirred when pump lasers up to about 1 MW are used, or continuously circulated from a reservoir when higher-energy or high-repetition-rate pump lasers are used. Suitable dye cells are available from Precision Cells. Cells can be constructed from stainless steel with Brewster window faces when end-on pumping with a high-energy laser is used (as shown in Figure 4.94). The dye can be magnetically stirred by placing a small

Teflon-coated stirring button in the bottom of the dye cell. When continuous circulation of dye is used, suitable pumps are available from Micropump Corporation. Once a narrow-linewidth pulsed-dye-laser oscillator has been constructed, its output power can be boosted by passing the laser beam through one or more additional dye cells, which can be pumped with the same laser as the oscillator. In a practical oscillator-amplifier configuration 10% of the pump power will be used to drive the oscillator and 90% to drive the amplifier(s). Further details of Hansch-type dye-laser construction are available in several publications.[76-78]

Two other approaches to achieving narrow-linewidth laser oscillation are worthy of note. The intracavity telescope beam expander can be replaced by a multiple-prism beam expander (see Section 4.3.4), which expands the beam onto the grating in one direction only. Such an arrangement does not need to be focused. An alternative, simple approach described by Littman[79,80] is to eliminate the beam expander and instead use a Littrow-mounted holographic grating in grazing incidence as shown in Figure 4.95. Linewidths below 0.08 cm^{-1} can be obtained without the additional use of an intracavity etalon.

4.6.4 Laser Radiation

Laser radiation is highly monochromatic in most cases, although flashlamp-pumped dye lasers in a worst case may have linewidths as large as 100 cm^{-1} (2.5 nm at 500 nm). The spectral characteristics of the radiation vary from laser to laser but will be specified by the manufacturer. Lasers are usually specified as single-mode or multimode.

A *single-mode* laser emits a single-frequency output. Continuous wave CO_2 lasers and other middle- and far-infrared lasers usually operate in this way, and helium-neon and argon-ion lasers can also be caused to do so. The single output frequency will itself fluctuate randomly, typically over a bandwidth of 100 kHz–1 MHz, unless special precautions are taken to stabilize the laser—usually by stabilizing the spacing between the two mirrors that constitute its resonant cavity.

Multimode lasers emit several modes, called longitudinal modes, spaced in frequency by $c_0/2L$, where c_0 is the velocity of light *in vacuo* and L is the optical path length between the resonator mirrors. In a medium of refractive index n and geometric length l, $L = nl$. There may, in fact, be several superposed combs of equally

Figure 4.95 Littman-type dye laser.

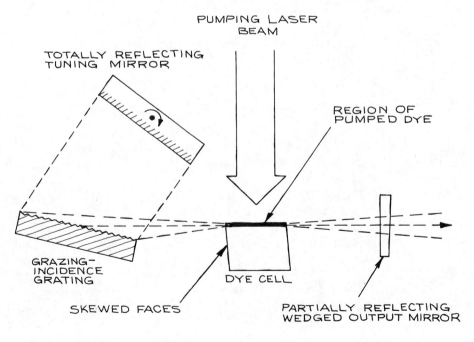

spaced modes in the output of a multimode laser if it is not operating in a single transverse mode. The transverse modes specify the different spatial distributions of intensity that are possible in the output beam. The most desirable such mode, which is standard in most good commercial lasers, is the fundamental TEM_{00} mode, which has a Gaussian radial intensity distribution (see Section 4.2.4). A laser will generally operate on multiple longitudinal modes if the linewidth $\Delta \nu$ of the amplifying transition is much greater than $c_0/2L$. In this case the output frequencies will span a frequency range of the order of $\Delta \nu$. In short-pulse lasers the radiation oscillating in the laser cavity may not be able to make many passes during the duration of the laser pulse, and (particularly if the linewidth of the amplifying transition is very large) the individual modes may not have time to become well characterized in frequency. This is the situation that prevails in pulsed dye and exciplex lasers. Additional frequency-selective components must be included in such lasers, such as etalons and/or diffraction gratings, to obtain narrow output linewidths.

Laser radiation is highly collimated; beam divergence angles as small as 1 mrad are quite common. This makes a laser the ideal way to align an optical system, far superior to traditional methods using point sources and autocollimators. As mentioned previously, lasers are highly coherent. Single-mode lasers have the highest temporal coherence—with coherence lengths that can extend to hundreds of kilometers. Fundamental-transverse-mode lasers have the highest spatial coherence.

In multimode lasers, the many output frequencies lie underneath the gain profile, as shown in Figure 4.87. Laser oscillation can be restricted to a single transverse mode by appropriate choice of mirror radii and spacing. Then the available output frequencies are uniformly spaced longitudinal modes a frequency $c_0/2L$ apart, where L is the optical spacing of the laser mirrors. Laser oscillation can be restricted to a single one of these longitudinal modes by placing an etalon inside the laser cavity. The etalon should be of such a length l that its free spectral range $c/2l$ is greater than the width of the gain profile, $\Delta \nu$. Its finesse should be high enough so that $c/2lF \leqslant c_0/2L$. The etalon acts as a filter that allows only one longitudinal mode to pass without incurring high loss.

Many lasers can be operated in a *mode-locked* fashion. Then the longitudinal modes of the laser cavity become locked together in phase and the output of the laser becomes a train of uniformly spaced, very narrow pulses. The spacing between these pulses corresponds to the round-trip time in the laser cavity, $c_0/2L$. The temporal width of the pulses is inversely proportional to the gain bandwidth; thus, the broader the gain profile, the narrower the output pulses in mode-locked operation. For example, an argon ion laser with a gain profile 5 GHz wide can yield mode-locked pulses 200 ps wide. A mode-locked dye laser with a linewidth of 100 cm^{-1} (3×10^{12} Hz) can in principle give mode-locked pulses as short as about 0.3 ps.

4.6.5 Optical Modulators

In many optical experiments, particularly those involving the detection of weak light signals or weak electrical signals generated by some light-stimulated phenomenon, the signal-to-noise ratio can be considerably improved by the use of phase-sensitive detection. The principles underlying this technique are discussed in Section 6.8.3. To modulate the intensity of a weak light signal falling on a detector, the signal must be periodically interrupted. This is most easily done with a mechanical chopper. Weak electrical signals that result from optical stimulation of some phenomenon can also be modulated in this way by chopping the radiation from the stimulating source.

Modulation of narrow beams of light such as laser beams, or of extended sources that can be focused onto a small aperture with a lens, is easily accomplished with a tuning-fork chopper. Such devices are available from Bulova. The region chopped can range up to several millimeters wide and a few centimeters long at frequencies from 5 Hz to 3 kHz. For chopping emission from extended sources, or over large apertures, rotating chopping wheels are very convenient. The chopping wheel can be made of any suitable rigid, opaque material and should have radially cut apertures of one of the forms indicated in Figure 4.96. This form of aperture ensures that the mark-to-space ratio of the modulated intensity is independent of where the light passes through the

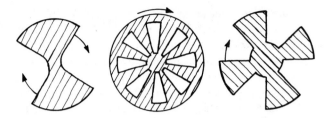

Figure 4.96 Some chopping-wheel designs.

wheels. By cutting very many slots in the wheel, very high modulation rates can be achieved—for example, up to 100 kHz with a 20,000-rpm motor and a wheel with 300 slots. The chopping wheel should usually be painted or anodized matt black. However, for chopping intense laser beams (in excess of perhaps 1 watt), it may be better to make the wheel reflective so that unwanted beam energy can be reflected into a beam dump. Reflective glass chopping wheels can be used at low speeds in experiments where a light beam must be periodically routed along two different paths.

To obtain an electrical reference signal that is synchronous with a mechanical chopping wheel, a small portion of the transmitted beam, if the latter is intense enough, can be reflected onto a photodiode or phototransistor. Alternatively, an auxiliary tungsten filament lamp and photodiode can be mounted on opposite sides of the wheel. If these are mounted on a rotatable arm, the phase of the reference signal they provide can be adjusted mechanically. Chopping wheels with built-in speed control and electrical rotation reference signals are available commercially from Laser Precision, Princeton Applied Research, and Ithaco.

Electro-optic modulators (Pockel's cells) can also be used in special circumstances, particularly for modulating laser beams or when mechanical vibration is undesirable. The operation of these devices is shown schematically in Figure 4.97. Light passing through the electro-optic crystal is plane-polarized before entry and then has its state of polarization altered by an amount that depends on the voltage applied to the crystal. The effect is to modulate the intensity of the light transmitted through a second linear polarizer placed behind the electro-optic crystal. By choosing the correct orientation of the input polarizer relative to the axes of the

electro-optic crystal and omitting the output polarizer, the device becomes an optical phase modulator. For further details of these electro-optic amplitude and phase modulators the reader is referred to the books by Yariv[10] and Kaminow.[81] Electro-optic modulators are available from Cleveland Crystals, Coherent (Modulator Division), Inrad, Lasermetrics, Quantum Technology, and II–VI Inc., among others. Acousto-optic modulators which operate by diffraction effects induced by sound-wave-produced periodic density variations in a crystal are also available.[15,82]

Very many liquids become optically active on application of an electric field; that is, they rotate the plane of polarization of a linearly polarized beam passing through them. This phenomena is the basis of the Kerr cell, which can be used as a modulator but is more commonly used as an optical shutter (for example, in laser Q-switching applications as discussed in Section 4.6.3). A typical Kerr cell uses nitrobenzene placed between two plane electrodes across which a high voltage is applied. This voltage is typically several kilovolts and is sufficient to rotate the plane of polarization of the incident light passing between the plates by 45 or 90°. Kerr cells are available from Korad/Hadron.

4.6.6 How to Work Safely with Light Sources

Light sources, whether coherent or incoherent, can present several potential safety hazards in the laboratory. The primary hazard associated with the use of home-built optical sources is usually their power supply. Lasers and flashlamps, in particular, generally operate with potentially lethal high voltages, and the usual safety considerations for constructing and operating such power supplies should be followed in their design:

1. Provide a good ground connection to the power supply and light-source housing.

2. Screen all areas where high voltages are present.

3. Install a clearly visible indicator that shows when the power supply is activated.

4. All power supplies, particularly those that operate

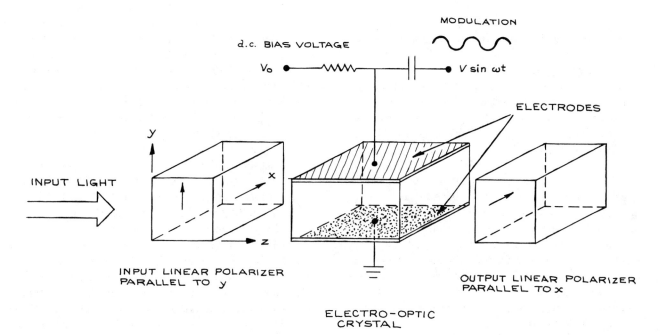

Figure 4.97 An electro-optic amplitude modulator using a transversely operated electro-optic crystal. The d.c. bias voltage can be used to adjust the operating point of the modulator so that, in the absence of modulation, the output intensity is a maximum, a minimum, or some intermediate value.

with pulsed power, can remain dangerous even after the power is turned off, unless energy-storage capacitors are automatically shunted to ground. Always short the capacitors in such a unit to ground after the power supply is turned off before working on the unit.

5. As a rule of thumb, keep a gap of about 1 inch for every 10 kV between high-voltage points and ground.

6. To avoid excessive corona at voltages above about 20 kV, make sure that high-voltage components and connections have no sharp edges. Where such points must be exposed, corona can be reduced by installing a spherical metal corona cap on exposed items such as bolts or capacitor terminals. Resistor and diode stacks can be potted in epoxy or silicone rubber to prevent corona.

7. Sources that are powered inductively or capacitively with r.f. or microwave power can burn fingers that come too close to the power source even without making contact.

Commercial optical sources are generally fairly safe electrically and are likely to be equipped with safety features, such as interlocks, which the experimentalist may not bother to incorporate into home-built equipment. However, *caveat emptor*. Remember the maxim "it's the volts that jolts but it's the mils that kills."

Other general precautions with regard to electric shock include the following:

1. Avoid wearing metallic objects such as watches, watchbands, and rings.

2. If any operations must be performed on a line circuit, wear well-insulated shoes and, if possible, use only one hand.

3. Keep hands dry; do not handle electrical equipment if you are sweating.

4. Learn rescue and resuscitation procedures for victims of electric shock: Turn off the equipment, remove the victim by using insulated material; if the victim is not breathing, start mouth-to-mouth resuscitation; if there is no pulse, begin CPR procedures immediately; summon medical assistance; continue resuscitation procedures until relieved by a physician. If the victim is conscious but continues to show symptoms of shock, keep him warm.

Other hazards in the use of light sources include the possibility of eye damage, direct burning (particularly, by exposure to the beam from a high-average-power laser), and the production of toxic fumes. The last is most important in the use of high-power CW arc lamps, which can generate substantial amounts of ozone. The lamp housing should be suitably ventilated and the ozone discharged into a fume hood or into the open air. Some lasers operate using a supply of toxic gas, such as hydrogen fluoride and the halogens. Workers operating such systems must be sufficiently experienced to work safely with these materials. The *Matheson Gas Data Book*, published by Matheson Gas Products (East Rutherford, N.J.), is a comprehensive guide to laboratory gases, detailing the potential hazards and handling methods appropriate to each.

The potential eye hazard presented by most incoherent sources is not great. If a source appears very bright, one should not look at it, just as one should not look directly at the sun. Do not look at sources that emit substantial ultraviolet radiation; at the least, severe eye irritation will result—imagine having your eyes full of sand particles for several days. Long-term exposure to UV radiation should be kept below 0.5 μW/cm^2. Ordinary eyeglasses will protect the eyes from ultraviolet exposure to some extent, but plastic goggles that wrap around the sides are better. Colored plastic or glass provides better protection than clear material. The manufacturer's specification of ultraviolet transmission should be checked, since radiation below 320 nm must be excluded from the cornea.

The use of lasers in the laboratory presents an optical hazard of a different order. Because a laser beam is generally highly collimated and at least partially coher-

ent, if the beam from a visible or near-infrared laser enters the pupil of the eye it will be focused to a very small spot on the retina (unless the observer is very short-sighted). If this focused spot happens to be on the optic nerve, total blindness may result: if it falls elsewhere on the retina, an extra blind spot may be generated. Although the constant motion of the human eye tends to prevent the focused spot from remaining at a particular point on the retina for very long, always obey the following universal rule: *Never look directly down any laser beam either directly or by specular reflection.* In practice laser beams below 1 mW CW are probably not an eye hazard, but they should still be treated with respect. A beam must be below about 10 μW before most workers would regard it as really safe.

The safe exposure level for CW laser radiation depends on the exposure time. For pulsed lasers, however, it is the maximum energy that can enter the eye without causing damage that is the important parameter. In the spectral region between 380 and 1.5 μm, where the interior material of the eye is transparent, the maximum safe dose is on the order of 10^{-7} J/cm^2 for Q-switched lasers and about 10^{-6} J/cm^2 for non-Q-switched lasers. If any potential for eye exposure to a laser beam or its direct or diffuse reflection exists, it is advisable to carry out the experiment in a lighted laboratory: in a darkened laboratory, the fully dark-adapted human eye with a pupil area of about 0.5 cm^2 presents a much larger target for accidental exposure.

Infrared laser beams beyond about 1.5 μm are not a retinal hazard, as they will not penetrate to the retina. Beams between 1.5 and 3 μm penetrate the interior of the eye to some degree. Lasers beyond 3 μm are a burn hazard, and eye exposure must be avoided for this reason. CO_2 lasers, for example, which operate in the 10-μm region, are less hazardous than equivalent-intensity argon-ion or neodymium lasers. Neodymium lasers represent a particularly severe hazard: they are widely used, emit substantial powers and energies, and operate at the invisible wavelength of 1.06 μm. This wavelength easily penetrates to the retina, and the careless worker can suffer severe eye damage without any warning. The use of safety goggles is strongly recommended when using such lasers. These goggles absorb or reflect 1.06

μm but allow normal transmission in at least part of the visible spectrum. They are available commercially from several sources, such as Fish-Schurman, Glendale Optical, Korad/Hadron, Spectra Optics, and Ultra-Violet Products. The purchaser must usually specify the laser wavelength(s) for which the goggles are to be used.

Unfortunately, one problem with goggles prevents their universal use. Because the goggles prevent the laser wavelength from reaching the eyes, they prevent the alignment of a visible laser beam through an experiment by observation of diffuse reflection of the beam from components in the system. When such a procedure must be carried out, we recommend caution and the operation of the laser at its lowest practical power level during alignment. Frequently a weak, subsidiary alignment laser can be used to check the potential path of a high-power beam before this is turned on—a strongly recommended procedure. For infrared laser beams Optical Engineering manufactures a series of thermal-image plates that allow observation of the beam path. The plates are exposed to a small ultraviolet source and fluoresce; where an infrared laser beam strikes the plate, the fluorescence is partially quenched and a dark spot appears. Used Polaroid film and thermally sensitive duplicating paper are also useful for such infrared-beam tracking.

The beam from an ultraviolet laser can generally be tracked with white paper, which fluoresces where the beam strikes it. A potential burn and fire hazard exists for lasers with power levels above about 1 W/cm^2, although the fire hazard will depend on the target. Black paper burns most easily. Very intense beams can be safely dumped onto pieces of firebrick. Pulsed lasers will burn at output energies above about 1 J/cm^2.

In the United States, commercial lasers are assigned a rating by the Bureau of Radiological Health, which identifies their type, power output range, and potential hazard. A convenient brief discussion of these safety standards has been given by Franks and Sliney in the 1981 LFBG.[15] It must be stressed, however, that lasers are easy to use safely: accidents of any sort have been rare. Their increasing use in laboratories requires that workers be well-informed of the standard safety practices. For further discussion of this and all aspects of laser safety and related topics, we recommend both the book by Sliney and Wolbarsht[83] and a series of articles on laser safety in the CRC *Handbook of Laser Science and Technology*, Vol. 1.[84]

4.7 OPTICAL DISPERSING INSTRUMENTS

Optical dispersing instruments allow the spectral analysis of optical radiation or the extraction of radiation in a narrow spectral band from some broader spectral region. In this general category we include interference filters, prism and grating monochromators, spectrographs and spectrophotometers, and interferometers. Interferometers have additional uses over and above direct spectral analysis—including studies of the phase variation over an optical wavefront, which allow the optical quality of optical components to be measured. Spectrometers, or *monochromators* as they are generally called, are optical filters of tunable center wavelength and bandwidth. The output narrow-band radiation from these devices is generally detected by a photon or thermal detector, which generates an electrical signal output. *Spectrographs*, on the other hand, record the entire spectral content of an extended bandwidth region photographically. *Spectrophotometers* are complete commercial instruments, which generally incorporate a source or sources, a dispersing system (which may involve interchangeable prisms and/or gratings), and a detector. They are designed for recording ultraviolet, visible, or infrared, absorption and emission spectra.

Before giving a discussion of important design considerations in the construction of various spectrometers and interferometers, it is worthwhile mentioning two important figures of merit that allow the evaluation and comparison of the performance of different types of optical dispersing instruments. The first is the resolving power, $\mathcal{R} = \lambda/\Delta\lambda$, which has already been discussed in connection with diffraction gratings (Section 4.3.5). The second is the *luminosity*, which is the flux collected by a detector at the output of a spectrometer when the

Figure 4.98   Basic elements of a prism monochromator.

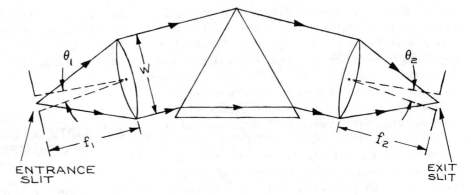

source at the input has a radiance of unity. The interrelation between resolving power and luminosity for spectrometers using prisms, gratings, and Fabry-Perot etalons has been dealt with in detail by Jacquinot.[85]

Figure 4.98, which shows the essential components of a prism monochromator, will serve to illustrate the points made here. High resolution is clearly obtained in this arrangement by using narrow entrance and exit slits. The maximum light throughput of the spectrometer results when the respective angular widths W_1, W_2 of the input and output slits θ_1, θ_2 satisfy

$$W_1 = W_2, \tag{150}$$

where

$$W_1 = \frac{\theta_1}{(d\alpha/d\lambda)_\delta}, \qquad W_2 = \frac{\theta_2}{(d\delta/d\lambda)_\alpha}. \tag{151}$$

The angles α, δ are the same as those used in Figure 4.28. The input and output dispersions $(d\alpha/d\lambda)_\delta$ and $(d\delta/d\lambda)_\alpha$ are only equal in a position of minimum deviation, or with a prism (or grating) used in a Littrow arrangement. If diffraction at the slits is negligible, the intensity distribution at the output slit when a monochromatic input is used is a triangular function as shown in Figure 4.99, where W is the spectral width of the slits given by

$$W = W_1 = W_2. \tag{152}$$

The limit of resolution of the monochromator is

$$\delta\lambda = W = \frac{\theta_2}{(d\delta/d\lambda)_\alpha}. \tag{153}$$

The maximum flux passing through the output slit is

$$\Phi = TE_e(\lambda)Sl\theta_2/f_2, \tag{154}$$

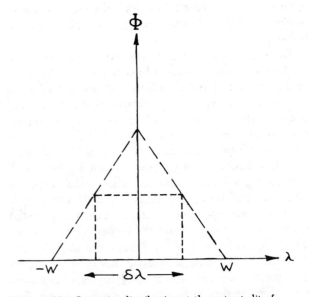

Figure 4.99   Intensity distribution at the output slit of a monochromator whose entrance and exit slits have the same spectral width W.

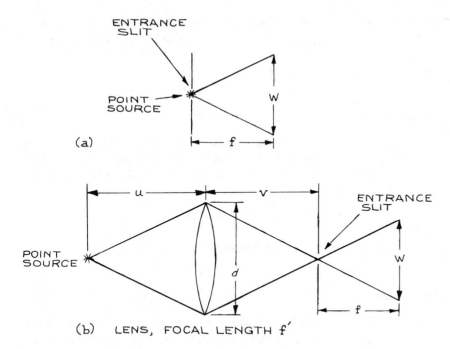

(a)

(b) LENS, FOCAL LENGTH f'

Figure 4.100 Schematic diagrams illustrating geometrical factors involved in optimizing light-collection efficiency by a spectrometer of aperture W: (a) point source directly at entrance slit; (b) remote point source imaged on entrance slit with a lens.

where S is the normal area of the output beam, T is the transmittance of the prism (or efficiency of the grating in the order used) at the wavelength being considered, l is the height of the entrance and exit slits (equal), and $E_e(\lambda)$ is the radiance of the monochromatic source of wavelength λ illuminating the entrance slit. $l\theta_2/f_2$ is the solid angle that the exit slit subtends at the output focusing lens. Equation (154) can be written as

$$\Phi = \frac{TE_e(\lambda)Sl\lambda(d\delta/d\lambda)_\alpha}{f_2\mathcal{R}}, \qquad (155)$$

which clearly shows that the output flux is inversely proportional to the resolving power [for a continuous source at the entrance slit, Equation (155) is multiplied by an additional factor $\lambda/\mathcal{R}$]. The efficiency E of the monochromator is defined by the equation

$$E = \frac{TSl}{f}\left(\frac{d\delta}{d\lambda}\right)_\alpha, \qquad (156)$$

where we have assumed the usual case in which $f_1 = f_2 = f$.

To collect all the light from the collimating optics in the arrangement of Figure 4.98, the prism or grating should have a normal area at least as large as the aperture of the collimating optics. Thus, the height of the prism (or diffraction grating) should be $h \geqslant W$. The ratio f/W, which is a measure of the light gathering power of the monochromator, is called the f/number. The most efficient way to use a spectrometer of any kind is to send the input radiation into the entrance slit within the cone of angles defined by the f/number. The maximum resolution is obtained if the input f/number matches the spectrometer f/number. If radiation enters the spectrometer with an f/number that is too small, this radiation overfills the dispersing element and some is wasted. This can be illustrated with the aid of Figure 4.100, which shows the use of a point source directly at the entrance slit. In Figure 4.100(b) a point source is imaged on the entrance slit using a lens that matches the input radiation to the f/number of the spectrometer. In Figure 4.100(a) the fractional useful light collection from the source is

$$\phi_1 \simeq \frac{\pi W^2}{4f^2} = \frac{\pi}{4F^2}, \tag{157}$$

where F is the f/number of the spectrometer. In contrast, in Figure 4.100(b) it is approximately

$$\phi_2 \simeq \frac{\pi d^2}{4u^2}, \tag{158}$$

which can be written as

$$\phi_2 = \frac{\pi d^2 (v-f)^2}{4f^2 v^2}, \tag{159}$$

and substituting $v/d = F$, $f/d = F'$ (the f/number of the lens), we have

$$\phi_2 = \frac{\pi (F/F' - 1)^2}{4F^2} \quad (F' \leqslant F). \tag{160}$$

Thus, equally efficient light gathering to that which would be obtained with the point source at the entrance slit results if

$$F' = F/2. \tag{161}$$

More efficient light gathering results if F' is less than this value. However, if the lens and point source are incorrectly positioned, so that $v/d < F$, then not all the light collected by the lens reaches the dispersing element of the spectrometer.

4.7.1   Comparison of Prism and Grating Spectrometers

Following Jacquinot,[85] we use Equation (155) to compare prism and grating instruments. Assume identical values of l and f_2, since for a given degree of aberration of the system their values are identical for prisms and gratings. Blazed-grating efficiencies can be high, so T is assumed to be similar for prism and grating. Thus, the comparison of luminosity under given operating conditions of wavelength and resolution depends solely on the quantity $S(d\delta/d\lambda)_\alpha$ for a prism and $S(d\beta/d\lambda)_\alpha$ for a grating.

For the prism,

$$\frac{d\delta}{d\lambda} = \frac{t}{W}\frac{dn}{d\lambda}, \tag{162}$$

where for maximum dispersion the whole prism face is illuminated, so that t is the width of the prism base. $S = hW$ and $th = A$, the area of the prism base, and so

$$S\frac{d\delta}{d\lambda} = A\frac{dn}{d\lambda}. \tag{163}$$

For the grating,

$$\frac{d\beta}{d\lambda} = \frac{m}{d\cos\beta}. \tag{164}$$

$S = A\cos\beta$, where A is the area of the grating. If it is assumed that the grating is used in Littrow, then $m\lambda = 2d\sin\beta$, and therefore

$$S\frac{d\beta}{d\lambda} = \frac{2A\sin\beta}{\lambda}. \tag{165}$$

The ratio of luminosities for a prism and grating of comparable size is therefore

$$\rho = \frac{\Phi(\text{prism})}{\Phi(\text{grating})} = \frac{\lambda\, dn/d\lambda}{2\sin\beta}. \tag{166}$$

This ratio can be improved by a factor of 2 if the prism is also used in Littrow. For a typical value of 30° for β, Equation (166) predicts that a grating is always superior to a prism instrument. Except for a few materials in restricted regions of wavelength where $dn/d\lambda$ becomes large and ρ may reach 0.2–0.3, a grating is more luminous than a prism by a factor of 10 or more. The only potential advantage of a prism instrument over a grating is the absence of overlapping orders. A grating instrument illuminated simultaneously with 300 nm and 600 nm, for example, will transmit both at the same angular position; a prism instrument will not. However, this minor problem is easily solved with an appropriate order-sorting color or interference filter.

Table 4.6   MAIN CHARACTERISTICS OF OPTICAL DISPERSING ELEMENTS

Element	Advantages	Disadvantages
Multilayer dielectric interference filter	High throughput at center frequency, $\simeq 50\%$. Tunable to some extent by tilting.	Low resolution—minimum bandwidth $\simeq 1$ nm. None available at short wavelengths ≤ 200 nm.
Prism spectrometer	Dispersion and resolution increase near absorption edge of prism (however, transmission falls). No ghosts. Scattered light can be lower than in grating spectrometer. No order sorting necessary.	Cannot be used below 120 nm (with LiF prism). Resolution not as good as grating spectrometer, best $\simeq 0.01$ nm. Resolution particularly poor in infrared.
Grating spectrometer	Resolution can be high (≥ 0.001 nm). Can be used from vacuum UV to far IR. Dispersion independent of wavelength. Luminosity depends on blaze angle, but generally much higher than for prism spectrometer.	Ghosts can be a problem. Expensive if really low scattered light required—may need double- or triple-grating instrument. Need to remove unwanted orders (not necessarily a serious problem).
Fabry-Perot interferometer	Very high light throughput —much higher than for grating spectrometer. Very high resolution—tens of MHz at optical frequencies ($\geq 10^{-5}$ nm).	Many orders generally transmitted simultaneously. Cannot be used at short wavelengths (≤ 200 nm). Transmission maximum can be scanned only over narrow range. Most difficult dispersing element to use.

Jacquinot[85] has demonstrated that a Fabry-Perot etalon or interferometer (see Section 4.7.4) used at a high or moderate resolving power is superior in luminosity to a grating instrument by a factor that can range from 30 to 400 or more. In general, an etalon cannot be used alone, because of its many potential overlapping orders. It is usual to couple it with a grating or prism monochromator in high-resolution spectroscopic applications, except in those cases where the source is already highly monochromatic.

Table 4.6 gives a comparison of the performance characteristics of prism and grating monochromators and Fabry-Perot interferometers. From the vacuum ultraviolet (prism instruments cannot be used below about 1200 Å) to the far infrared, grating instruments are much superior to comparable-size prism instruments in both resolving power and luminosity. Apart from the minor inconvenience of potential overlapping orders, grating instruments are almost always to be preferred over prism instruments. There is one exception worth noting: when a monochromator is being used to observe some weak optical emission in the simultaneous presence of a strong laser beam, unless the optical emission is too close in wavelength to the laser for prism resolution to be adequate. In this application, a prism avoids the problems of ghost emission or grating scattering. A good prism has very low scattered light. It is common to predisperse a light beam with a prism before sending the beam into the entrance slit of a monochromator, which avoids both the overlapping-order problem and spurious signals from a very strong light signal at another wavelength. Fabry-Perot interferometers have very high resolution and luminosity, but require careful and regular adjustment to maintain high performance. They are used only where their very high resolution is essential, frequently in conjunction with a grating monochromator for preisolation of a narrow spectral region.

4.7.2   Design of Spectrometers and Spectrographs

The construction of monochromators and spectrographs is a specialized undertaking. The availability of good commercial instruments generally makes it unnecessary for anyone but the enthusiast to contemplate their construction in the laboratory. However, the principles that underlie good design in a dispersing instrument are not complex. Some essential features of such instruments can be illustrated with reference to particular designs that are used both in laboratory and commercial instruments. Attention will be restricted to monochromators; spectrographs are similar, apart from having a photographic plate instead of an exit slit, as shown in Figure 4.101. Spectrographs are also simpler in construction because wavelength scanning is not required

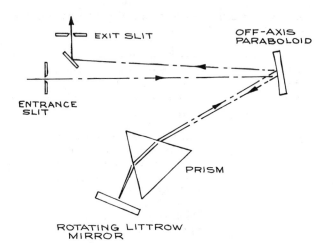

Figure 4.102   Littrow monochromator. (From R. J. Meltzer, "Spectrographs and Monochromators," in *Applied Optics and Optical Engineering*, Vol. 5, R. Kingslake, Ed., Academic Press, New York, 1969; by permission of Academic Press.)

and all optical components of the system remain fixed. The discussion here should also allow intelligent evaluation and comparison of commercial instruments. For further details of the many different instrument designs that have been used for the construction of monochromators, spectrographs and spectrophotometers, References 86 and 87 should be consulted.

Figure 4.102 shows a Littrow prism monochromator. The light from the input slit is collimated with an off-axis paraboloidal mirror. The output wavelength is changed by rotating the Littrow mirror. If the output wavelength must change uniformly with rotation of a drive shaft, then the rotation of the mirror requires a cam arrangement. Figure 4.103 shows a Czerny-Turner prism arrangement. The collimating mirrors are spherical. In the arrangement of Figure 4.103(*a*) the coma of one mirror compensates that of the other, while in the arrangement of Figure 4.103(*b*) these comas are additive. The wavelength is changed by rotating the prism.

In any of these spectroscopic instruments the optical arrangement should be enclosed in a light-tight enclosure painted on its interior with black nonreflective paint such as Nextel.[88] This helps to minimize scattered light. It is essential that light passing through the entrance slit not be able to reach the exit slit without

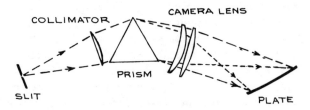

Figure 4.101   Schematic diagram of a simple prism spectrograph. (From R. J. Meltzer, "Spectrographs and Monochromators," in *Applied Optics and Optical Engineering*, Vol. 5, R. Kingslake, Ed., Academic Press, New York, 1969; by permission of Academic Press.)

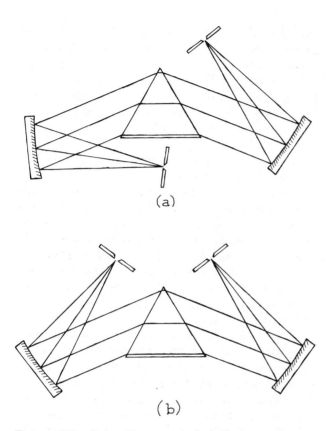

Figure 4.103 Czerny-Turner monochromator arrangements: (*a*) coma of one mirror compensates that of the other; (*b*) comas are additive.

going through the illuminating optics and prism. To this end it is common practice to incorporate internal black-painted baffles in the monochromator enclosure. The most likely source of scattered light is overillumination of the input collimator: input radiation should not enter with an *f*/number smaller than that of the monochromator. To check for stray light, visual observation through the exit slit in a darkened room when the entrance slit is illuminated with an intense source will reveal where additional internal baffles would be helpful. Stray light is more likely to be a problem in a Littrow arrangement than in a straight-through type such as the Czerny-Turner.

In the visible and near infrared, lens collimating optics can be used, as, for example, in the design shown

in Figure 4.101. Good lenses, such as camera lenses, should be used. Mirror collimating optics are useful over a much broader wavelength region. For ultraviolet use they should be aluminum overcoated with MgF_2, and they are frequently of this type even in instruments designed for longer wavelengths.

Slit design is important in achieving high resolution. Entrance and exit slits of equal width are generally used. If the source is small, having a height much less than the height of the prism or grating, parallel-sided slits are adequate. Adjustable versions are available from Ealing and Melles Griot. Fixed slits can be easily made using razor blades. If the source is high compared to the height of the dispersing element, a straight entrance slit produces a curved image at the exit slit. In the case of a prism, this happens because light passing through the prism at an angle inclined to the meridian plane suffers slightly more deviation than light in the meridian plane. Light striking a grating in a plane that is not perpendicular to the grating grooves sees an apparently smaller groove spacing, and is consequently deviated more than light in the plane perpendicular to the grating. To maintain high resolution in the presence of image curvature, curved entrance and exit slits should be used, each with a radius equal to the distance of the slit from the axis of the system. This slit arrangement is shown in Figure 4.104, which also shows a very good, and widely used, grating monochromator design due to Fastie[89] (based on a design originally described by Ebert[90] for use with a prism). It is also quite common for the single spherical mirror in Figure 4.104 to be replaced by a pair, in which case the design becomes the Czerny-Turner one. In either case the wavelength is tuned by rotating the grating. The wavelength variation at the exit slit is much more nearly linear with grating rotation than with prism rotation. If linearity is required, a cam, sine-drive, or other such arrangement is used for rotating the grating or prism.[87]

Grating monochromators for use in the vacuum ultraviolet generally have a concave grating. Such a grating, which has equally spaced grooves on the chord of a spherical mirror, combines both dispersion and focusing in one element. For maximum resolution both the object and image formed by a concave grating must be on a circle called the *Rowland circle*, whose diameter is the

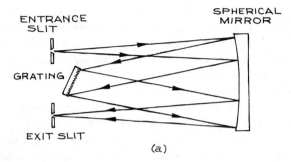

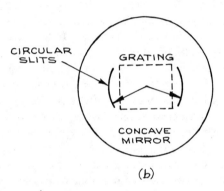

Figure 4.104 (*a*) Ebert-Fastie monochromator; (*b*) curved entrance- and exit-slit configuration for use with the above.

principal radius of the spherical surface on which the grating is ruled, as shown in Figure 4.105. The use of a concave-grating monochromator (or spectrograph) entails mounting grating and slits (or grating, slit, and photographic plate) on a Rowland circle. Adjustment of the instrument will involve moving the grating, slits, or plate on the circle. Various mounting arrangements—such as the Rowland, Abney, Paschen, Eagle, Wadsworth, and grazing-incidence mounting—have been used to accomplish this.

A convenient concave-grating design widely used in vacuum-ultraviolet monochromators is the Seya-Namioka,[91,92] shown in Figure 4.106. In this arrangement, by judicious choice of slit distances from the grating and slit angular separation, the only adjustment needed for wavelength adjustment is rotation of the grating. In comparison with an ideal Rowland-circle mounting, only a small resolution loss results over a wide wavelength range.

Vacuum-ultraviolet instruments possess the added complication that all interior adjustments of the instrument must be transmitted through the wall of the vacuum chamber that houses the optical arrangement. Various kinds of rotary-shaft and bellows seals and magnetically coupled drives exist for this purpose.[93]

Multiple monochromators use multiple dispersing ele-

Figure 4.105 Rowland circle of a concave diffraction grating. For maximum resolution, both object and image must be on the Rowland circle. R = object (entrance-slit) distance; R_2 = image (exit-slit) distance. (From R. J. Meltzer, "Spectrographs and Monochromators," in *Applied Optics and Optical Engineering*, Vol. 5, R. Kingslake, Ed., Academic Press, New York, 1969; by permission of Academic Press.)

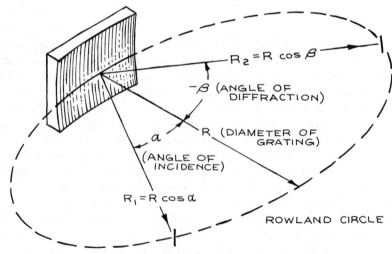

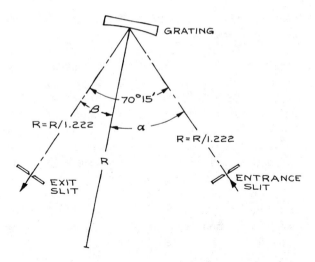

Figure 4.106   Layout of the Seya-Namioka monochromator.

ments in series to obtain greater dispersion and/or lower scattered light. For example, in a double-grating monochromator two diffraction gratings move simultaneously; the desired diffracted wavelength from the first grating is diffracted once again from the second grating before being focused on the exit slit. The considerable reduction in scattered light that can be obtained with multiple monochromators makes them particularly useful for analyzing weak emissions close in wavelength to a strong emission, as in Raman spectroscopy.

Prism and grating monochromators and spectrographs are available from several suppliers, such as Bausch and Lomb, Cary Instruments, Gaertner Scientific, Rank-Hilger (formerly Hilger and Watts), Jarrell-Ash, Jobin-Yvon, McPherson Instruments, Minuteman Laboratories, Pye-Unicam, and Spex. Jarrell-Ash and Rank-Hilger are noted for their large high-resolution spectrographs; Jarrell-Ash, Jobin-Yvon, Minuteman Laboratories, and Spex for their compact grating monochromators; and McPherson for vacuum-ultraviolet instruments. Cary and Spex are noted for their double and triple monochromators. Complete spectrophotometer systems are available from several companies, such as Baird-Atomic, Beckman, Cary, Rank-Hilger, and Pye-Unicam.

Optical multichannel analyzers (OMA) are a new type of very useful spectroscopic tool. They are essentially spectrographs where, instead of a photographic plate, an array of photodetector elements, or a photoelectric imaging tube such as a vidicon, is used to allow the simultaneous photoelectric recording in real time of a whole spectral region. These instruments are available from Princeton Applied Research and Tracor.

4.7.3 Calibration of Spectrometers and Spectrographs

To calibrate the wavelength scale of a spectrometer or spectrograph, the entrance slit is illuminated with a reference source that has characteristic spectral features of accurately known wavelength. It is quite common to mount a small reference lamp—mercury, neon, and hollow-cathode iron lamps are the most popular—near the entrance slit and use a small removable mirror or beamsplitter to superimpose the spectrum of the reference on the unknown spectrum under study. To calibrate a grating instrument through its various orders, a helium-neon laser is ideal. If the slits are set to their narrowest position and a low-power (≤ 1 mW) He-Ne laser is directed at the entrance slit, the various grating settings that transmit a signal (which can be safely observed by eye at the exit slit) correspond to the wavelengths 0, 632.8 nm, 632.8×2 nm, 632.8×3 nm, and so on. By interpolation between the observed values a reliable calibration curve can be obtained even for instruments operating far into the infrared.

The efficiency of the instrument at a given wavelength can be determined with any suitable detector. A laser or other narrow-band source at the desired wavelength illuminates the entrance slit, or a duplicate of it, and the transmitted power is measured with the detector placed directly behind the slit. A similar measurement is then made at the exit slit of the whole instrument. The relative spectral efficiency can be determined by illuminating the entrance slit with a blackbody source. A lens should not be used for focusing light on the entrance slit, as its collection and focusing efficiency will vary with wavelength. The signal at the output slit is then measured as a function of wavelength with a spectrally flat detector, such as a pyroelectric detector, thermopile, or Golay cell (see Section 4.8.5), and normalized with respect to the blackbody distribution.

4.7.4 Fabry-Perot Interferometers and Etalons

Fabry-Perot interferometers and etalons are optical filters that operate by multiple-beam interference of light reflected and transmitted by a pair of parallel flat or coaxial spherical reflecting interfaces. In its simplest form, an *etalon* consists of a flat, parallel-sided slab of transparent material, which may or may not have semi-reflecting coatings on each face. An etalon may also consist of a pair of parallel, air-spaced flat mirrors of fixed spacing. If the reflective surfaces have adjustable spacing, or if their effective optical spacing can be adjusted by changing the gas pressure between the surfaces, the device is called a *Fabry-Perot interferometer*. These devices can be analyzed by the impedance concepts discussed in Section 4.2.3, but it is instructive

to discuss their operation from the standpoint of multiple-beam interference.

Figure 4.107 shows the successive reflected and transmitted field amplitudes of a plane electromagnetic wave striking a plane-parallel Fabry-Perot etalon or interferometer at angle of incidence θ'. Although the reflection coefficients at the two reflecting interfaces may be different, only the case where they are equal will be analyzed. Almost all practical devices are built this way. The optical-path difference between successive transmitted waves is $2nl\cos\theta$, where l is the interface spacing and n is the refractive index of the medium between these interfaces (air, glass, quartz, or sapphire, for example). If refraction occurs at the reflecting interfaces, θ and θ' will be related by Snell's law. The phase difference between successive transmitted waves is

$$\delta = 2kl\cos\theta + 2\epsilon, \tag{167}$$

where

$$k = \frac{2\pi}{\lambda} = \frac{2\pi n}{\lambda_0} = \frac{2\pi n\nu}{c_0},$$

and ϵ is the phase change (if any) on reflection.

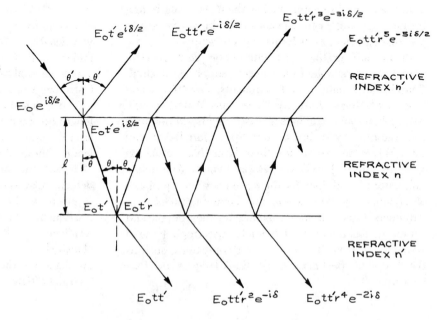

Figure 4.107 Paths of transmitted and reflected rays in a Fabry-Perot etalon. The complex amplitude of the fields associated with the rays at various points is shown.

The total resultant transmitted complex amplitude is

$$E_T = E_0 tt' = E_0 tt' r^2 e^{-i\delta} + E_0 tt' r^4 e^{-2i\delta} + \cdots$$

$$= \frac{E_0 tt'}{1 - r^2 e^{-i\delta}}. \tag{168}$$

Here r and t are the reflection and transmission coefficients of the waves passing from the medium of refractive index n to that of refractive index n' at each interface, r' and t' are the corresponding coefficients for passage from n' to n, and E_0 is the amplitude of the incident electric field.

The total transmitted intensity is

$$I_T = \frac{|E_T|^2}{2Z} = \frac{I_0 |tt'|^2}{|1 - r^2 e^{-i\delta}|^2}. \tag{169}$$

Here $|tt'| = T$ and $|r|^2 = |r'|^2 = R$, where T and R are the transmittance and reflectance of each interface.

If there is no energy lost in the reflection process,

$$T = 1 - R. \tag{170}$$

If the etalon had slightly absorbing reflective layers, this would no longer be true; if each layer absorbed a fraction A of the incident energy, then

$$T = 1 - R - A. \tag{171}$$

If $A = 0$, Equation (169) gives

$$\frac{I_T}{I_0} = \frac{1}{1 + \dfrac{4R}{(1-R)^2} \sin^2 \dfrac{\delta}{2}}. \tag{172}$$

This variation of transmitted intensity with δ is called an *Airy function* and is illustrated in Figure 4.108 for several values of R. The variation of reflected intensity with δ is just the inverse of Figure 4.108, since

$$\frac{I_R}{I_0} = 1 - \frac{I_T}{I_0}.$$

If $A \neq 0$, Equation (172) becomes

$$\frac{I_T}{I_0} = \frac{\left(\dfrac{T}{1-R}\right)^2}{1 + \dfrac{4R}{(1-R)^2} \sin^2 \dfrac{\delta}{2}}. \tag{173}$$

In either case, maximum transmitted intensity results when

$$\delta = 2m\pi, \tag{174}$$

where m is an integer. If the phase change on reflection is neglected, this reduces to

$$2l \cos\theta = m\lambda, \tag{175}$$

where λ is the wavelength of the light in the medium between the two reflecting surfaces (plates). In normal incidence, transmission maxima result when

$$l = m\lambda/2. \tag{176}$$

Even if the phase change on reflection is included, the adjustment in spacing necessary to go from one transmission maximum to the next is one half wavelength.

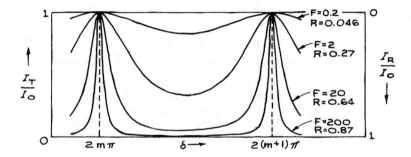

Figure 4.108 Transmission and reflection characteristics of a Fabry-Perot etalon. (From M. Born and E. Wolf, *Principles of Optics*, 3rd edition, Pergamon Press, Oxford, 1965; by permission of Professor Emil Wolf.)

Thus, for example, the transmitted intensity as a function of plate separation when a Fabry-Perot interferometer is illuminated normally with plane monochromatic light is also as shown in Figure 4.108. In an ideal Fabry-Perot device, the overall transmittance at a transmission-intensity maximum is unity, whereas in a practical device, which will invariably have some losses, the maximum overall transmittance is reduced by a factor $[T/(1-R)]^2$. As is clear from Figure 4.108, the transmission peaks of the device become very sharp as R approaches unity.

(a) Free Spectral Range, Finesse, and Resolving Power. If a particular transmission maximum for fixed l occurs for normally incident light of frequency ν_0, then

$$\nu_0 = \frac{mc}{2l}, \qquad (177)$$

where c is the velocity of light in the material between the plates. Of course, the device will also show transmission maxima at all frequencies $\nu_0 \pm pc/2l$ as well, where p is an integer. The frequency between successive transmission maxima is called the *free spectral range*:

$$\Delta\nu = \frac{c}{2l}. \qquad (178)$$

When R is close to unity, all phase angles δ within a transmission maximum differ from the value $2m\pi$ by only a small angle. Thus, we can write

$$\delta = \frac{4\pi\nu l}{c} = \frac{4\pi\nu_0 l}{c} + \frac{4\pi(\nu-\nu_0)l}{c}, \qquad (179)$$

which can be written in the form

$$\delta = 2m\pi + \frac{2\pi(\nu-\nu_0)}{\Delta\nu}. \qquad (180)$$

Equation (173) becomes

$$\frac{I_T}{I_0} = \frac{1}{1 + \dfrac{4R\pi^2}{(1-R)^2}\dfrac{(\nu-\nu_0)^2}{\Delta\nu}}. \qquad (181)$$

Writing $\pi\sqrt{R}/(1-R) = F$, where F is called the *finesse*, the shape of a narrow transmission maximum can be written as

$$\frac{I_T}{I_0} = \frac{1}{1 + \left[2(\nu-\nu_0)/\Delta\nu_{1/2}\right]^2}. \qquad (182)$$

Here $\Delta\nu_{1/2}$, the full width at half maximum transmission of the transmission peak, is given by

$$\Delta\nu_{1/2} = \Delta\nu/F. \qquad (183)$$

The *resolving power* of a Fabry-Perot device is a measure of its ability to distinguish between two closely spaced monochromatic signals. A good criterion for determining this is the *Rayleigh criterion*, which recognizes two closely spaced lines as distinguishable if their half-intensity points on opposite sides of the two lineshapes are coincident, as shown in Figure 4.109. Thus, for a Fabry-Perot device the resolving power is

$$\mathcal{R} = \frac{\nu}{\Delta\nu_{1/2}} = \frac{F\nu}{\Delta\nu}. \qquad (184)$$

It would appear that very high resolving powers can be obtained by the use of high-reflectance plates (and consequent high finesse values, as illustrated in Table

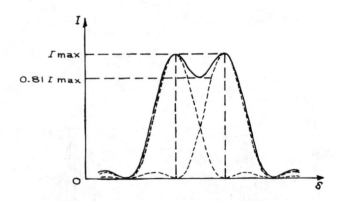

Figure 4.109 Two monochromatic spectral lines just resolved according to the Rayleigh criterion. (From M. Born and E. Wolf, *Principles of Optics*, 3rd edition, Pergamon Press, Oxford 1965; by permission of Professor Emil Wolf.)

Table 4.7   PROPERTIES OF FABRY-PEROT ETALONS

Plate Separation l (cm)	Reflectance R (%)	Free Spectral Range $\Delta\nu = c/2Z$ (Hz)	Finesse $F = \pi\sqrt{R}/(1-R)$	Resolving Power $\mathcal{R} = F\nu/\Delta\nu$
0.1	80	1.5×10^{11}	14	5.6×10^4
0.1	90	1.5×10^{11}	30	1.2×10^5
0.1	95	1.5×10^{11}	61	2.44×10^5
0.1	99	1.5×10^{11}	313	1.25×10^6
1.0	80	1.5×10^{10}	14	5.6×10^5
1.0	90	1.5×10^{10}	30	1.2×10^6
1.0	95	1.5×10^{10}	61	2.44×10^6
1.0	99	1.5×10^{10}	313	1.25×10^7
10.0	80	1.5×10^9	14	5.6×10^6
10.0	90	1.5×10^9	30	1.2×10^7
10.0	95	1.5×10^9	61	2.44×10^7
10.0	99	1.5×10^9	313	1.25×10^8

Note: Properties shown are at 500 nm (600 THz).

4.7) and/or large plate spacings (and consequent small free spectral ranges). However, additional practical considerations limit how far these approaches to high resolution can actually be used.

(b) Practical Operating Configurations and Performance Limitations of Fabry-Perot Systems.   The use of very high reflectance values to obtain improved resolving power is limited by the ability of the optical polisher to achieve ultraflat optical surfaces. Very good-quality plates for use in the visible can be obtained with flatnesses of $\lambda/200$ over regions a few centimeters in diameter. Some claims of flatness in excess of $\lambda/200$ are also seen. These should be viewed with skepticism—such a degree of flatness is extremely difficult to verify. If the plates have surface roughness Δs, the average error in plate spacing is $\sqrt{2}\,\Delta s$. This gives rise to a spread in the frequency of transmission maxima equal to

$$\Delta\nu_s \simeq \frac{mc\,\Delta s}{\sqrt{2}\,l^2} \simeq \sqrt{2}\,\nu_0\frac{\Delta s}{l}. \qquad (185)$$

Thus the flatness-limited finesse is

$$F_s = \frac{\Delta\nu}{\Delta\nu_s} = \frac{1}{\sqrt{2}}\left(\frac{\Delta\nu}{\nu_0}\right)\frac{l}{\Delta s} = \frac{\lambda}{2\sqrt{2}\,\Delta s}. \qquad (186)$$

For example, two plates with surface roughness of $\lambda/200$ would have a flatness-limited finesse of about 71. In practice Fabry-Perot plates may not be randomly rough, and their stated flatness figure may represent the average deviation from parallelism of the two plates. In that case, the parallelism-limited finesse is

$$F_\rho = \frac{\lambda}{2\,\Delta s}, \qquad (187)$$

where Δs is the deviation from parallelism over the used aperture of the system. Thus, for example, with a 1-cm aperture, a $0.01''$ deviation from parallelism would imply $\Delta s = 4.84$ nm and a parallelism-limited finesse of only 56 at 546 nm. Consequently, Fabry-Perot plates must be held very nearly parallel to achieve high resolution. Also, their separation should not be allowed to drift. A random variation of Δs will limit the finesse as predicted by Equation (187). A fractional length change of 10^{-6} in a 1-cm-spacing Fabry-Perot device will limit the finesse to only about 27. Thus it must be isolated from vibrational interference to achieve high finesse. Further, if the frequencies of the transmission maxima are not to drift, the change in plate spacing due to thermal expansion must be minimized by constructing the device of low-expansion materials such as Invar,[94] Super-Invar,[95] fused silica,[96] or special ceramics.[97]

Figure 4.110 (*a*) Operating scheme for a Fabry-Perot interferometer, using collimated light; (*b*) detail of typical plate configuration (the wedge angle of the reflecting plates is exaggerated).

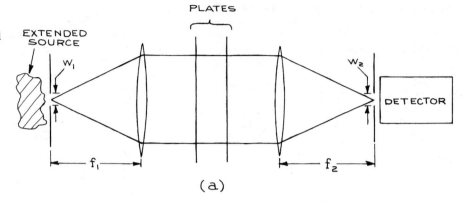

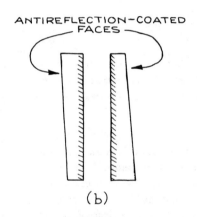

The resolving power of a Fabry-Perot system is also limited by the range of angles that can be transmitted through the instrument. One popular configuration for using it is shown in Figure 4.110. A small source, or light from an extended source that passes through a small aperture, is collimated by a lens, passes through the interferometer, and is focused onto a second small aperture in front of a detector. The maximum angular spread of rays passing through the system will be governed by the aperture sizes W_1 and W_2 and the two focal lengths f_1 and f_2. In the paraxial-ray approximation, the angular width $2\,\Delta\theta$ of the paraxial ray bundle that comes from, or is delivered to, an aperture of diameter W is equal to the angular width $2\,\Delta\theta'$ that traverses the system, as illustrated in Figure 4.111. In

this case,

$$\Delta\theta = W/2f. \qquad (188)$$

In Figure 4.110 the resolving power will be limited by W_1/f_1 or W_2/f_2, whichever is the larger. The transmitted frequency spread associated with an angular spread $\Delta\theta$ around the normal direction is, from Equations (175) and (177),

$$\Delta\nu_{1/2} = \nu_0(\Delta\theta)^2/2. \qquad (189)$$

The corresponding aperture-limited finesse is

$$F_A = \frac{\lambda}{l(\Delta\theta)^2} = \frac{4\lambda f^2}{lW^2}. \qquad (190)$$

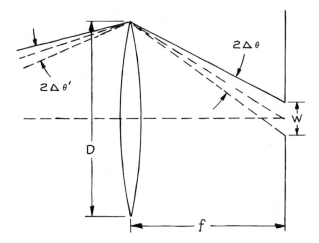

Figure 4.111 Angular factors involved in the collection of light transmitted through a Fabry-Perot interferometer by a circular aperture. In the paraxial-ray approximation, $\Delta\theta' = \Delta\theta$.

For example, using a 1-mm-diameter aperture, a 10-cm-focal-length lens, and a 1-cm air-spaced Fabry-Perot device at 546.1 nm, the aperture-limited finesse is 218. The operating conditions of a Fabry-Perot system should be arranged so that F_A is at least three times larger than the desired operating finesse, which usually is ultimately limited by the flatness-limited finesse. The apertures in Figure 4.110 should be circular and accurately coaxial with both lenses, or the aperture finesse will be further reduced. If the limiting aperture (either a lens or the Fabry-Perot device itself) is D, there is also a diffraction limit to the finesse, which on axis is

$$F_D = \frac{2D^2}{\lambda nl}. \tag{191}$$

At the pth fringe off axis,

$$F_D = \frac{D^2}{2p\lambda nl}, \tag{192}$$

where n is the refractive index of the material between the plates. The overall finesse of a Fabry-Perot system, F_I, is related to the individual contributions to the finesse, F_i, by

$$F_I^{-2} = \sum_i F_i^{-2}. \tag{193}$$

In the operating configuration of Figure 4.110, if the plate spacing is fixed, then with a broad-band source, all frequencies that satisfy Equation (177) will be transmitted. These frequencies become more closely spaced as the plate separation l is increased. There is a concomitant increase in resolution, but this may not prove useful if any of the spectral features under study are broader than the free spectral range. In this case simultaneous transmission of two broadened spectral features always occurs. For example, for two lines of wavelength λ_1 and λ_2 and of FWHM $\Delta\lambda$, it is usually possible to find two integers m_1 and m_2 such that

$$\begin{aligned} m_1\lambda_1' = 2l, \qquad |\lambda_1' - \lambda_1| \lesssim \Delta\lambda, \\ m_2\lambda_2' = 2l, \qquad |\lambda_2' - \lambda_2| \lesssim \Delta\lambda. \end{aligned} \tag{194}$$

Thus, for isolated, high-resolution studies of the spectral feature at λ_1, say, all potential interference features such as λ_2 must be filtered out. This is done with a color or interference filter, or a prism or grating monochromator, before the signal is sent to the interferometer. The choice of prefiltering element will depend on the relative wavelength spacing of λ_1 and λ_2. The optical throughput of filters can be very high, 50% or more, but will not allow prefiltering of lines closer than a few nanometers. When very high throughput coupled with high resolution is essential, one or more additional Fabry-Perot devices of appropriate free spectral range can be used. For example, an etalon of spacing 0.01 mm has a free spectral range of about 15 mm. If it has a finesse of 50, its useful transmission bandwidth is 0.3 m. Such an etalon could be coupled with etalons of spacing 0.1 m and 1 nm with similar finesse to isolate a band only 0.003 nm wide from an original relatively broad frequency band transmitted through a filter. The final transmitted wavelength of such a combination of etalons can be tuned by tilting.

The wavelength of maximum transmission of a single etalon varies as

$$\lambda = \lambda_0 \cos\theta, \tag{195}$$

where λ_0 is the wavelength for maximum transmission in normal incidence. In the interferometer arrangement shown in Figure 4.110, wavelength scanning can be accomplished in two ways: by pressure scanning, or by adjusting the axial position of one of the plates with a piezoelectric transducer.[98] To accomplish pressure scanning, the interferometer is mounted in a vacuum- or pressure-tight housing into which gas can be admitted at a slow steady rate. The resultant change of refractive index of the gas between the plates, which is approximately linear with pressure, causes a continuous, unidirectional scanning of the center transmitted frequency. The interferometer should be mounted freely within the housing so that the pressure differential between the interior and exterior of the housing causes no deformation of the interferometer. Pressure scanning is not convenient where rapid or bidirectional scanning is required. However, because no geometrical dimensions of the interferometer change during the scan, no change in finesse should occur during scanning.

Piezoelectric scanning is very convenient, but the transducer material must be specially selected to give uniform translation of the moving plate without any tilting—which would cause finesse reduction during scanning. If separate piezoelectric transducers[98] are mounted on the kinematic-mounted alignment drives of the plate holders, it is possible to trim the finesse electronically after good mechanical alignment has been achieved by micrometer adjustment. The alignment of the plates during scanning can be maintained with a servo system.[99,100]

A Fabry-Perot etalon can be used in the configuration shown in Figure 4.112 to give a high-resolution display of a narrow-band spectral feature that cannot be displayed with a scanning interferometer because (for example) the optical signal is a short pulse. In the arrangement shown in Figure 4.112, all rays from a monochromatic source that are incident at angle θ_m on the plates will be transmitted and brought to a focus in a bright ring on a screen if

$$\cos \theta_m = \frac{m\lambda}{2l}. \tag{196}$$

In the paraxial approximation, θ_m is a small angle and the radius ρ_m of a ring on the screen is

$$\rho_m = f\theta_m, \tag{197}$$

which can be written as

$$\rho_m = f\sqrt{2 - \frac{m\lambda}{l}}. \tag{198}$$

The radius of the smallest ring corresponds to the largest integer m, for which $m\lambda/2l \leqslant 1$. Successive

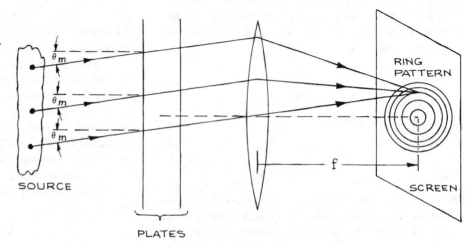

Figure 4.112 Arrangement for observing a ring interference pattern with a Fabry-Perot etalon.

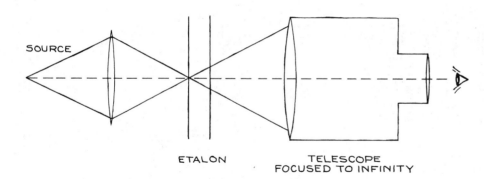

SOURCE

ETALON   TELESCOPE
                FOCUSED TO INFINITY

Figure 4.113 Arrangement for visual observation of Fabry-Perot fringes.

rings going out from the center of the pattern correspond to integers $m-1$, $m-2$, and so on.

If there is a bright spot at the center of the ring pattern, the radius of the pth ring from the center is

$$\rho_p = f\sqrt{\frac{p\lambda}{l}}. \qquad (199)$$

This ring will be diffuse because of the finite finesse of the etalon, even if the source is highly monochromatic. The FWHM at half maximum intensity of the pth ring in this case is

$$\Delta\rho_p = \frac{\rho_p}{2pF}, \qquad (200)$$

which can be verified by expanding Equation (173) in the angle θ. This ring width is the same as would be produced by a source of FWHM $\Delta\lambda = \lambda^2/2lF$.

Equation (200) predicts that the total flux through each ring of the pattern will be identical, since the area of each ring is

$$\Delta A = 2\pi\rho_p\Delta\rho_p = \pi f^2\lambda/2lF. \qquad (201)$$

The ratio of ring width to ring spacing for the pth ring in the pattern is

$$\frac{\Delta\rho_p}{\rho_{p+1}-\rho_p} = \left[2Fp\left(1-\sqrt{1+\frac{1}{p}}\right)\right]^{-1}, \qquad (202)$$

which is close to $1/F$ except for the first few rings. Thus

observation of the ring pattern from a quasimonochromatic source will yield its approximate bandwidth $\Delta\lambda$ as

$$\Delta\lambda \simeq \frac{\lambda}{2l}\frac{\Delta\rho}{(\rho_{p+1}-\rho_p)}. \qquad (203)$$

A convenient way of making a qualitative visual observation of this kind is to use an etalon in conjunction with a small telescope as shown in Figure 4.113.

(c) Luminosity, Throughput, Contrast Ratio, and Étendue. The luminosity of a Fabry-Perot system is, as already mentioned, much higher than that of any prism or grating monochromator. In the arrangement of Figure 4.110, the luminosity is

$$\Phi = \left(\frac{T}{1-R}\right)^2 E_e(\lambda)S\frac{\pi W_2^2}{4f_2}, \qquad (204)$$

where $\pi W_2^2/4f_2$ is the solid angle subtended by the output circular aperture at the output focusing lens, and S is the illuminated area of the plates. If the system has an aperture-limited finesse, then from Equation (190) the luminosity becomes

$$\Phi = \left(\frac{T}{1-R}\right)^2 E_e(\lambda)\frac{\pi\lambda S}{\rho F_A}, \qquad (205)$$

which can be written in terms of the resolving power $\Re$ as

$$\Phi = \left(\frac{T}{1-R}\right)^2 E_e(\lambda)\frac{2\pi S}{\Re}. \qquad (206)$$

The factor $[T/(1-R)]^2$ is called the *throughput* of the system. To compare the luminosity predicted by Equation (206) with that of a grating, we shall assume that the throughput of the Fabry-Perot system is equal to the corresponding efficiency factor T of the grating in Equation (143). In this case, for identical resolving powers $\mathscr{R}$, from Equations (206) and (155), we have

$$\frac{\Phi(\text{F.P.})}{\Phi(\text{grating})} = \frac{\pi S}{A \sin \beta \, (l/f_2)}, \qquad (207)$$

where l/f_2 is the angular slit height at the output-focusing element of the grating monochromator. For a blaze angle of 30° and equal Fabry-Perot aperture and grating area,

$$\frac{\Phi(\text{F.P.})}{\Phi(\text{grating})} = \frac{\pi}{l/f_2}. \qquad (208)$$

For a 0.3-m-focal-length grating instrument and a slit height of 3 cm (which is a typical maximum value for a small instrument of this size), the luminosity ratio is 71.4. This probably represents an optimistic comparison as far as the grating instrument is concerned, since high-resolution grating monochromators typically have focal lengths in excess of 0.75 m.

The *contrast ratio* of a Fabry-Perot system is defined as

$$C = \frac{(I_t/I_0)_{\text{max}}}{(I_t/I_0)_{\text{min}}}, \qquad (209)$$

which from Equation (173) clearly has the value

$$C = \left(\frac{1+R}{1-R}\right)^2 = 1 + \frac{4F^2}{\pi^2}. \qquad (210)$$

The *étendue* U of a Fabry-Perot system is a measure of its light-gathering power for a given frequency bandwidth $\Delta\nu_{1/2}$. It is defined by

$$U = \Omega S, \qquad (211)$$

where A is the area of the plates and Ω is the solid angle within which incident rays can travel and be transmitted with a specified frequency bandwidth. From

Equation (189), since $\Omega = \pi \Delta \theta^2$,

$$\Delta\nu_{1/2} = \frac{\nu_0 \Omega}{2\pi}. \qquad (212)$$

Therefore

$$U = \Omega S = \frac{2\pi \Delta\nu_{1/2}}{\nu_0} \frac{\pi D^2}{4}, \qquad (213)$$

which can be written in the form

$$U = \Omega S = \frac{\pi^2 D^2 \lambda}{4 l F_I}. \qquad (214)$$

(d) Confocal Fabry-Perot Interferometers. Fabry-Perot interferometers with spherical concave mirrors are generally used in a confocal arrangement, as shown in Figure 4.114. In this configuration the various characteristics of the interferometer can be summarized as follows:

$$\left(\frac{I_T}{I_0}\right)_{\text{max}} = \frac{T^2}{2(1-R)^2}, \qquad (215)$$

$$\Delta\nu(\text{FSR}) = c/4l, \qquad (216)$$

$$\text{reflectivity-limited finesse} = \frac{\pi R}{(1-R)^2}. \qquad (217)$$

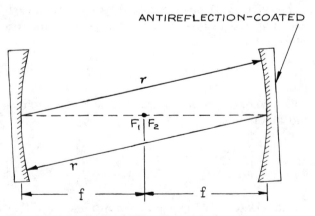

Figure 4.114   Confocal Fabry-Perot interferometer.

The approximate optimum radius of the apertures used in an arrangement similar to Figure 4.110 is $1.15 \, (\lambda r^3 / F_I)^{1/4}$, where r is the radius of the spherical mirrors.

The characteristics of all the possible combinations of two spherical mirrors that can be used to form a Fabry-Perot interferometer have been very extensively studied in connection with the use of such mirror arrangements to form the resonant cavities of laser oscillators.[11] In that application, these spherical-mirror resonators support Gaussian beam modes, and it is with such Gaussian beams that highest performance can be achieved with a spherical-mirror Fabry-Perot interferometer. Thus, a confocal Fabry-Perot interferometer is ideal for examining the spectral content of a Gaussian laser beam. The beam to be studied should be focused into the interferometer with a lens that matches the phase-front curvature and spot size of the laser beam to the radius of the Fabry-Perot mirror and the spot size of the resonant modes. (See Section 4.2.4.)

4.7.5 Design Considerations for Fabry-Perot Systems

Fabry-Perot interferometers and etalons are extremely sensitive to angular misalignment and fluctuations in plate spacing. Consequently, their design and construction are more demanding than for other laboratory optical instruments. The plates must be extremely flat, as already mentioned, and must be held without any distortion in high-precision alignment holders. The plate spacing must be held constant to about 1 nm in a typical high-finesse device for use in the visible, and the angular orientation must be maintained to better than 0.01 second of arc. Small-plate-spacing etalons are generally made commercially by optically contacting two plates to a precision spacer, or spacers, made of quartz. Such etalons are available from ICOS Ltd., Burleigh, Virgo, CVI, and Klinger, among others. Etalons of thickness in excess of 1 or 2 mm for use in the visible can be made from solid plane-parallel pieces of quartz or sapphire with dielectric coatings on their faces, or from materials such as germanium or zinc selenide for infrared use. Fabry-Perot interferometer plates are al-

most always slightly wedge-shaped and antireflection-coated on their outside faces to eliminate unwanted reflections, as shown in Figure 4.110(b). To prevent distortion of the plates they should be kinematically mounted in their holders. A good way to do this is to use three quartz or sapphire balls cemented to the circumference of the plates and held in ball, slot, and plane locations with small retaining clips.

For the construction of Fabry-Perot interferometers, low-expansion materials should be used. The plate holders can be held at a fixed spacing by having them spring-loaded against three quartz or ceramic spacer rods or by the use of an Invar or Super-Invar spacer structure.

After a Fabry-Perot interferometer is aligned mechanically, it will creep slowly out of alignment, because of the easing of micrometer threads, for example. The alignment can be retuned by having a piezoelectric element[98] incorporated in each plate-orientation adjustment screw.

Excellent Fabry-Perot interferometers are made commercially by Burleigh and Tropel.

4.7.6 Double-Beam Interferometers

In a double-beam interferometer, waves from a source are divided into two parts at a beamsplitter and then recombined after traveling along different optical paths. The most practically useful of these interferometers are the Michelson and the Mach-Zehnder.

(a) **The Michelson Interferometer.** The operation of a Michelson interferometer can be illustrated with reference to Figure 4.115. Light from an extended source is divided at a beamsplitter and sent to two plane mirrors M_1 and M_2. If observations of a broad-band (temporally incoherent) source are being made, a compensating plate of the same material and thickness as the beamsplitter is included in arm 2 of the interferometer. This plate compensates for the two additional traverses of the beamsplitter substrate made by the beam in arm 1. Because of the dispersion of the beamsplitter material, it produces a phase difference that is wavelength-dependent, and without the com-

Figure 4.115 Michelson interferometer.

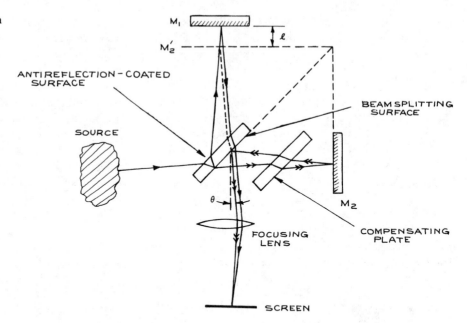

pensating plate, interference fringes would not be observable with broad-band illumination (such as white light). A maximum of illumination results at angle θ in Figure 4.115 if the phase difference between the two beams coming from M_1 and M_2 is an integral multiple of 2π. If M_2' represents the location of the reflection of M_2 in the beamsplitter, the condition for a maximum at angle θ is

$$2l\cos\theta = m\lambda. \tag{218}$$

The positioning of M_1 at M_2' corresponds to $l = 0$, $m = 0$. If the output radiation is focused with a lens, a ring pattern is produced. A clear ring pattern is only visible if

$$2l\cos\theta \leq l_c, \tag{219}$$

where l_c is the coherence length of the radiation from the source. The clearness of the rings is usually described in terms of their *visibility* V, defined by the relation

$$V_{\max} = \frac{I_{\max} - I_{\min}}{I_{\max} + I_{\min}}. \tag{220}$$

To obtain a ring-shaped interference pattern, M_1 and M_2' must be very nearly parallel; otherwise a pattern of curved dark and light bands will result. To study the fringes, mirror M_1 must be scanned in a direction perpendicular to its surface. This can be done with a precision micrometer, or a piezoelectric transducer if only small motion is desired.

In practice, Michelson interferometers are rarely used in the configuration shown in Figure 4.115, but are used instead with *collimated* illumination (a laser beam, or parallel light obtained by placing a point source at the focal point of a converging lens) as shown in Figure 4.116. In this case, the output illumination is (almost) perfectly uniform, being a maximum if

$$2l = m\lambda, \tag{221}$$

where $2l$ is the path difference between the two arms.

If monochromatic illumination of angular frequency ω is used we can write the fields of the waves from arms 1 and 2 as

$$\begin{aligned} V_1 &= V_0 e^{i(\omega t - \phi_1)}, \\ V_2 &= V_0 e^{i(\omega t - \phi_2)}, \end{aligned} \tag{222}$$

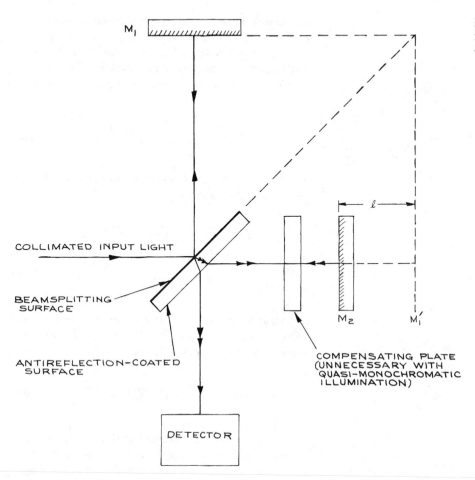

where it is assumed that the beamsplitter divides the incident beam into two equal parts. ϕ_1 and ϕ_2 are the optical phases of these two beams, which are dependent on the optical path lengths traveled in arms 1 and 2.

The output intensity is

$$I \propto (V_1 + V_2)^* (V_1 + V_2), \qquad (223)$$

which gives

$$I = I_0 \left[1 + \cos(\phi_2 - \phi_1) \right]$$

$$= I_0 \left[1 + \cos\left(\frac{4\pi \nu l}{c} \right) \right], \qquad (224)$$

in which I_0 is the intensity of the light from one arm alone. Suppose an optical component such as a prism or lens is placed in one arm of the interferometer, as shown in Figure 4.117, with the plane mirror replaced by an appropriate-radius spherical mirror in the case of the lens, and quasi-monochromatic illumination is used. Then the output fringe pattern will reveal inhomogeneities and defects in the inserted component. A Michelson interferometer used in this way for testing optical components is called a *Twyman-Green interferometer*.[7]

Equation (224) illustrates why Michelson (and all other double-beam) interferometers are not directly suitable for spectroscopy. The distance the movable mirror must be adjusted to go from one maximum to the next is

$$\Delta l = \lambda / 2.$$

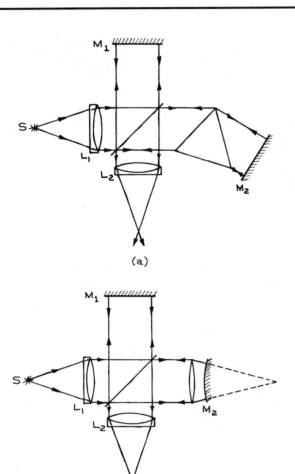

Figure 4.117   Twyman-Green interferometer: (*a*) arrangement for testing a prism; (*b*) arrangement for testing a lens. S is a monochromatic source, L_1 is a collimating lens, and L_2 is a focusing lens. (From M. Born and E. Wolf, *Principles of Optics*, 3rd edition, Pergamon Press, Oxford, 1965; by permission of Professor Emil Wolf.)

The distance required to go from a maximum- to a half-maximum-intensity point is

$$\Delta l = \lambda / 8.$$

Thus the effective finesse is only 4: sharp spectral peaks with monochromatic illumination are not obtained as they are in multiple-beam interferometers. However, if

the movable mirror is scanned at a steady velocity v, the resultant intensity as a function of position can be used to obtain considerable information about the spectrum of the source. If the source has intensity $I(\nu)$ at ν, the contribution to the intensity from the small band $d\nu$ at ν is

$$I_\nu(l) = I(\nu)\left(1 + \cos\frac{4\pi\nu l}{c}\right) d\nu, \qquad (225)$$

where $l = vt$. The total observed output from the full spectrum of the source is

$$
\begin{aligned}
I(l) &= \int_{\nu=0}^{\infty} I(\nu)\left(1 + \cos\frac{4\pi\nu l}{c}\right) d\nu \\
&= \frac{I_0}{2} + \int_{\nu=0}^{\infty} I(\nu)\cos\frac{4\pi\nu l}{c}\, d\nu. \qquad (226)
\end{aligned}
$$

The second term on the second line is essentially the cosine transform of $I(\nu)$. Thus $I(\nu)$ can be found as the cosine transform of $I(l)-(I_0/2)$:

$$I(\nu) = \frac{2}{c}\int_{l=0}^{\infty}\left(I(l) - \frac{I_0}{2}\right)\cos\frac{4\pi l\nu}{c}\, dl. \qquad (227)$$

Further details of Fourier-transform spectroscopy, as this technique is called, will not be given here; the interested reader should refer to the specialized literature.[32,87,101,102] Note, however, that Fourier spectrometers have high luminosity and collect spectral information very efficiently: the whole emission spectrum is in reality sampled at once, rather than one feature at a time as with a prism or grating spectrometer. Commercial Fourier-transform spectrometers are available from Bomem, Digilab, Beckmann Instruments (England), Sir Howard Grubb Parsons (England), and Nicolet.

(b) The Mach-Zehnder Interferometer. The Mach-Zehnder interferometer, shown in Figure 4.118, is widely used for studying refractive-index distributions in gases, liquids, and solids. It is particularly widely used for studying the density variations in compressible gas flows,[103] the thermally induced density (and consequent refractive-index) changes behind shock fronts, and ther-

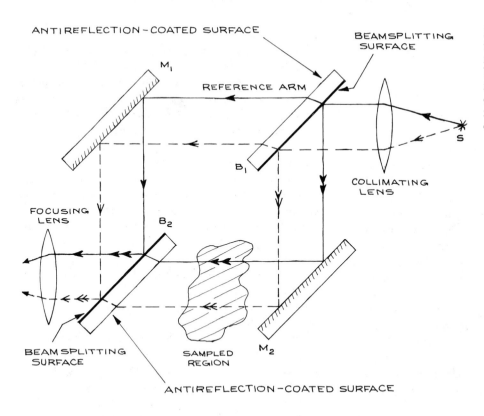

Figure 4.118 Mach-Zehnder interferometer. M_1, M_2 are totally reflecting mirrors; B_1, B_2 are beamsplitters (usually equally reflective and transmissive). The distribution of output illumination reflects the refractive-index distribution in the sampled region.

mally induced density changes produced by laser beams propagating through transparent media.[104,105] If the light entering the interferometer is perfectly collimated, uniform output illumination results, which is a maximum if the path difference between the two arms is an integral number of wavelengths. If the image of M_1 in beamsplitter B_2 is not parallel to M_2, the output fringe pattern is essentially the interference pattern observed from a wedge—a series of parallel bright and dark bands. If the medium in arm 2 is perturbed in some way, so that a spatial variation in refractive index results, the output fringe pattern changes. Analysis of the new fringe pattern reveals the spatial distribution of refractive-index variation along the beam path in arm 2. Various design variations on the Mach-Zender interferometer exist, notably the Jamin and the Sirks-Pringsheim.[7]

(c) Design Considerations for Double-Beam Interferometers. Michelson and Mach-Zehnder interferom-

eters are much easier to build in the laboratory than multiple-beam interferometers such as the Fabry-Perot. Because the beams only reflect off the interferometer mirrors once, mirrors of flatness $\lambda/10$ or $\lambda/20$ are quite adequate for building a good instrument. The usual precautions should be taken in construction: stable rigid mounts for the mirrors and beamsplitters and, if long-term thermal stability is required, low-expansion materials. Piezoelectric control of mirror alignment is not generally necessary, as these instruments are less sensitive to small angular misalignments than Fabry-Perot interferometers. To achieve freedom from alignment disturbances along one or two orthogonal axes, the plane mirrors of a Michelson interferometer can be replaced with right-angle prisms or corner-cube retroreflectors, respectively. Some care should be taken to avoid source-polarization effects in these instruments; the beamsplitters are generally designed to reflect and transmit equal amounts only for a specific input polarization. If polarizing materials are placed in either arm of the interferometer, no fringes will be seen if, by any chance, the beams from the two arms emerge orthogonally polarized.

4.8   DETECTORS

The development of optical detectors has occurred, in common with other branches of electronics, by a series of advances through the use of gas-filled tubes and vacuum tubes to various semiconductor devices. However, whereas in general electronics the vacuum tube is now reserved for specialized applications, vacuum-tube optical detectors such as the photomultiplier or vacuum photodiode are still in widespread use and are significantly better than semiconductor detectors in very many circumstances, most notably the detection of weak or very short-pulse-duration radiation with a wavelength below about 1 μm. Thus, the designer of an optical system is confronted with a very broad range of detector types: vacuum-tube, semiconductor, and thermal. The last category includes thermopiles, pyroelectric detectors, Golay cells, and bolometers, each with its own advantages and disadvantages.

The choice of detector for a given application will almost invariably involve a choice from among the numerous commercially available devices in each of these categories. The construction of optical detectors is rarely undertaken in the laboratory except by the specialist. The choice of detector will be governed by factors such as the following:

1. The wavelength region of the radiation to be detected.

2. The intensity of the radiation to be detected.

3. The time response required to resolve high-speed events.

4. The environmental conditions under which the detector is to be used.

5. The cost.

The choice of detector should not, however, be an exercise performed in isolation. The other components of the optical system may influence it, and vice versa. This is particularly true if very weak radiation is to be detected. In such a situation action should be taken to maximize the light signal actually reaching the detector; this will involve a careful choice of optical components to be placed between source and detector—for example, suitable light collection and focusing optics. If the light from the source must pass through a dispersing element (monochromator, etalon, etc.) before reaching the detector, then the choices of dispersing element and detector are likely to be interrelated. Some dispersing elements have high resolution (see Section 4.6) but low light throughput; others the reverse. Still others have high throughput and resolution but can only be scanned in a restricted way, if at all. In a given experimental situation the various requirements of resolution, light throughput, scannability, and available detector sensitivities must be offset one against another in arriving at a sensible compromise.

Detectors can be classified as photon detectors or thermal detectors. In *photon detectors*, individual incident photons interact with electrons within the detector material. This leads to detectors based on *photoemission*, where the absorption of a photon frees an electron from a material; *photoconductivity*, where absorption of photons increases the number of charge carriers in a material or changes their mobility; the *photovoltaic effect*, where absorption of a photon leads to the generation of a potential difference across a junction between two materials; the *photon-drag effect*, where absorbed photons transfer their momentum to free carriers in a heavily doped semiconductor; and the photographic plate.

In *thermal detectors* the absorption of photons leads to a temperature change of the detector material, which may be manifest as a change in resistance of the material, as in the *bolometer*; the development of a potential difference across a junction between two different conductors, as in the *thermopile*; or a change of internal dipole moment in a temperature-sensitive ferroelectric crystal, as in a *pyroelectric* detector. With the exception of the last, thermal detectors have slow time response when compared to photon detectors. They do, however, respond uniformly at all wavelengths (in principle).

Photon detectors can be further classified by the

spectral region to which they are sensitive. Photoemissive detectors operate from the vacuum ultraviolet through to about 1.2 μm; they rapidly become insensitive beyond 1 μm. However, work to develop improved infrared-sensitive devices is continuing. Photoconductive and photovoltaic detectors based on different materials are used right from the visible into the very far infrared, at least to 1000 μm. Photographic films, which are also photon detectors, generally respond best in the visible, although films marginally sensitive out to about 1.2 μm are available.

This brief survey is not intended to be exhaustive. Many other detection mechanisms have been described, but not all are commonly used or are commercially available. For example, no further mention will be made of photoelectromagnetic detectors, or superconducting bolometers, though these are used in specialized applications. For more information and details about detectors in general than will be found here, the reader should consult References 36, 37, 106, and 107.

4.8.1   Figures of Merit for Detectors

Before discussing the characteristics of commonly used and commercially available detectors individually, the various figures of merit used by manufacturers to specify their products will be described. These are the noise-equivalent power (NEP), detectivity (D^*), responsivity ($\mathcal{R}$), and time constant (τ), which allow such questions to be answered as: What is the minimum light intensity falling on the detector that will give rise to a signal voltage equal to the noise voltage from the detector? What signal will be obtained for unit irradiance? How does the electrical signal from the detector vary with the wavelength of the light falling on it? What is the modulation frequency response of the detector or its ability to respond to short light pulses?

(a) Noise-Equivalent Power. The *noise-equivalent power* (NEP) is the rms value of a sinusoidally modulated radiant power falling upon a detector that gives an rms signal voltage equal to the rms noise voltage from the detector. The NEP is usually specified in terms of a blackbody source at 500 K, the reference band-

width for the detection of signal and noise (usually 1 or 5 Hz), and the modulation frequency of the radiation (usually 90, 400, 800 or 900 Hz). For example, a noise-equivalent power written NEP(500 K, 900, 1) implies a blackbody source at 500 K, a 900-Hz modulation frequency, and a 1-Hz detection bandwidth. Thus, if a radiant intensity I (W m^{-2}) falls on a detector of sensitive area A (m^2), and if signal and noise voltages V_s and V_n are measured with bandwidth Δf (Hz) (which is small enough so that the noise-voltage frequency spectrum is flat within it), then the NEP measured in W Hz$^{-1/2}$ is

$$\text{NEP} = \frac{IA}{\sqrt{\Delta f}} \frac{V_N}{V_S}. \qquad (228)$$

(b) Detectivity. At one time, detectivity was defined as the reciprocal of the NEP. However, most detectors exhibit an NEP that is proportional to the square root of the detector area; so a detector-area-independent *detectivity* D^* is now used, specified by

$$D^* = \frac{\sqrt{A}}{\text{NEP}}. \qquad (229)$$

D^* is specified in the same way as NEP: for example, $D^*(500 \text{ K}, 900, 1)$. To specify the variations in response of a detector with wavelength, the *spectral detectivity* is used. Thus, the symbol $D^*(\lambda, 900, 1)$ would specify the response of the detector to radiation of wavelength λ, modulated at 900 Hz and detected with a 1-Hz bandwidth. The units of D^* are cm Hz$^{1/2}$ W^{-1}. Some curves of $D^*(\lambda)$ for various detectors are shown in succeeding figures.

(c) Responsivity. The *responsivity* $\mathcal{R}$ of a detector specifies its response to unit irradiance:

$$\mathcal{R} = V_S / IA. \qquad (230)$$

A similar figure of merit, usually used to characterize photoemissive detectors, is the *radiant sensitivity S*, which is the current per unit area of the photoemissive surface produced by unit irradiance:

$$S = \frac{i_S}{P}, \qquad (231)$$

Figure 4.119 Wavelength dependence of the radiant sensitivity of several commercially available photocathode materials. (From *Handbook of Lasers*, R. J. Pressley, Ed., CRC Press, Cleveland, 1971; by permission of the CRC Press.)

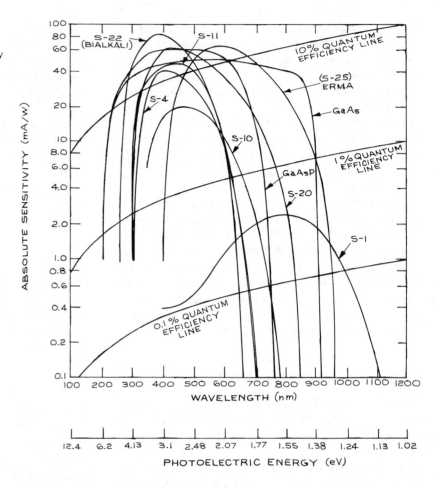

where i_S is the total current from the detector and P the total radiant power falling on it. Some curves showing the spectral variation of radiant sensitivity for different photoemissive surfaces are given in Figure 4.119.

(d) Frequency Response and Time Constant. The frequency response of a detector is the variation of responsivity or radiant sensitivity as a function of the modulation frequency of the incident radiation. The signal from the detector should be a.c.-coupled; otherwise the generated d.c. signal that is produced as the detector begins to fail to respond to the modulation will also be detected. The frequency variation of $\mathcal{R}$ and the time constant τ of the detector are generally related

according to

$$\mathcal{R}(f) = \frac{\mathcal{R}(0)}{\left(1 + 4\pi^2 f^2 \tau^2\right)^{1/2}}, \qquad (232)$$

as shown in Figure 4.120.

Thus the responsivity has decayed to a value of $\mathcal{R}(0)/\sqrt{2}$ at a frequency $1/2\pi\tau$. Hence τ is, in reality, a simple measure of the ability of the detector to respond to a sharply rising or falling optical signal.

(e) Noise. The random fluctuations in the output voltage or current from a detector set a lower limit to the radiant power that can be detected, given the

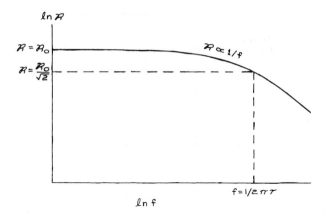

Figure 4.120 Typical frequency dependence of detector responsivity $\mathscr{R}$.

associated with random conductance fluctuations; and *generation-recombination* (gr) noise, also important in semiconductors, which results from the random recombination of electrons and holes in the material. Thermal noise and shot noise have so-called white spectra—they contribute equal noise signals in a given detection bandwidth at any frequency. Current noise generally shows a $1/f$ *frequency dependence* (it is frequently referred to as $1/f$ noise) and is therefore the dominant source of noise to be dealt with at low frequencies. As shown in Figure 4.121, gr noise contributes at frequencies up to a value of the order of the reciprocal of the carrier lifetime, and then falls off rapidly with increasing frequency.

In a practical detection system the detector is generally coupled to various forms of electronic processing systems, preamplifiers, amplifiers, filters, lock-in detectors, and so on. This electronic system necessarily creates additional noise; in a well-designed system this is kept to a minimum.

It will be seen in the following sections that there are various ways to reduce, and even essentially eliminate, detector noise. However, there is a fundamental fluctuation component in the signal from the detector that cannot be removed. This results from fluctuations in the arrival rate of photons from the source that is being observed. Thus, the ultimate performance of any detector is *photon-noise-limited*. The photon-noise limit is different for thermal detectors, which are sensitive to

detector temperature and operating conditions, source modulation frequency, and electronic-detection-system bandwidth. These fluctuations are from a variety of sources, some of which can be reduced by operating the detector in an appropriate way. The most important forms of noise are thermal noise, which results from the random motion of charge carriers in the detector; shot noise, which is primarily important in vacuum and gas-filled tubes and arises from the random, discrete nature of the photoemission process; current noise, which is important in semiconductor detectors and is

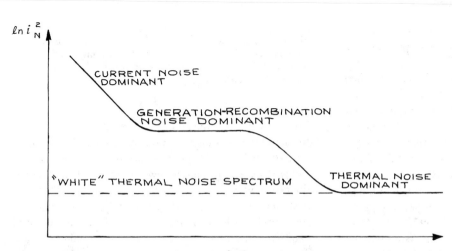

Figure 4.121 Schematic frequency dependence of noise in semiconductor detectors; i_N is the noise current.

total absorbed radiation, and photon detectors, which respond, at least in a microscopic way, to the absorption of individual quanta.

For a thermal detector of absorptivity and emissivity ϵ, at absolute temperature T_1, completely enclosed by an environment at temperature T_2, the photon-noise-limited detectivity can be shown to be

$$D^* = \frac{4.0 \times 10^{16} \epsilon^{1/2}}{\left(T_1^5 + T_2^5\right)^{1/2}} \ \mathrm{cm\,Hz^{1/2}W^{-1}}. \qquad (233)$$

For a photon detector, the limiting spectral detectivity is

$$D^*(\lambda) = \frac{c_0 \eta(\nu)}{2h\nu\pi^{1/2}\left[\displaystyle\int_{\nu_0}^{\infty} \frac{\eta(\nu')\nu'^2 e^{h\nu'/kT_2}d\nu'}{\left(e^{h\nu'/kT_2}-1\right)^2}\right]^{1/2}}, \qquad (234)$$

where $\eta(\nu)$ is the *quantum efficiency* (the average number of charge carriers produced for each incident photon absorbed), $\lambda = c_0/\nu$, and ν_0 is the lowest frequency to which the detector is sensitive. Some evaluations from Equation (234) have been given by Kruse, McGlauchlin, and McQuistan,[36] who also give a detailed discussion of noise in optical detectors.

4.8.2 Photoemissive Detectors

Photoemissive detectors include gas-filled and vacuum photodiodes, photomultiplier tubes, and photo-channeltrons. These are all photon detectors that utilize the photoelectric effect. When radiation of frequency ν falls upon a metal surface, electrons are emitted, provided the photon energy $h\nu$ is greater than a minimum critical value ϕ, called the *work function*, which is characteristic of the material being irradiated. A simplified energy-level diagram illustrating this effect for a metal-vacuum interface is shown in Figure 4.122(a). For most metals, ϕ is in a range from 4–5 eV (1 eV $\leftrightarrow$ 1.24 μm), although for the alkali metals it is lower, for example, 2.4 eV for sodium and 1.8 eV for cesium. Pure metals or alloys, particularly beryllium-copper, are used as photoemissive surfaces in ultraviolet and vacuum-ultraviolet detectors.

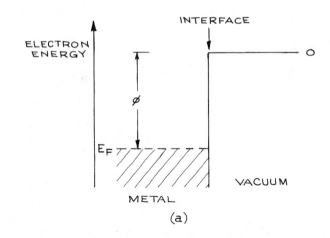

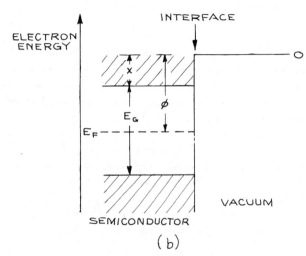

Figure 4.122 Band structure of (a) a metal vacuum interface and (b) a pure-semiconductor-vacuum interface (ϕ = work function; χ = electron affinity; E_g = band-gap energy; E_F = Fermi level).

Lower work functions, and consequently sensitivities that can be extended into the infrared, can be obtained with special semiconductor materials. Figure 4.122(b) shows a schematic energy-level diagram of a semiconductor-vacuum interface. In this case the work function is defined as $\phi = E_{\mathrm{vac}} - E_F$, where E_F is the energy of the Fermi level. In a pure semiconductor, the Fermi level is in the middle of the band gap, as illustrated in Figure 4.122(b). In a p-type doped semiconductor, E_F

moves down toward the top of the valence band, while in n-type material it moves up toward the bottom of the conduction band. Consequently, ϕ is not as useful a measure of the minimum photon energy for photoemission as it is for a metal. The electron affinity χ is a more useful measure of this minimum energy for a semiconductor. Except at absolute zero, photons with energy $h\nu > \chi$ cause photoemission. Photons with energy $h\nu > E_g$ lead to the production of carriers in the conduction band; this leads to intrinsic photoconductivity, which is the operative detection mechanism in various infrared detectors, such as InSb.

(a) **Vacuum Photodiodes.** Once electrons are liberated from a photoemissive surface (a photocathode), they can be accelerated to an electrode positively charged with respect to the cathode—the anode—and generate a signal current. If the acceleration of photoelectrons is directly from cathode to anode through a vacuum, the device is a vacuum photodiode. Because the electrons in such a device take a very direct path from anode to cathode and can be accelerated by high voltages—up to several kilovolts in a small device—vacuum photodiodes have the fastest response of all photoemissive detectors. Risetimes of 100 ps or less can be achieved. External connections and electronics are generally the limiting factors in obtaining short risetimes from such devices. However, vacuum photodiodes are not very sensitive, since at most one electron can be obtained for each photon absorbed at the photocathode. In principle, of course, the limiting sensitivity of the device is set by the *quantum efficiency* η of the photoemissive surface, defined as

$$\eta = \frac{\text{electrons emitted}}{\text{photons absorbed}}. \tag{235}$$

Practical quantum efficiencies for photoemissive materials range up to about 0.4.

If the space between photocathode and anode is filled with a noble gas, photoelectrons will collide with gas atoms and ionize them, yielding secondary electrons. Thus, an electron multiplication effect occurs. However, because the mobility of the electrons moving from cathode to anode through the gas is slow, these devices have a long response time, typically about 1 ms. Gas-filled photocells are no longer competitive with solid-state detectors in laboratory applications.

(b) **Photomultipliers.** If photoelectrons are accelerated *in vacuo* from the photocathode and allowed to strike a series of secondary electron emitting surfaces, call *dynodes*, held at progressively more positive voltages, a considerable electron multiplication can be achieved and a substantial current can be collected at the anode. Such devices are called *photomultipliers*. Practical gains of 10^9 (anode electrons per photoelectron) can be achieved from these devices for short light pulses. Continuous gains must be held lower, to perhaps 10^7, a limit imposed primarily by the ability of the final dynodes in the chain to withstand the thermal loading produced by continuous electron impact. Because of their very high gain, photomultipliers can generate substantial signals when only a single photon is detected: for example, an anode pulse of 2-ns duration containing 10^9 secondary photoelectrons produced from a single photoelectron will generate a voltage pulse of 4 V across 50 ohms. This, coupled with their low noise, makes photomultipliers the only effective single-photon detectors available. Photomultiplier D^*-values can range up to 10^{16} cm Hz$^{1/2}$ W^{-1}. Only the dark-adapted human eye, which can detect bursts of about 10 photons in the blue, comes close to this sensitivity.

The time response characteristics of photomultiplier tubes depend to a considerable degree on their internal dynode arrangement. The response of a given device can be specified in terms of the output signal at the anode that results from a single photoelectron emission at the photocathode. This is illustrated in Figure 4.123. Because electrons passing through the dynode structure can generally take slightly different paths, secondary electrons arrive at the anode at different times. The anode pulse has a characteristic width called the *transit-time spread*, which usually ranges from 0.1 to 20 ns. The time interval between photoemission at the cathode and the appearance of an anode pulse is called the *transit time*, and is usually a few tens of nanoseconds. The transit time and transit-time spread also

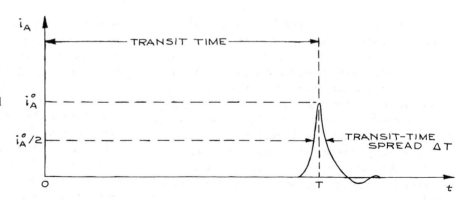

Figure 4.123 Typical anode pulse produced by a single photoelectron emission at the cathode of a photomultiplier tube. t is the time following photoemission. The transit-time T, the transit-time spread ΔT, and the peak anode current i_A^o all fluctuate from pulse to pulse.

fluctuate slightly from one single-photoelectron-produced anode pulse to the next. Although this is a fine point, it may be something to worry about in precision-timing experiments.

There are four main types of dynode structure in common usage in photomultiplier tubes; these are illustrated in Figure 4.124. The circular cage structure (used, for example, in the RCA 7102 photomultiplier tube) is compact and can be designed for good electron-collection efficiency and small transit-time spread. This dynode structure works well with opaque photocathodes, but is not very suitable for high-amplification requirements where a larger number of dynodes is required. The box-and-grid and Venetian-blind structures (used, for example, in the EMI 9684 and EMI 6256, respectively) offer very good electron-collection efficiency. Because they collect multiplied electrons independently of their path through the dynode structure, a wide range of secondary- electron trajectories is possible, leading to a large transit-time spread and slow response. Typical response times of these types of tube are 10–20 ns. Venetian-blind tubes can easily be extended to many dynode stages and have a very stable gain in the presence of small power supply fluctuations. These tubes also have an optically opaque dynode structure, which contributes to their exhibiting very low dark-current noise when operated under appropriate conditions.

The focused dynode structure (used, for example, in the Phillips 56 AVP and other 56-series tubes) is designed so that electrons follow paths of similar length through the dynode structure. To accomplish this, electrons that deviate too much from a specified range of trajectories are not collected at the next dynode. These types of tube offer short response times, typically 1–2 ns; some recently developed tubes are even faster.

Other types of multiplier structures, not shown in Figure 4.124, are also available. Traveling-wave phototubes and various types of crossed-field photomultiplier have very short response times, down to 0.1 ns. Some designs of crossed-field photomultiplier are available commercially.[108] These tubes use a combination of static electric and magnetic fields to direct high-energy electrons along long paths and low-energy electrons along appropriately shorter paths between dynodes. This minimizes transit-time variations between secondary electrons of different energy. Channel multipliers (channeltrons) are made in the form of a tube of high-resistivity material whose inside surface forms a continuous dynode.[109] Electrons emitted from a photocathode at one end of the structure are accelerated along the tube, making multiple reflections from its walls and generating secondary electrons. These devices have high secondary-electron collection efficiency, and can have high gain and fairly low noise; their response time is typically about 10 ns. Unfortunately, the gain of these devices tends to change continuously with use, so they are inconvenient to use when quantitative detector calibration must be maintained.

(c) **Photocathode and Dynode Materials.** The performance of a photomultiplier depends not only on its

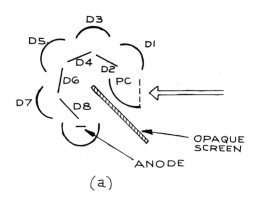

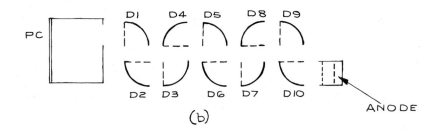

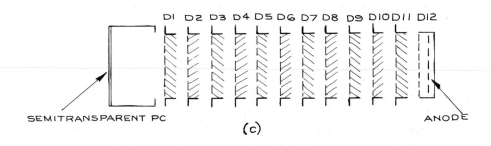

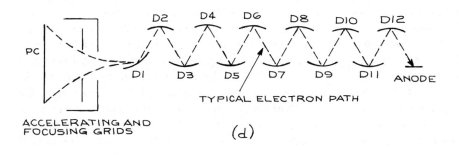

Figure 4.124 Schematic diagram of the internal structure of various types of photomultiplier tube: (*a*) squirrel cage; (*b*) box and grid; (*c*) venetian blind; (*d*) focused dynode. D = dynode; PC = photocathode.

Table 4.8 CHARACTERISTICS OF PHOTOEMISSIVE SURFACES

Cathode	Radiant Sensitivity (mA/W)			Peak Quantum Efficiency (%)	λ_{peak} (μm)	Dark Current[a] (A/cm^2)
	515 nm	694 nm	1.06 μm			
S-1 Cs-O-Ag	0.6	2	0.4	0.08	800	9×10^{-13}
S-10 Cs-O-Ag-Bi[b]	20	2.7	0	5	470	4×10^{-16}
S-11 Cs$_3$Sb on MnO[c]	39	0.2	0	13	440	10^{-16}
S-20 (Cs)Na$_2$KSb (tri-alkali)	53	20	0	18	470	3×10^{-16}
S-22 (bi-alkali)	42	0	0	26	390	$1{-}6 \times 10^{-18}$
GaAs[d-g]	48	28	0	14	560[h]	10^{-16}
GaAsP[d]	60	30	0	19	400[h]	3×10^{-15}
InGaAs[d,e]	—	—	4.3	—	—[h]	3×10^{-14}
InGaAsP[d,f]	—	—	—	47	300[h]	2.5×10^{-13}
S-25 (ERMA)[i]	53	26	0	25	430	1×10^{-15}
Cs-Te (solar blind)	—	—	—	15	254	2.5×10^{-17}

Note: Table shows typical values, but these can vary greatly from one manufacturer to another.

[a] At room temperature.

[b] Cathode designated S-3 is similar.

[c] Several types of CsSb photocathodes exist where the CsSb is deposited on different opaque and semitransparent substrates and various window materials are used. These photocathodes have the designations S-4, 5, 13, 17, and 19 as well as S-11.

[d] NEA photoemitters.

[e] Available from RCA Electronic Components, Harrison, N.J.

[f] Available from Varian (LSE Division).

[g] Available from Hamamatsu Corporation, Middlesex, N.J.

[h] May show no wavelength of maximum quantum efficiency; quantum efficiency falls with increasing wavelength. However, exact spectral characteristics will depend on thickness of photoemitter and whether it is used in transmission or not.

[i] Extended-red S-20.

internal structure, but also on the photoemissive material of its photocathode and the secondary-electron-emitting material of its dynodes. The selection of a photodiode or photomultiplier for a particular spectral response depends on an appropriate selection of photocathode material. The wavelength dependence of various commercially available photocathode materials is shown in Figure 4.119; some radiant sensitivities and quantum efficiencies are given in Table 4.8. The short-wavelength cutoff of a given material depends on its work function. This cutoff is not sharp because, except at absolute zero, there are always a few electrons high up in the conduction band available for photoexcitation by low-energy photons. Some few of these electrons, because of their thermal excitation, will be emitted without any photostimulation. This contributes the major part of the dark current observed from the photocathode. Materials that have low work functions, and are consequently more red- and infrared-sensitive, have higher—often much higher—dark currents than materials that are optimized for visible and ultraviolet sensitivity. Unless a phototube is to be used primarily for detecting long-wavelength radiation, we recommend selecting a photocathode with the shortest possible wavelength response.

Photocathodes are available in both opaque and semitransparent forms, depending on the model of phototube. In the semitransparent form the photoelectrons are ejected from the thin photoemissive layer on the opposite side from the incident light. In both types the photocathode has to perform two important functions: it must absorb incident photons and allow the emitted photoelectron to escape. The latter event is inhibited if photons are absorbed too deep in a thick photoemissive layer, or if the photoelectron suffers energy loss from scattering in the layer. If the emitted photoelectron has too much energy, it can excite a further electron across the band gap. This pair-production phenomenon inhibits the release of photoelectrons from the photoemissive layer, and accounts for the ultraviolet cutoff characteristics of the different materials shown in Figure 4.119. With reference to Figure 4.122(b) it can be shown that for pair production to occur the incident photon energy must be greater than $2E_g$. Photoelectrons have the best chance of escaping, and the photocathode its highest quantum efficiency, for materials where $\chi < E_g$.

Practical photoemissive materials fall into two main categories: classical photoemitters and negative-electron-affinity (NEA) materials. Classical photoemitters generally involve an alkali metal or metals, a group-V element such as phosphorus, arsenic, antimony, or bismuth, and sometimes silver and/or oxygen. Examples are the Ag-O-Cs (S1) photoemitter, which has the highest quantum efficiency beyond about 800 nm of any classical photoemitter, and Na_2KSbCs, the so-called tri-alkali (S-20) cathode.

NEA photoemitters have been developed only within the last ten years. These materials utilize a photoconductive single-crystal semiconductor substrate with a very thin surface coating of cesium and usually a small amount of oxygen. The cesium (oxide) layer lowers the electron affinity below the value it would have in the pure semiconductor, achieving an effectively negative value. Examples of such NEA photoemitters are GaAs (CsO) and InP (CsO). NEA emitters can offer very high quantum efficiency and extended infrared response. GaAs (CsO), for example, has higher quantum efficiency in the near infrared then an S-1 photocathode. Commercial photomultipliers with NEA photocathodes are available from RCA and EMR. Development of NEA photoemitters is continuing, and the experimentalist seeking long-wavelength response would be well advised to follow the literature and check for the introduction of new commercial tubes. For further details about NEA materials, the interested reader should consult the article by Zwicker.[110]

The performance of the dynode material in photo-multiplier tubes is specified in terms of the secondary-emission ratio δ as a function of energy. For a phototube with n dynodes the gain is δ^n. In the past, the commonest dynode materials were CsSb, AgMgO, and BeCuO. The last is also used as the primary photoemitter in windowless photomultipliers that are operated *in vacuo* for the detection of vacuum-ultraviolet radiation. BeCuO can be reactivated after exposure to air. The above materials have δ-values of 3–4. Newer NEA dynode materials have much higher δ-values; in particular, that of GaP can range up to 40 for an incident-electron input energy of 800 eV. With such high δ-values a photomultiplier tube needs fewer dynodes for a given gain, which means a more compact and faster-response tube can be built. In very many commercial photomultipliers, the first dynode at least is now frequently made of GaP. This offers improved characterization of the single-photoelectron response of the tube, which is important in designing a system for optimum signal-to-noise ratio.

(d) Practical Operating Considerations for Photomultiplier Tubes. Detailed advice is now given on the use of photomultipliers in experimental apparatus.

(i) *Dynode chains.* The accelerating voltages are supplied to the dynodes of a photomultiplier by a resistive voltage divider called a *dynode chain*. The relative resistance values in the chain determine the distribution of voltages applied to the dynodes. The total chain resistance R determines the chain current at a given total photocathode-anode applied voltage. Some phototubes are supplied with an integral dynode chain, but most are not. Adequate dynode-chain designs are generally supplied by the manufacturer with each tube. However, alternative designs intended for certain types of response, such as maximum gain for short pulses or highest linearity, can frequently be found by consulting the literature. Two examples of such dynode-chain designs for use with Phillips 56-series tubes are given in Figure 4.125. The total dynode-chain resistance is chosen so that the chain current is at least 100 times greater than the average d.c. anode current to be drawn from the tube. This current is specified for each

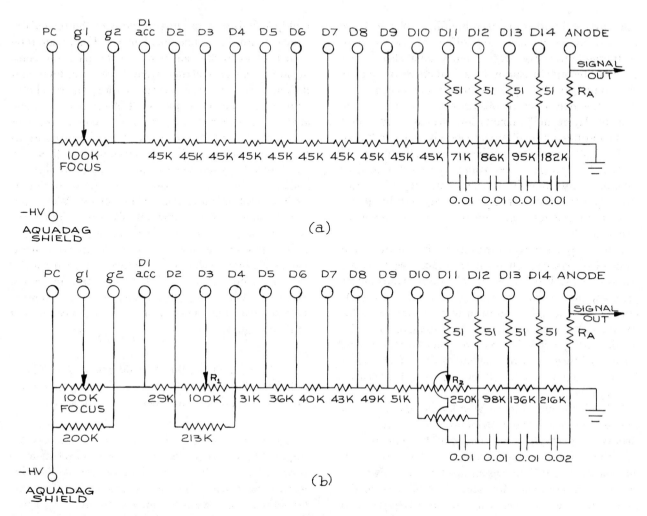

Figure 4.125 Dynode chains for Phillips 56-series photomultiplier tubes: (*a*) chain for high gain and fast time response; (*b*) chain for high linearity (R_1 and R_2 adjusted for optimum performance). PC = photocathode; g1, g2 = grids; acc = accelerator grid; D1–D14 = dynodes. Resistance values are in ohms, capacitance values in μF. For very low-level light detection, all resistances can be scaled upward to reduce overall chain current.

tube and may range as high as 1 mA for high-gain tubes. However, the operation of photomultiplier tubes at such high average currents for very long is not recommended. Long-term d.c. anode currents should be kept between 1 and 10 μA for most tubes.

There are several interesting features of the dynode chains shown in Figure 4.125. The photocathode is generally operated at negative potential with the anode near ground. This makes it easy to couple the output signal to other electronics. The tube can be operated with the cathode grounded, but in this case the signal from the anode, which is now at high positive potential,

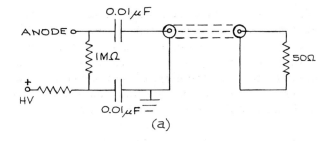

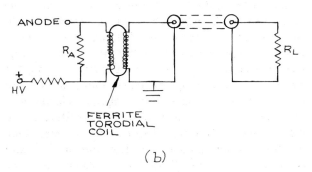

Figure 4.126 Anode signal coupling methods for a photomultiplier operated with the anode at high positive potential: (*a*) capacitor coupling for fast-risetime, short-pulse operation (the component sizes are typical; the capacitors should be high-frequency, high-voltage types); (*b*) transformer coupling for lower-frequency, modulated operation (the high-voltage winding on the transformer should be sufficiently well insulated for isolation from the core; the size of R_A will depend on various factors, such as transformer primary impedance and operating frequency).

must be coupled through high-voltage capacitors or an insulated transformer, as shown in Figure 4.126. The photocathode is also often connected to a conducting shield, which is either a painted coat of colloidal graphite (Aquadag) on the outside of the tube envelope, or a metal foil wrapped around the tube envelope. If the photocathode is at negative potential, appropriate insulation must be used around the tube if the tube is in close proximity to any grounded metal parts inside its housing.

Many photomultiplier tubes have one or more focusing electrodes between the photocathode and the first dynode. The voltage on these electrodes can be adjusted to optimize the collection of photoelectrons from the photocathode. The last three or four dynodes are decoupled with high-voltage, high-frequency capacitors (disc ceramics work well in this application). These capacitors prevent depression of the dynode-chain voltage on the last few dynodes when a large pulse of electrons passes through the tube (the secondary-electron current drawn from each dynode must be supplied by the dynode chain). For high-frequency applications the inclusion of small damping resistors (typically $50\,\Omega-1\,\mathrm{k}\Omega$) between the decoupling capacitors and the dynodes is recommended. The resistor R_A, the anode resistor, determines the actual voltage that will be produced by a given anode current, since the photomultiplier serves as a current source. R_A should not be so large that the voltage developed across it is significant compared with the voltage between anode and last dynode. A value of R_A between 1 and 10 MΩ is usual. Lower values than this can be obtained by terminating the signal lead from the anode with a second resistor. In many cases, the effective value of the anode load is set by the input impedance of the electronics to which the tube is connected. In high-frequency applications, such as single-photon counting, the signal from R_A should be coupled out through a 50-Ω coaxial cable terminated in 50 Ω. In such applications the cable connection to the anode should be shielded as close to the anode as possible, even going so far as to insert R_A under the grounded screen of the coaxial cable. Some high-frequency tubes are supplied with an integral coaxial connector on the anode. In any case, the aim is to reduce parasitic inductance and capacitance associated with the anode connection. For example, even with a 50-Ω load a parasitic capacitance greater than 20 pF will limit the response time of the tube to 1 ns.

The actual response-time behavior of the photomultiplier can be determined by observing its single-photoelectron response. This is done by looking at the anode pulses with a fast oscilloscope. The photocathode should not need to be illuminated for this to be done; sufficient noise pulses will usually be observed. The pulses should appear as in Figure 4.123. If they have too much of an exponential tail they are probably limited by the anode resistor and parasitic capacitance. These anode pulses reflect the time distribution and number of secondary

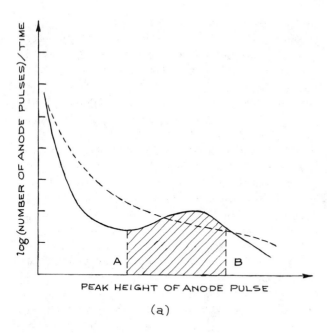

(a)

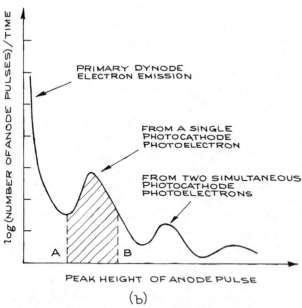

(b)

Figure 4.127 Schematic photomultiplier anode pulse-height distributions: (*a*) two forms likely to be observed in practice; (*b*) idealized form from tube with dynodes having a high and well-defined secondary emission coefficient. The best signal/noise ratio would be obtained in a photon-counting experiment by collecting only anode pulses in a height range roughly indicated by the shaded region *AB*.

electrons reaching the anode following single (or multiple) photoelectron emissions from the photocathode. If the height distribution of anode pulses is measured, a distribution such as is shown in Figure 4.127(*a*) will probably be seen. The idealized distribution shown in Figure 4.127(*b*) may be seen from newer tubes with GaP dynodes, which have a very high δ-value.

(ii) *Mounting photomultiplier tubes.* Photomultiplier tubes are generally mounted by clamping their plastic socket inside a cylindrical tube housing. The housing should be light-tight: a photomultiplier should never be exposed to ambient lighting when its high-voltage dynode chain is on, for the resultant large secondary-electron current will disintegrate the last dynodes and may strip the photocathode layer itself. If the total power dissipation of the dynode chain is low, the whole dynode chain may be mounted directly on the tube socket. However, if space is not at a premium, the dynode chain can be remote from the tube base and connected to it with high-voltage-insulated wire. The damping resistors, decoupling capacitors, and anode resistor should, however, be kept in close proximity to the tube. The photomultiplier should be mounted so there is no optical path from dynode chain to photocathode.

Photomultipliers are fragile and should be mounted so they are not subject to stress. This is particularly important if a tube is to be cooled. Rugged (and expensive) photomultipliers that can withstand severe vibrations and accelerations are available from EMR.

(e) Noise in Photomultiplier Tubes. Noise in photomultipliers comes from several sources:

1. Thermionic emission from the photocathode.

2. Thermionic emission from dynodes.

3. Field emission from dynodes (and photocathode) at high interdynode voltages.

4. Radioactive materials in the tube envelope (for example, ^{40}K in glass).

5. Electrons striking the tube envelope and causing fluorescence.

6. Electrons striking the dynodes and causing fluorescence.

7. Electrons colliding with residual atoms of vapor in the tube (cesium for example) and causing fluorescence.

8. Cosmic rays.

Noise from photomultipliers is always greater after they have been exposed to light (without high voltage applied). They gradually become quieter after operation under dark conditions for an extended period.

Noise from thermionic emission can be reduced considerably by cooling the tube (particularly for S-1 tubes). A factor of 10–1000 reduction in noise can be obtained by cooling to about $-20°C$. S-1 tubes need to be cooled to low temperatures, but temperatures below about $-70°C$ are not recommended. Too much cooling has a marginal effect on reducing thermionic emission—which will already be negligible for most tubes at $-20°C$—and may have deleterious effects. At low temperatures the conductivity of the photocathode layer falls, which causes it to no longer be an equipotential. The resultant distorted field distribution between cathode and dynodes will increase the possibility of noise from sources such as item 5 above. Two convenient designs of photomultiplier tube-coolers are shown in Figure 4.128. In the design shown in Figure 4.128(a), a copper or aluminum cylinder encloses the tube (without touching it). This tube is in thermal contact with a chamber containing Dry Ice (or liquid nitrogen). Condensation on the face of the photomultiplier tube is minimized, since moisture preferentially condenses and freezes on the much colder metal tube. In the second design, shown in Figure 4.128(b), cold nitrogen gas boiled off from a liquid-nitrogen Dewar flows through a copper tube surrounding the photomultiplier, and finally sprays over its photocathode surface to minimize water-vapor condensation. The rate of supply of cold nitrogen gas can be adjusted to provide a range of temperatures.

Noise from field emission may be a problem with certain high-voltage, high-gain tubes; one solution is to reduce the voltage. In single-photon-counting experiments, larger than normal anode pulses may result from field emission and can be rejected with an appropriate window discriminator.[111]

Noise from radioactive envelope materials should generally be negligible. Noise from electrons striking the tube envelope can be reduced by a shield around the tube held at photocathode potential. Noise from electrons striking the dynodes and causing fluorescence causes the fewest problems in Venetian-blind tubes, which have little or no direct optical path from dynodes to photocathode. Noise from gas in the tube can be a problem in old tubes, which frequently become gassy; solution: buy a new tube. Noise from cosmic rays is usually unimportant; it can be reduced by shielding.

Of all these sources of noise, thermionic emission from the photocathode is generally by far the most important. For a given photocathode material the total noise will depend on the photocathode area. Thus, under comparable conditions tubes with small photocathode area like the S-20 ITT FW-130, with a photocathode diameter of 0.25 cm, are less noisy than tubes with large photocathodes, like the 5-cm photocathode of the S-20 Phillips 56 TVP. In many experiments a large photocathode area is unnecessary—light coming through a monochromator slit illuminates a very small area. The effective photocathode area, and consequently its thermionic noise, can be reduced by wrapping a magnetic coil around the photocathode, which prevents electrons other than those from the center of the photocathode from reaching the first dynode. Under circumstances where photomultipliers must be operated in close proximity to magnetic fields, magnetic shields to enclose the tube are available from MμShield Company, from Perfection Mica Company, or from the various tube suppliers.

In the anode pulse-height distributions shown in Figure 4.127, small anode pulses are much more likely to correspond to noise than are pulses in the middle of the distribution. In photon-counting experiments these small pulses should be rejected with a discriminator. A window discriminator provides simultaneous rejection of these pulses and those which are much larger than average.[111]

(f) Ultraviolet and Vacuum-Ultraviolet Detection with Photodiodes and Photomultipliers. For detection of radiation down to about 105 nm, phototubes with windows of LiF are available, as are tubes with windows of MgF_2 or sapphire. For detection of even shorter

Figure 4.128
Photomultiplier-tube coolers: (*a*) using Dry Ice reservoir in thermal contact with a metal tube close to, but electrically insulated from, the photocathode; (*b*) using cold nitrogen gas circulated around tube and sprayed onto photocathode surface (a thermistor monitors temperature and allows control of the rate of supply of cold gas).

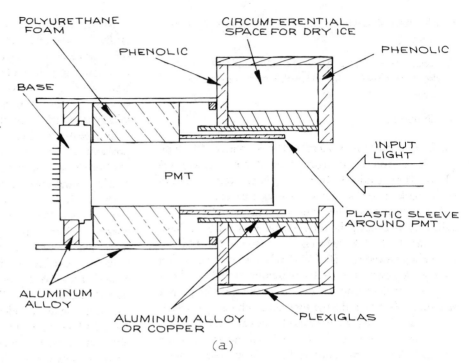

(a)

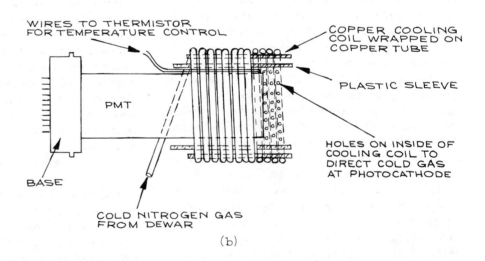

(b)

wavelengths windowless phototubes or channeltrons are used. For detection of radiation down to 58 nm or beyond, conventional photomultiplier tubes sensitive to visible light (S-11) can be used if the outer surface is coated with a layer of sodium salicylate—or if a sodium-salicylate-coated disc is placed close to the pho-

tocathode. Sodium salicylate fluoresces in the visible with almost unity quantum efficiency and converts incident vacuum-ultraviolet radiation to a wavelength where a conventional tube can detect it. The response of the sodium salicylate is also fast, so that 1 ns ultraviolet detection is possible. To prepare a sodium-salicylate-coated window a solution of sodium salicylate in reagent-grade methanol (80 g liter^{-1}) is atomized and sprayed onto the window, which is placed about 10 cm above the atomizer. Commercial nasal sprays pressurized with dry nitrogen are suitable for this purpose. The atomizer should be far enough from the window so that no large drops can be deposited. The method works better if the window is kept warm (50–70°C). Spraying should be continued until the density of the sodium salicylate layer reaches about 20 g m^{-2}, which can be determined by weighing.

4.8.3 Photoconductive Detectors

Photoconductive detectors can operate through either intrinsic or extrinsic photoconductivity. The physics of intrinsic photoconductivity is illustrated in Figure 4.129(a). Photons with energy $h\nu > E_g$ excite electrons across the band gap. The electron-hole pair that is thereby created for each photon absorbed leads to an increase in conductivity—which comes mostly from the electrons. Semiconductors with small band gaps respond to long-wavelength infrared radiation but must be cooled accordingly; otherwise thermally excited electrons swamp any small photoconductivity effects. Table 4.9 lists commonly available intrinsic photoconductive detectors together with their usual operating temperature and the limit of their long-wavelength response, λ_0, together with some representative figures for detectivities and time constants. Note that silicon and germanium

are also operated in both photovoltaic and avalanche modes (see Section 4.8.4).

If a semiconductor is doped with an appropriate material, impurity levels are produced between the valence and conduction bands as shown in Figure 4.129(b). Impurity levels that are able to accept an electron excited from the conduction band are called *acceptor levels*, whereas impurity levels that can have an electron excited from them into the conduction band are called *donor levels*. Thus, in Figure 4.129 photons with energy $h\nu > E_A$ excite an electron to the impurity level, leaving a hole in the valence band and thereby giving rise to p-type extrinsic photoconductivity. Photons with energy $h\nu > E_D$ will excite an electron into the conduction band, giving n-type extrinsic photoconductivity. For example, gold-doped germanium has an acceptor level 0.15 eV above the valence band and is an extrinsic p-type photoconductor, as is copper-doped germanium, which has an acceptor level 0.041 eV above the valence band. These are the two most commonly used extrinsic photoconductive detectors, responding out to about 9 μm and 30 μm, respectively. Curves showing the variation of their D^* with wavelength are given in Figure 4.130. Table 4.10 lists the operating characteristics of these and some other commercially available extrinsic photoconductive detectors. Detectors listed with operating temperatures below 28 K will in practice frequently be operated at 4 K, since liquid helium is a safe, available coolant, although its use involves more complicated technology than the use of liquid nitrogen.

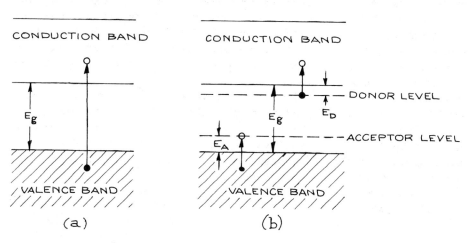

Figure 4.129 Mechanism for (a) intrinsic photoconductivity; (b) extrinsic photoconductivity.

Table 4.9 INTRINSIC PHOTOCONDUCTIVE DETECTORS

Semiconductor	T (K)	E_g (eV)	λ_0 (μm)	D^*(max) (cm Hz$^{1/2}$ W^{-1})	τ
CdS	295	2.4	0.52	3.5×10^{14}	$\simeq$ 50 ms
CdSe	295	1.8	0.69	2.1×10^{11}	$\simeq$ 10 ms
Si	295[a]	1.12	1.1	$\leqslant 2 \times 10^{12}$	—[b]
Ge	295[a]	0.67	1.8	$\leqslant 10^{11}$	10 ns[c]
PbS	295	0.42	2.5	$\leqslant 2 \times 10^{11}$	—[d]
	195	0.35	3.0	$\leqslant 5 \times 10^{11}$	—[d]
	77	0.32	3.3	$\leqslant 8 \times 10^{11}$	—[d]
PbSe	295	0.25	4.2	$1 \times 10^9 - 5 \times 10^9$	1 μs
	195	0.23	5.4	$1.5 - 4 \times 10^{10}$	30–50 μs
	77	0.21	5.8	$\leqslant 3 \times 10^{10}$	50 μs
InSb[e]	77	$\simeq$ 0.23	5.5–7.0	$\leqslant 3 \times 10^{10}$	0.1–1 μs
Hg$_{0.8}$Cd$_{0.2}$Te	77	$\simeq$ 0.1	12–25	$10^9 - 10^{11}$	> 1 ns

[a] Increased sensitivity can be obtained by cooling.
[b] Detectors with time constants from 50 ps upward and various detectivities are available.
[c] Detectors operated in a photovoltaic mode with time constants from 120 ps upward and various detectivities are available.
[d] Detectors with time constants ranging from about 100 μs to 10 ms and varying detectivities are available.
[e] More commonly operated in a photovoltaic mode.

All photoconductive detectors, whether intrinsic or extrinsic, are operated in essentially the same way, although there are wide differences in packaging geometry. These differences arise from differing operating temperatures and speed-of-response considerations. A schematic diagram which shows the main construction features of a liquid-nitrogen-cooled photoconductive or photovoltaic detector is given in Figure 4.131. Uncooled detectors can be of much simpler construction—for example, in a transistor or flat solar-cell package.

One feature of the cooled detector design shown in Figure 4.131 is worthy of note. The field of view of the detector is generally restricted by an aperture, which is kept at the temperature of the detection element. This shields the detector from ambient infrared radiation, which peaks at 9.6 μm. For detection of low-level narrow-band infrared radiation the influence of background radiation can be further reduced by incorporating a cooled narrow-band filter in front of the detector element. The filter will only radiate beyond the cutoff wavelength of the detector, and it restricts transmitted ambient radiation to a narrow band. Liquid-helium-cooled detectors generally have a double Dewar; an outer one for liquid nitrogen surrounds the inner for liquid helium. Liquid-nitrogen-cooled detectors, such as HgCdTe, are recommended in preference to liquid-helium-cooled detectors, such as Ge:Cu, whenever there is a choice. Some miniature-package commercial detectors can be cooled with Joule-Thompson refrigeration units driven with compressed gas, which eliminates the need for externally supplied liquified-gas coolant. Thermoelectrically cooled (Peltier effect) infrared detector packages are available from Santa Barbara Research Center for the operation of PbS or PbSe detectors down to temperatures of 193 K. Further details of the operating principles behind these and other refrigeration techniques used with infrared detectors are given by Hudson.[37]

Figure 4.132 shows a basic biasing circuit commonly used for operating photoconductive detectors. R_d is the detector dark resistance. It is easy to see that the change in voltage, ΔV, that appears across the load resistor R_L for a small change ΔR in the resistance of the detector is

$$\Delta V = \frac{-V_0 R_L \Delta R}{R_d + R_L}. \qquad (236)$$

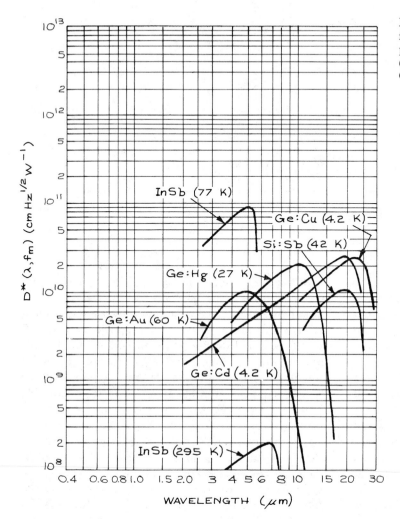

Figure 4.130 $D^*(\lambda)$ as a function of wavelength for various photoconductive detectors. (Courtesy of Hughes Aircraft Company.)

This is at a maximum when $R_L = R_d$. Thus it is common practice to bias the detector with a load resistance equal to the detector's dark resistance. The bias voltage is selected to give a bias current through the detector that gives optimum detectivity. This bias current will generally be specified by the manufacturer.

To obtain fast response from a photoconductive detector, great care must be taken to minimize the stray capacitance C_S in the input circuit to the preamplifier. Otherwise, the time constant of the detector will be limited by $R_L C_S$. The ultimate limits to the speed of an actual detector are set by its internal capacitance C_d and

the majority-carrier lifetime. Many commercial detectors are manufactured so that the stray capacitance of the detector and its connection leads is very small. This is generally true of high-speed commercial detectors packaged in all-metal Dewars with integral coaxial bias connections. Metal Dewar packages are preferable to glass ones; although the latter are cheaper, they are very fragile.

If a detector is not specifically packaged to minimize stray inductance, it is possible to reduce the stray capacitance of the detector package and improve its speed of response by a technique called "bootstrapping,"

Table 4.10   EXTRINSIC PHOTOCONDUCTIVE DETECTORS

Semiconductor	Impurity	T (K)	λ_0 (μm)	D^*	τ	Suppliers[a]
Ge:Au[b]	p-type	77	8.3	$3 \times 10^9 - 10^{10}$	30 ns	A, B, S
Ge:Hg[c]	p-type	<28	14	$1-2 \times 10^{10}$	> 0.3 ns	A, M, S, T
Ge:Cd	p-type	<21	21	$2-3 \times 10^{10}$	10 ns	S
Ge:Cu[c]	p-type	<15	30	$1-3 \times 10^{10}$	> 0.4 ns	A, M, S, T
Ge:Zn[b,c]	p-type	<12	38	$1-2 \times 10^{10}$	10 ns	S
Ge:Ga	p-type	4	115	2×10^{10}	10 ns	A
Ge:In	p-type	4	111	—	< 1 μs	A
Si:Ga	p-type	<25	17	$10^9 - 10^{10}$	0.1 μs	S
Si:As	n-type	<20	23	$1-3.5 \times 10^{10}$	0.1 μs	S

[a] A: Advanced Kinetics; B: Barnes Engineering; M: Mullard; S: Santa Barbara Research Center; T: Texas Instruments.
[b] Sometimes also contain silicon.
[c] Sometimes also contain antimony.

which is illustrated in Figure 4.133. All bias leads to the detector and the leads to the the preamplifier are double-shielded. The preamplifier should ideally have a high input impedance, a low input capacitance, a wide bandwidth, and 50-ohm output impedance. Such preamplifiers can be constructed, or bought from detector suppliers or Perry Amplifier. The inner shield is not grounded, but is connected directly to the low-impedance output of the unit-gain preamplifier. Thus the inner shield is "bootstrapped" to the signal voltage and stray capacitance is effectively eliminated. By this means the speed of response of a detector can be improved substantially, say from 1 μs down to 50 ns. An alternative way to improve the speed of response, at the expense of signal (but not detectivity), is to reduce the value of the load resistor.

A few photoconductive detectors are worthy of brief

extra comment. Silicon and germanium are much more commonly used for photodiodes, frequently in an avalanche mode. These devices are discussed in Section 4.7.4. Lead sulfide detectors have high impedance, 0.5 to 100 MΩ, and slow response, but are the most sensitive detectors in the spectral region between 1.2 and 3 μm and can be used uncooled. $D^*(\lambda)$ curves for these detectors are shown in Figure 4.134. They are available from Sanders Associates, Hamamatsu, Infrared Industries, Optoelectronics, and Santa Barbara Research Center. Lead selenide is sensitive to longer wavelengths than lead sulfide, but InAs operated in a photovoltaic mode is to be preferred in this spectral region (3–4 μm). Gold-doped germanium is a simple detector to use in the spectral region between 1 and 9 μm; it is superior to cooled PbSe in this region. InSb operated in a photovoltaic mode is probably the detector of choice between

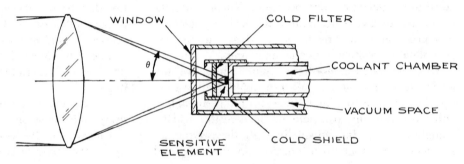

Figure 4.131   Radiation-shielded liquid-nitrogen-cooled photoconductive or photovoltaic infrared detector assembly.

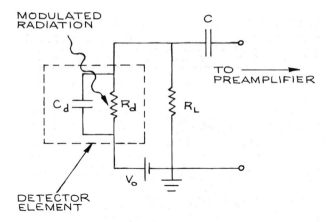

Figure 4.132 Simple biasing circuit for operating a photoconductive detector with modulated radiation.

3 and 5.6 μm, but is somewhat more difficult to operate. Beyond about 8 μm, HgCdTe is the detector of choice and is to be preferred to detectors such as Ge:Hg and Ge:Cu, which must be operated below liquid-nitrogen temperature. Gold-doped germanium is still sufficiently responsive at 10.6 μm to be used as a CO_2 laser detector. Ge:Au detectors are available from Santa Barbara Research Center, Advanced Kinetics, Barnes Engineering, and Judson Infrared. HgCdTe detectors are available from Infrared Associates, Santa Barbara

Research Center, Spectronics, Honeywell, Société Anonyme Télécommunications (SAT), Princeton Infrared Equipment, and New England Research. For further details of the above detectors, the reader should consult References 36, 37, and 112. Suppliers of other detectors are listed in Reference 15.

The use of extrinsic photoconductivity for the detection of far-infrared radiation requires the introduction of appropriate doping material into a semiconductor in order to generate an acceptor or donor impurity level extremely close to the valence or conduction bands, respectively. Well-characterized impurity levels can be generated in germanium in this way using gallium,[113] indium,[114] or boron,[115] doping, but the long-wavelength sensitivity limit is restricted to about 120 μm. Longer-wavelength-sensitive, extrinsic photoconductivity can be observed in appropriately doped InSb in a magnetic field. However, bulk InSb can be used more efficiently for infrared detection in a different photoconductive mode entirely.[116] Even at the low temperature at which far-infrared photoconductive detectors operate (≤ 4 K), there are carriers in the conduction band. These free electrons can absorb far-infrared radiation efficiently and move into higher-energy states within the conduction band. This change in energy results in a change of mobility of these free electrons, which can be detected as a change in conductivity, using a circuit such as the

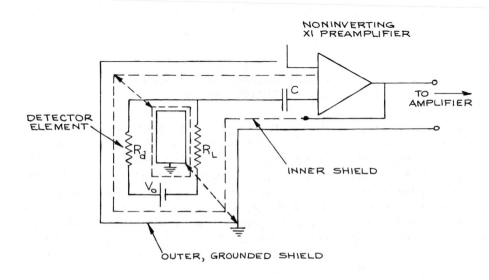

Figure 4.133 Arrangement for "bootstrapping" an infrared detector to minimize effects of parasitic capacitance.

Figure 4.134 D^* as a function of wavelength for various lead-salt photoconductive detectors. (Courtesy of Hughes Aircraft Company.)

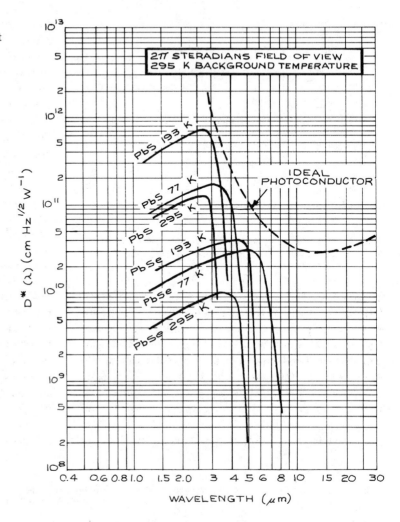

one shown previously in Figure 4.132. Some typical wavelength response curves are shown in Figure 4.135. These *hot-carrier-effect* photoconductive detectors can be used successfully over a wavelength range extending from 50 to 10^4 μm; they have detectivities up to 2×10^{12} cm Hz$^{1/2}$ W^{-1} and response times down to 10 ns or less. They are frequently operated in a large magnetic field (several hundred kA/m or more). These detectors are available from Mullard, Texas Instruments, Raytheon, and Advanced Kinetics. These detectors can also be tuned in wavelength response by operating them in a magnetic field.[116, 117]

4.8.4 Photovoltaic Detectors (Photodiodes)

In a photovoltaic detector photoexcitation of electron-hole pairs occurs near a junction when radiation of energy greater than the band gap is incident on the junction region. Extrinsic photoexcitation is rarely used in photovoltaic photodetectors. The internal energy barrier of the junction causes the electron and hole to separate, creating a potential difference across the junction. This effect is illustrated for a *p-n* junction in Figure 4.136. Other types of structure are also used, such as

4.2 K
1.45 K
7.08
3.5
8.83
8.9
0
0
500
1000
1500
WAVELENGTH (μm)
ℓn ℛ (ARBITRARY UNITS)

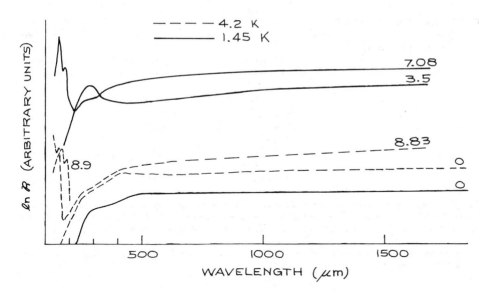

Figure 4.135 Responsivity as a function of wavelength of a hot-carrier InSb photoconductive detector operated at two different temperatures and various magnetic field strengths. The number on each curve is the magnetic field strength in units of 10^5 A/m.

p-i-n, Schottky-barrier (a metal deposited onto a semiconductor surface) and heterojunction (a junction between two different semiconductors). The *p-n* and *p-i-n* structures are the most commonly used. All these devices are commonly called photodiodes. The characteristics of some important photodiodes are listed in Table 4.11. Important photodiodes include silicon for detection of radiation between 0.1 and 1.1 μm, germanium for use between 0.4 and 1.8 μm, indium arsenide between 1 and 3.8 μm, indium antimonide between 1 and 7 μm, lead-tin telluride between 2 and 18 μm, and mercury-cadmium telluride between 1 and 12 μm. Some

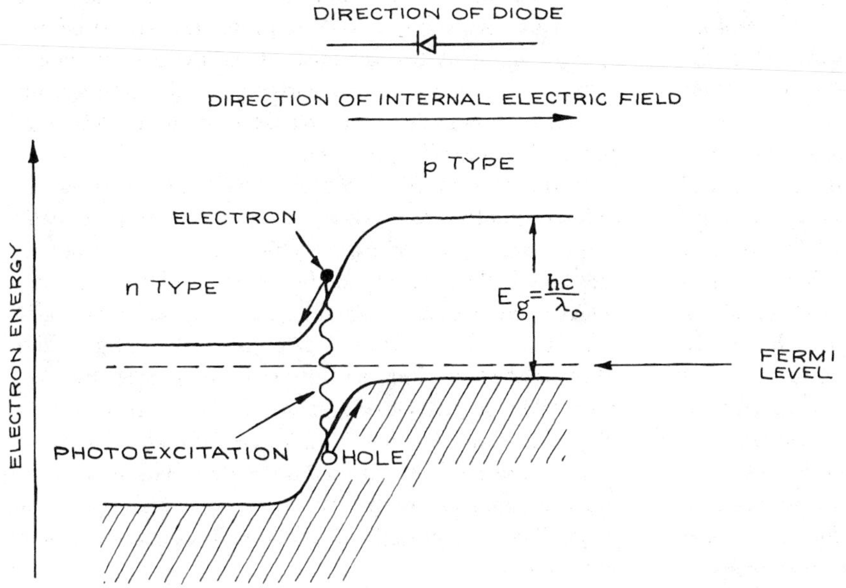

DIRECTION OF DIODE
DIRECTION OF INTERNAL ELECTRIC FIELD
p TYPE
ELECTRON
n TYPE
$E_g = \frac{hc}{\lambda_0}$
FERMI LEVEL
PHOTOEXCITATION
HOLE
ELECTRON ENERGY

Figure 4.136 Photoexcitation at a *p-n* junction.

Table 4.11 PHOTOVOLTAIC DETECTORS (PHOTODIODES)

Semiconductor	T (K)	Wavelength Range (μm)	D^*(max)	τ	Suppliers[a]
Si	300	0.2–1.1	$\leqslant 2 \times 10^{13}$	—	—[b]
Ge	300	0.4–1.8	10^{11}	0.3 ns	(1)
InAs	300	1–3.8	$\leqslant 4 \times 10^{9}$	5 ns–1 μs	(2)
InAs	77	1–3.2	4×10^{11}	0.7 μs	(2)
InSb	300	1–7	1.5×10^{8}	0.1 μs	(3)
InSb	77	1–5.6	$\leqslant 2 \times 10^{11}$	> 25 ns	(4)
PbSnTe	77	2–18[c]	$\leqslant 10^{11}$	20 ns–1 μs	(5)
HgCdTe	77	1–25[c]	10^{9}–10^{11}	> 1.6 ns	(6)

[a](1) Available from Judson Infrared and Ford Aerospace, among others; (2) Judson Infrared and Barnes Engineering; (3) Optoelectronics; (4) Santa Barbara Research Center (SBRC), Spectronics, SAT, Ford Aerospace, Hamamatsu, and Advanced Kinetics; (5) SBRC, Barnes Engineering, and Plessey Optoelectronics; (6) Infrared Associates, New England Research, Ford Aerospace, Honeywell Electrooptics Center, SBRC, Spectronics, and SAT.
[b]A very wide range of silicon photoconductive and photovoltaic detectors are available with NEP figures down to 10^{-16} W Hz$^{1/2}$ and time constants down to 6 ps. Reference 15 contains a list of suppliers.
[c]Range varies from one supplier to another.

typical curves of $D^*(\lambda)$ are shown in Figure 4.137. These spectral response regions are not all necessarily covered by a detector operating at the same temperature; for example, InSb responds to 7 μm at 300 K but to wavelengths no longer than 5.6 μm at 77 K. The wavelength response of PbSnTe and HgCdTe depends also on the stoichiometric composition of the crystal. All these photodiodes have very high quantum efficiency, defined in this case as the ratio of photons absorbed to mobile electron-hole pairs produced in the junction region. Values in excess of 90% have been observed in the case of silicon.

When a photodiode detector is illuminated with radiation of energy greater than the band gap, it will generate a voltage and can be operated in the very simple circuit illustrated in Figure 4.138(a). However, it is much better to operate a photodiode detector in a reverse-biased mode, as shown in Figure 4.138(b), where positive voltage is applied to the n-type side of the junction and negative to the p-type. In this case, the observed photosignal is seen as a change in current through the load resistor. The difference between the two modes of operation can be easily seen from Figure 4.139(a), which shows the I-V characteristic of a photodiode in the dark and in the presence of illumination. A photodiode responds much more linearly to changes in light intensity and has greater detectivity when operated in the reverse-biased mode. Ideal operation is obtained when the diode is operated in the current mode with an operational amplifier that effectively holds the photodiode voltage at zero—its optimum bias point. A simple practical circuit which can be used to operate a photodiode in this way is shown in Figure 4.139(b). In this circuit, the bias voltage V_B is not necessary, but for many photodiodes will improve the speed of response, albeit at the expense of an increase in noise. Integrated packages incorporating a photodiode and operational amplifier are available from EG&G and Silicon Detector Corporation.

If the reverse bias voltage on a photodiode is increased, photoinduced charge carriers can acquire sufficient energy transversing the junction region to produce additional electron-hole pairs. Such a photodiode exhibits current gain and is called an *avalanche photodiode*. It is in some respects the solid-state analog of the photomultiplier. The avalanche process does not increase the signal-to-noise ratio of the detector—it is likely to reduce it—but its internal gain allows the use of higher-signal-level external amplifiers, which are very much easier to construct than low-noise high-gain preamplifiers for conventional photodiode operation at low light levels.

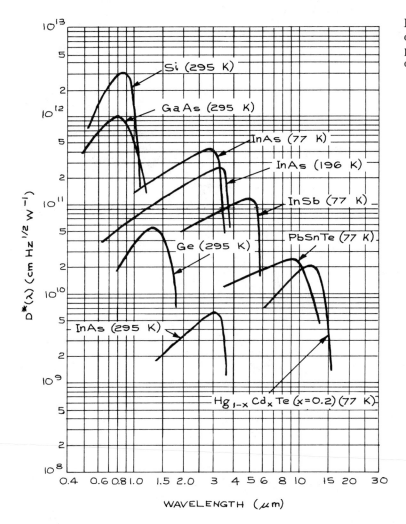

Figure 4.137 D^* as a function of wavelength for various photovoltaic detectors. (Courtesy of Hughes Aircraft Company.)

4.8.5 Thermal Detectors

Thermal detectors, in principle, have a detectivity that is independent of wavelength from the vacuum ultraviolet upward, as shown in Figure 4.140. However, the absorbing properties of the "black" surface of the detector will, in general, show some wavelength dependence, and the necessity for a protective window on some detector elements may limit the useful spectral bandwidth of the device. Most commonly available thermal detectors, although by no means as sensitive as various types of photon detectors, achieve spectral response very far into the infrared—to the microwave region, in fact—while conveniently operating at room temperature. The operating characteristics of some commercially available thermal detectors are given in Table 4.12. Each of these detectors is discussed briefly below. Putley[118] gives a more detailed discussion.

(a) Thermopile. Thermopiles, although they are one of the earliest forms of infrared detector, are still widely

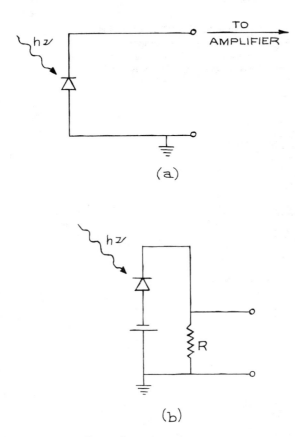

Figure 4.138 Photovoltaic detector operated in (a) open-circuit mode; (b) reverse-biased mode.

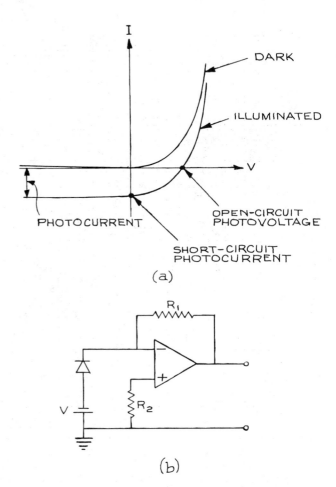

Figure 4.139 (a) Current-voltage characteristics of a photodiode; (b) current-mode-operated photodiode.

used. Their operation is based on the Seebeck effect, where heating the junction between two dissimilar conductors generates a potential difference across the junction. An ideal device should have a large Seebeck coefficient, low resistance (to minimize ohmic heating), and a low thermal conductivity (to minimize heat loss between the hot and cold junctions of the thermopile). These devices are usually operated with an equal number of hot (irradiated) and cold (dark) junctions, the latter serving as a reference to compensate for drifts in ambient temperature. Both metal (copper-constantan, bismuth-silver, antimony-bismuth) and semiconductor junctions are used as the active elements. The junctions can take the form of evaporated films, which improves the robustness of the devices and reduces their time

constant, although this is still slow (0.1 ms at best). Because a thermopile has very low output impedance, it must be used with a specially designed low-noise amplifier, or with a step-up transformer, as shown in Figure 4.141. Such transformers can be conveniently built in the laboratory by winding the primary and secondary coils on a small ferrite torus.

Although thermopiles, along with other thermal detectors, do not have absolutely flat spectral response over unlimited wavelength regions, they can be remarkably flat within restricted regions—in the visible or from 1 to 10 μm, for example. In addition, such detec-

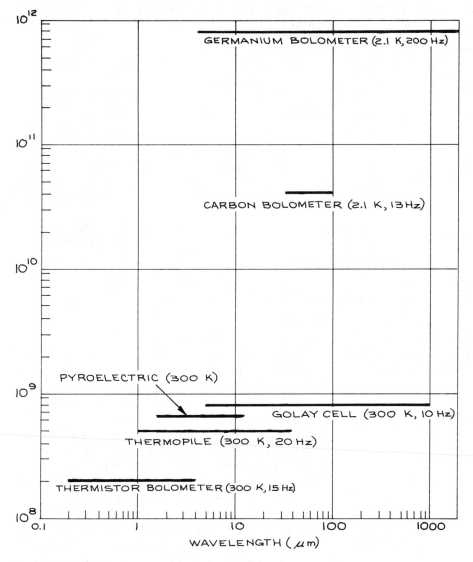

Figure 4.140 Typical $D^*(\lambda)$ curves for various thermal detectors assuming total absorption of incident radiation. The operating temperature, modulation frequency, and typical useful wavelength range are shown for each detector. In each case, the detector is assumed to view a hemispherical surround at a temperature of 300 K.

tors with calibrations traceable to the national Bureau of Standards are available. Thus they are invaluable in the absolute calibration (both for radiant sensitivity and for spectral response) of light sources, detectors, and spectrometers.

(b) Pyroelectric Detectors. These are detectors that utilize the change in surface charge that results when

certain asymmetric crystals (ones that can possess an internal electric dipole moment) are heated. The crystalline material is fabricated as the dielectric in a small capacitor, and the change in charge is measured when the element is irradiated. Thus, these devices are inherently a.c. detectors. If the chopping frequency of the input radiation is slow compared to the thermal relaxation time of the crystal, the crystal remains close to thermal equilibrium and the current response is small.

Table 4.12   CHARACTERISTICS OF COMMERCIALLY AVAILABLE THERMAL DETECTORS

Device	D^* ($\mathrm{cm\,Hz}^{1/2}\mathrm{W}^{-1}$)	Time Constant	Suppliers [a]
Thermopile	$(1-4)\times 10^8$	20 μs–60 ms	Barnes Engineering, Oriel, Horiba Instruments, Eppley
Pyroelectric	10^6-10^9	> 100 ps	Molectron, Barnes Engineering, Oriel, Laser Precision, Plessey Opto-electronics, New England Research
Bolometer	$2.5-10^8$	1 ms	Barnes Engineering, Servo Corporation
Golay Cell	NEP ($\mathrm{W\,Hz}^{1/2}$) 7×10^{-11}	$\simeq$ 10 ms	Oriel

[a] Not intended to be an exhaustive list; for further information, see Reference 15.

When the chopping period becomes shorter than the thermal relaxation time, much greater heating and current response results. The responsivity of the detector in this case can be written as

$$R = \frac{p(T)}{\rho C_p d} \quad (\mathrm{A\,W}^{-1}), \qquad (237)$$

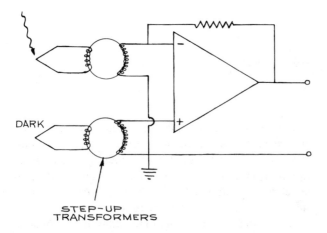

Figure 4.141   Operating circuit for a thermopile using a dark junction (or junctions) and a differential amplifier for compensation.

where $p(T)$ is the pyroelectric coefficient at temperature T, d is the spacing of the capacitor electrodes, and ρ and C_p are the density and specific heat of the crystal, respectively.

The equivalent circuit of a pyroelectric detector is a current source in parallel with a capacitance, which can range from a few to several hundred picofarads. For optimum performance, the resultant high impedance must be matched to a high-input-impedance, low-output-impedance amplifier. Two examples of such circuits are given in Figure 4.142. Complete detector-amplifier packages are available from Molectron. For fastest response the capacitance of the detector must be shunted with a small resistor (though this reduces the detectivity). Response times as short as 2 ps have been seen in detectors operated in this way without amplification when illuminated with high-intensity mode-locked laser pulses.

Pyroelectric detectors are robust and have frequency responses extending from a few hertz to 100 GHz or so. Their detectivities are comparable with those of thermopiles, and they also have flat spectral response. Consequently, they can replace the thermopile in many applications as a convenient, room temperature, wide-spectral-sensitivity detector of infrared and visible light.

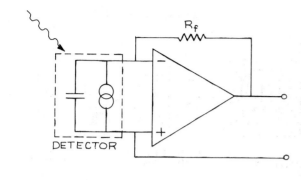

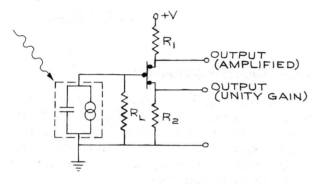

Figure 4.142 Two examples of operating circuits for pyroelectric detectors.

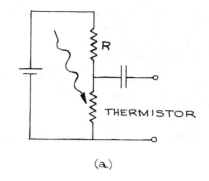

(a)

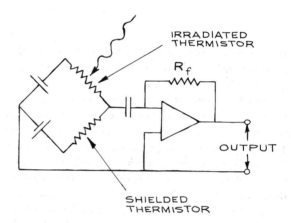

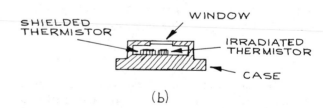

(b)

Figure 4.143 Operating circuits for thermistor bolometers: (*a*) simple bias circuit; (*b*) bridge circuit using compensating shielded thermistor, with device construction shown.

Pyroelectric detectors are widely used for the detection of very short-duration infrared laser pulses. They are much more sensitive than photon-drag detectors[119] for this purpose. However, high-energy pulses generate acoustic waves in the detector crystal, which give rise to spurious signals. These acoustic signals are more of a problem in the observation of long laser pulses (≥ 100 ns) than they are with short pulses.

(c) Bolometer. The resistance of a solid changes with temperature according to a relation of the form

$$R(T) = R_0 \left[1 + \gamma (T - T_0) \right], \qquad (238)$$

where γ is the temperature coefficient of resistance, typically about 0.05 K^{-1} for a metal, and R_0 is the resistance at temperature T_0.

A *bolometer* is constructed from a material with a large temperature coefficient of resistance. Absorbed

radiation heats the bolometer element and changes its resistance. Bolometers utilize metal, semiconductor, or almost superconducting elements. Metal bolometers utilize fine wires (platinum or nickel) or metal films. The mass of the element must be kept small in order to maximize its temperature rise. Even so, the response time is fairly long (≥ 1 ms). Semiconductor bolometer

Figure 4.144 Schematic design of the Golay detector. The top half of the line grid is illuminated by the LED and imaged back on the lower half of the grid by the flexible mirror and meniscus lens. Any radiation-induced deformation of the flexible mirror moves the image of the line grid and changes the illumination reaching the photodiode.

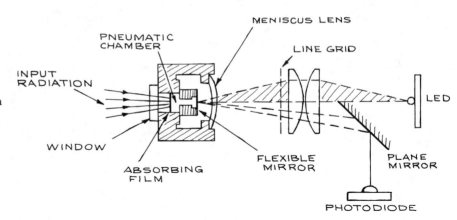

elements (thermistors) have larger absolute values of γ and have largely replaced metals except where very long-term stability is required.

Bolometer elements can be operated in several ways; a simple bias circuit for a single element is shown in Figure 4.143(a). However, is is usual to operate the elements in pairs in a bridge circuit, as shown in Figure 4.143(b). One element is irradiated, while the second serves as a reference and compensates for changes in ambient temperature. Thermistors have a negative I-V characteristic above a certain current and will exhibit destructive thermal runaway unless operated with a bias resistor. It is therefore usually best to operate the thermistor at currents below the negative-resistance part of its I-V characteristic.

(d) **The Golay Cell.** In a Golay cell (named for its inventor, M.J.E. Golay),[120] radiation is absorbed by a metal film that forms one side of a small sealed chamber containing xenon (used because of its low thermal conductivity). Another wall of the chamber is a flexible membrane, which moves as the xenon is heated. The motion of the membrane is used to change the amount of light reflected to a photodetector. The operating principle and essential design features of a modern Golay cell are shown in Figure 4.144. Although these detectors are fragile, they are very sensitive and are still widely used for far-infrared spectroscopy.

4.8.6 Detector Calibration

In certain cases the data supplied by a detector manufacturer on parameters such as $D^*(\lambda)$ or τ will only indicate a range of values within which the detector's characteristics will fall, or the data may not be sufficiently reliable. If necessary, more accurate detector calibration can be carried out.

To determine $D^*(\lambda)$ for a detector, its response must be measured with a source giving a known irradiance $E_e(\lambda)$ at the detector. This can be done either with a calibrated source such as a blackbody, or by comparison with another detector, such as a thermopile, which has a calibrated response. To determine the temporal response function of a detector operated in a given configuration, all that is necessary is to irradiate the detector with a very short light pulse and record the output pulse shape on a fast oscilloscope. For response functions below about 1 ns a sampling oscilloscope should be used in conjunction with repetitive short-light-pulse irradiation of the detector. For detectors with response faster than about 1 ns, sufficiently short light pulses can be obtained from a mode-locked Nd: YAG, Nd: glass, ruby, or dye laser. Pulses from mode-locked argon or CO_2 lasers may not be short enough to calibrate a very fast detector. To obtain short light pulses outside conveniently available wavelength regions, nonlinear harmonic generation or mixing schemes can be used, but the difficulties of such techniques can be severe. To determine the time response of a detector that responds slower than 1–2 ns, any pulsed light source of much shorter duration than the detector response can be used. Short-duration ($\simeq$ 10 ns) flashlamps are available from EG & G and Xenon Corporation.

Very-short-duration, low-energy, pulsed light sources are easily made in the laboratory. A very simple design is shown in Figure 4.145. A high-voltage discharge between two electrodes spaced by a few millimeters or less in a high-pressure gas is used. Hydrogen works very well, and even air at atmospheric pressure is satisfactory. To obtain a short-duration flash, only the self-capacitance C_s of the electrodes must be allowed to discharge. To accomplish this a small charging resistor must be placed very close to the high-voltage electrode; a larger series charging resistor can be placed farther from the electrode. The tips from ballpoint pens make excellent electrodes in this application. The flash duration is proportional to L/p, where L is the electrode spacing and p the gas pressure. The breakdown voltage is proportional to L and also approximately to p: the capacitance of the electrode gap is proportional to L, so the flash energy $\frac{1}{2}C_s V^2$ is proportional to $L^3 p^2$. Thus, for short-duration, high-energy flashes, p should be high.

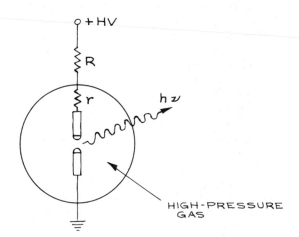

Figure 4.145 Simple design for high-pressure, nanosecond-duration pulsed light source. Generally $R \gg r$, where r ranges from 1 kΩ to 1 MΩ.

The repetition frequency of the lamp, which is most easily operated in a free-running mode, will depend on the charging time constant.

CITED REFERENCES

1. *Proceedings of the Symposium on Quasi-Optics*, New York, June 8–10, 1964, J. Fox, Ed., Polytechnic Press of the Polytechnic Institute of Brooklyn, N.Y., 1964.
2. *Proceedings of the Symposium on Submillimeter Waves*, New York, March 31–April 2, 1970, J. Fox, Ed., Polytechnic Press of the Polytechnic Institute of Brooklyn, N.Y. 1964.
3. K. M. Baird, D. S. Smith, and B. G. Whitford, "Confirmation of the Currently Accepted Value 299792458 Metres per Second for the Speed of Light," Opt. Commun., 31, 367–368, 1979.
4. Recommendation M2 of the comité consultatif pour la définition du mètre, June 1979.
5. G. W. C. Kaye and T. H. Laby, *Tables of Physical and Chemical Constants*, 14th edition, Longman, New York, 1972.
6. C. D. Coleman, W. R. Bozman, and W. F. Meggers, *Table of Wavenumbers*, U.S. Nat'l. Bur. Std. Monograph 3, Vols. 1–2, 1960.
7. M. Born and E. Wolf, *Principles of Optics*, 3rd edition, Pergamon Press, Oxford, 1965.
8. E. E. Wahlstrom, *Optical Crystallography*, 3rd edition, Wiley, New York, 1960.
9. W. A. Shurcliff, *Polarized Light*, Harvard University Press, Cambridge, Mass., 1962.
10. A. Yariv, *Introduction to Optical Electronics*, 2nd edition, Holt, Rinehart and Winston, New York, 1976.
11. H. Kogelnik and T. Li, "Laser Beams and Resonators," *Proc. IEEE*, 54, 1312–1329, 1966.
12. S. Ramo, J. R. Whinnery, and T. Van Duzer, *Fields and Waves in Communication Electronics*, Wiley, New York, 1967.
13. R. W. Ditchburn, *Light*, 2nd edition, Interscience, New York, 1963.
14. Diffraction-limited spherical lenses are available from Melles Griot, Optics for Research, Oriel, Special Optics, and J. L. Wood Optical Systems, among others (see following reference).

15. *Laser Focus Buyers Guide*, published annually by Advanced Technology Publications Inc., 1001 Watertown Street, Newton, MA 02165, lists a large number of suppliers of a wide range of optical components and systems.

16. The "float" method for manufacturing plate glass involves the drawing of the molten glass from a furnace, where the molten glass floats on the surface of liquid tin. The naturally flat surface of the liquid metal ensures the production of much larger sheets of better-flatness glass than was possible by earlier techniques.

17. L. Levi, *Applied Optics*, Wiley, New York, Vol. 1, 1968.

18. A. B. Meinel, "Astronomical Telescopes," in *Applied Optics and Optical Engineering*, Vol. 5, R. Kingslake, Ed., Academic Press, New York, 1969.

19. W. Brouwer and A. Walther. "Design of Optical Instruments," in *Advanced Optical Techniques*, A. C. S. Van Heel, Ed., North-Holland, Amsterdam, 1967.

20. Stimulated Raman scattering is a very useful technique for generating new wavelength from high-intensity, pulsed visible or ultraviolet lasers. A high-pressure gas cell specifically designed for this purpose is available from Quantaray. For further information about the technique, the following articles, and references therein, can be consulted: M. Bierry, R. Frey, and F. Pradère, "Multimegawatt Tunable Infrared Source up to 7.7 μm," Rev. Sci. Instr., 48, 733–737, 1977; T. R. Loree, R. C. Sze, D. L. Barker, and P. B. Scott, "New Lines in the UV: SRS of Excimer Laser Wavelengths," IEEE J. Quant. Electron., QE-15, 337–342, 1979; T. R. Loree, R. C. Sze, and D. L. Barker, "Efficient Raman Shifting of ArF and KrF Laser Wavelengths," Appl. Phys. Lett., 31, 37–39, 1977; H. Komine and E. A. Stappaerts, "Efficient Higher-Stokes-Order Raman Conversion in Molecular Gases," Opt. Lett., 4, 398–400, 1979. In a paper presented at the 1979 Conference on Laser Engineering and Applications in Washington, D.C., J. A. Paisner and R. L. Hargrove reported tunable, efficient vacuum-UV and IR generated from 166.5 nm to 1.9 μm using hydrogen gas as the Raman-scattering medium and a variety of pump lasers, including Nd: YAG, dye, and ArF.

21. A. Girard and P. Jacquinot, "Principles of Instrumental Methods in Spectroscopy," in *Advanced Optical Techniques*, A. C. S. Van Heel, Ed., North-Holland, Amsterdam, 1967.

22. J. D. Strong, *Procedures in Experimental Physics*, Prentice-Hall, Englewood Cliffs, N. J., 1938.

23. *Handbook of Chemistry and Physics*, 62nd edition, R. C. Weast, Ed., CRC Press, Boca Raton, Fla., 1981–82.

24. U. Hochuli and P. Haldemann, "Indium sealing techniques," Rev. Sci. Instr. 43, 1088–1089, 1972.

25. Available from Oakite Products.

26. T. C. Poulter "A Glass Window Mounting for Withstanding Pressure of 30,000 Atmospheres," Phys. Rev. 35, 297, 1930.

27. W. Paul, W. W. DeMeis, and J. M. Besson, "Windows for Optical Measurements at High Pressures and Long Infrared Wavelengths," Rev. Sci. Instr., 39, 928–930, 1968.

28. *High Pressure Technology*, I. L. Spain and J. Paawe, Eds., Dekker, New York, 1967.

29. *Compilation of ASTM Standard Definitions*, 3rd edition, American Society for Testing and Materials, Philadelphia, 1976.

30. Glass, quartz, and sapphire vacuum window assemblies are available from Adolf Meller, Ceramaseal, Varian, and Vacuum Generators.

31. *Spectroscopic Techniques for Far Infra-Red, Submillimetre and Millimetre Waves*, D. H. Martin, Ed., North-Holland, Amsterdam, 1967.

32. K. D. Möller and W. G. Rothschild, *Far-Infrared Spectroscopy*, Wiley-Interscience, 1971.

33. A. Hadni, *Essentials of Modern Physics Applied to the Study of the Infrared*, Pergamon Press, Oxford, 1967.

34. G. A. Fry, "The Eye and Vision," in *Applied Optics and Optical Engineering*, Vol. 2, R. Kingslake, Ed., Academic Press, 1965.

35. C. C. Davis and R. A. McFarlane, "Lineshape Effects in Atomic Absorption Spectroscopy," J. Quant. Spect. Rad. Trans., 18, 151–170, 1977.

36. P. W. Kruse, L. D. McGlauchlin, and R. B. McQuistan, *Elements of Infrared Technology: Generation, Transmission and Detection*, Wiley, New York, 1962.

37. R. D. Hudson, Jr., *Infrared System Engineering*, Wiley-Interscience, New York, 1969.

38. J. A. R. Samson, *Techniques of Vacuum Ultraviolet Spectroscopy*, Wiley, New York, 1967.

39. Available from Candela, EG & G, ILC, and Xenon Corporation.

40. Available from ILC.

41. J. P. Markiewicz and J. L. Emmett, "Design of Flashlamp Driving Circuits," IEEE J. Quant. Electron. QE-2, 707–711, 1966.

42. J. G. Edwards, "Some Factors Affecting the Pumping Efficiency of Optically Pumped Lasers," Appl. Opt., 6, 837–843, 1967.

43. H. J. Baker and T. A. King, "Optimization of Pulsed UV Radiation from Linear Flashtubes," J. Phys. E.: Sci. Instr., 8, 219–223, 1975.

44. B.A. Lengyel, *Lasers*, 2nd edition, Wiley-Interscience, New York, 1971.

45. A.E. Siegman, *An Introduction to Lasers and Masers*, McGraw-Hill, New York, 1971.

46. *Quantum Electronics*, H. Rabin and C.L. Tang, Eds., Vols. 1A and 2A, Nonlinear Optics, Academic Press, New York, 1975.

47. *Nonlinear Optics*, Proceedings of the Sixteenth Scottish Universities Summer School in Physics, 1975, P.G. Harper and B.S. Wherrett, Eds., Academic Press, London, 1977.

48. N. Bloembergen, *Nonlinear Optics*, Benjamin, New York, 1965.

49. *Lasers, Light Amplifiers and Oscillators*, D. Röss, Academic Press, New York, 1969.

50. *Dye Lasers*, F.P. Schäfer, Ed., Topics in Applied Phys., Vol. 1, 2nd revised edition, Springer, Berlin, 1977.

51. H.W. Furumoto and H.L. Ceccon, "Optical Pumps for Organic Dye Lasers," Appl. Opt., 8, 1613–1623, 1969.

52. J.F. Holzrichter and A.L. Schawlow, "Design and Analysis of Flashlamp Systems for Pumping Organic Dye Lasers," Ann. N.Y. Acad. Sci. 168, 703–714, 1970.

53. T.B. Lucatorto, T.J. McIlrath, S. Mayo, and H.W. Furumoto, "High-Stability Coaxial Flashlamp-Pumped Dye Laser," Appl. Opt., 19, 3178–3180, 1980.

54. C.C. Davis and T.A. King, "Gaseous Ion Lasers," in *Advances in Quantum Electronics*, Vol. 3, D.W. Goodwin, Ed., Academic Press, London, 1975, pp. 169–454.

55. W.B. Bridges, "Ion Lasers," in *Handbook of Laser Science and Technology*, Vol. 1: *Lasers in All Media*, M.J. Weber, Ed., CRC Press, Boca Raton, Fla., 1982.

56. C.C. Davis, "Neutral Gas Lasers," in *Handbook of Lasers Science and Technology*, Vol. 1: *Lasers in All Media*, M.J. Weber, Ed., CRC Press, Boca Raton, Fla., 1982.

57. D.C. Tyte, "Carbon Dioxide Lasers," in *Advances in Quantum Electronics*, Vol. 1, D.W. Goodwin, Ed., Academic Press, London, 1970.

58. H. Seguin and J. Tulip, "Photoinitiated and Photosustained Laser," Appl. Phys. Lett., 21, 414–415, 1972.

59. J.D. Cobine, *Gaseous Conductors*, Dover, N.Y., 1958.

60. H.J. Seguin, K. Manes, and J. Tulip, "Simple Inexpensive Laboratory-Quality Rogowski TEA Laser," Rev. Sci. Instr., 43, 1134–1139, 1972.

61. A.J. Beaulieu, "Transversely Excited Atmospheric Pressure CO_2 Laser," Appl. Phys. Lett., 16, 504–505, 1970.

62. D. Basting, F.P. Schäfer and B. Steyer, "A Simple, High Power Nitrogen Laser," Opto-electron., 4, 43–49, 1972.

63. P. Schenck and H. Metcalf, "Low Cost Nitrogen Laser Design for Dye Laser Pumping," Appl. Opt., 12, 183–186, 1973.

64. C.P. Wang, "Simple Fast-Discharge Device for High-Power Pulsed Lasers," Rev. Sci. Instr., 47, 92–95, 1976.

65. A.J. Schwab and F.W. Hollinger, "Compact High-Power N_2 Laser: Circuit Theory and Design," IEEE J. Quant. Electron., QE-12, 183–188, 1976.

66. C.L. Sam, "Small-Size Discrete-Capacitor N_2 Laser," Appl. Phys. Lett., 29, 505–506, 1976.

67. M. Feldman, P. Lebow, F. Raab, and H. Metcalf, "Improvements to a Home-Built Nitrogen Laser," Appl. Opt., 17, 774–777, 1978.

68. L.F. Mollenauer, "Dyelike Lasers for the 0.9–2 μm Region Using F_2^+ Centers in Alkali Halides," Opt. Lett., 1, 164, 1977 (see also Opt. Lett. 3, 48–50, 1978; 4, 247–299, 1979; 5, 188–190, 1980).

69. P.F. Moulton, A. Mooradian, and T.B. Reed, "Efficient CW Optically-Pumped Ni:MgF_2 Laser," Opt. Lett., 3, 164–166, 1978.

70. P.F. Moulton and A. Mooradian, "Broadly Tunable CW Operation of Ni:MgF_2 and Co:MgF_2 Lasers," Appl. Phys. Lett., 35, 838–840, 1979.

71. R.C. Greenhow and A.J. Schmidt, "Picosecond Light Pulses," in *Advances in Quantum Electronics*, Vol. 2, D.W. Goodwin, Ed., Academic Press, London, 1973.

72. S.L. Shapiro, Ed., *Ultrashort Light Pulses, Picosecond Techniques and Applications*, Topics in Applied Physics, Vol. 18, Springer, Berlin, 1977.

73. M.B. Panish, "Heterostructure Injection Lasers," Proc. IEEE, 64, 1512–1540, 1976.

74. E.D. Hinkley, K.W. Nill, and F.A. Blum. "Infrared Spectroscopy with Tunable Lasers," in *Topics in Applied Physics*, Vol. 2, H. Walther, Ed., Springer, Berlin, 1976, pp. 125–196.

75. T.W. Hänsch, "Repetitively Pulsed Tunable Dye Laser for High Resolution Spectroscopy," Appl. Opt., 11, 895–898, 1972.

76. R. Wallenstein and T.W. Hänsch, "Linear Pressure Tuning of a Multielement Dye Laser Spectrometer," Appl. Opt., 13, 1625–1628, 1974.

77. R. Wallenstein and T.W. Hänsch, "Powerful Dye Laser Oscillator—Amplifier System for High-Resolution Spectroscopy," Opt. Commun., 14, 353–357, 1975.

78. G.L. Eesley and M.D. Levenson, "Dye-Laser Cavity Employing a Reflective Beam Expander," IEEE J. Quant. Electron. QE-12, 440–442, 1976.

79. M.G. Littman, "Single-Mode Operation of Grazing-Incidence Pulsed Dye Laser," Opt. Lett., 3, 138–149, 1978.

80. M.G. Littman and H.J. Metcalf, "Spectrally Narrow

Pulsed Dye Laser without Beam Expander," Appl. Opt., 17, 2224–2227, 1978.

81. I. P. Kaminow, *An Introduction to Electro-optic Devices*, Academic Press, New York, 1974.

82. Available from Andersen Laboratories, Coherent Associates, Harris Corp., Intra Action Corp., and Isomet, among others.

83. D. Sliney and M. Wolbarsht, *Safety with Lasers and Other Optical Sources: A Comprehensive Handbook*, Plenum, New York, 1980.

84. M. J. Weber, Ed., *CRC Handbook of Laser Science and Technology*, Vol. 1, Lasers and Masers, CRC Press, Boca Raton, Fla., 1982.

85. P. Jacquinot, "The Luminosity of Spectrometers with Prisms, Gratings, or Fabry-Perot Etalons," J. Opt. Soc. Am., 44, 761–765, 1954.

86. *Applied Optics and Optical Engineering*, Vol. 5: *Optical Instruments Part II*, R. Kingslake, Ed., Academic Press, New York, 1969.

87. J. F. James and R. S. Sternberg, *The Design of Optical Spectrometers*, Chapman and Hall, London, 1969.

88. Available from 3M, Decorative Products Division.

89. W. G. Fastie, "A Small Plane Grating Monochromator," J. Opt. Soc. Am., 42, 641–647, 1952.

90. H. Ebert, "Zwei Formen von Spectrographen," Annalen der Physik und Chemie, 38, 489–493, 1889.

91. M. Seya, "A New Mounting of Concave Grating Suitable for a Spectrometer," Sci. Light (Tokyo), 2, 8–17, 1952.

92. T. Namioka, "Constitution of a Grating Spectrometer," Sci. Light (Tokyo), 3, 15–24, 1952.

93. Available from numerous suppliers of vacuum equipment—for example, Ceramaseal, Edwards, Ferrofluidics Corp., Perkin Elmer (Norwalk), Vacuum Generators, Varian, and Veeco.

94. Available from Carpenter Technology.

95. Available from Burleigh.

96. Available from Heraeus-Amersil, Dynasil, Esco, and Quartz Products Corp., among others.

97. Available from Corning Glass Works, Optical Products Department.

98. Piezoelectric transducers are available from Burleigh, Lansing Research Corporation, Jodon, and Gulton Industries, among others.

99. C. F. Bruce, "On Automatic Parallelism Control in a Scanning Fabry-Perot Interferometer," Appl. Opt., 5, 1447–1452, 1966.

100. Commercial Fabry-Perot Interferometers that can incorporate this feature are available from Burleigh.

101. G. A. Vanasse and H. Sakai, "Fourier Spectroscopy," in *Progress in Optics*, Vol. 6, E. Wolf, Ed., North-Holland, Amsterdam, 1967.

102. G. W. Chantry, *Submillimetre Spectroscopy*, Academic Press, London, 1971.

103. R. Ladenburg and D. Bershader, "Interferometry," in *Physical Measurements in Gas Dynamics and Combustion*, R. Ladenburg, Ed., Vol. 9 of High Speed Aerodynamics and Jet Propulsion, Princeton University Press, Princeton, N.J., 1954.

104. P. R. Longaker and M. M. Litvak, "Perturbation of the Refractive Index of Absorbing Media by a Pulsed Laser Beam," J. Appl. Phys., 40, 4033–4041, 1969.

105. D. C. Smith, "Thermal Defocusing of CO_2 Laser Radiation in Gases," IEEE J. Quant. Electron., QE-5, 600–607, 1969.

106. "Optical and Infrared Detectors," in *Topics in Applied Physics*, R. J. Keyes, Ed., Springer, Berlin, 1977.

107. *Infrared Detectors*, R. D. Hudson, Jr., and J. W. Hudson, Eds., Benchmark Papers in Optics, Vol. 2, Dowden, Hutchinson and Ross, Stroudsburg, Pa., 1975.

108. Static cross-field photomultipliers are available from Varian (LSE Division).

109. Available from EMR, Galileo Electro Optics Corp., Mullard, and Varian (LSE Division).

110. H. R. Zwicker, "Photoemissive Detectors," in *Optical and Infrared Detectors*, R. S. Keyes, Ed., Topics in Applied Physics, Vol. 19, pp. 149–196, Springer, Berlin, 1977.

111. "Window" discriminators are available from EG & G and LeCroy.

112. D. Long, "Photovoltaic and Photoconductive Infrared Detectors" in *Optical and Infrared Detectors*, R. J. Keyes, Ed., Topics in Applied Physics, Vol. 19, pp. 101–147, Springer, Berlin, 1977.

113. W. J. Moore and H. Shenker, "A High-Detectivity Gallium-Doped Germanium Detector for the 40–120 μ Region," Infrared Phys., 5, 99–106, 1965.

114. F. J. Low, "Low Temperature Germanium Bolometer," J. Opt. Soc. Am., 51, 1300–1304, 1961.

115. H. Shenker, W. J. Moore, and E. M. Swiggard, "Infrared Photoconductive Characteristics of Boron-Doped Germanium," J. Appl. Phys., 35, 2965–2970, 1964.

116. E. H. Putley, "Indium Antimonide Submillimeter Photoconductive Detectors," Appl. Opt., 4, 649–656, 1965.

117. M. A. C. S. Brown and M. F. Kimmitt, "Far-Infrared Resonant Photoconductivity in Indium Antimonide," Infrared Phys., 5, 93–97, 1965.

118. E. H. Putley, "Thermal Detectors," in *Optical and Infrared Detectors*, R. J. Keyes, Ed., Topics in Applied Physics, Vol. 19, Springer, Berlin, 1977.

119. A. F. Gibson, M. F. Kimmitt, and A. C. Walker, "Photon Drag in Germanium," Appl. Phys. Lett. 17, 75–77, 1970.
120. M. J. E. Golay, "A Pneumatic Infra-Red Detector," Rev. Sci. Instrum., 18, 357–362, 1947.

GENERAL REFERENCES

Comprehensive (General) Optics Texts

M. Born and E. Wolf, *Principles of Optics*, Macmillan, New York, 1964.
R. W. Ditchburn, *Light*, Academic Press, New York, 1976.
F. A. Jenkins and H. E. White, *Fundamentals of Optics*, McGraw-Hill, New York, 1957.
M. V. Klein, *Optics*, Wiley, New York, 1970.
R. S. Longhurst, *Geometrical and Physical Optics*, Longman, London, 1973.
F. G. Smith and J. H. Thomson, *Optics*, Wiley, London, 1971.
J. Strong, *Concepts of Classical Optics*, Freeman, San Francisco, 1958.

Applied Optics

Applied Optics and Optical Engineering, R. Kingslake, Ed., Academic, New York, Vol. 1, 1965; Vol. 2, 1965; Vol. 3, 1965; Vol. 4, 1967; Vol. 5, 1969.
L. Levi, *Applied Optics*, Wiley, New York, Vol. 1, 1968; Vol. 2, 1980.

Electro-optic Devices

I. P. Kaminow, *An Introduction to Electro-optics*, Academic Press, New York, 1974.
A. Yariv, *Optical Electronics*, 2nd edition, Holt, Rinehart and Winston, New York, 1976.

Far-Infrared Techniques

A. Hadni, *Essentials of Modern Physics Applied to the Study of the Infrared*, Pergamon Press, Oxford, 1967.
K. D. Möller and W. G. Rothschild, *Far-Infrared Spectroscopy*, Wiley-Interscience, New York, 1971.

L. C. Robinson, *Physical Principles of Far-Infrared Radiation*, Methods in Experimental Physics, Vol. 10, L. Marton, Ed., Academic Press, New York, 1973.

Fiber Optics

Introduction to Integrated Optics, M. K. Barnoski, Ed., Plenum, New York, 1974.
Optical Fiber Technology, D. Gloge, Ed., IEEE, New York, 1976.

Filters

Handbook of Chemistry and Physics, 62nd ed., R. C. Weast, Ed., CRC Press, Boca Raton, Fla., 1981.
Handbook of Lasers, R. J. Pressley, Ed., CRC Press, Cleveland, 1971.
L. Levi, *Applied Optics*, Vol. 2, Wiley, New York, 1980.
H. A. Macleod, *Thin-Film Optical Filters*, American Elsevier, New York, 1969.

Incoherent Light Sources

Advanced Optical Techniques, A. C. S. Van Heel, Ed., North-Holland, Amsterdam, 1967.
Applied Optics and Optical Engineering, R. Kingslake, Ed., Vol. 1, Academic Press, New York, 1965.
Handbook of Lasers, R. J. Pressley, Ed., CRC Press, West Palm Beach, Fla., 1971.
L. Levi, *Applied Optics*, Vol. 1, Wiley, New York, 1968.

Infrared Technology

A. Hadni, *Essentials of Modern Physics Applied to the Study of the Infrared*, Pergamon Press, Oxford, 1967.
R. D. Hudson, *Infrared System Engineering*, Wiley-Interscience, New York, 1969.
P. W. Kruse, L. D. McGlauchlin, and R. B. McQuistan, *Elements of Infrared Technology*, Wiley, New York, 1962.

Interferometers and Interferometry

M. Born and E. Wolf, *Principles of Optics*, Macmillan, New York, 1964.

M. Francon, *Optical Interferometry*, Academic Press, New York, 1966.

W.H. Steel, *Interferometry*, Cambridge University Press, Cambridge, 1967.

S. Tolansky, *An Introduction to Interferometry*, Longman, London, 1955.

Lasers

Handbook of Laser Science and Technology, Vol. I: *Lasers in all Media*, M. Weber, Ed., CRC Press, Boca Raton, Fla., 1982.

B.A. Lengyel, *Lasers*, 2nd edition, Wiley-Interscience, New York, 1971.

A. Maitland and M.H. Dunn, *Laser Physics*, North-Holland, Amsterdam, 1969.

D.C. O'Shea, W.R. Callen, and W.T. Rhodes, *Introduction to Lasers and Their Applications*, Addison-Wesley, Reading, Mass., 1977.

A.E. Siegman, *An Introduction to Lasers and Masers*, McGraw-Hill, New York, 1971.

J.T. Verdeyen, *Laser Electronics*, Prentice-Hall, Englewood Cliffs, N.J., 1981.

A. Yariv, *Optical Electronics*, 2nd edition, Holt, Rinehart and Winston, New York, 1976.

A. Yariv, *Quantum Electronics*, 2nd edition, Wiley, New York, 1975.

Nonlinear Optics

N. Bloembergen, *Nonlinear Optics*, Benjamin, New York, 1965.

Nonlinear Optics, P.G. Harper and B.S. Wherrett, Eds., Academic Press, New York, 1977.

Quantum Electronics, Vols. 1A and 2A, *Nonlinear Optics*, C.L. Tang and H. Rabin, Eds., Academic Press, New York, 1975.

F. Zernike and J.E. Midwinter, *Applied Non-linear Optics*, Wiley, New York, 1973.

Optical Component and Instrument Design

Advanced Optical Techniques, A.C.S. Van Heel, Ed., North-Holland, Amsterdam, 1967.

Applied Optics and Optical Engineering, R. Kingslake, Ed.,

Academic Press, New York, Vol. 3, 1965; Vol. 4, 1967; Vol. 5, 1969.

Optical Detectors

J.B. Dance, *Photoelectronic Devices*, Iliffe, London, 1969.

P.W. Kruse, L.D. McGlauchlin, and R.B. McQuistan, *Elements of Infrared Technology*, Wiley, New York, 1962.

L. Levi, *Applied Optics*, Vol. 2, Wiley, New York, 1980.

Optical and Infrared Detectors, R.S. Keyes, Ed., Topics in Applied Physics, Vol. 19, Springer, Berlin, 1977.

Optical Materials

Applied Optics and Optical Engineering, R. Kingslake, Ed., Vol. 1, Academic Press, New York, 1965.

P.W. Kruse, L.D. McGlauchlin, and R.B. McQuistan, *Elements of Infrared Technology*, Wiley, New York, 1962.

A.J. Moses, *Optical Material Properties*, IFI/Plenum, New York, 1971.

Optical Safety

Handbook of Laser Science and Technology, Vol. 1: *Lasers and Masers*, M.J. Weber, Ed., CRC Press, Boca Raton, Fla., 1982.

D. Sliney and M. Wolbarsht, *Safety with Lasers and Other Optical Sources*: *A Comprehensive Handbook*, Plenum, New York, 1980.

Polarized Light and Crystal Optics

R.M.A. Azzam and N.M. Basham, *Ellipsometry and Polarized Light*, North-Holland, Amsterdam, 1977.

W.A. Shurchuff, *Polarized Light*, Harvard University Press, Cambridge, Mass., 1962.

E. Wahlstrom, *Optical Crystallography*, 3rd edition, Wiley, New York, 1960.

Spectrometers

J.F. James and R.S. Sternberg, *The Design of Optical Spectrometers*, Chapman and Hall, London, 1969.

H.S. Strobel, *Chemical Instrumentation*, Addison-Wesley, Reading, Mass., 1973.

Spectroscopy

Atomic Absorption Spectrometry, M. Pinta, Ed., Wiley, New York, 1975.

J. R. Edisbury, *Practical Hints on Absorption Spectrometry*, Hilger and Watts, London, 1966.

R. J. Reynolds and K. Aldous, *Atomic Absorption Spectroscopy*, Barnes and Noble, New York, 1970.

R. A. Sawyer, *Experimental Spectroscopy*, Prentice-Hall, Englewood Cliffs, N.J., 1951.

Spectroscopy, D. Williams, Ed., Methods of Experimental Physics, Vol. 13, Parts A and B, Academic Press, New York, 1968.

S. Walker and H. Straw, *Spectroscopy*, Vol. I: *Microwave and Radio-Frequency Spectroscopy*; Vol. II: *Ultraviolet, Visible, Infrared and Raman Spectroscopy*, Macmillan, New York, 1962.

Submillimeter Wave Techniques

G. W. Chantry, *Submillimeter Spectroscopy*, Academic Press, New York, 1971.

Spectroscopic Techniques, D. H. Martin, Ed., North-Holland, Amsterdam, 1967.

Tables of Physical and Chemical Constants

G. W. C. Kaye and T. H. Laby, *Tables of Physical and Chemical Constants*, 14th edition, Longman, London, 1978.

Tables of Spectral Lines

G. R. Harrison, *MIT Wavelength Tables*, MIT Press, Cambridge, Mass., 1969.

A. R. Striganov and N. S. Sventitskii, *Tables of Spectral Lines of Neutral and Ionized Atoms*, IFI/Plenum, New York, 1968.

A. N. Zaidel', V. K. Prokof'ev, S. M. Raiskii, V. A. Slavnyi, and E. Ya. Shreider, *Tables of Spectral Lines*, IFI/Plenum, New York, 1970.

Ultraviolet and Vacuum-Ultraviolet Technology

The Middle Ultraviolet: Its Science and Technology, A. E. S. Green, Ed., Wiley, New York, 1966.

J. A. R. Samson, *Techniques of Vacuum Ultraviolet Spectroscopy*, Wiley, New York, 1967.

MANUFACTURERS AND SUPPLIERS

Advanced Kinetics Inc.
1231 Victoria St.
Costa Mesa, CA 92626

Aerotech Inc.
101 Zeta Dr.
Pittsburgh, PA 15238

Allied Chemical Corp.
Electro Optical Products Group
P.O. Box 4901
7 Powder Horn Dr.
Warren, NJ 07060

American Acoustical Products
9 Cochituate St.
Natick, MA 01760

American Instrument Co.
8030 Georgia Ave.
Silver Spring, MD 20910

American Laser Corp.
1832 S. 3850 W
Salt Lake City, UT 84104

Amperex Electronic Corp.
230 Duffy Ave.
Hicksville, NY 11802

Andersen Laboratories Inc.
1280 Blue Hills Ave.
Bloomfield, CT 06002

Apollo Lasers Inc.
6357 Arizona Circle
Los Angeles, CA 90045

Ardel Kinamatic Corp.
125-20 18th Ave.
College Point, NY 11356

Baird Atomic, Inc.
125 Middlesex Turnpike
Bedford, MA 01730

Barnes Engineering Co.
Systems & Instruments Div.
30 Commerce Rd.
Stamford, CT 06904

Bausch & Lomb Inc.
Analytical Systems Div.
820 Linden Ave.
Rochester, NY 14625

Beckmann Instruments
2500 Harbor Blvd.
Fullerton, CA 92634

Bomen
910 Pl. Dufour
Ville Vanier, Quebec
Canada G1M 3B1

Bond Optics Inc.
Etna Rd.
Lebanon, NH 03766

Broomer Research Corp.
23 Sheer Plaza
Plainview, NY 11803

Bruker Instruments
Manning Park
Billerica, MA 01821

Bulova Watch Co. Inc.
Electronics Div.
61-20 Woodside Ave.,
Woodside, NY 11377

Burleigh Instruments Inc.
Burleigh Park
Fishers, NY 14453

Candela Corp.
96 South Ave.
Natick, MA 01760

Canrad Hanovia Inc.
100 Chestnut St.
Newark, NJ 07105

Carpenter Steel Div.
Carpenter Technology Corp.
Reading, PA 19600

Cary
Varian Associates, Instrument Div.
611 Hansen Way
Palo Alto, CA 94303

Ceramaseal Inc.
P.O. Box 25
New Lebanon Center, NY 12126

Chance Pilkington
St. Asaph
Flutshire, U.K.

Chromatix Inc.
560 Oakmead Pkwy.
Sunnyvale, CA 94086

CILAS—Compagnie Industrielle des Lasers
Route de Nozay
91460 Marcoussis, France

Cleveland Crystals Inc.
P.O. Box 17157
Euclid, OH 44117

Coherent Inc.
3210 Porter Dr.
Palo Alto, CA 94304

Coherent Modulator Division
14 Finance Dr., Commerce Pk.
Danbury, CT 06810

Commercial Crystal Laboratories Inc.
111 Chevalier Ave.
South Amboy, NJ 08879

Continental Optical Corp.
15 Power Dr.
Hauppauge, NY 11787

Control Laser Corp.
11222 Astronaut Blvd.
Orlando, FL 32809

Corning Glass Works
Optical Products Dept.
Corning, NY 14830

Crystal Systems Inc.
35 Congress St.
Shetland Ind. Pk.
Salem, MA 01970

Crystal Technology Inc.
1035 E. Meadow Cir.
Palo Alto, CA 94303

CSI Technologies Inc., Capacitor Div.
P.O. Box 2052
Escondido, CA 92025

CVI Laser Corp.
P.O. Box 11308
290 Dorado Pl., SE
Albuquerque, NM 87123

Diffraction Products Inc.
P.O. Box 645
Woodstock, IL 60098

Digilab, Inc.
237 Putnam Ave.
Cambridge, MA 02139

Dynasil Corp. of America
Cooper Rd.
Berlin, NJ 08009

Ealing Corp.
22 Pleasant St.
South Natick, MA 01760

Eastman Kodak Co.
343 State St.
Rochester, NY 14650

Edinburgh Instruments Ltd.
Riccarton, Currie
Edinburgh EH14 4AP, U.K.

Edmund Scientific Co.
Product Development Div.
101 E. Gloucester Pike
Barrington, NJ 08007

Edwards High Vacuum Inc.
3279 Grand Island Blvd.
Grand Island, NY 14072

EG&G Inc.
Electro-Optics Div.
35 Congress St.
Salem, MA 01970

EMI Electronics Ltd.
Electron Tube Div.
253 Blyth Rd.
Hayes, Middlesex UB3 1HJ, U.K.

EMI Gencom Inc.
80 Express St.
Plainview, NY 11803

EMR Photoelectric
P.O. Box 44
Princeton, NJ 08540

Eppley Laboratory Inc.
12 Sheffield Ave.
Newport, RI 02840

Esco Products Inc.
171 Oak Ridge Rd.
Oak Ridge, NJ 07438

Exotic Materials
2968 Randolph Ave.
Costa Mesa, CA 92626

Ferrofluidics Corp.
40 Simon St.
Nashua, NH 03061

Fish-Schurman Corp.
75 Portman Rd.
New Rochelle, NY 10802

Ford Aerospace & Communications Corp.
Aeronutronic Div.
Ford Rd.
Newport Beach, CA 92663

Gaertner Scientific Co.
1201 Wrightwood Ave.
Chicago, IL 60614

Galileo Electro-Optics Corp.
Galileo Park
Sturbridge, MA 01518

General Electric Co.
Space Sciences Laboratory
P.O. Box 8555
Philadelphia, PA 19101

General Photonics Corp.
2255F Martin Ave.
Santa Clara, CA 95050

Gen-Tec Inc.
Electro-Optics Div.
2625 Dalton St.
Ste.-Foy, Quebec, Canada G1P 3S9

Glendale Optical Inc.
130 Crossways Park Dr.
Woodbury, NY 11797

GP Instrumentation
Whitley Rd., Longbenton
Newcastle-upon-Tyne
NE12 9SP, U.K.

Sir Howard Grubb Parsons and Co. Ltd.
(see also GP Instrumentation)
Shields Road
Walkergate
Newcastle-upon-Tyne
NE6 2YB, U.K.

GTE Sylvania
Laser Products Dept.
P.O. Box 1-8
Mountain View, CA 94042

Gulton Industries, Inc.
212 Durham Ave.
Metuchen, NJ 08840

Hamamatsu Corp.
420 South Ave.
Middlesex, NJ 08846

Harris Corp.
Government Comm. Systems Div.
P.O. Box 37
Melbourne, FL 32901

Harshaw Chemical Co.
Crystal & Electronic Products Dept.
6801 Cochran Rd.
Solon, OH 44139

Heraeus-Amersil Inc.
650 Jernee Mill Rd.
Sayreville, NJ 08872

Hi Voltage Components Inc.
P.O. Box 851
Largo, FL 33540

Hibshman Optical Lab Inc.
860 Capitolio Way
P.O. Box 1243
San Luis Obispo, CA 93401

Hilger and Watts
Westwood, Margate
Kent CT9 4JL, U.K.

Holobeam Laser Inc.
(Control Laser)
11222 Astronaut Blvd.
Orlando, FL 32809

Holotron Corp.
2400 Stevens Dr.
Richland, WA 99352

Honeywell Inc.
Electro-Optics Center
2 Forbes Rd.
Lexington, MA 02173

Horiba Instruments, Inc.
Optoelectronics Division
1021 Durgea Ave.
Irvine, CA 92714

Hughes Aircraft Co.
Industrial Products Div.
6155 El Camino Real
Carlsbad, CA 92008

IC Optical Systems Ltd.
Franklin Rd.
London SE20 8HW, U.K.

ILC Technology Inc.
399 Java Dr.
Sunnyvale, CA 94086

Illumination Industries Inc./PEK
825 East Evelyn Ave.
Sunnyvale, CA 94086

Infrared Associates Inc.
14A Jules Ln.
New Brunswick, NJ 08901

Infrared Industries Inc., Eastern Div.
62 Fourth Ave.
Waltham, MA 02154

Infrared Optics
170-17 Central Ave.
E. Farmingdale, NY 11735

Inrad (Interactive Radiation) Inc.
181 Legrand Ave.
Northvale, NJ 07647

Instruments SA Inc.
173 Essex Ave.
Metuchen, NJ 08840

IntraAction Corp.
3719 Warren Ave.
Bellwood, IL 60104

Isomet Corp.
5263 Port Royal Rd.
Springfield, VA 22151

Ithaco Inc.
735 W. Clinton St.
Ithaca, NY 14850

ITT
Electro-Optical Products Division
3700 East Pontiac St.
Fort Wayne, IN 46803

Janos Optical Corp.
Rt. 35
Townshend, VT 05353

Jarrell-Ash Div.
Fisher Scientific Co.
590 Lincoln St.
Waltham, MA 02154

JEC Associates Inc.
253 Crooks Ave.
Paterson, NJ 07503

JK Lasers Ltd.
Somers Rd.
Rugby, Warwickshire CV22 7DG, U.K.

Jobin-Yvon
Instruments SA Inc.
173 Essex Ave.
Metuchen, NJ 08840

Jodon Engineering Associates Inc.
145 Enterprise Dr.
Ann Arbor, MI 48103

Judson Infrared Inc.
565 Virginia Dr.
Ft. Washington, PA 91034

KGM Electronics, Hivotronics Div.
Wella Rd.
Basingstoke, Hampshire, U.K.

Klinger Scientific Corp.
110-20 Jamaica Ave.
Richmond Hill, NY 11418

Korad/Hadron Inc.
1700 Old Meadow Road
McLean, VA 22102

Laakman Electro-Optics Inc.
33052 Calle Aviador
San Juan Capistrano, CA 92675

Karl Lambrecht Corp.
4204 N. Lincoln Ave.
Chicago, IL 60618

Lansing Research Corp.
P.O. Box 730
705 Willow Ave.
Ithaca, NY 14850

Lasag AG
Bernstrasse 11
CH-3600 Thun, Switzerland

Laser Inc. (Coherent)
Laser Lane
P.O. Box 537
Sturbridge, MA 01566

Laser Analytics Inc. (Spectra Physics)
25 Wiggins Ave.
Bedford, MA 01730

Laser Consultants Inc.
344 W. Hills Rd.
Huntington, NY 11743

Laser Nucleonics Inc.
123 Moody St.
Waltham, MA 02154

Laser Optics Inc.
P.O. Box 127
Danbury, CT 06810

Laser Precision Corp.
1231 Hart St.
Utica, NY 13502

Lasermetrics Inc.
111 Galway Pl.
Teaneck, NJ 07666

LeCroy Research Systems Corp.
1800 Embarcadero Rd.
Palo Alto, CA 94303

Lexel Corp.
928 E. Meadow Dr.
Palo Alto, CA 94303

Leybold-Heraeus Vacuum
    Products Inc.
200 Seco Rd.
Monroeville, PA 15146

Liconix
1390 Borregas Ave.
Sunnyvale, CA 94086

Lumonics Research Ltd.
105 Schneider Rd.
Kanata, Ontario, Canada K2K 1Y3

3M
Decorative Products Div.
3M Center
St. Paul, MN 55101

Magnetic Shield Div.
(Perfection Mica Co.)
740 N. Thomas Dr.
Bensenville, IL 60106

Mathematical Sciences NorthWest Inc.
2755 Northrup Way
Bellevue, WA 98004

Maxwell Laboratories
8835 Balboa Ave.
San Diego, CA 92123

McPherson Instrument
530 Main St.
Acton, MA 01720

Adolf Meller Co.
P.O. Box 6001
120 Corliss St.
Providence, RI 02904

Melles Griot
1770 Kettering St.
Irvine, CA 92714

Minuteman Laboratories Inc.
916 Main St.
Acton, MA 01720

MμShield Co.
121 Madison St.
Malden, MA 02148

Modern Optics
2207 Merced Ave.
El Monte, CA 91733

Molectron Corp.
177 N. Wolfe Rd.
Sunnyvale, CA 94086

Mullard Ltd.
Mullard House
Torrington Pl.
London, WC1, U.K.

Murata Corp. of America
1148 Franklin Rd. S.E.
Marietta, GA 30062

National Research Group Inc.
P.O. Box 5321
Madison, WI 53705

New England Research Center Inc.
Minuteman Dr., Longfellow Center
Sudbury, MA 01776

Newport Research Corp.
18235 Mt. Baldy Cir.
Fountain Valley, CA 92708

Nicolet Analytical Instruments
5225-1 Verona Rd.
Madison, WI 53711

J.A. Noll Co.
P.O. Box 312
Monroeville, PA 15146

Oakite Products, Inc.
50 Valley Rd.
Berkeley Heights, NJ 07922

Opthos Instrument Co.
9600 Overlea Dr.
Rockville, MD 20850

Optical Coating Laboratory Inc.
Technical Products Div.
P.O. Box 1599
Santa Rosa, CA 95402

Optical Engineering Inc.
P.O. Box 696
3300 Coffey Ln.
Santa Rosa, CA 95402

Optics Plus Inc.
1351 E. Edinger Ave.
Santa Ana, CA 92705

Optics for Research Inc.
P.O. Box 82
Caldwell, NJ 07006

OptoElectronics Inc.
1309 Dynamic St.
Petaluma, CA 94952

Optovac Inc.
E. Brookfield Rd.
North Brookfield, MA 01535

Oriel Corp.
15 Market St.
Stamford, CT 06902

Oxford Lasers Ltd.
18 Croft Ave.
Kidlington, Oxford OX5 2HU, U.K.

The Perkin-Elmer Corp.
1751 Kettering St.
Irvine, CA 92705

The Perkin-Elmer Corp.
Main Ave.
Norwalk, CT 06856

The Perkin-Elmer Corp.
Applied Optics Div.
2930 Bristol
P.O. Box 2218
Costa Mesa, CA 92626

Perry Amplifier
138 Fuller St.
Brookline, MA 02146

Phase-R Corp.
Box G-2, Old Bay Rd.
New Durham, NH 03855

Photon Sources Inc.
Laser Div.
37100 Plymouth Rd.
Livonia, MI 48150

Plessey Optoelectronics & Microwave Ltd.
Wood Burcote Way
Towcester, Northants, U.K.

Precision Cells Inc.
560 S. Broadway
Hicksville, NY 11801

Precision Tool & Instrument Co.
Coombe Rd.
Hill Brow, Liss, Hants, U.K.

Princeton Applied Research (EG&G)
P.O. Box 2565
Princeton, NJ 08540

Princeton Infrared Equipment, Inc.
248 U.S. Rt. 1
Monmouth Junction, NJ 08852

PTR Optics Corp.
145 Newton St.
Waltham, MA 02154

Pye Unicam Ltd.
York St.
Cambridge CB1 2PX, U.K.

Quanta-Ray Inc. (Spectra-Physics)
2134 Old Middlefield Way
Mountain View, CA 94043

Quantel International Inc.
385 Reed St.
Santa Clara, CA 95050

Quantronix Corp.
225 Engineers Rd.
Smithtown, NY 11787

Quantum Technology Inc.
2620 Iroquois Ave.
Sanford, FL 32771

Quartz Products Corp.
(Quartz & Silice)
P.O. Box 1347
Plainfield, NJ 07061

Quartz & Silice
8 rue d'Anjou,
75008 Paris, France

Rank Hilger
Westwood, Margate,
Kent CT9 4JL, U.K.

Raytheon Co.
Laser Center
Fourth Ave.
Burlington, MA 01803

Raytheon Co.
Research Div.
28 Seyon St.
Waltham, MA 02154

RCA Corp.
Electro-Optics & Devices
New Holland Ave.
Lancaster, PA 17604

Rofin Ltd.
Winslade House
Egham Hill
Egham, Surrey TW20 0AZ, U.K.

Rolyn Optics Co.
P.O. Box 148
300 Rolyn Pl.
Arcadia, CA 91006

RPC Industries
P.O. Box 3306
Hayward, CA 94540

Sanders Associates
Defensive Systems Div.
95 Canal St.
Nashua, NH 03061

Santa Barbara Research Center
Hughes Aircraft Co.
75 Coromar Dr.
Goleta, CA 93017

SAT (Société Anonyme de Télécommunications)
41 Rue Cantagrel
75624 Paris Cedex 13, France

Schott Optical Glass Inc.
400 York Ave.
Duryea, PA 18642

Servo Corp. of America
111 New South Rd.
Hicksville, NY 11802

Silicon Detector Corp.
855 Lawrence Dr.
Newbury Park, CA 91320

Sopra
68 Rue Pierre Joigneaux
Bois Columbes, France 92270

Space Optics Research Labs
7 Stuart Rd.
Chelmsford, MA 01824

Spawr Optical Research Inc.
1527 Pomona Rd.
Corona, CA 91720

Special Optics
P.O. Box 163
101 E. Main St.
Little Falls, NJ 07424

Spectra Optics
12317 Gladstone Ave.
Sylman, CA 91342

Spectra-Physics Inc.
Laser Instruments Div.
1250 W. Middlefield Rd.
Mountain View, CA 94042

Spectronics Div. (Honeywell)
830 E. Arapaho Rd.
Richardson, TX 75081

Spex Industries Inc.
3880 Park Ave.
Metuchen, NJ 08840

Sprague Elec. Co.
Marshall St.
North Adams, MA 01247

Systems, Science & Software
(RPC Industries)
P.O. Box 3306
Hayward, CA 94540

Tachisto Inc.
13 Highland Circle
Needham, MA 02194

Texas Instruments, Inc.
Equipment Group
P.O. Box 6015
Dallas, TX 75222

Tracor Northern Inc.
2551 W. Beltline Hwy.
Middleton, WI 53562

Tropel (Coherent)
P.O. Box 164
1000 Fairport Pk.
Fairport, NY 14450

II-VI Inc.
Saxonburg Blvd.
Saxonburg, PA 16056

Ultra-Violet Products Inc.
5100 Walnut Grove Ave.
San Gabriel, CA 91778

Unique Optical Co.
P.O. Box 585
Farmingdale, NY 11735

Vacuum Generators Ltd.
Charlwoods Rd.
East Grinstead, Sussex, U.K.

Valtec Corp.
Optical Thin Film Products
211 Second Ave.
Waltham, MA 02154

Varian Associates, Instrument Div.
611 Hansen Way
Palo Alto, CA 94303

Varian Associates, LSE Div.
601 California Ave.
Palo Alto, CA 94303

Varian Associates, Vacuum Div.
611 Hansen Way
Palo Alto, CA 94303

Veeco Instruments, Inc.
Terminal Dr.
Plainview, NY 11803

Velmex Inc.
P.O. Box 38
East Bloomfield, NY 14443

Virgo Optics, Inc.
33 Poplar Dr.
Stirling, NJ 07980

WEC Corp.
4399 Gollihar Rd.
Corpus Christi, TX 88411

J.L. Wood Optical Systems
1361 E. Edinger Ave.
Santa Ana, CA 92705

Xenon Corp.
66 Industrial Way
Wilmington, MA 01887

Suppliers of Optical Windows

Several of the suppliers listed will supply lenses, prisms, and other components fabricated from these materials:

Arsenic trisulphide: Infrared Optics, Optics for Research, Servo, Unique Optical

Barium fluoride: Barnes, Harshaw, Infrared Optics, Janos, Optovac

Cadmium telluride (Irtran 6): Cleveland Crystals, Eastman Kodak, Adolf Meller, II-VI

Calcium carbonate (calcite): Harshaw, Inrad, Karl Lambrecht Corp.

Calcium fluoride (Irtran 3): Barnes, Harshaw, Janos, Laser Optics, II-VI

Cesium bromide: Barnes, Harshaw, Janos

Cesium chloride: Harshaw

Cesium fluoride: Harshaw

Cesium iodide: Barnes, Harshaw, Janos, Adolf Meller

Diamond: Optics for Research, Oriel

Gallium arsenide: Laser Optics, Adolf Meller, II-VI, Unique Optical

Germanium: Continental Optical, Exotic Materials, Infrared Optics, Janos, Laser Optics, II-VI, Unique Optical

Glasses: Coherent, Ealing, Edmund Scientific, NRC, Rolyn

Lithium fluoride: Harshaw, Janos, Optovac, Unique Optical

Magnesium fluoride (Irtran 1): Harshaw, Janos, Optovac

Magnesium oxide (Irtran 5): Eastman Kodak

Potassium bromide: Barnes, Harshaw, Janos, Optovac

Potassium chloride: Harshaw, Janos, Optovac

Potassium iodide: Harshaw

Quartz (crystalline): Continental Optical, ESCO, Janos, Infrared Optics, Optics for Research, Oriel, Adolf Meller, Unique Optical

Sapphire: Crystal Systems, Janos, Adolf Meller

Silica (fused): Continental Optical, ESCO, Janos, Optical Coating Lab (OCLI), Unique Optical

Silicon: Exotic Materials, Infrared Optics, Janos, Laser Optics, Unique Optical

Silver bromide: Barnes, Harshaw

Silver chloride: Barnes, Harshaw

Sodium bromide: Harshaw

Sodium chloride: Barnes, Harshaw, Janos, Optovac

Sodium fluoride: Harshaw, Optovac

Strontium fluoride: Harshaw, Janos, Optovac

Strontium titanate: Commercial Crystal Labs, Hibshman Optical Labs

Tellurium: Inrad

Thallium bromide: Harshaw

Thallium bromoiodide (KRS-5): Barnes, Harshaw, Infrared Optics, Unique Optical

Thallium chlorobromide (KRS-6): Harshaw

Titanium dioxide (rutile): Commercial Crystal Labs, Adolf Meller

Zinc selenide: Cleveland Crystals, Infrared Optics, Laser Optics, II-VI, Unique Optical

Zinc sulphide (Irtran 2): Barnes, Cleveland Crystals, Eastman Kodak, II-VI, Unique Optical

CHAPTER 5   CHARGED-PARTICLE OPTICS

Forty years ago, devices employing charged-particle beams were confined to the purview of a small group of physicists studying elementary processes. However, we are now in an era in which beams of ions or electrons are used by chemists, biologists, and engineers to probe various materials and investigate discrete processes. Physicists are constructing beam machines to control the momentum of interacting particles with energies from a few tenths of an electron volt to billions of electron volts. Chemists routinely use mass spectrometers as analytical tools, and various electron spectrometers to probe molecular structures. The electron microscope has become a major tool of the modern biologist. Furthermore, charged-particle beam technology has spread to industry, where electron-beam machines are used for cleaning surfaces, and welding and ion-beam devices are used in the preparation of semiconductors.

The properties of charged-particle beams are analogous in many respects to those of photon beams: hence the appellation *charged-particle optics*. In the following sections, the laws of geometrical optics will be covered insofar as they apply to charged-particle beams. The

consequences of the coulombic interaction of charged particles will be considered. In addition, we shall discuss the design of electron and ion sources, as well as the design of electrodes that constitute optical elements for manipulating beams of charged particles. We shall consider primarily electrostatic focusing by elements of cylindrical symmetry, and restrict discussion to particles of sufficiently low kinetic energies that relativistic effects can be ignored.

A number of excellent books and review articles are available to the reader who wishes to pursue the topic of electron and ion optics in greater detail.[1-6]

5.1 BASIC CONCEPTS OF CHARGED-PARTICLE OPTICS

5.1.1 Brightness

The *brightness* β of a point on a luminous object is determined by the differential of current dI, which passes through an increment of area dA about the point, and shines into a solid angle $d\Omega$:

$$\beta = \frac{dI}{dA \, d\Omega} \qquad (\mu \text{A cm}^{-2} \text{sr}^{-1}).$$

In electron or ion optics, a luminous object is usually defined by an aperture, called a *window*, which is

The science of charged-particle optics owes a great deal to the principles of design formulated by C. E. Kuyatt of the National Bureau of Standards. The authors are especially indebted to Dr. Kuyatt for his advice and for permission to present parts of the contents of his "Electron Optics Lectures" in the beginning of this chapter.

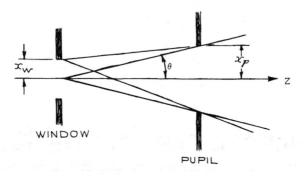

Figure 5.1 A pupil determining the half angle θ of rays from each point on an object defined by a window.

uniformly illuminated from one side by a stream of charged particles. The angular spread of particles emanating from the window is limited by a second aperture called a *pupil*. This situation, in the xz plane, is illustrated in Figure 5.1. As will usually be the case in this chapter, it will be assumed that the system is cylindrically symmetric about the z-axis and the distance between the window and pupil is sufficiently great that the angular spread of rays emanating from each point on the object is the same. Then the integrated brightness of the object outlined by the window is

$$\beta = \frac{I}{\pi^2 x_w^2 \theta^2},$$

where x_w is the window radius and θ is the half angle defined by the pupil relative to a point at the window.

5.1.2 Snell's Law

When a beam of charged particles enters an electric field, the particles will be accelerated or decelerated, and the trajectory will depend on the angle of incidence with respect to the equipotential surfaces of the field. This effect is analogous to the situation in optics when a light ray passes through a medium in which the refractive index changes. Figure 5.2 illustrates the behavior of a charged-particle beam as it passes from a region of uniform potential V_1 to a region of uniform potential V_2. The initial and final energies of a particle of charge q

that originated at ground potential with no kinetic energy are $E_1 = qV_1$ and $E_2 = qV_2$, respectively. α_1 and α_2 are the angles of incidence and refraction with respect to the normals to the equipotential surfaces that separate the field-free regions.

The quantity in charged-particle optics which corresponds to the index of refraction is the particle velocity, which is proportional to the square root of the particle energy. Thus, the charged-particle analog of Snell's law is

$$\sqrt{E_1}\,\sin\alpha_1 = \sqrt{E_2}\,\sin\alpha_2.$$

Clearly, this property can be exploited in charged-particle optics, as in light optics, to make lenses by shaping the equipotential surfaces. This is equivalent to varying the refractive index and shape of lenses for light.

5.1.3 The Helmholtz-Lagrange Law

Consider the situation illustrated in Figure 5.3, in which an object defined by a window at z_1 is imaged at z_2. The rays emanating from a point (x_1, z_1) at the edge of the object are limited by a pupil to fall within a cone defined by a half angle θ_1, which is called the *pencil angle*. The angle of incidence α_1 of the central ray from (x_1, z_1) on the plane of the pupil is referred to as the *beam angle*. The rays emanating from each point on the

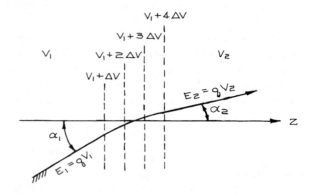

Figure 5.2 Charged-particle refraction at a potential gradient.

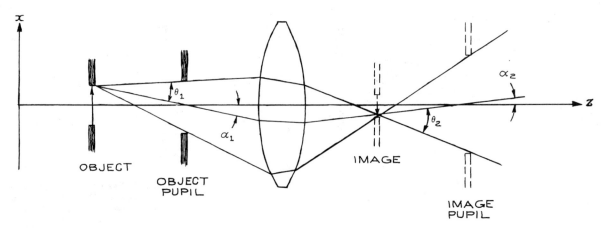

Figure 5.3 Relation of object pencil angle θ_1 and beam angle α_1 to image pencil angle θ_2 and beam angle α_2.

image appear to be limited by an aperture that corresponds to the image of the object pupil. An image-pencil angle θ_2 and beam angle α_2 from a point (x_2, z_2) at the edge of the image are defined with respect to this image pupil.

For most practical electron-optical systems, the pencil and beam angles are small and the approximation $\sin \theta \simeq \theta$ is valid. The treatment of optics based on this approximation is called *Gaussian* or *paraxial* optics.

The image pencil angle is determined by the object pencil angle. This relation is given to first order by the Helmholtz-Lagrange law

$$x_1 \theta_1 \sqrt{E_1} = x_2 \theta_2 \sqrt{E_2} ,$$

$$\frac{E_1}{E_2} = Mm; \qquad M = \frac{x_2}{x_1} , \qquad m = \frac{\theta_1}{\theta_2} ,$$

where M and m are the linear and angular magnifications, respectively.

The current I_1 through the object window that falls within the object pupil is the same as the current I_2 through the image window and pupil. Setting $I_1 = I_2 = I$, it follows that

$$\frac{I}{E_1 \theta_1^2 x_1^2} = \frac{I}{E_2 \theta_2^2 x_2^2} .$$

Furthermore, if β_1 and β_2 are the brightnesses of the object and image, respectively, then this equality can be written

$$\frac{\beta_1}{E_1} = \frac{\beta_2}{E_2} ,$$

demonstrating that *the ratio of brightness to energy is conserved from object to image.*

5.1.4 Vignetting

The foregoing discussion assumes that the object radiates uniformly; to the extent that the pencil angles are the same for all points on the object, it follows that the image is uniformly illuminated. However, if, as illustrated in Figure 5.4, a second pupil is added to the system, the illumination will vary across the image. Such a situation is called *vignetting*. The one case in which an aperture, in addition to the object window and pupil, does not produce vignetting is when an aperture is placed at the location of the image of the original window or pupil. Such an aperture is sometimes employed to skim off stray current resulting from scattering from slit edges and from aberrations. Also, an aperture is placed at the exit window of an energy or momentum analyzer to define its resolution. Except for

Figure 5.4 Vignetting.

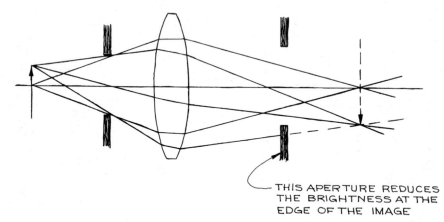

THIS APERTURE REDUCES
THE BRIGHTNESS AT THE
EDGE OF THE IMAGE

"spatter apertures" or resolving apertures, an electron-optical system should have only two apertures.

5.2 ELECTROSTATIC LENSES

It is a simple matter to produce axially symmetric electrodes that, when electrically biased, will produce equipotential surfaces with shapes similar to those of optical lenses. A charged particle passing across these surfaces will be accelerated or decelerated, and its path will be curved so as to produce a focusing effect. The chief difference between such a charged-particle lens and an optical lens is that the quantity analogous to the refractive index, namely the particle velocity, varies continuously across an electrostatic lens, whereas a discontinuous change of refractive index occurs at the surfaces of an optical lens. Charged-particle lenses are "thick" lenses, meaning that their axial dimensions are comparable to their focal lengths.

5.2.1 Geometrical Optics of Thick Lenses

In the case of a thick lens it is not correct to measure focal distances from a plane perpendicular to the axis through the center of the lens. The focal points of a

thick lens are located by *focal lengths*, f_1 and f_2, measured from *principal planes* H_1 and H_2, respectively. As shown in Figure 5.5, the principal planes of a charged-particle lens are always crossed, and they are both located on the low-voltage side of the central plane M of the lens. The locations of the focal points with respect to the central plane are given by F_1 and F_2, and hence the distances from the central plane to the principal planes are $F_1 - f_1$ and $F_2 - f_2$, respectively. The object and image distances with respect to the principal planes are p and q, and with respect to the central plane are P and Q, respectively. All distances are positive in the direction indicated by the arrows in Figure 5.5.

As in light optics, it is possible to graphically construct the image produced by a lens if the cardinal points of the lens are known. This procedure is illustrated in Figure 5.6. It is only necessary to trace two principal rays. From a point on the object draw a ray through the first focal point and thence to the first principal plane. From the point of intersection with this plane draw a ray parallel to the axis. Draw a second ray through the object point and parallel to the axis. From the point of intersection of this ray with the second principal plane, draw a ray through the second focal point. The intersection of the first and second principal rays gives the location of the image point corresponding to the object point.

A number of important relationships can be derived geometrically from Figure 5.6. The linear and angular

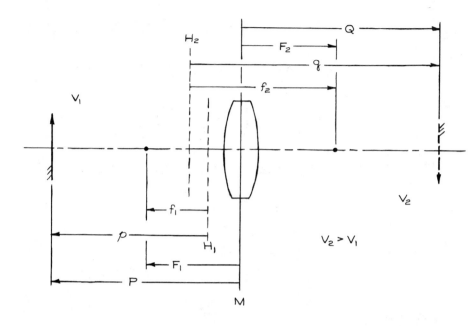

Figure 5.5 Lens parameters.

magnifications are given by

$$M = \frac{f_2 - q}{f_2} = \frac{f_1}{f_1 - p}$$

and

$$m = \frac{f_1 - p}{f_2} = \frac{f_1}{f_2 - q},$$

where negative magnification implies an inverted image.

The object and image distances from the central plane are

$$P = F_1 - \frac{f_1}{M}$$

and

$$Q = F_2 - Mf_2.$$

The object and image distances from the principal

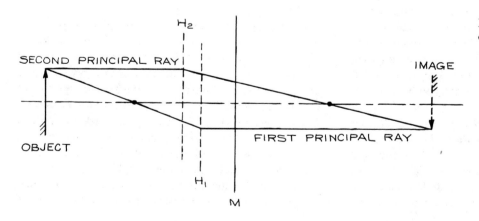

Figure 5.6 Graphical construction of an image.

planes are given by Newton's law,

$$(p - f_1)(q - f_2) = f_1 f_2,$$

and Newton's formula,

$$\frac{f_1}{p} + \frac{f_2}{q} = 1.$$

For lenses that are not very strong, the principal planes are close to the central plane. Then

$$p \to P,$$

$$q \to Q,$$

and

$$f_1 \simeq f_2 \simeq f,$$

and Newton's formula takes the approximate form

$$\frac{1}{P} + \frac{1}{Q} = \frac{1}{f},$$

which is useful for initial design work. Furthermore, Spangenberg[1] has observed that for the lenses discussed in succeeding sections,

$$M \simeq -0.8 \frac{Q}{P}.$$

5.2.2   Cylinder Lenses

The most widely used lens for focusing charged particles with energies of a few eV to several keV is that produced by two cylindrical coaxial electrodes biased at voltages corresponding to the desired initial and final particle energies. Figure 5.7 illustrates the equipotential surfaces associated with the field produced by a pair of cylinders biased at V_1 and V_2.

The focal properties of a *two-cylinder lens* depend upon the diameters of the cylinders, the spacing between them, and the ratio E_2/E_1 of the final to initial kinetic energies of the transmitted particles. The lens properties scale with the diameter, so all dimensions are taken in units of the diameter D of the larger cylinder.

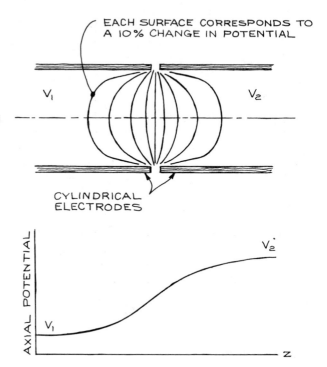

Figure 5.7   Potential distribution in a two-cylinder lens.

If the gap between cylinders is large, the focal properties become sensitive to the cylinder wall thickness and external fields are liable to penetrate to the lens. Most lenses are constructed with a gap of $0.1D$. The particle energies depend upon the bias potentials on the elements. In most cases it is assumed that the bias voltage supplies are referenced to the particle source, so that $V_2/V_1 = E_2/E_1$.

Focal properties are usually presented as graphs (Figure 5.8) or tables of F_1, f_1, F_2, and f_2 as a function of V_2/V_1. For simple lenses it is also convenient to plot P versus Q for different V_2/V_1 ratios as in Figure 5.9. The chief sources of information on the focal properties of electrostatic lenses are the book *Electrostatic Lenses* by Harting and Read[7] and various journal articles by Read and his coworkers.[8] These data are determined from numerical solutions of the equation of motion of an electron in a lens field and are thought to be accurate to 1–3%. Unfortunately, these sources contain no information applicable to systems with virtual objects or images.

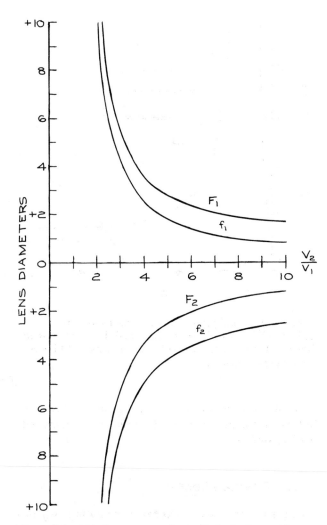

Figure 5.8 Focal properties of a two-cylinder lens with equal-diameter cylinders and a gap of $0.1D$. (From E. Harting and F. H. Read, *Electrostatic Lenses*, Elsevier, New York, 1976; by permission of Elsevier Publishing Company.)

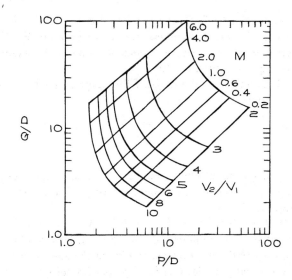

Figure 5.9 *P-Q* curves for a two-cylinder lens with equal diameter cylinders and a gap of $0.1D$. (From E. Harting and F. H. Read, *Electrostatic Lenses*, Elsevier, New York, 1976; by permission of Elsevier Publishing Company.)

For two-cylinder lenses, Natali, DiChio, Uva, and Kuyatt[9] have computed *P-Q* curves that extend to negative values of the object and image distances.

To obtain desired focal properties with a given acceleration ratio, two or more two cylinder lenses can be used in series. The lens field extends about one diameter D on either side of the midplane gap. When two gaps

are sufficiently close that their lens fields overlap, it is convenient to treat the combination as a single lens. Such a *three-cylinder lens* has the advantage that it can produce an image of a fixed object at a fixed image plane for a range of final-to-initial energy ratios. Alternatively, it can produce a variable image location with a fixed energy ratio.

A three-cylinder lens for which the initial and final energies differ is called an *asymmetric lens*. It can be used with an electron or ion source to produce a variable-energy beam. Such lenses are also used to focus an image of the exit slit of a monochromator on a target while still allowing the particle acceleration between monochromator and target to be varied.

The focal properties of three-cylinder lenses are not conveniently presented graphically, since each focal length is a function of two independent variables. These variables are usually taken to be V_2/V_1 and V_3/V_1 where V_1, V_2, and V_3 are the potentials on the first, second, and third lens elements. Harting and Read[7] and Heddle[10] present data on three-element lenses for V_3/V_1 values between 1.0 and 30.0 in the form of tables of lens parameters (f_1, f_2, F_1, F_2) as a function of V_2/V_1. They also plot "zoom-lens curves," which are graphs of V_2/V_1 versus V_3/V_1 for selected values of P and Q. An example is given in Figure 5.10.

Read has defined a length f that approximately satisfies the relation

$$\frac{1}{f} = \frac{1}{P} + \frac{1}{Q}$$

for three-element lenses. For use in initial design work, V_2/V_1 is graphed as a function of f/D for various values of V_3/V_1.[3]

A three-element lens for which $V_3/V_1 = 1$ is called an *einzel lens* or *unipotential lens*. Such a lens produces focusing without an overall change in the energy of the transmitted particle. The focal properties are symmetrical: $f_1 = f_2 = f$ and $F_1 = F_2 = F$. Figure 5.11 illustrates the focal properties of a typical einzel lens. For any desired object and image distance there are two focusing conditions: a decelerating mode with $V_2/V_1 < 1$ and an accelerating mode with $V_2/V_1 > 1$. The accelerating mode is preferred, because deceleration in the middle

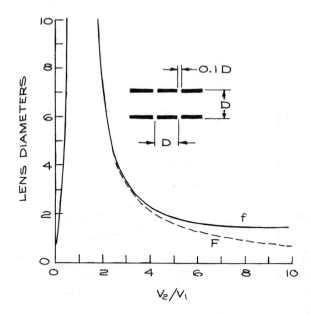

Figure 5.11   Focal properties of an einzel lens, with $A = D$ and $g = 0.1D$. (From E. Harting and F. H. Read, *Electrostatic Lenses*, Elsevier, New York, 1976; by permission of Elsevier Publishing Company.)

element causes expansion of the transmitted beam, resulting in aberrations and unwanted interactions with the lens surfaces.

5.2.3 Aperture Lenses

An aperture in a plane electrode separating two uniform-field regions will have a focusing effect if the field on one side is greater than on the other. As illustrated in Figure 5.12, this lens results from a bulge in the equipotential surfaces caused by field penetration through the aperture. A lens of this type is called a *Calbick lens*. Single-aperture focusing is important in electron-gun and ion-source design because these devices usually have an aperture at their output.

The Calbick lens is (Figure 5.12) the analog of a thin lens. For a circular aperture,

$$f_1 = f_2 = F_1 = F_2 = \frac{4V_2}{E_B - E_A},$$

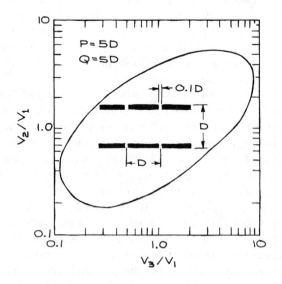

Figure 5.10   Typical "zoom-lens curve." (From A. Adams and F. H. Read, "Electrostatic Cylinder Lenses III," J. Phys., E5, 156, 1972; copyright 1972 by The Institute of Physics.)

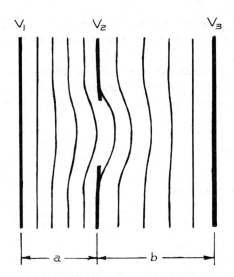

Figure 5.12 Penetration of equipotential surfaces through an aperture separating two different field regions.

where

$$E_A = \frac{V_2 - V_1}{a}, \qquad E_B = \frac{V_3 - V_2}{b};$$

the potentials are referred to the source, and the ratio of aperture potential to radius is much larger than E_A or E_B. For a long slot in the xy plane,

$$f_x = \frac{2V_2}{E_B - E_A}$$

and

$$f_y \simeq \infty$$

when the short dimension of the slot is in the x-direction. The focal properties of a slot are important in the design of electron guns that employ a ribbon filament.[11]

Lenses can also be constructed using two or more apertures. Read has computed the properties of two-aperture and three-aperture lenses.[7,8]

In practice, circular-aperture electrodes of an aperture lens are supported by concentric cylindrical mounts. The focusing effects of these cylindrical elements can be ignored if the cylinder diameter is at least three times greater than the aperture diameter.

5.2.4 Matrix Methods

A useful description of the charged-particle trajectories through a focusing system can be formulated in using the matrix methods introduced in Section 4.2.2. A particle trajectory through a system that is cylindrically symmetrical about the z-axis is determined by (r, $dr/dz = r'$, z), where r is the radial distance of the trajectory from the axis at z. A trajectory through a plane perpendicular to the central axis at z is represented by a vector

$$\begin{pmatrix} r \\ r' \end{pmatrix}.$$

The effect of a displacement in a field-free region from z_1 to z_2 is given by

$$\begin{pmatrix} r_2 \\ r_2' \end{pmatrix} = \mathbf{M}(z_1 \to z_2)\begin{pmatrix} r_1 \\ r_1' \end{pmatrix},$$

where the *ray transfer matrix* is

$$\mathbf{M}(z_1 \to z_2) = \begin{pmatrix} 1 & \Delta z \\ 0 & 1 \end{pmatrix}, \qquad \Delta z = z_2 - z_1.$$

A lens can be represented as if its effect were confined to the region between the principal planes. The matrix operating between the principal planes is

$$\mathbf{M}(H_1 \to H_2) = \begin{pmatrix} 1 & 0 \\ -\dfrac{1}{f_2} & \dfrac{f_1}{f_2} \end{pmatrix}.$$

Consider, for example, a ray originating at a point on an object at z_1 and passing through a point on an image at z_2. The trajectory at z_2 can be determined from the trajectory at z_1 by

$$\begin{pmatrix} r_2 \\ r_2' \end{pmatrix} = \mathbf{M}(H_2 \to z_2)\mathbf{M}(H_1 \to H_2)\mathbf{M}(z_1 \to H_1)\begin{pmatrix} r_1 \\ r_1' \end{pmatrix}.$$

In terms of the parameters defined in Figure 5.5, the

displacement between the object and the first principal plane is $P - F_1 + f_1$; thus

$$\mathbf{M}(z_1 \rightarrow H_1) = \begin{pmatrix} 1 & P - F_1 + f_1 \\ 0 & 1 \end{pmatrix},$$

and similarly

$$\mathbf{M}(H_2 \rightarrow z_2) = \begin{pmatrix} 1 & Q - F_2 + f_2 \\ 0 & 1 \end{pmatrix}.$$

Substitution yields

$$\begin{pmatrix} r_2 \\ r_2' \end{pmatrix} = \begin{pmatrix} \dfrac{F_2 - Q}{f_2} & \dfrac{f_1 f_2 - (Q - F_2)(P - F_1)}{f_2} \\ -\dfrac{1}{f_2} & \dfrac{F_1 - P}{f_2} \end{pmatrix} \begin{pmatrix} r_1 \\ r_1' \end{pmatrix}.$$

This transformation matrix is generally applicable to the problem of tracing a trajectory through the field of a lens. For the two-cylinder lens, simple analytical expressions for the elements of the matrix have been derived by DiChio et al.[12] When r_1 and r_2 refer to points in the object and image planes, respectively, the diagonal elements of the transformation matrix are the linear and angular magnification. Furthermore, the focusing condition given by Newton's formula requires the numerator of the upper right element to be zero. Thus, the transformation matrix between object and image is

$$\mathbf{M}(\text{object} \rightarrow \text{image}) = \begin{pmatrix} M & 0 \\ -1/f_2 & m \end{pmatrix}.$$

The matrix formulation is a desirable alternative to graphical ray tracing for a system of lenses that are so close to one another that an image does not appear between each lens and the next. Consider, for example, the system of two lenses in Figure 5.13. In this case the transformation matrix is

$$\mathbf{M}(z_1 \rightarrow z_2) = \mathbf{M}(H_2' \rightarrow z_2)\mathbf{M}(H_1' \rightarrow H_2')$$
$$\times \mathbf{M}(H_2 \rightarrow H_1')\mathbf{M}(H_1 \rightarrow H_2)\mathbf{M}(z_1 \rightarrow H_1).$$

In general, the image distance L_3 will be unknown

initially. However, it can be determined from the focusing condition, which requires the upper right element of $\mathbf{M}(z_1 \rightarrow z_2)$ to be zero.

It is occasionally necessary to determine a particle trajectory in a region of uniform field. The transformation matrix from z_1 to z_2 in a uniform field in the z-direction is

$$\mathbf{M}(\text{uniform field}) = \begin{pmatrix} 1 & \dfrac{2\,\Delta z\, V_1}{V_2 - V_1}\left(\sqrt{\dfrac{V_2}{V_1}} - 1\right) \\ 0 & \sqrt{\dfrac{V_1}{V_2}} \end{pmatrix},$$

where V_1 and V_2 are the potentials of the planar equipotential surfaces perpendicular to the z-axis at z_1 and z_2.

5.2.5 Aberrations

There are three types of aberrations that must be considered in designing charged-particle optics:

1. *Geometrical aberrations* due to deviations from the paraxial assumption that the angle θ between a ray and the central axis is sufficiently small that $\theta = \sin \theta = \tan \theta$.

2. *Chromatic aberrations* caused by variations in the kinetic energies of transmitted particles.

3. *Space charge* due to electron-electron interactions.

The *geometrical aberration* associated with a lens is defined in terms of the radius Δr_2 of a spot in the image plane formed by rays within the pencil angle θ_1 that emanate from an object point on the axis, as shown in Figure 5.14. The important geometrical errors can be expressed in terms of the coefficients of a power series in θ_1:

$$\Delta r_2 = b\theta_1^2 + c\theta_1^3 + \cdots.$$

An imaging system is said to be *second-order-focusing* if

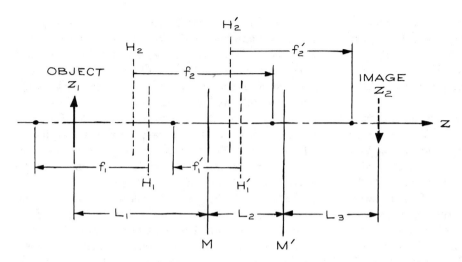

Figure 5.13 A compound lens system.

$b = 0$, and *third-order-focusing* if $c = 0$. An axially symmetric system is always at least second-order-focusing, since only terms of odd order appear in the expansion of Δr_2. The coefficient c is referred to as the *third-order-aberration coefficient*. Higher-order coefficients can usually be ignored. For an object point on a lens axis, the error represented by c is spherical aberration. Harting and Read[7] have computed spherical-aberration coefficients C_s defined by

$$c = MC_s$$

for all of the lenses for which they give focal properties. For object points off axis, other types of aberration must be considered (coma, field curvature, distortion, and astigmatism). Most off-axis aberrations can be related to the magnitude of C_s. When the object is small compared to the lens diameter, the spherical-aberration coefficient gives a reasonable measure of the total geometrical aberration.

As can be seen from Figure 5.14, the pencil of rays emanating from an object point achieves a minimum diameter at a point slightly in front of the image plane.

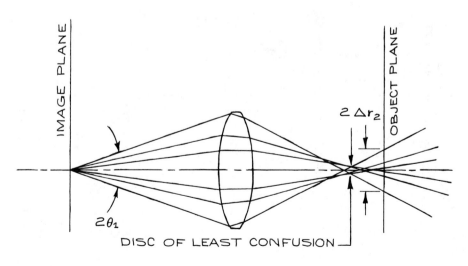

Figure 5.14 Geometrical aberration of an image formed from an object point on axis.

This minimum diameter is the *circle of least confusion*. To achieve the sharpest focus it is often possible to vary the lens voltages so as to place the circle of least confusion on the desired image plane. The radius of this circle is $\frac{1}{4}\Delta r_2$, and thus C_s can in practice be reduced by a factor of 4.

The diameter of the bundle of rays emanating from an object is usually determined by a pupil, which limits the portion of the lens that is illuminated by rays from the object. The extent of illumination is specified by a *filling factor*, which is the ratio of the diameter of the bundle of rays near the lens gap to the lens diameter. Because the spherical aberration depends upon the third power of the angle between the limiting ray and the axis, it is obvious that aberration increases rapidly with increasing filling factor. In practice, a lens system should not be designed with a filling factor in excess of about 50%.

For an object of finite extent, the worst aberration occurs for off-axis image points. The magnitude of the aberration depends upon the maximum value of the angle between a ray from the edge of the object and the axis. From Figure 5.3 it can be seen that this angle is the sum of the object pencil angle and beam angle. In many lens systems, the image produced by one lens serves as an object for succeeding lenses. The pencil angle associated with this intermediate image is determined by the Helmholtz-Lagrange law from the pencil angle associated with the initial object. There are, however, no such physical restrictions on the beam angle from the intermediate image. In fact, as shown in Figure 5.15, if an object pupil is placed at the first focal point of a lens, then the corresponding image pupil is at infinity and the image beam angle is zero. Obviously such an arrangement reduces aberrations produced by lenses that treat this image as an object.

The effect of *chromatic aberration* depends upon the relative spread in particle energies, $\Delta E/E$, passing through a lens. Variations in particle energies can occur because of conditions in the particle source or within the lens. An energy spread of 0.2 to 0.4 eV is characteristic of electron sources. Some ion sources yield ions in a bandwidth as great as 30 eV. Energy variations may also be caused by fluctuations in the voltage from the power supplies that establish both source and lens-element potentials.

For a given lens, the extent of chromatic aberration can best be determined from a calculation of the image distances for particles with energies at the extremes of the anticipated energy distribution. Consider for exam-

Figure 5.15   Placement of an object pupil at the first focal point of a lens. This moves the image pupil to infinity and reduces the beam angle from the image to zero.

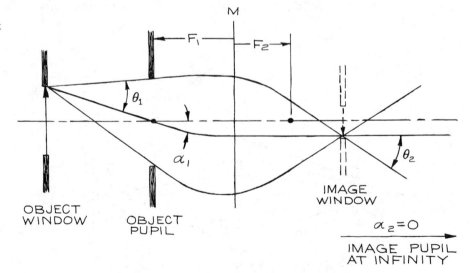

ple the case of a source at ground potential that produces particles with energies between 0 and ΔE. For a lens of voltage ratio V_2/V_1 a distance P from the source, determine the corresponding image position, that is, $Q = f(P, V_2/V_1)$. Determine also

$$Q + \Delta Q = f\left(P, \frac{V_2 + \Delta E/q}{V_1 + \Delta E/q}\right).$$

Then a point on axis at the source yields an image at Q of radius

$$\Delta r_2 = \theta_2 \Delta Q,$$

or, by the Helmholtz-Lagrange law,

$$\Delta r_2 = \frac{\theta_1 \Delta Q}{M}\sqrt{\frac{V_2}{V_1}},$$

where θ_1 and θ_2 are the pencil angles at the source and the image, respectively.

In electron-beam devices, the main cause of chromatic aberration is the spread of electron kinetic energies emitted from a thermionic cathode. The energy of electrons at the maximum of the current distribution from a cathode is

$$E_{max} = 8.6 \times 10^{-5} T \text{ eV}.$$

The width of the distribution is

$$\Delta E \simeq E_{max}.$$

For a bare tungsten filament a temperature of about 3000 K is required to produce significant emission. At this temperature $E \simeq 0.25$ eV. An oxide-coated cathode (available from Electron Technology) can be operated at temperatures as low as 1200 K, where $\Delta E \simeq 0.1$ eV.

The *space-charge* effect arises from the mutual repulsion of particles of like charge. The effect increases with current density. As a result, even a focused beam will give a diffuse image. Furthermore, the beam will diverge away from the image more rapidly than predicted by geometrical optics.

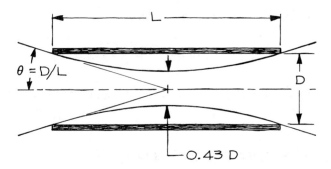

Figure 5.16 A beam focused to give maximum current through a tube.

Space charge places a limit on the current in a charged-particle beam. As an example of practical value, consider the problem of transmitting maximum current through a cylindrical element of length L and diameter D. As shown in Figure 5.16, the maximum current is achieved by focusing the beam on a point at the center of the tube with a pencil angle $\theta = D/L$. It can be shown that the space-charged-limited current of electrons of energy $E = qV$ is

$$I_{max}(\text{electrons}) = 38.5 V^{3/2}\left(\frac{D}{L}\right)^2 \mu\text{A},$$

and the maximum ion current is

$$I_{max}(\text{ions}) = 0.90\left(\frac{M}{n}\right)^{1/2} V^{3/2}\left(\frac{D}{L}\right)^2 \mu\text{A},$$

where V is in volts, M is in amu, and n is the charge state.[2] Furthermore, the minimum beam diameter at the space-charge limit is $0.43D$.

Space charge also limits the current available from a surface that emits electrons or ions. The typical source geometry is the plane diode consisting of a planar anode at potential V a distance d from the cathode. Charged particles produced by thermionic emission from the cathode are accelerated toward the anode. The maximum electron current density at the anode is

$$J_{max}(\text{electrons}) = 2.34 \frac{V^{3/2}}{d^2} \mu\text{A cm}^{-2},$$

while for an ion-emitting cathode

$$J_{max}(\text{ions}) = 0.054 \left(\frac{n}{M}\right)^{1/2} \frac{V^{3/2}}{d^2} \, \mu\text{A cm}^{-2},$$

with V in volts, d in cm, M in amu, and n the charge state.

5.2.6 Lens Design Example

Consider the problem of producing the image of the anode aperture of an electron source on the entrance plane of an energy analyzer as illustrated in Figure 5.17. The anode potential is $V_1 = 100$ volts, and the analyzer entrance-plane potential is $V_2 = 10$ volts. The anode aperture radius is $r_1 = 1.0$ mm, and the distance from anode to analyzer is $l = 10$ cm. It is desired that the image of the anode aperture serve as the entrance aperture for the analyzer with radius $r_2 = 0.5$ mm. So as not to overfill the analyzer, one must also have the image pencil angle $\theta_2 = 0.14$ and the image beam angle $\alpha_2 = 0$.

The lens to be designed is a decelerating lens with $V_2/V_1 = 0.1$ and magnification $M = 0.5$. Customarily only the focal properties of accelerating lenses are tabulated. The focal properties for the desired lens are just the reverse of those of the accelerating lens with $V_2/V_1 = 10$. From Harting and Read,[7] the focal properties of the $V_2/V_1 = 10$ lens (identified by a prime) are

$$f_1' = 0.80D, \qquad F_1' = 1.62D,$$
$$f_2' = 2.54D, \qquad F_2' = 1.19D.$$

The corresponding parameters for the $V_2/V_1 = 0.1$ lens are

$$f_1 = 2.54D, \qquad F_1 = 1.19D,$$
$$f_2 = 0.80D, \qquad F_2 = 1.62D.$$

From the P-Q curve of Figure 5.9 for $V_2/V_1 = 10$ and $M' = 1/M = 2$, it can be estimated that

$$P = Q' = 6.20D,$$
$$Q = P' \simeq 2.05D.$$

Taking $P = 6.20D$ and $p = P - F_1 = 5.01D$, Newton's law gives

$$q = \frac{f_1 f_2}{P} = 0.41D$$

and

$$Q = q + F_2 = 2.03D.$$

The overall length of the lens is

$$l = P + Q = 6.20D + 2.03D = 10 \text{ cm},$$

and thus

$$D = 1.22 \text{ cm}.$$

In order for the image beam angle to be zero, the object pupil must be placed at the first focal point of the lens, so that the image pupil is at infinity. The object pupil must define a pencil angle θ_1, which, in terms of the image pencil angle, is given by the Helmholtz-Lagrange

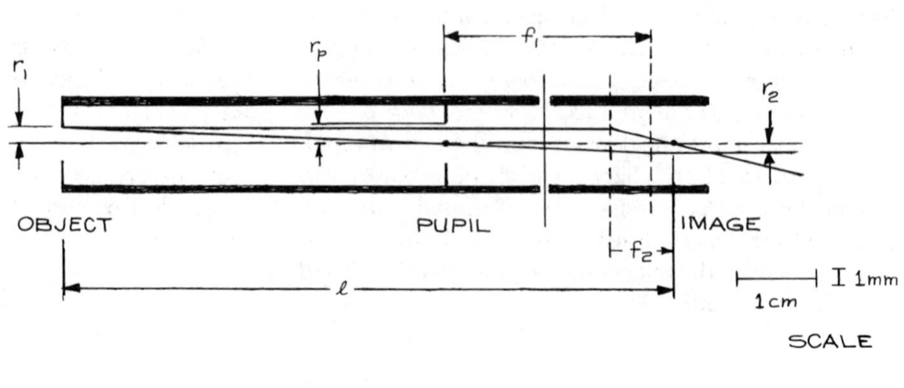

Figure 5.17 A simple lens design.

law as

$$\theta_1 = M\theta_2\sqrt{\frac{V_2}{V_1}} = 0.022.$$

The radius r_p of the aperture at the first focal point, which defines θ_1, is given by

$$r_p = \theta_1(P - F_1) = 0.11 \text{ cm}.$$

To estimate the geometrical aberration of this lens, calculate the spherical aberration for the lens used in reverse. This is necessary because spherical-aberration coefficients are available only for lenses with $V_2/V_1 > 1$. For $V_2/V_1 = 10$, Harting and Read give

$$\frac{M'C_s}{D} = 20.$$

Thus for the lens used in reverse, the image of a point is a spot of diameter

$$\Delta r' = M'C_s\theta_2^3 = 0.067 \text{ cm}.$$

For the lens used as designed, the size of the image of an object point is

$$\Delta r_2 = M\Delta r' = 0.033 \text{ cm}.$$

Thus spherical aberration will enlarge the image by 67%. If, however, the lens voltage is adjusted slightly to bring the circle of least confusion onto the image plane, the spherical aberration can be reduced to 17%.

For a typical electron source, $\Delta E = 0.3$ eV. This energy spread will result in chromatic aberration at the image. To estimate this effect, calculate the image position $Q + \Delta Q$ for electrons at the extreme of the energy distribution where the deceleration ratio of the lens is $(V_2 + \Delta E/q)/(V_1 + \Delta E/q) = 0.103$. As before, the focal properties can be determined from those of the corresponding accelerating lens. For the lens with $V_2/V_1 = 0.103$,

$$f_1 = 2.58D, \qquad F_1 = 1.22D,$$
$$f_2 = 0.82D, \qquad F_2 = 1.65D.$$

The image position in this case is

$$q_{\Delta E} = \frac{f_1 f_2}{P} = \frac{f_1 f_2}{P - F_1} = 0.42D,$$

and

$$Q + \Delta Q = q_{\Delta E} + F_2 = 2.07D,$$
$$\Delta Q = 0.04D = 0.05 \text{ cm}.$$

This displacement of the image plane will cause the image of a point object to appear as a disc of radius

$$\Delta r_2 = \Delta Q\,\theta_2 = 0.007 \text{ cm}$$

in the original image plane.

The overall aberration of the lens as designed is about 30%. The filling factor is

$$\frac{2(\theta_1 + \alpha_1)P}{D} = 54\%.$$

From the aberration calculations it can be seen that the aberrations would become serious if the filling factor were to exceed this value.

5.3 CHARGED-PARTICLE SOURCES

An electron or ion gun consists of a source of charged particles, such as a hot metal filament or a plasma, and an electrode structure that gathers particles from the source and accelerates them in a particular direction to form a beam. There are a wide range of practical electron or ion guns. Some simple devices and their principles of operation will be described in this section. In addition to these simple devices, which can be conveniently constructed for laboratory use, there are a number of very sophisticated guns that have been developed commercially. These include electron guns for cathode-ray tubes and electron-beam welders, and ion guns for sputter-cleaning apparatus and for ion-implantation doping of semiconductors.

5.3.1 Electron Guns

Electrons are produced by thermionic emission, field emission, photoelectric emission, and electron-impact ionization. Thermionic sources are most common.[13] These sources typically consist of a wire filament of some refractory metal such as tungsten or tantalum, which is heated by an electrical current passing through it. In some cases thermionic sources consist of a metal cup or button, which is indirectly heated by an electrical heater or by electron bombardment of the back surface. As explained in Section 5.2.5, the electrons from a hot filament possess energies of a few tenths of an electron volt. These electrons will be lost because of space charge if they are not immediately accelerated before being admitted to a field-free region. Many accelerator structures are used. Most are either diode or triode geometries, although some TV tubes employ a pentode geometry. The Pierce diode geometry is most commonly employed in laboratory devices.

Consider the diode made up of a plane emissive surface and a parallel plane anode as illustrated in Figure 5.18(a). Electrons leave the cathode with energy E_k. The anode is biased at a positive potential V_a relative to the cathode, so that electrons from a spot on the cathode will appear at a spot on the anode with energies of approximately E_a. To admit the accelerated electrons to the system beyond, a hole is made in the anode. If the cathode and anode were infinite in extent, then the space-charge-limited current density given in Section 5.2.5 could be achieved at the anode, and the electron beam emerging from the anode hole would be characterized by a beam angle

$$\alpha = \frac{r_a}{3d},$$

where r_a is the radius of the anode hole and d the cathode-to-anode spacing. Since the most divergent electron arriving at the anode would be one emitted parallel to the cathode with energy E_k, it can be seen that the pencil angle characterizing the beam from the anode hole would be

$$\theta = \sqrt{\frac{E_k}{E_a}}.$$

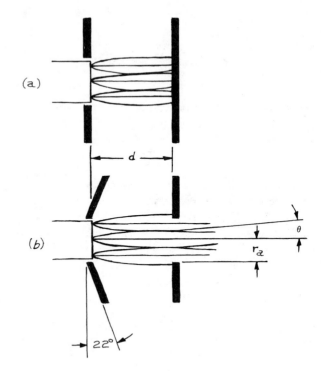

Figure 5.18   (a) The plane diode; (b) the Pierce geometry that simulates the field of the plane diode.

For a cathode of finite extent, the space-charge interaction causes the beam to spread laterally within the gap between cathode and anode. Pierce[2] has shown that the electric field in an infinite space-charge-limited diode can be reproduced in the region of a finite cathode by means of the conical cathode structure illustrated in Figure 5.18(b). The maximum current of electrons is as follows:

$$I_{\max} = \pi r_a^2 J_{\max} = 7.35 \left(\frac{r_a}{d}\right)^2 V_a^{3/2} \ \mu\text{A}.$$

The corresponding ratio of brightness to energy (which is conserved) is

$$\frac{\beta}{E_a} = \frac{I}{E_a \pi^2 r_a^2 \theta^2}$$

$$= 0.74 \frac{E_a^{3/2}}{E_k d^2} \ \mu\text{A}\,\text{cm}^{-2}\text{sr}^{-1}\text{eV}^{-1}.$$

The design of multielectrode guns is a complicated process involving time-consuming experimentation. A number of special guns have been developed for such purposes as producing low-energy electron beams or very narrow beams. Many of these are described by Klemperer and Barnett,[3] and most can be constructed easily if they are really needed. In general, however, it is best to use the simplest device that will fulfill one's requirements.

The guns developed for TV tubes are the product of much industrial development. These guns will produce a sharply focused spot at a distance of 10 to 20 cm. Cathode-ray-tube guns can be obtained quite inexpensively from commercial suppliers (such as Cliftronics) or from shops that specialize in rebuilding TV picture tubes.

5.3.2 Electron-Gun Design Example

Consider the problem of constructing a gun consisting of a Pierce diode and a lens system that can inject a beam of 20–200-eV electrons into a gas cell, as illustrated in Figure 5.19. It is desired that the beam current approach the space-charge limit at 20 eV. The overall length of the gun is to be about 10 cm, and the cell is to be located 4 cm in front of the gun. The cell is 2 cm long, with an input aperture of radius $r_2 = 0.1$ cm and an exit aperture of radius 0.15 cm. Treat the exit aperture as though its radius r_4 were 0.1 cm, so as to make the beam tight enough to minimize backscattering of electrons from the edges of this aperture.

The maximum current through the cell can be taken as the space-charge-limited current through a cylinder with a diameter D equal to that of the gas-cell apertures and length L equal to the length of the gas cell. From Section 5.2.5, the maximum current of electrons with energy $E_3 = 20$ eV is in this case

$$I_{\max} = 38.5 V_3^{3/2} \left(\frac{D}{L} \right)^2$$

$$= 34.4 \; \mu\text{A}.$$

To approach this limit, focus an image of the anode aperture on the center of the cell. Make the image radius $r_3 = 0.43 r_2$, the minimum size of the space-

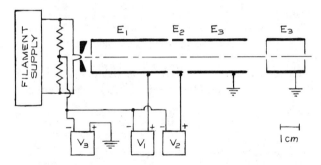

Figure 5.19   An electron-gun system for injecting a variable-energy beam into a gas cell.

charge-limited beam.[14] The ratio of brightness to energy at this image is

$$\frac{\beta_3}{E_3} = \frac{I_{\max} / \pi^2 r_3^2 \theta_3^2}{V_3}$$

$$= 9500 \; \mu\text{A cm}^{-2}\text{sr}^{-1}\text{eV}^{-1},$$

where the image pencil angle is $\theta_3 = D/L = 0.1$.

At the anode of the Pierce diode (Section 5.3.1), the ratio of brightness to energy is given by

$$\frac{\beta_1}{E_1} = 0.74 \frac{V_1^{3/2}}{V_k d^2},$$

where V_k is the mean potential (relative to the cathode) of electrons emitted by the cathode, and V_1 is the potential of the anode relative to the cathode. Since this ratio is a conserved quantity,

$$\frac{\beta_3}{E_3} = \frac{\beta_1}{E_1}.$$

Substituting from the previous two equations yields

$$V_1 = \left[\left(\frac{\beta_3}{E_3} \right) \frac{V_k d^2}{0.74} \right]^{2/3}.$$

For $V_k = 0.25$ volts (Section 5.2.5) and a cathode-to-anode spacing $d = 0.5$ cm, the anode potential corresponding to a final electron energy of 20 eV is

$$V_1(E_3 = 20 \text{ eV}) = 86 \text{ volts}.$$

A similar calculation yields

$$V_1(E_3 = 200 \text{ eV}) = 185 \text{ volts}.$$

The current from the diode depends upon the size of the anode hole r_1 and the current density J at the anode:

$$I = J\pi r_1^2.$$

For $E_3 = 20$ eV and $I_{max} = 34.4$ μA,

$$J_{max} = \frac{2.34}{d^2} V_1^{3/2}$$

$$= 7500 \text{ } \mu\text{A}.$$

Then

$$r_1 = \left(\frac{I_{max}}{\pi J_{max}}\right)^{1/2} = 0.04 \text{ cm}.$$

A similar calculation for $E_3 = 200$ eV would suggest using a larger anode aperture. However, in that case an excess of current would be injected into the lens if the space-charge-limited current were extracted from the diode when the system was tuned to produce a low-energy beam. This situation would be undesirable because of the large numbers of stray electrons that would result.

Note that when the system is designed to approach the space-charge limit at 20 eV, the pencil angle at the anode is consistent with the pencil angle defined by the cell apertures. The pencil angle characteristic of the Pierce diode at the low-energy limit is

$$\theta_1 = \sqrt{\frac{V_k}{V_1}} = 0.05,$$

and the anode pencil angle that gives the space-charge-limited pencil angle in the cell is, according to the Helmholtz-Lagrange law,

$$\theta_1 = \sqrt{\frac{V_3}{V_1}}\left(\frac{r_3}{r_1}\right)\theta_3 = 0.05.$$

For final energies greater than 20 eV, these pencil angles will be less than the angle defined by the cell apertures, and thus scattering from the edges of the apertures will be minimized.

It remains to design a lens system that will image the anode aperture on the center of the gas cell. A variable-ratio lens is required, and in this case a three-cylinder asymmetric lens seems appropriate. The voltage ratios at the extremes of the desired operating conditions are

$$\frac{V_3}{V_1}(E_3 = 20 \text{ eV}) = 0.23$$

and

$$\frac{V_3}{V_1}(E_3 = 200 \text{ eV}) = 1.1.$$

The desired magnification is

$$M = \frac{r_3}{r_1} \simeq 1.$$

The lens diameter must be chosen to give an acceptable filling factor. As a starting point, recall that

$$M \simeq 0.8\frac{Q}{P}.$$

It is desired that $P + Q = 10$ cm, and thus the distance from the anode to the middle of the lens will be

$$P = \left(\frac{P}{P + Q}\right) \times 10 \text{ cm}$$

$$= \left(\frac{P}{P + 1.25P}\right) \times 10 \text{ cm}$$

$$= 4.4 \text{ cm}.$$

The maximum diameter of the beam through the lens system will be approximately

$$2\theta_1 P = 0.44 \text{ cm}$$

for a final energy of 20 eV. To achieve a filling factor less than 30%, choose $D = 1.5$ cm. Then

$$P = 3.0D,$$
$$Q = 3.7D,$$

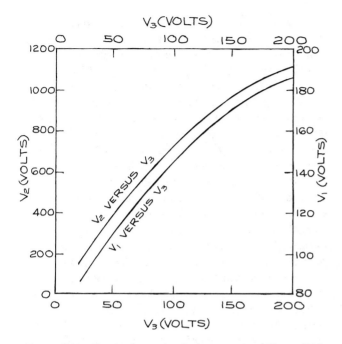

Figure 5.20   Lens voltages for the lens system of Figure 5.19.

and the nominal focal length of the desired lens (see Section 5.2.2) is

$$f = \left(\frac{1}{P} + \frac{1}{Q} \right)^{-1} = 1.66D.$$

The calculations of Adams and Read[8] or Harting and Read[7] can then be used to determine the lens voltages that correspond to this nominal focal length. The lens voltages and voltage ratios for a three-cylinder asymmetric lens with a central cylinder of length $A = 0.5D$ and lens gaps $g = 0.1D$ are presented graphically in Figure 5.20.

5.3.3   Ion Sources

An extensive range of parameters must be considered in choosing or designing an ion source. Foremost is the desired type of ionic species and charge state. The physical and chemical properties of the corresponding parent material determine to a large extent the means of ion production. Other important parameters are the desired current, brightness, and energy distribution. In some cases it is also necessary to consider the efficiency of utilization of the parent material. Considering the number of variables involved, it is not surprising that many different types of ion sources have been developed to meet various scientific and industrial demands.[15]

Ion sources can be classified according to the ion production mechanism employed. The two most common means are surface ionization and electron impact. The electron-impact sources include both electron-bombardment sources and plasma sources. More exotic ion sources employ ion impact, charge exchange, field ionization, and photoionization. Three typical sources, illustrating different ion production mechanisms, will be discussed below.

In practice, the simplest means of producing ions is by thermal excitation of neutral species. This process, known as *surface ionization*, occurs efficiently when an atom or molecule is brought in contact with a heated surface whose work function exceeds the ionization potential of the atom or molecule. This requirement is fulfilled for alkali atoms (Li, Na, K, Rb, Cs) on surfaces of tungsten, iridium, or platinum. In order for ionization to compete with vaporization it is necessary that the surface be sufficiently hot that the substrate surface is only partially covered with neutrals, so that the work function of the surface is characteristic of the substrate rather than the material to be ionized.

In a surface-ionization ion gun, the source of neutral atoms can either be remote from, or integral with, the ionizing surface. In the remote type, the alkali metal is contained in an oven. A stream of metal vapor from the oven is directed toward the ionizer, which is a heated metal filament located within an electrode structure similar to that of an electron gun. The electrodes are negatively biased to accelerate positive ions into a beam. The most convenient alkali-ion emitter consists of a porous tungsten disc that has been infused with an alkali-containing mineral and mounted on an electrical heater. When heated, alkali atoms are generated, the atoms diffuse to the surface of the disc, and they are ionized as they leave the surface. These emitters with integral heaters are commercially available (e.g. from Spectra-Mat). As shown in Figure 5.21, a source is

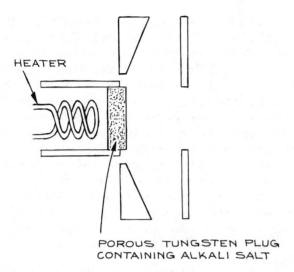

Figure 5.21   An alkali-ion source consisting of a heated emitter in a Pierce diode.

constructed by inserting the emitter into the cathode of a Pierce diode or into some other extractor electrode structure.

The majority of laboratory ion sources employ *electron-impact* ionization. The simplest configuration is illustrated in Figure 5.22. In this source, an electron beam from a simple diode gun is injected into an

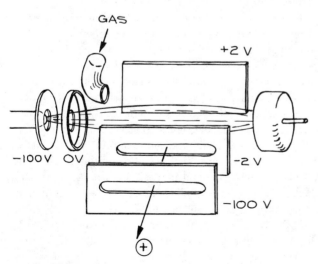

Figure 5.22   A simple electron-impact ion source.

ionization chamber containing an appropriate parent gas. A transverse electric field of a few V/cm causes ions produced by electron impact to drift across the chamber and out through a slit in the side of the chamber. An extractor electrode system accelerates the ions into a beam. The ionic species produced will depend upon the parent gas, the gas pressure, and the electron energy. Singly charged ions are produced with maximum efficiency at electron energies of about 70 eV. Higher electron energies favor multiply charged ions. Lower electron energies yield only singly charged ions and reduce the extent of fragmentation in the event that the parent gas is a molecular species. Gas pressures are typically 1 to 10 mtorr. Higher pressures often place an unreasonable gas load on the vacuum system and impede the flow of ions from the ionization chamber.

The efficiency of this simple source can be improved by imposing a magnetic field of a few hundred gauss coaxial with the electron beam. The field confines the electron beam to a spiral around the beam axis, thus increasing the path length and maximizing the probability of collision with the gas molecules. Electron-bombardment sources of this type yield currents of only a few tens of microamperes. The chief advantages are simplicity of construction and an energy spread in the product ions of only a few eV.

Relatively high ion densities can be achieved in an electrical discharge through a gas. The ion density in a discharge can be further increased if the discharge is confined and compressed by a magnetic field. The duoplasmatron source illustrated in Figure 5.23 is the prototypical discharge-type ion source.[16] A discharge through a gas at about 100 mtorr is produced by applying a potential of 300 to 500 volts between the heated cathode and the anode. After the arc is struck, the discharge is maintained by passing a current 0.5 to 2 A through the ionized gas. In the duoplasmatron, the intermediate element, known as the *zwischen*, is one pole of a magnet. The axial magnetic field at the tip of the zwischen confines the discharge to a dense plasma bubble at the anode. Plasma leaks through a hole in the anode to fill a cup on the front face of the anode. An extractor electrode, biased at about 10 kV relative to the anode, withdraws ions from the surface of the plasma. Ion currents of 1 to 10 mA are easily obtained. In operation, several hundred watts of electrical power are

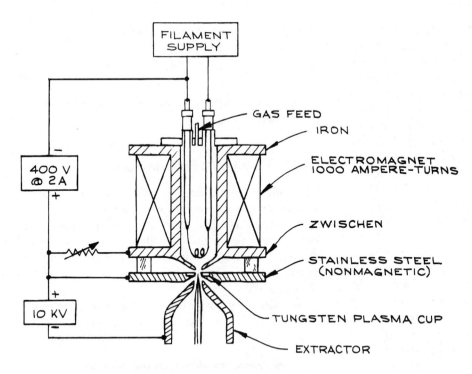

Figure 5.23 A duoplasmatron ion source.

GAS FEED

IRON

ELECTROMAGNET 1000 AMPERE-TURNS

ZWISCHEN

STAINLESS STEEL (NONMAGNETIC)

TUNGSTEN PLASMA CUP

EXTRACTOR

FILAMENT SUPPLY

400 V @ 2A

10 KV

dissipated in the arc and the electromagnet. In its original form this source was liquid-cooled. However, the author has found air cooling to be acceptable for slightly smaller versions. The duoplasmatron is the brightest of ion sources. Copious quantities of singly charged atomic and diatomic ions can be obtained for these species for which an appropriate gaseous parent can be found. It is also possible to obtain a few micro-amperes of doubly charged ions, and in addition nega-tive ions can be obtained by reversing the extraction potential.[17] The chief disadvantages of plasma sources are their complicated construction and the fact that stable operation is only possible for a few days before cleaning and filament replacement are necessary.

5.4 ENERGY ANALYZERS

There are three basic means of measuring the energy of charged particles in a beam. These involve measuring the time of flight over a known distance, the retarding

potential required to stop the particles, or the extent of deflection in an electric or magnetic field.

Because of the great velocities involved, the flight time of a charged particle over any reasonable distance is very short. Determination of the kinetic energy of a particle from the time required to cover a distance of a few centimeters typically requires electronics with a response time of a few nanoseconds. Time-of-flight analyzers are used for energy analysis of electrons with energies less than 10 eV and ions below 1 keV.

An energy analysis of the particles in a beam can be performed by placing a grid or aperture in front of a particle collector and varying the potential on this ele-ment.[18] The current at the collector is the integrated current of particles whose energy exceeds the potential established by the grid. As the grid potential is reduced from that at which all current is cut off, the collector current increases. To obtain the energy distribution the integrated current as a function of retarding potential must be differentiated. One drawback to this method is that only the component of velocity normal to the retarding grid is selected. There are also a number of practical difficulties. The ratio of the initial energy to the energy at the potential barrier varies rapidly for particles near the threshold for penetrating the retarding barrier. Thus focusing effects vary rapidly near threshold, and particles approaching slightly off axis are often deflected away from the collector. The result is that the transmission of the analyzer is unpredictable near threshold. Another problem with retarding-potential analyzers is that low-energy particles near the retarding grid are seriously affected by space charge and by stray electric and magnetic fields. Retarding-potential analyzers are very easily constructed and very compact, but the vagaries of their performance suggests that their use for high-resolution energy analysis be avoided if possible.

Finally, the energies of particles in a beam can be determined by passing the beam through an electric or magnetic field, so that the deflection of the particle paths is a function of the particle energy per unit charge or momentum per unit charge. In these dispersive-type analyzers, the shape of the deflection field must be carefully controlled. Since it is generally easier to produce a shaped electric field than a shaped magnetic field, dispersive analyzers for particle energies up to several keV are usually of the electrostatic variety. For very high-energy particles, magnetic analyzers are preferred, since production of the required electric fields would demand inconveniently large electrical potentials. The following, however, is restricted to a discussion of electrostatic analyzers.

The energy passband ΔE of an analyzer may be defined as the full width at half maximum (FWHM) of the peak that appears in a plot of transmitted current versus energy. Ideally this transmission function is triangular, although for any real analyzer it resembles a Gaussian. We shall take ΔE to be half the full width of the transmission function; such an approximation overestimates the passband of a real analyzer.[19] To first order the passband is established by entrance and exit slits or apertures, which are usually of equal width w. The transmission function also depends upon the maximum angular extent to which particles can deviate from the central path that leads from the entrance to the exit slit. This angular deviation is defined by angles $\Delta \alpha$ in the plane of deflection and $\Delta \beta$ in the perpendicular plane. If E is the central energy of particles transmitted

through an analyzer, then the *resolution* is

$$\frac{\Delta E}{E} = aw + b(\Delta \alpha)^2 + c(\Delta \beta)^2,$$

where a, b, and c are constants characteristic of the particular analyzer.

5.4.1 Parallel-Plate Analyzers

The simplest electrostatic analyzer employs a uniform field created by placing a potential difference across a pair of plane parallel plates as shown in Figure 5.24. With the entrance and exit slits in one of the plates, first-order focusing in the deflection plane is obtained when the angle of incidence of entering particles is about $\alpha = 45°$.

The deflection potential in relation to the incident energy E, the plate spacing d, and the slit separation L

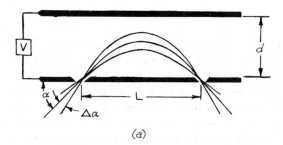

(a)

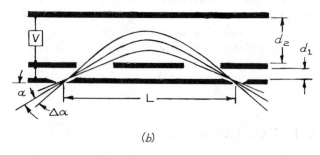

(b)

Figure 5.24 Parallel-plate analyzers with (a) the energy-resolving slits in one of the plates and (b) the slits in a field-free region.

is given by

$$V = E\frac{2d}{L},$$

with the proviso that $d > L/2$ to prevent particles from striking the back plate. The resolution is

$$\frac{\Delta E}{E} = \frac{w}{L} + (\Delta\alpha)^2 + \tfrac{1}{2}(\Delta\beta)^2.$$

A point at the entrance aperture is focused into a line of length $2\sqrt{2}\,\Delta\beta$, and thus the length of the exit slit must exceed that of the entrance slit by this amount.

When the slits are placed in a field-free region[20] as in Figure 5.24(b), optimum performance is obtained with the angle of incidence $\alpha = 30°$ and the distance from the slits to the entrance plate of the analyzer $d_1 = \tfrac{1}{4}d_2 E/V$, where d_2 is the plate spacing. The deflection potential is given by

$$V = 2.6\left(\frac{d_2}{L}\right)E,$$

and L should be no more than 3 times greater than d_2. This arrangement gives second-order focusing in the plane of deflection. The resolution is

$$\frac{\Delta E}{E} = 1.5\frac{w}{L} + 4.6(\Delta\alpha)^3 + 0.75(\Delta\beta)^2,$$

and the image of a point on the entrance slit is a line of length $4L\,\Delta\beta$ at the exit slit.

Although the parallel plate is an attractive design because of its simple geometry, there are several problems with it. The entrance apertures or slits in the front plate are at the boundary of a strong field and act as lenses to produce unwanted aberrations. For the design with the energy-resolving slits in a field-free region this problem can be alleviated by placing a fine wire mesh over these entrance apertures to mend the field. A large electrical potential must be applied to the back plate, creating a strong electrostatic field outside the analyzer. The apparatus in which the analyzer is installed must often be shielded from this field. In addition, the fringing field at the edges of the plates can penetrate into the

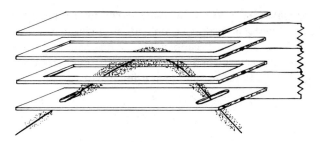

Figure 5.25 Guard electrodes at the edges of a parallel-plate analyzer to offset the field distortion caused by fringing.

deflection region, since the gap between the plates is large. This problem can be solved by extending the edges of the plate well beyond the deflection region, or by placing compensating electrodes at the edges of the gap as in Figure 5.25.

5.4.2 Cylindrical Analyzers

There are two well-known electrostatic analyzers that employ cylindrical electrodes. These are the radial cylindrical analyzer and the cylindrical mirror analyzer.

In the *radial cylindrical analyzer* shown in Figure 5.26, a radial electric field is produced by an electrical potential placed across concentric cylindrical electrodes. Particles are injected along tangents to the cylindrical

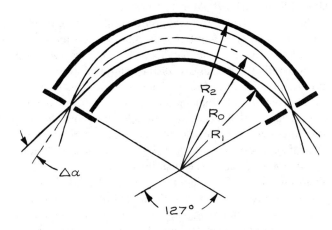

Figure 5.26 The radial cylindrical analyzer.

equipotential surfaces. First-order focusing is obtained if the cylindrical electrodes subtend an angle of $\pi/\sqrt{2} = 127°$. If the electrical potential V is related to the incident particle energy by $E = qV$, the potentials to be applied to the outer and inner cylinders are

$$V_{\text{outer}} = V + 2V \ln \frac{R_2}{R_0}$$

and

$$V_{\text{inner}} = V + 2V \ln \frac{R_1}{R_0},$$

where R_2, R_0, and R_1 are the radius of the outer cylinder, the midradius, and the inner radius, respectively. The resolution of this analyzer is

$$\frac{\Delta E}{E} = \frac{w}{R_0} + \tfrac{2}{3}(\Delta \alpha)^2 + \tfrac{1}{2}(\Delta \beta)^2.$$

It is good practice to limit the angle of divergence in the plane of deflection so that

$$\Delta \alpha < \frac{2\sqrt{2}}{\pi} \frac{R_2 - R_1}{R_0}$$

in order to keep the filling factor below 50%. The radial-field analyzer gives focusing only in the plane of deflection. A point at the entrance slit is imaged as a line of length $\sqrt{2}\,\pi R_0 \Delta \beta$ at the exit slit.

The *cylindrical-mirror analyzer* is similar to the parallel-plate analyzer except the deflection plates are coaxial cylinders. In fact, the parallel-plate analyzer can be considered as a special case of the cylindrical mirror. In the usual geometry, shown in Figure 5.27, the source is located on the axis, and particles emitted into a cone defined by polar angle α pass through an annular slot in the inner cylinder. Particles of energy E are deflected so that they pass through an exit slot and are focused to an image on the axis. The cylindrical mirror is double-focusing, that is, focusing occurs in both the deflection plane and the perpendicular plane, so that the image of a point at the source appears as a point at the detector. An obvious advantage of this analyzer is that particles at any azimuthal angle can be collected.

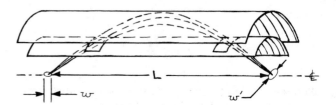

Figure 5.27   The axial-focusing cylindrical-mirror analyzer.

For optimum performance the entry angle is $\alpha = 42.3°$, in which case the distance from source to detector is $L = 6.12R_1$.[21] The inner cylindrical plate is at the same potential as the source, and the potential on the outer cylinder is

$$V_{\text{outer}} = 0.763V \ln\left(\frac{R_2}{R_1}\right)$$

for transmission of particles of energy $E = qV$. It is wise to choose $R_2 > 2.5R_1$ so that particles are not scattered from the outer electrode. The resolution of the axial-focusing cylindrical mirror analyzer is approximately

$$\frac{\Delta E}{E} = 1.09 \frac{w}{L}$$

for a source of axial extent w and an energy-resolving aperture of diameter $w' = w \sin \alpha$ perpendicular to the axis (as shown in Figure 5.27).

If the source is not small and well defined, the entry and exit slots in the inner cylinder can be used to define the resolution of the cylindrical-mirror analyzer as in Figure 5.28. Although the axial-focusing geometry gives focusing to second order, this arrangement is only first-order-focusing.

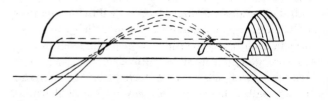

Figure 5.28   Cylindrical-mirror analyzer with energy-resolving slits in the inner electrode.

5.4.3 Spherical Analyzers

For many applications the most desirable analyzer geometry employs an inverse-square-law field, which is created by placing a potential across a pair of concentric spherical electrodes. Focusing in both the deflection plane and the perpendicular plane can be obtained using any sector portion of a sphere. As shown in Figure 5.29, the object and image lie on lines that are perpendicular to the entrance and exit planes, respectively, and tangent to the circle described by the midradius R_0. Furthermore, by Barber's rule, the object, the center of curvature of the spheres, and the image lie on a common line. As with the cylindrical mirror, the spherical analyzer can be used to collect all particles emitted from a point source at or near a particular polar angle.

For energy analysis of a beam, the 180° spherical sector (Figure 5.30) is often used because of the compact geometry that results from folding the beam back onto a line parallel to its original path. As with any spherical sector, the electrical potentials for transmission of a particle of energy $E = qV$ are given by

$$V_{\text{outer}} \approx V \left[2 \frac{R_0}{R_2} - 1 \right]$$

and

$$V_{\text{inner}} \approx V \left[2 \frac{R_0}{R_1} - 1 \right].$$

The resolution is

$$\frac{\Delta E}{E} = \frac{w}{2R_0} + \tfrac{1}{2} (\Delta \alpha)^2,$$

and the maximum deviation of a trajectory from the central path within the analyzer is

$$w_m \simeq \frac{\Delta E}{E} + \Delta \alpha + \frac{1}{2\,\Delta \alpha} \left(\frac{w}{2R_0} + \frac{\Delta E}{E} \right)^2.$$

For spherical-sector analyzers in general, the resolution is approximately

$$\frac{\Delta E}{E} = \frac{w}{R(1 - \cos \phi) + l \sin \phi},$$

where, as illustrated in Figure 5.29, ϕ is the angle substended by the analyzer sector and l is the distance from the exit boundary of the analyzer to the exit

Figure 5.29 Focusing with a spherical analyzer.

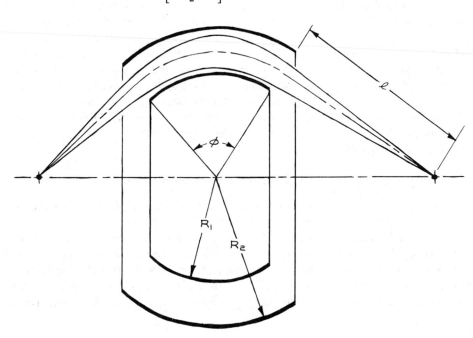

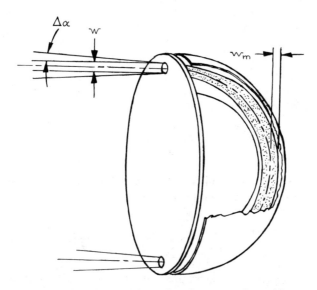

Figure 5.30   The hemispherical analyzer: w is the width of the entrance and exit apertures, $\Delta\alpha$ the maximum angular deviation of an incident trajectory with respect to the central path, and w_m the maximum deviation of a trajectory from the central path within the analyzer.

aperture. Although it would appear that the resolution can be increased arbitrarily by increasing l, the aberrations increase rapidly as the system becomes asymmetric. It is best to employ the symmetric geometry with ϕ in the range 60–180°.

In comparison with the parallel-plate or cylindrical-mirror analyzer, the spherical analyzer has the advantage of requiring relatively low electrical potentials on the electrodes. Because the electrodes are closely spaced in the spherical analyzer, fringing fields are less of a problem and more easily controlled. The chief drawback of this type of analyzer is the difficulty of fabrication and mounting.

5.4.4   Preretardation

In all electrostatic-deflection analyzers, the passband ΔE is a linear function of the transmitted energy. Thus, the absolute energy resolution of these devices can be improved by retarding the incident particles prior to their entering the analyzer. In principle, the passband of an analyzer can be reduced arbitrarily by preretardation; in practice, there are limitations on this technique. A beam of particles incident on the entrance aperture of an analyzer can be slowed by a decelerating lens as in the example given in Section 5.2.6. However, it would be very difficult to design a lens system to be used with analyzers that have an annular entrance slit. Space charge and stray fields must be considered. Because the flight path through an analyzer is long, it is usually not possible to reduce the energy of transmitted particles below about 5 eV before these effects result in severe aberrations. Finally, it is important to recall that if all particles of a particular energy in a beam are to be transmitted through an analyzer, the ratio of brightness to energy must be conserved (Section 5.1.3). As the energy of particles incident on an analyzer is decreased, the current is ultimately reduced because the pencil angle of the beam exceeds the acceptance angle of the analyzer.

When using a decelerating lens with an analyzer, it is good practice to place an aperture on the high-energy side of the lens and design the lens so that the image of this aperture appears at the entrance plane of the analyzer. The need for a real entrance aperture is thereby eliminated, and there are no metal surfaces in the vicinity of the low-energy beam. Since space charge will cause the low-energy beam to expand at the entrance plane of an analyzer, current would be lost if a real entrance aperture were used. As demonstrated by Kuyatt and Simpson, space-charge expansion at a virtual aperture is compensated by the spreading of the beam, with the result that the beam appears to originate at an aperture of about the same size as that in the absence of space charge[22] (see Figure 5.16).

5.4.5   The Energy-Add Lens

An electrostatic energy selector can be used as a monochromator to select the particles in a beam whose energies fall within a narrow range, or as an analyzer to determine the energy distribution of particles originating from a process under study. In the latter application the passband of the analyzer must be scanned over the energy range of interest. Scanning can be accomplished by varying the potential difference between the analyzer electrodes; however, neither the transmission nor the energy resolution will be constant. A more satisfactory method for determining the energy distribution of particles in a beam is to fix the analyzer potential at that voltage which allows transmission of particles of the highest desired energy and preaccelerate the incident particles by a variable amount. The energy that must be added to the particles so that they pass through the analyzer is then a measure of the difference between their initial energy and the energy necessary to transmit them with no preacceleration. In a scattering experiment, for example, the analyzer might be set to transmit elastically scattered particles, and the distribution of energy lost by inelastically scattered particles would be scanned by monitoring the transmitted current as a function of energy added.

An *energy-add lens* must add back energy without disturbing the optics of the analyzer. Kuyatt[23] has shown that the electron-optical analog of the *field lens* is suited to this application. This lens is positioned so that its first principal plane coincides with the particle source to be examined. The source is then imaged onto the second principal plane with unit magnification. For a reasonably strong lens ($V_2 / V_1 > 3$), the position of the principal planes is nearly independent of the voltage ratio and the separation of the planes is small. Thus particles from the source can be accelerated by a variable amount without changing the apparent position of the source. An energy-add lens used in conjunction with a fixed-ratio decelerating lens for preretardation at the input of an electron energy analyzer is schematically illustrated in Figure 5.31. For transmission of electrons with energy A (eV) less than the maximum energy to be transmitted, the potential of the second element of the energy-add lens and *every electrode thereafter* is increased by A (volts). For a positive ion analyzer, the potential would be decreased by A / n (volts), where n is the charge state of the ions.

A problem with the energy-add lens is that the object pupil moves as the lens voltages change. Thus, while the apparent source remains stationary, the beam angle may vary over a considerable range as the lens is tuned. As a result, the lenses following the energy-add lens may be

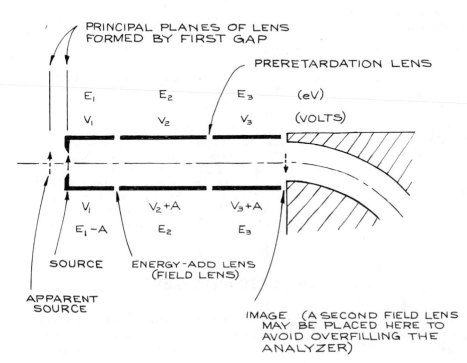

Figure 5.31 An energy-add lens and a fixed-ratio retarding lens at the input of an analyzer. The energies and corresponding lens voltages for particles (assumed to be electrons) that are transmitted without having energy added are given above the lens. The energies and voltages for transmission of particles that have lost energy A are shown below. The "source" is defined by an aperture that is finally imaged onto the entrance plane of the analyzer.

overfilled. This can be avoided by placing a second field lens downstream from the first at an intermediate image of the apparent source. The second field lens is then tuned to give a zero beam angle. So that there is no net change of energy as the beam angle is manipulated, an einzel lens (Section 5.2.2) can be used.

5.4.6 Fringing-Field Correction

For the radial-field cylindrical analyzer or the spherical analyzer, particles must pass through the fringing field at the edge of the electrode gap as they enter the deflection region. The electric field at the edge bulges out of the gap, and this curvature produces a focusing effect, which causes undesirable aberrations. Herzog[24] and Wollnik and Ewald[25] have shown that this effect can be largely eliminated by placing a diaphragm at the entrance of the condenser gap. As illustrated in Figure 5.32, the diaphragm, when properly located, produces a field that has the same effect as a field abruptly terminated at a distance d^* in front of the condenser gap. The appropriate location of the diaphragm as a function of the dimensions given in Figure 5.32 can be determined for either a thick or thin diaphragm from the graphs in Figure 5.33.

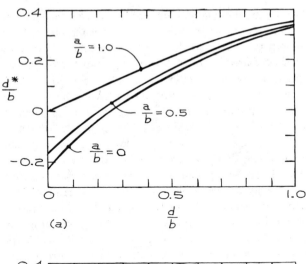

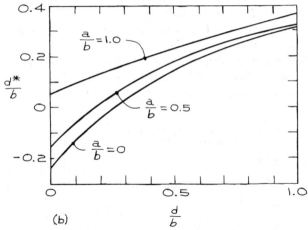

(a)

(b)

Figure 5.33 The distance of the field boundary from the edge of the condenser electrodes as a function of the dimensions given in Figure 5.32 for (a) a thick diaphragm and (b) a thin diaphragm. (From H. Wollnik and H. Ewald, "The Influence of Magnetic and Electric Fringing Fields on the Trajectories of Charged Particles," Nucl. Instr. Meth., 36, 93, 1965; by permission of North-Holland Publishing Company.)

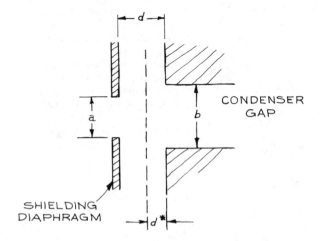

Figure 5.32 A shielding diaphragm to correct for the fringing field at the edge of an electrostatic condenser. The field appears as if it were terminated abruptly at a distance d^* from the gap.

5.5 MASS ANALYZERS

Mass analysis is more complicated than energy analysis of an isotopically pure beam, since energy and momentum are independent variables in a mixed beam. A detailed discussion of the considerable range of mass analyzers that has been developed is beyond the scope of this book. The principles of operation of a few representative analyzers will be described.

5.5.1 Magnetic Sector Analyzers

The deflection of ions in a perpendicular magnetic field is proportional to the particle momentum per unit charge. If all particles entering a magnetic field have the same energy per unit charge, then the field will separate particles according to their masses. In the usual configuration a magnetic field is produced between the two parallel, sector-shaped polefaces of an electromagnet as illustrated in Figure 5.34. The focusing properties of this field are the same as those of the spherical electrostatic analyzer, and the locations of the entrance and exit slits are given by Barber's rule (Section 5.4.3). The radius of curvature of ions of energy E in a magnetic field B is

$$R = \frac{144}{Bn}\sqrt{mE} \quad \text{cm},$$

with B in gauss, E in eV, m in amu, and n the charge state. In terms of the parameters specified in Figure 5.34, the resolution is

$$\frac{\Delta m}{m} = \frac{w}{R\left(1 - \cos\phi\right) + l\sin\phi}.$$

To first order, there is no focusing in the plane perpendicular to the plane of deflection. However, when the curvature of the fringing field is taken into consideration, some focusing in the perpendicular plane results. Incident ion trajectories should be normal to the entrance plane to avoid focusing effects.

5.5.2 Wien Filter

The Wien filter makes use of mutually perpendicular electric and magnetic fields normal to the ion trajectory[26] (Figure 5.35). Ions of a unique velocity v will experience equal and opposite forces through interaction with the two fields (the Coulomb force and the Lorentz force) and be transmitted in a straight line. If

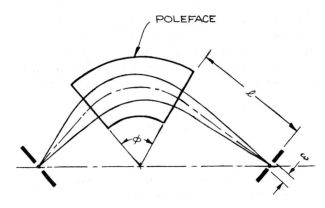

Figure 5.34   Focusing properties of the magnetic sector.

all of the ions injected into the fields of the Wien filter have been accelerated through the same potential relative to their source, then only ions of a unique mass-to-charge ratio are transmitted. In most Wien filters the magnetic field is supplied by a pair of permanent magnets, and the electric field is produced by placing an electrical potential V_d between a pair of parallel electrodes separated by a distance d. As suggested in Figure 5.35, these electrodes are often shaped to ensure that the electric field lines are parallel near the center of the device. The velocity of ions transmitted straight through the Wien filter is

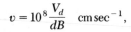

$$v = 10^8\frac{V_d}{dB} \quad \text{cm sec}^{-1},$$

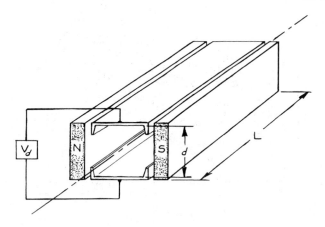

Figure 5.35   The Wien filter.

with V_d in volts, d in cm, and B in gauss. Obviously, the mass spectrum of ions entering the filter can be scanned by varying V_d. If slits of width w are placed at the entrance and exit of a Wien filter of length L, the mass resolution for ions that have been accelerated through a potential V is

$$\frac{\Delta m}{m} = \frac{2dVw}{V_d L^2}.$$

Incident ion trajectories should be normal to the entrance plane to avoid focusing effects.

5.5.3   Dynamic Mass Spectrometers

A class of analyzers known as *dynamic mass spectrometers* use time-varying electric or magnetic fields or timing circuits to disperse ions according to their masses.[27] The quadrupole mass analyzer and the linear time-of-flight mass spectrometer are probably the two most successful designs of this type.

The *quadrupole mass analyzer* illustrated in Figure 5.36 employs a time-varying electric quadrupole field. For a particular field intensity and frequency, only ions of a unique mass-to-charge ratio will follow a stable path

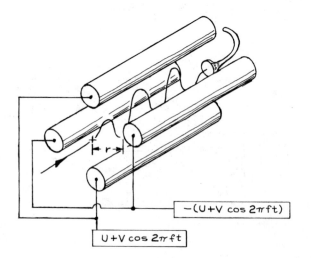

Figure 5.36   The quadrupole mass analyzer.

and pass through the field. As shown in the figure, the quadrupole field is approximated by a square array of four cylindrical electrodes parallel to the axis along which ions are injected. The potential applied to the vertical pair of electrodes is

$$V_v = U + V\cos 2\pi ft,$$

while for the horizontal pair

$$V_h = -U + V\cos(2\pi ft + \pi),$$

where f refers to an r.f. frequency. For optimum performance the ratio of the d.c. field to the r.f. field is adjusted so that $U/V \simeq 0.17$. For singly charged ions, the mass of the ions that are transmitted is

$$m = 0.14\frac{V}{r^2 f^2} \text{ amu,}$$

with V in volts, r in cm, and f in MHz. Mass scanning is usually accomplished by varying the r.f. amplitude V, while fixing the ratio of d.c. to r.f. voltage (U/V) and the frequency (f). Quadrupole analyzers are compact and offer the advantages of high transmission, fast scanning, and insensitivity to the initial ion energy. These instruments are available commercially as residual-gas analyzers for high-vacuum systems and are suitable for use in many scattering experiments.

The *time-of-flight mass spectrometer* depends upon the fact that the velocity of ions of the same kinetic energy is a function of mass. In this spectrometer, ions that have been accelerated to an energy of about 1 keV are admitted to a long drift tube in short bursts by a negative electrical pulse applied to a grid at the entrance of the drift region. The lightest ions travel most rapidly and are the first to arrive at the detector, located at the end of the drift tube. Heavier ions arrive in the order of their masses. The obvious advantage of this design is that the entire mass spectrum is scanned in a few microseconds. Furthermore, the spectrum can be scanned thousands of times each second, making this system ideal for studying rapidly varying processes. Its chief drawback is its size, since the length of the flight tube required to achieve good resolution may be greater than a meter.

5.6 ELECTRON- AND ION-BEAM DEVICES: CONSTRUCTION

To realize the design of a charged-particle optical system, it is necessary to create an environment where a stream of particles can travel without loss of momentum and to make electrodes and pole pieces that faithfully produce the electrical and magnetic fields necessary to deflect the particles.

5.6.1 Vacuum Requirements

Charged particles lose energy through interactions with gas molecules; thus ions or electrons can only be transported in a vacuum. The required pressure depends upon the distance they must travel. As noted in Section 3.1.2, the mean free path at a pressure of 1 mtorr is a few centimeters, while at 10^{-5} torr this increases to a few meters. It follows that high vacuum is required for the operation of a charged particle optical system.

Conducting surfaces must be kept clean, since contamination with an insulating material will result in the buildup of surface charges, which cause an unpredictable deflection of charged particles passing nearby. Clean electrode surfaces are particularly important for particles of energy less than 100 eV and when high spatial resolution is required. All electrodes must be clean and free of hydrocarbon contamination before installation in the vacuum system (Section 3.6.3). A vacuum system with oil-free pumps such as mercury diffusion pumps, ion pumps, or sorption pumps (Section 3.4.3) is most desirable, although a system evacuated with a properly trapped oil diffusion pump (Sections 3.5.5 and 3.4.2) is adequate in many cases. Polyphenyl ether diffusion-pump fluids such as Convalex-10 (Consolidated Vacuum) or Neovac Sy (Varian) have been found to be the least offensive diffusion-pump oils for electron-optical systems. In low-energy beam devices a daily bake to 300–400°C will keep electrode surfaces clean and ensure stable operation. Hydrocarbons on aperture surfaces exposed to charged particles create a particularly obnoxious problem. Bombardment of the adsorbed hydrocarbons produces an insulating carbon polymer, which adheres tenaciously to the underlying metal. This carbon material can only be removed by high-temperature baking (400°C), by vigorous application of an abrasive, or by etching with a strong NaOH solution.

When baking is impractical, electrode surface quality and stability are often improved by a coating of carbon black. A surface can be blacked by brushing with a sooty acetylene flame; however, the coating produced in this manner does not adhere well. A superior coating can be produced by spraying or brushing the surface with a thin water or ethanol slurry of colloidal graphite[28] (Aquadag, made by Acheson Colloids Co.). The tenacity of this coating is improved by preheating the metal surface to about 100°C.

5.6.2 Materials

The materials used to construct lens elements must be such that the equipotential surfaces near the electrodes faithfully follow the contours of the electrode surfaces. As mentioned above, these surfaces must be clean, and thus the electrode material must withstand periodic cleansing by baking, bead blasting, electropolishing, or harsh chemical action. In addition, the electrodes of an ion or electron source often must resist high temperatures as well as erosion through ion or electron bombardment.

The *refractory metals* such as tungsten, tantalum, and particularly molybdenum are probably the best electrode materials. These metals have a low, uniform surface potential, they do not oxidize, and they are bakeable. Refractory metals are hard and rather brittle. Only the wrought material can be machined or formed.[29] Stock produced by sintering cannot be used.

Oxygen-free high-conductivity (OFHC) *copper* (Section 1.2.4) is often used for electrode fabrication, although Kuyatt has used commercial, half-hard copper in some applications. Exposed to air, copper forms a surface oxide, but this oxide is conducting. Copper is bakeable and reasonably machinable.

Stainless steels (Section 1.2.3) contain domains of different composition, some of which may be ferromagnetic. The magnetic properties of stainless steels vary considerably even between samples of the same net composition, and as a result these materials are unsuitable for use with low-energy charged particles unless they are carefully annealed to remove residual magnetism (Section 1.2.3). On the other hand, stainless steels have the advantage of being bakeable, strong, and easy to machine.

Aluminum (Section 1.2.5) is unsuitable because its surface is rapidly attacked by oxygen in the air to form a hard oxide insulting layer. However, aluminum has a low density, and most aluminum alloys are easy to machine or form by bending, spinning or rolling. When these properties are important, the surface can be plated with copper or gold in order to overcome the problem of oxidation.

Electrodes and lenses must be mounted on nonconducting materials. Glass or ceramic (Section 1.2.8) or, in some cases, plastic (Section 1.2.7) can be used for this purpose. Of the ceramic materials, *alumina* is the best insulator and the strongest. Precision-ground rods and balls are available commercially from McDanel and from Industrial Tectonics, respectively. Alumina can only be shaped by grinding; however, there are machinable ceramics from which complicated shapes can be formed.[30] *Plastics* are machinable and inexpensive. Unfortunately, they are not bakeable; a further disadvantage is that they contain hydrocarbons that can contaminate the vacuum environment.

5.6.3 Lens and Lens-Mount Design

Most electron-optical lenses consist of a coaxial series of cylindrically symmetric electrodes. These electrodes must be designed so that the electric fields associated with the lens mounts and the vacuum-container walls do not penetrate the gap between electrodes. The involuted design illustrated in Figure 5.37 ensures that the lens gap is shielded. The step on the shoulder of the lens elements is designed so that the inner gap is of the correct size when the outer gap is adjusted to some standard width. In this way all the lenses in a system can be correctly positioned with the aid of a single gauge.

There are two widely used, semikinematic schemes for mounting lens elements. These are the *rod mount*

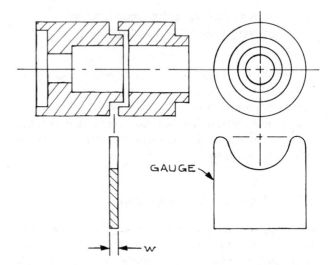

Figure 5.37   The construction of lens electrodes so that external fields cannot penetrate the lens gap. When mounting the lens elements, a gauge of width *w* is inserted to ensure correct spacing at the invisible lens gap.

and the *ball mount*, illustrated in Figures 5.38 and 5.39, respectively. In the rod mount, cylindrical lens elements rest on ceramic rods, which insulate the elements from one another and from a grounded mounting plate. The critical dimensions are the diameters of the lens elements and the width of the channel in which the rods rest. In order for the lens elements to be coaxial, their outer diameters must be identical. This requirement is met by making all of the elements from a single piece of rod stock, which has been carefully turned to a uniform diameter. To ensure that the elements are mounted coaxially, it is then only necessary that the sides of the channel in which the rods rest be parallel. This is easily accomplished in a milling operation. If the vertical position of the lens axis is important, then the width of the channel becomes a critical dimension. Of course, the diameter of the rods is important, but, as noted elsewhere (Section 1.2.8), alumina rod, centerless-ground to high precision, is commercially available. For maximum strength the rods should be about 90° apart around the circumference of the lens element.

In the ball-mounting scheme, lens elements are in-

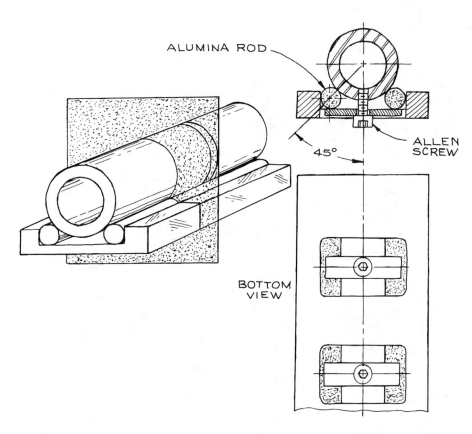

ALUMINA ROD

45°

ALLEN
SCREW

BOTTOM
VIEW

Figure 5.38 Cylindrical lens elements mounted on ceramic rods.

sulated from one another and positioned by ceramic balls that rest in holes near the edges of the lenses. This system is preferred when mounting very thin elements, as in an aperture lens system. Ideally only three balls should be used. The critical dimensions are the locations of the holes and their diameter. The holes should be bored, or else drilled and reamed, in a milling machine or jig borer using a dividing head or a precision rotary table. For balls of diameter d, the hole diameter should be $d\sin 45°$, in which case the spacing between lens elements will be $d\cos 45°$. A lens system is assembled by stacking lens elements alternating with balls. The stack must be clamped. The clamp should have some spring and should bear on the topmost element only at one or two points near the center of the circle around which the balls are located. A rigid clamp will tend to drive the balls into their holes and thus may crush the

edges of the holes. If the clamp bears too near the edge, there is a danger that the stack may become cocked off axis.

5.6.4 Charged-Particle Detection

For particle energies up to a few tens of keV there are two detection schemes in wide use. The charged particles can be collected at a metal surface and the resultant electrical current measured directly, or they can be detected by collecting the slower secondary electrons that are ejected from a metal surface by impact of the primary particle. Charged particles can also be detected with photographic emulsions, scintillators, and various solid-state devices. Emulsions and scintillators have largely been supplanted by the methods

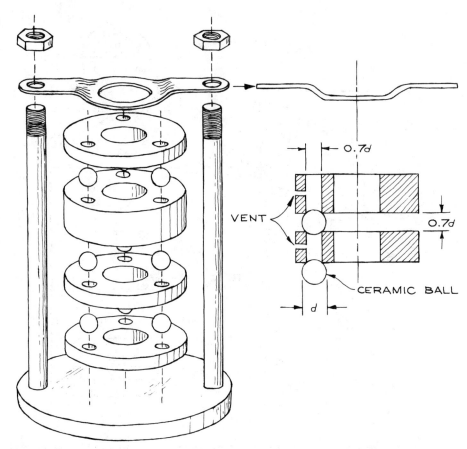

Figure 5.39 Lens elements mounted on ceramic balls.

mentioned above, and solid-state devices are only suitable for the detection of high-energy particles (above 30 keV). Table 6.20 lists the properties of common particle detectors.

The collector for the direct detection of a current of charged particles is called a *Faraday cup*. Typical designs are illustrated in Figure 5.40. These simple collectors are connected directly to a current-measuring device and are useful at currents down to the detection limits of modern electrometers—about 10^{-14} A. A properly designed Faraday cup does not permit secondary electrons to escape. For a positive-ion collector, the loss of a secondary electron appears to the current-measuring instrument as an additional ion, while for an electron collector the loss of each secondary cancels the effect of an incident primary electron. To prevent the

escape of secondary electrons, the depth of a Faraday cup should be at least five times its diameter. A suppressor aperture or grid biased to about -30 V in front of a Faraday cup effectively prevents the escape of most secondaries. A grounded grid in front of the suppressor as illustrated in Figure 5.40 (*b*) prevents field penetration from the suppressor in the direction of the incident current source. When a particle beam must be aligned and sharply focused, a concentric pair of collectors [Figure 5.40(*c*)] is useful. With this arrangement the beam-focusing elements are adjusted to maximize the ratio of current to the inner cup in relation to the current to the outer cup.

Charged-particle currents of less than 10^{-14} A can be detected with an *electron multiplier*. As illustrated in Figure 5.41, these devices consist of a series of elec-

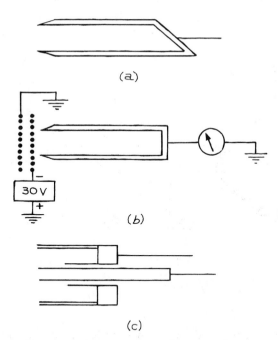

Figure 5.40   Faraday-cup designs. The design illustrated in (*b*) employs a grid to suppress the emission of secondary electrons. Design (*c*) is a double cup for aligning and focusing a beam.

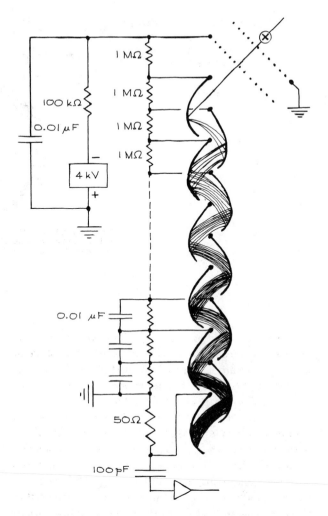

Figure 5.41   An electron multiplier. Electrical connections shown are appropriate for detecting positive ions. Grids at the entrance of the multiplier prevent the escape of secondary electrons.

trodes known as *dynodes*. Secondary electrons produced by impact of a particle on the first dynode are accelerated towards the second dynode through a potential drop of 100–300 volts. The impact of these electrons results in a number of secondaries, which are in turn accelerated into the third dynode, and so on. The secondary emission coefficient of the dynodes is typically about 3 electrons per incident particle, so that for a multiplier with n electrodes, the impact of a single particle results in a pulse of about 3^n electrons. Electron multipliers are available with 10 to 20 dynodes, which provide gains of 10^6 to 10^9. The dispersion in time of the shower of electrons from the last dynode is of the order of 10 ns. The corresponding current through a 50-Ω measuring resistor produces a pulse of 500 μV to 500 mV, which is easily detected by modern pulse-counting electronics. Furthermore, if time resolution is not required, the output current from an electron multiplier can be measured directly with an electrometer.

The electrodes of an electron multiplier are fabricated of a material that has a high work function—typically a Be-Cu alloy that has been "activated" by some proprietary process. The high work function is desirable because it inhibits thermionic emission, which would result in noise pulses at the multiplier output. On the other hand, this requires that incident particles have energies in excess of a few hundred eV to assure effi-

Figure 5.42   A channeltron electron multiplier. Electrical connections are typical for detection of electrons.

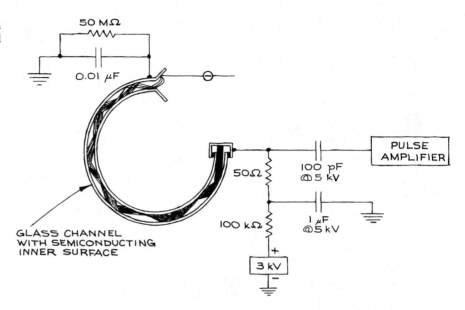

GLASS CHANNEL
WITH SEMICONDUCTING
INNER SURFACE

cient detection. When working with lower-energy particles it is necessary to provide an acceleration stage at the multiplier input. For incident particles with energies greater than 300 eV the detection efficiency of an electron multiplier is essentially 100%.

An important variant of the electron multiplier uses a continuous dynode as illustrated in Figure 5.42. These devices are known as *channeltrons* and are made by Galileo Electro-Optics. They are glass or ceramic tubes with semiconducting inner surfaces. The end-to-end resistance is about 10^9 ohms. In operation, a potential of about 3 kV is applied across them, resulting in a gain of about 10^8. Channeltrons have the advantage of small size (about 2 in. long), low cost, and ruggedness.

For the detection of positive ions an electron multiplier is usually operated with its first dynode at a high negative potential and its last dynode near ground potential, as in Figure 5.41. This arrangement provides acceleration of incident ions. For electrons or negative ions the cathode is usually biased positive by a few hundred volts and the last dynode is at a high positive potential relative to ground, as in Figure 5.42. In this latter configuration, it is only convenient to operate in a pulse-counting mode, since the detection electronics must be electrically insulated from the measuring resis-

tor. Ordinarily, the multiplier output is coupled to the pulse-counting electronics *via* a high-voltage disc ceramic capacitor of about 100 pF. A problem sometimes encountered with this arrangement is that occasionally sparks between dynodes or across dynode resistors, or switching noise, give rise to large high-frequency transients, which are transmitted through the coupling capacitor and cause damage to the electronics. This problem can be cured by coupling the output of the multiplier to the electronics through a transformer that will attenuate large pulses because of saturation. A suitable transformer[31] can be produced by making a ten-turn bifilar winding of well-insulated wire on a ferrite core (e.g. Ferroxcube), as illustrated in Figure 5.43. The number of turns and their position must be carefully adjusted to assure a proper impedance match between the multiplier and the electronics.

The output current from an electron multiplier is limited by the current available from the resistive voltage divider that establishes the potential of the dynodes. To ensure that the gain of a multiplier is not reduced because of depletion of the dynode charge, the maximum output current should be at least an order of magnitude less than the current drawn by the dynode resistor chain.

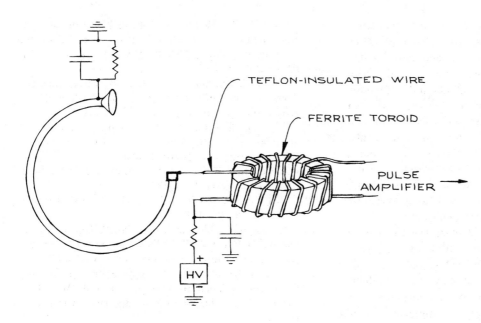

Figure 5.43 A coupling transformer for use between an electron multiplier and a sensitive amplifier. Properly constructed, such a transformer will faithfully transmit signal pulses while attenuating large noise pulses.

5.6.5 Magnetic-Field Control

A magnetic field can cause serious aberrations in a charged-particle optical system. The earth's magnetic field can significantly alter the trajectories of low-energy electrons and light ions. (It is about 0.6 gauss, at an angle of elevation in the north-south plane roughly equal to the local latitude.) For example, the radius of curvature of the path of a 10-eV electron moving perpendicular to the earth's field is only 15 cm.

In order to eliminate the effect of the earth's magnetism, it is practical to surround a sensitive apparatus with an electromagnet whose field approximately cancels that of the earth. This usually takes the form of a pair of coaxial coils of wire, referred to as *Helmholtz coils*. The appropriate spacing is illustrated in Figure 5.44. The field along the axis of a round Helmholtz pair a distance z above the bottom coil is given by

$$B = 0.32 \frac{NI}{R} \left\{ \left[1 + \left(\frac{z}{R} \right)^2 \right]^{-3/2} + \left[1 + \left(1 - \frac{z}{R} \right)^2 \right]^{-3/2} \right\} \text{ gauss,}$$

where N is the number of turns, I is in amperes, and R

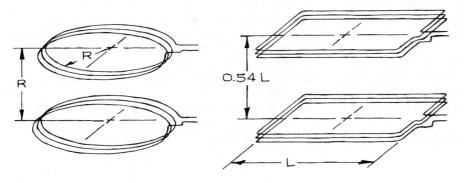

Figure 5.44 The geometry of round and square Helmholtz coils.

and Z are in cm. For square coils, a similar expression with R replaced by $L/2$ applies. The orientation of the coils and the exact value of the current must be determined with the aid of a magnetometer. The field produced by a Helmholtz pair is of high uniformity only in a small volume midway between the coils and near the axis. For example, a uniformity of 0.1% is achieved in a central volume of 2×10^{-2} cubic radii, a uniformity of 1% in a volume of 1.5×10^{-1} cubic radii, and a uniformity of 5% in a volume of 0.6 cubic radii.[32] Thus, the coil dimensions must exceed those of the apparatus to be shielded by at least a factor of 10 in order to reduce the magnetic field to a few milligauss.

A sensitive apparatus may be shielded from the earth's field by placing the instrument within an enclosure made of a high-permeability magnetic material. Such shields have been used in conjunction with a Helmholtz coil to achieve a residual field less than a milligauss. The proper design of these shields is quite complicated[33] and will not be dealt with here.

CITED REFERENCES

1. K. R. Spangenberg, *Vacuum Tubes*, McGraw-Hill, New York, 1948.
2. J. R. Pierce, *Theory and Design of Electron Beams*, 2nd edition, Van Nostrand, New York, 1954.
3. O. Klemperer and M. E. Barnett, *Electron Optics*, 3rd edition, Cambridge University Press, Cambridge, 1971.
4. V. E. Cosslett, *Introduction to Electron Optics*, Oxford University Press, Oxford, 1946.
5. D. Roy and J. D. Carette, "Design of Electron Spectrometers for Surface Analysis," in *Electron Spectroscopy*, H. Ibach, Ed., Topics in Current Physics, Springer, Berlin, 1977, Chapter 2.
6. *Methods of Experimental Physics*, Vol. 4A, *Atomic Sources and Detectors*, V. W. Hughes and H. L. Schulz, Eds., Academic Press, New York, 1967.
7. E. Harting and F. H. Read, *Electrostatic Lenses*, Elsevier, New York, 1976.
8. (a) F. H. Read, J. Phys., E2, 165, 1969; (b) F. H. Read, J. Phys., E2, 679, 1969; (c) F. H. Read, A. Adams, and J. R. Soto-Montiel, J. Phys., E4, 625, 1971; (d) A. Adams and F. H. Read, J. Phys., E5, 150, 1972; (e) A. Adams and F. H. Read, J. Phys., E5, 156, 1972.
9. S. Natali, D. DiChio, E. Uva, and C. E. Kuyatt, Rev. Sci. Instr., 43, 80, 1972; D. DiChio, S. V. Natali, and C. E. Kuyatt, Rev. Sci. Instr., 45, 559, 1974.
10. D. W. O. Heddle, *Tables of Focal Properties of Three-Element Electrostatic Cylinder Lenses*, J. I. L. A. Report No. 104, University of Colorado, Boulder.
11. R. E. Collins, B. B. Aubrey, P. N. Eisner, and R. J. Celotta, Rev. Sci. Instr., 41, 1403, 1970.
12. D. DiChio, S. V. Natali, C. E. Kuyatt, and A. Galejs, Rev. Sci. Instr., 45, 566 1974.
13. J. A. Simpson, "Electron Guns," in *Methods of Experimental Physics*, Vol. 4A, V. W. Hughes and H. L. Schultz, Eds., Academic Press, New York, 1967, Section 1.15.
14. An estimation of image expansion at the space charge limit can be obtained from the data of W. Glaser, *Grundlagen der Elektronenoptik*, Springer, Vienna, 1952, p. 75.
15. R. G. Wilson and G. R. Brewer, *Ion Beams*, Wiley, New York, 1973.
16. C. D. Moak, H. E. Banta, J. N. Thurston, J. W. Johnson, and R. F. King, Rev. Sci. Instr., 30, 694, 1959; M. von Ardenne, *Tabellen der Electronenphysik, Ionenphysik, und Übermikroskopie*, Deutscher Verlag der Wissenschaften, Berlin, 1956.
17. W. Aberth and J. R. Peterson, Rev. Sci. Instr., 38, 745, 1967.
18. J. A. Simpson, Rev. Sci. Instr., 32, 1283, 1961.
19. M. E. Rudd, "Electrostatic Analyzers," in *Low Energy Electron Spectrometry*, by K. D. Sevier, Wiley-Interscience, New York, 1972, Chapter 2, Section 3; A. Poulin and D. Roy, J. Phys., E11, 35, 1978.
20. T. S. Green and G. A. Proca, Rev. Sci. Instr., 41, 1409, 1970; G. A. Proca and T. S. Green, Rev. Sci. Instr., 41, 1778, 1970.
21. V. V. Zashkvara, M. I. Korsunskii, and O. S. Kosmachev, Soviet Phys. Tech. Phys., 11, 96, 1966; H. Sar-el, Rev. Sci. Instr., 38, 1210, 1967, and 39, 533, 1968; J. S. Risley, Rev. Sci. Instr., 43, 95, 1972.
22. C. E. Kuyatt and J. A. Simpson, Rev. Sci. Instr., 38, 103, 1967.
23. C. E. Kuyatt, unpublished lecture notes; see also A. J. Williams, III, and J. P. Doering, J. Chem. Phys., 51, 2859, 1969; J. H. Moore, J. Chem. Phys., 55, 2760, 1971.

24. R. F. Herzog, Z. Phys., 89, 447, 1934; 97, 596, 1935; Phys. Z., 41, 18, 1940.
25. H. Wollnik and H. Ewald, Nucl. Instr. Meth., 36, 93, 1965.
26. R. L. Seliger, J. Appl. Phys., 43, 2352, 1972.
27. E. W. Blauth, *Dynamic Mass Spectrometers*, Elsevier, Amsterdam, 1966; P. H. Dawson, *Quadrupole Mass Spectrometry*, Elsevier, Amsterdam, 1976.
28. E. I. Lindholm, Rev. Sci. Instr., 31, 210, 1960.
29. Electrodes in cylindrical and spherical shapes, spun from sheet molybdenum, are available from Bomco, Inc.
30. MACOR, machinable glass-ceramic, Code 9658, Corning Glass Works, Corning, N.Y.
31. J. Millman and H. Taub, *Pulse, Digital, and Switching Waveforms*, McGraw-Hill, New York, 1965, Chapter 3; C. N. Winningstad, IRE Trans. Nucl. Sci., NS-6, 26, 1959; C. L. Ruthroff, Proc. IRE, 47, 1337, 1959.
32. R. K. Cacak and J. R. Craig, Rev. Sci. Instr., 40, 1468, 1969.
33. W. G. Wadley, Rev. Sci. Instr., 27, 910, 1956.

MANUFACTURERS AND SUPPLIERS

Acheson Colloids Co.
Port Huron, MI 48060

Bomco, Inc.
Rt. 128
Blackburn Circle
Gloucester, MA 01930

Cliftronics, Inc.
515 Broad St.
Clifton, NJ 07013

Corning Glass Works
Corning, NY 14830

Electron Technology
626 Schuyler Ave.
Kearny, NJ 07032

Ferroxcube
5083 Kings Highway
Saugerties, NY 12477

Galileo Electro-Optics Corp.
Galileo Park
Sturbridge, MA 01518

Industrial Tectonics
P.O. Box 1128
Ann Arbor, MI 48106

McDanel Refractory Porcelain Co.
510 Ninth Ave.
Beaver Falls, PA 15010

Spectra-Mat, Inc.
Watsonville, CA 95076

CHAPTER 6

ELECTRONICS

It is the purpose of this chapter to discuss electronics at a level somewhere between that of a handbook, which consists essentially of charts, tables, and graphs, and a textbook, where the interesting, important, and useful conclusions come only after well-developed discussions with examples. The aim here is a presentation that has sufficient continuity and readability that individual sections can be profitably read without having to refer to preceding sections. On the other hand, it has been deemed important to have useful and frequently referenced material in the form of readily accessible tables, graphs, and diagrams that are so self-explanatory that very little, if any, reference to the text material is necessary. Another important goal is vocabulary. A large amount of jargon in electronics is meaningless to the uninitiated, but when it is necessary to understand the properties of an electronic device from a written technical description, when writing the specifications for electronic equipment, or when talking to an electronics engineer, salesman, or technician, this vocabulary is vital. With this in mind, terms not current outside of electronics are italicized.

To be used to best advantage, this chapter should be supplemented with manufacturers' catalogs, data books, applications texts, handbooks, and more specialized texts that treat the topic of interest in depth. Many manufacturers of laboratory electronic equipment, discrete devices, and integrated circuits have publications that describe in clear practical terms the properties of their products and their application to a wide variety of tasks.

Since the goal in experimental work is apparatus that works reliably and predictably, such publications are very valuable, and references to them are included.

The material has been organized and written exactly as one explains it to a student or technician coming to work in a laboratory for the first time. The complexity of modern electronics is such that the cut-and-try approach is too inefficient and costly in terms of material and time. There are just too many possibilities to try when connecting devices and multiple-component circuits, and it is important to establish a systematic approach based on a limited number of simple, well-understood principles. It is probably not reasonable in the laboratory to expect quick solutions to problems that are entirely outside one's previous experience. However, the number of really new situations that can arise is limited—most problems being variations on a few basic situations. The ability to recognize this and to isolate the source of difficulty comes with experience and mastery of basic principles. When confronted with a new situation involving rack upon rack of complicated equipment, the tendency is to believe that an understanding of how everything works is beyond the capabilities of any but electronics engineers. This is far from the truth. At the operational level, present-day electronics is the most reliable, easy-to-use, and easy-to-understand element of most experiments; this, of course, explains the great proliferation of electronics, not only in physical-science laboratories but in life-science and social-science laboratories as well.

6.1   PRELIMINARIES

6.1.1   Circuit Theory

A knowledge of elementary circuit theory permits one to reduce complex circuits consisting of many elements to a few essentials, predict the behavior of complex circuits, specify the operation of components, and understand and use the electronics vocabulary. In routine laboratory work it is not necessary to be skillful with circuit theory. What is necessary is to be able to isolate the basic elements of a circuit and understand its behavior in terms of them. With that ability, when circuits fail to operate correctly, the causes of the malfunction can often be localized.

Linear circuit theory applies to devices whose output is directly proportional to the applied input. For example, if one increases the current through a resistor by a factor of two the voltage across it will double. An example of a nonlinear device is a switch that is either open or closed and whose state changes abruptly at some threshold. Nonlinear devices can often be treated by the linear theory by dividing the response of the device into separate regions over which the device behaves in a quasilinear manner. This is called *linearizing the response curve*. An example is the piecewise linearization of the current-voltage response of a diode

as shown in Figure 6.1. The exponentially rising forward current and constant reverse current are represented by straight lines of slope $1/R_f$ and $1/R_r$, which are joined at voltage V_γ.

We begin by considering only *passive* linear devices; that is, devices that either dissipate energy (resistors) or store energy in electric (capacitors) or magnetic (inductors, transformers) fields. *Active* devices such as transistors can supply energy to a circuit when appropriately powered by external sources. The analysis of circuits with such active devices is based on representations using equivalent circuits consisting of passive devices.

Conventional circuit analysis uses *lumped* circuit elements, of which there are three: resistors (R), capacitors (C), and inductors (L). This way of analyzing circuits is valid at frequencies f for which the wavelength λ is much larger than the physical dimensions of the circuit. Since $\lambda = c/f$, where c is the speed of light, this means that analysis in terms of lumped elements is valid up to frequencies of a few hundred megahertz.

This limitation also excludes waveforms with significant frequency components above a few hundred megahertz, even though the repetition rate of the waveform may be much less. A convenient way to estimate the frequency of the highest-frequency component of a nonsinusoidal waveform is to divide 0.3 by the *risetime* of the waveform, t_r, defined as the time between the

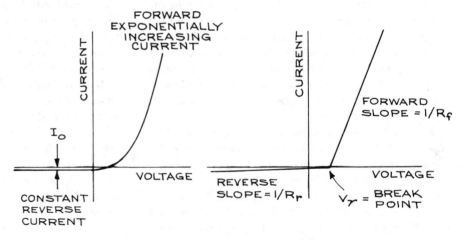

Figure 6.1   (*a*) Real and (*b*) piecewise linear representation of the current-voltage characteristics of a diode.

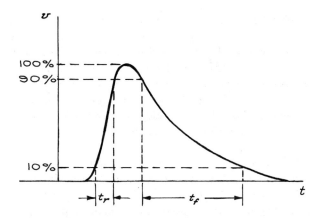

Figure 6.2 Risetime (t_r) and falltime (t_f) of a pulse.

Table 6.1 RATIO OF A.C. TO D.C. WIRE RESISTANCE

	$R_{\mathrm{a.c.}}/R_{\mathrm{d.c.}}$			
Wire Gauge	10^6 Hz	10^7 Hz	10^8 Hz	10^9 Hz
#22	6.86	21.7	68.6	217
#18	10.9	34.5	109	345
#14	17.6	55.7	176	557
#10	27.6	87.3	276	873

10%- and 90%-amplitude points on the leading edge of the waveform. By analogy, the *falltime*, t_f, is the time between the 10%- and 90%-amplitude points on the trailing edge. These relations are shown in Figure 6.2.

Even at low frequencies there are no ideal resistors, capacitors, or inductors. Actual resistors have some capacitance and inductance, while capacitors have resistance and inductance, and inductors have resistance and capacitance. These departures from ideality are largely a matter of construction.

At high frequencies stray capacitances and inductances become significant, and one commonly speaks of *distributed* parameters in contrast to the lumped parameters at low frequencies. Coaxial cable is an example of a type of distributed parameter circuit. The electrical properties of coaxial cable are normally given in terms of ohms per unit length and attenuation per unit length as a function of frequency. At high frequencies, the resistance of conductors, even connecting wires, increases due to what is termed *skin effect*. The magnitude of this effect for round-cross-section wires is given in Table 6.1 as a function of frequency. High-frequency connections are best made with leads having a high surface-area-to-volume ratio, the flat-ribbon geometry being best.

Conventional circuit theory is based on a few laws, principles, and theorems. In the equations that follow, lowercase letters represent instantaneous values for voltage and current, whereas uppercase letters indicate

effective or d.c. values. It is also convenient to distinguish between rms (root-mean-square), peak-to-peak, and average values of voltage and current for sinusoidally varying voltages. If $v = V\cos \omega t$, where v is the instantaneous value of voltage and V is the peak value, the rms value is $V/\sqrt{2}$, the peak-to-peak value is $2V$, and the average value is clearly zero. This is illustrated in Figure 6.3. Common U.S. line voltage is specified as 110 a.c., which is the rms value. The peak voltage is 156 volts, so the peak-to-peak voltage is 312 volts.

Under certain conditions the rms value is not sufficient for specifying the output of an a.c. source. A source producing voltage spikes of large amplitude but short duration superimposed on a small sinusoidally varying voltage will have an rms value very close to that without the spikes, but the spikes can have a large effect on circuits connected to the source. Therefore, when specifying the output of a d.c. power supply, not only is the rms value of any a.c. component of the output important but so also are the magnitude, frequency, and duration of nonsinusoidal waveforms that appear at the output.

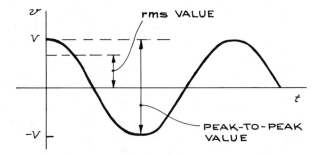

Figure 6.3 Relation of rms and peak-to-peak voltages for a sinusoidal waveform.

(a) Laws

(i) *Current-voltage relations.* For resistors, capacitors, and inductors we have, respectively,

$$v = iR, \qquad v = \frac{1}{C} \int i \, dt, \qquad v = L \frac{di}{dt},$$

where R is in ohms, C in farads, and L in henrys.

(ii) *Loops and nodes (Kirchhoff's laws).* In these, the sums are algebraic (signs taken into account):

1. Σ (voltage drops around a closed loop) $= \Sigma$ (voltage sources).

2. Σ (current into a node) $= \Sigma$ (current out of a node), where a node is a point where two or more elements have a common connection.

(b) Principles

(i) *Voltage.* The voltage between any two points in a circuit is equal to the algebraic sum of the voltages produced by each one of the voltage sources separately.

(ii) *Current.* The current past any point in a circuit is the algebraic sum of the currents due to each of the current sources in the circuit.

(c) Theorems

(i) *Thevenin's theorem.* A real voltage source in a circuit can always be replaced by an ideal voltage source in series with a generalized resistance. An ideal voltage source is one that can maintain a constant voltage across its terminals regardless of load. In other words, an ideal voltage source has zero internal resistance. An automobile battery, with an internal resistance of a few hundredths of an ohm, is a good approximation to an ideal voltage source at currents of a few amperes. Electronically regulated power supplies often have very low effective internal resistances when operated within their voltage and current ratings.

(ii) *Norton's theorem.* A real current source in a circuit can always be replaced by an ideal current source shunted by a generalized resistance. An ideal current source is one that supplies a constant current regardless of load. Such a source has an infinite internal resistance. Electron-multiplier devices provide currents, albeit very low, that are independent of load and approximate an ideal current source.

When connecting a source of current or voltage to other circuitry, it is often important to know the internal resistance of the source. This is accomplished by first measuring the open-circuit voltage of the source with a high-internal-resistance voltmeter and then connecting a variable resistance across the source and adjusting it until the voltmeter reading is one-half the open-circuit value. The source resistance is then equal to the value of the variable-resistance setting. If the source has a very high internal resistance, a current measurement can be substituted for the voltage measurement. In this case the output is shunted with an ammeter and the short-circuit current measured. Next, a variable resistance is placed in series with the ammeter and adjusted until the current through the ammeter has been reduced by a factor of two. The value of the variable resistor at this point is equal to the internal resistance of the source. Identical measurements can be made on a.c. sources by using either a.c. voltmeters and ammeters or an oscilloscope.

6.1.2 Circuit Analysis

For any given source, the choice of representation (Thevenin or Norton) is completely arbitrary and, in fact, the series resistance in the Thevenin representation is exactly equal to the parallel resistance in the Norton representation. Thevenin's and Norton's theorems simplify the application of the laws and principles discussed above.

The most general method for solving circuit problems is to apply Kirchhoff's laws using the appropriate current-voltage relations for each element in the circuit. This gives rise to one or more linear differential equations, which when solved with the proper boundary conditions give the general solution.

When dealing with sinusoidal sources of angular frequency ω, circuit analysis can be greatly simplified if only the steady-state solution is required. In this case circuit capacitances and inductances are replaced by *reactances*:

$$\text{capacitative reactance} = X_C = -\frac{j}{\omega C} \qquad (j=\sqrt{-1}),$$

$$\text{inductive reactance} = X_L = j\omega L.$$

The *impedance* Z of a circuit is obtained by combining reactances and resistances according to the formula $Z = R + X_L + X_C$. These quantities can be represented in the complex plane by vectors (Figure 6.4). The angle between Z and the real axis is ϕ. By analogy with the I-V relations for a pure resistance, X_C is the ratio of the a.c. voltage across a capacitor to the current through it, X_L is the ratio of the a.c. voltage across an inductor to the current through it, and Z is the net ratio of a.c. voltage to current in a circuit composed of resistors, capacitors, and inductors.

The fact that X_C and X_L are pure imaginary means that the voltage and current are 90° out of phase with each other. In the capacitor the voltage lags the current by 90°, while in the inductor the voltage leads the current by 90°.

Another quantity that is occasionally useful in circuit analysis is the *complex admittance* Y, which is the reciprocal of the impedance. The usefulness of the admittance arises in circuits with several parallel branches, where the net admittance is the sum of the admittances of the branches.

In carrying out circuit analysis, the following facts are useful:

(a) Series Circuits. At any instant the current is everywhere the same in a series circuit, and the algebraic sum of the voltage drops around a circuit equals the algebraic sum of the sources. For circuit elements of impedance $Z_1, Z_2, \ldots, Z_n$ in an N-element series circuit, the total impedance is $Z = Z_1 + Z_2 + \cdots + Z_N$. If all the elements are resistors, or inductors, or capacitors, the general expression reduces, respectively, to

$$R = R_1 + R_2 + \cdots + R_N,$$

$$\frac{1}{C} = \frac{1}{C_1} + \frac{1}{C_2} + \cdots + \frac{1}{C_N},$$

$$L = L_1 + L_2 + \cdots + L_N.$$

(b) Parallel Circuits. For circuit elements in a parallel configuration, the voltage drop across each branch is the same, while the current through each branch is inversely proportional to the impedance of the branch. The total current through all of the branches is the voltage across the network divided by the equivalent impedance for the network. The equivalent impedance Z and admittance Y for an N-branch parallel circuit are

$$\frac{1}{Z} = \frac{1}{Z_1} + \frac{1}{Z_2} + \cdots + \frac{1}{Z_N}$$

and

$$Y = Y_1 + Y_2 + \cdots + Y_N,$$

where $Z_1, Z_2, \ldots, Z_N$ are the impedances of the branches and $Y_1, Y_2, \ldots, Y_N$ are the admittances. In the special cases where all the circuit elements in the branches are

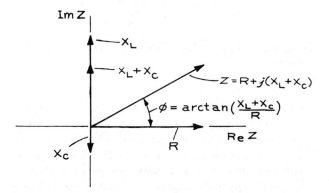

Figure 6.4 Relations between reactance, resistance, impedance, and phase angle.

of the same type,

$$\frac{1}{R} = \frac{1}{R_1} + \frac{1}{R_2} + \cdots + \frac{1}{R_N},$$

$$C = C_1 + C_2 + \cdots + C_N,$$

$$\frac{1}{L} = \frac{1}{L_1} + \frac{1}{L_2} + \cdots + \frac{1}{L_N},$$

where R, C, and L are the net resistance, capacitance, and inductance of the circuits.

(c) Voltage Dividers. The *voltage divider*, illustrated in Figure 6.5(a), is a very common circuit element. The instantaneous voltage across Z_N is $v_{in} \times Z_N/(Z_1 + Z_2 + Z_3 + \cdots + Z_N)$; that is, the fraction of v_{in} that appears across any circuit element is the impedance of that element divided by the total impedance of the series circuit. Voltage dividers provide an easy way to obtain a variable-voltage output from a fixed-voltage input, but there are limitations. To avoid drawing too much current from the voltage source, the impedance of the voltage divider string should not be too low. However, if, in the interest of conserving power, the impedance is made large, the output impedance of the circuit will be large and v_{out} will depend critically on the load. This can be seen from the Thevenin equivalent of the circuit given in Figure 6.5(b) when v_s is the instantaneous voltage of the ideal voltage source. When Z_S is large, the voltage across a load will depend critically on the value of the load. Such "loading" of a divider is to be avoided. Generally, for noncritical applications, Z_S should be at most $\frac{1}{10}$ of any load.

Precision, highly linear, multiturn potentiometers are commonly used for mechanical-position sensing. In this application a stable voltage source is connected across the fixed ends of the potentiometer and the ratio of the voltage from the variable contact to one end to the total voltage is measured with a voltmeter. The accuracy of this measurement is of the order R_S/R_V, where R_S is the total resistance of the potentiometer, and R_V is the internal resistance (input resistance) of the voltmeter.

(d) Equivalent Circuits. Two circuits are *equivalent* if the relationships between the measurable cur-

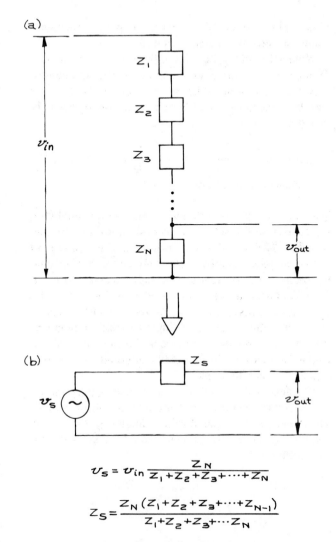

$$v_s = v_{in} \frac{Z_N}{Z_1 + Z_2 + Z_3 + \cdots + Z_N}$$

$$Z_S = \frac{Z_N(Z_1 + Z_2 + Z_3 + \cdots + Z_{N-1})}{Z_1 + Z_2 + Z_3 + \cdots Z_N}$$

Figure 6.5 (a) The voltage divider; (b) the Thevenin equivalent.

rents and voltages are identical. As has been seen, a circuit with two external terminals can be replaced by its Thevenin or Norton equivalent. Common equivalence transformations for circuits with three terminals (*Miller* and *Y-Δ transformations*) are shown in Figures 6.6 and 6.7. In the first Miller transformation of Figure 6.6 it is necessary to know the ratio of the voltages at nodes 1 and 2; in the second it is necessary to know the ratio of the currents through 1 and 2.

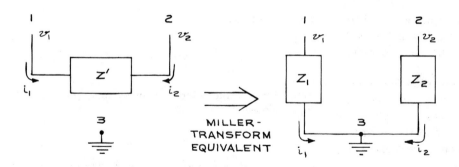

$$Z_1 = Z'/(1-K), \quad Z_2 = Z'K/(K-1), \quad K = v_2/v_1$$

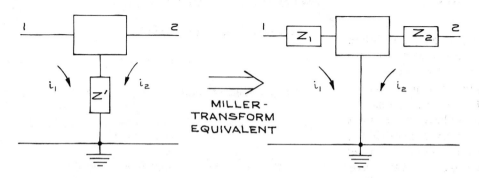

$$Z_1 = Z'(1-A_i), \quad Z_2 = Z'(A_i-1)/A_i, \quad A_i = -i_2/i_1$$

Figure 6.6 Miller transformations for a circuit with three terminals.

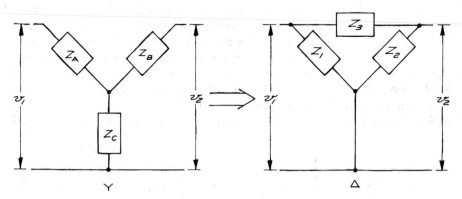

$$Z_1 = \frac{Z_A Z_B + Z_A Z_C + Z_B Z_C}{Z_B}$$

$$Z_2 = \frac{Z_A Z_B + Z_A Z_C + Z_B Z_C}{Z_A}$$

$$Z_3 = \frac{Z_A Z_B + Z_A Z_C + Z_B Z_C}{Z_C}$$

Figure 6.7 Y-Δ transformation for a circuit with three terminals.

Discussions of active circuits often refer to the *Miller effect*. This is nothing more than the multiplication of Z by $1 - A_i$ in the upper left-hand branch, where A_i is the current gain of the active element.

The term *Miller capacitance* refers to the first transformation in the case where Z is purely capacitive. The transformation results in Z being multiplied by $1/(1 - K)$, where K is the voltage gain with the transformed impedance inserted in the circuit between node 1 and the common or ground node.

6.1.3 High-Pass and Low-Pass Circuits

Analysis of the *high-pass* and *low-pass* circuits shown in Figure 6.8 illustrates some of the above circuit-analysis principles. The combination of v_s and R_s represents a real voltage source with instantaneous open-circuit voltage v_s and internal resistance R_s. For the high-pass or *differentiating circuit* the output voltage, v_o, is across the resistor R; for the low-pass or *integrating circuit* it is across the capacitor C. Very often the principal features of extremely complex circuits reduce to one of these two circuits, so it is useful to be acquainted with their characteristics.

These circuits can be analyzed by the *differential-equation method*. For either circuit,

$$v_s(t) = iR_s + \frac{1}{C}\int i\, dt + iR.$$

This equation makes use of the fact that the sum of the voltage drops in the circuit equals the sum of the voltage sources and the current is everywhere the same in a series circuit at any instant. Differentiating with

respect to time,

$$\frac{dv_s}{dt} = \frac{di}{dt}(R_s + R) + \frac{i}{C}.$$

The solution to the homogeneous equation ($dv_s/dt = 0$) is as follows:

$$i = Ae^{-t/R'C},$$

where $R' = R_s + R$ and A is the integration constant determined from the initial conditions. The general solution requires that the functional form of v_s be known. For this analysis, consider three cases:

1. An a.c. voltage of amplitude V,

$$v_s = V\cos(\omega t + \phi).$$

2. A step voltage of amplitude V,

$$v_s = \begin{cases} 0 & \text{for} \quad t < 0, \\ V & \text{for} \quad t > 0. \end{cases}$$

3. A rectangular pulse of amplitude V and duration T,

$$v_s = \begin{cases} V & \text{for} \quad 0 \leqslant t \leqslant T, \\ 0 & \text{for} \quad t < 0, \quad t > T. \end{cases}$$

For case 1 the output voltage is sinusoidal at the same frequency as the input voltage. The ratio of v_o to v_s as a function of normalized frequency is shown in Figure 6.9(a). At the frequencies $\omega = \omega_H$ and $\omega = \omega_L$ for the two circuits, v_o is $1/\sqrt{2}$ of the maximum value. These frequencies are called the *upper* and *lower corner frequencies*, respectively.

The maximum power that can be delivered to a load

Figure 6.8 (*a*) High-pass or differentiating circuit; (*b*) low-pass or integrating circuit.

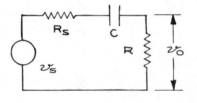

(a)

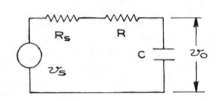

(b)

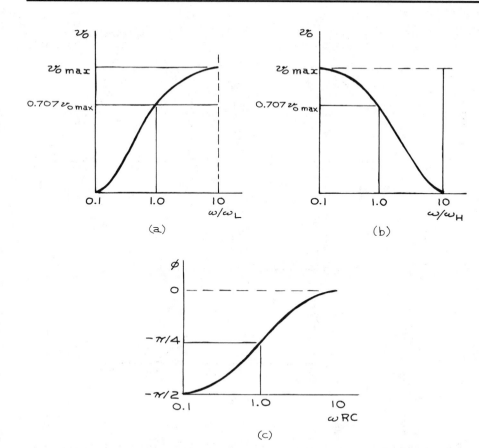

Figure 6.9 Output voltage as a function of frequency for (*a*) high-pass and (*b*) low-pass circuits; (*c*) voltage-current phase relationship in an RC circuit.

is proportional to the square of the output voltage, so that at $\omega = \omega_H$ and $\omega = \omega_L$ the maximum power that the circuits can deliver to a constant load is one-half the maximum possible value. The usual way of expressing this is in terms of decibels (dB), where

$$(\text{ratio in dB}) = 10\log_{10}\left(\frac{\text{power out}}{\text{power in}}\right)$$

$$= 10\log_{10}\frac{v_{\text{out}}^2 R_{\text{out}}}{v_{\text{in}}^2 R_{\text{in}}}.$$

If $R_{\text{out}} = R_{\text{in}}$, which is often assumed, then (ratio in dB) $= 20\log_{10}[v_{\text{out}}/v_{\text{in}}]$. When $v_{\text{out}}/v_{\text{in}} = \frac{1}{2}$, this comes out $= -3$, so that -3 dB represents a power reduction of a factor of about 2. Since the frequency response of amplifiers, filters, and transducers is routinely given in dB, it is important to keep in mind that the dB scale is

logarithmic. Human sensory preception is approximately logarithmic, and a 3-dB change in sound level or light level is barely perceptible at any level.

Because of the reactive element in the circuits (the capacitor), the voltage is not in phase with the current, as illustrated in Figure 6.9(*b*). These plots of phase and log(output voltage) as a function of log(frequency) are called *Bode plots*, after H. W. Bode.[1]

Often it is convenient to approximate the frequency-response curves by a piecewise linear function. Such idealized Bode plots are shown in Figure 6.10(*a*) and (*b*). The *corner frequencies* where ω/ω_L and $\omega/\omega_H = 1.0$ correspond to the -3-dB points on the unapproximated Bode plots. For most purposes, the simplified curves are an entirely satisfactory representation. From these curves, every tenfold reduction in frequency below ω_H for the high-pass circuit decreases the output

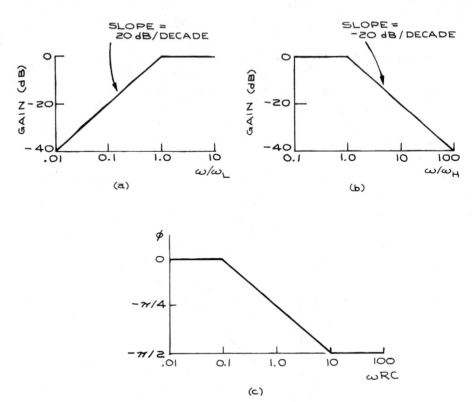

Figure 6.10 Idealized gain response for (a) high-pass and (b) low-pass circuits; (c) the idealized phase response.

voltage by 20 dB, and every twofold reduction decreases it by 6 dB. The low-pass circuit has just the opposite properties: a tenfold increase in frequency above ω_L results in a 20-dB decrease in output voltage, and a twofold increase results in a 6-dB decrease. Often one states these facts as "20 dB per decade and 6 dB per octave." The linearized phase-response curves are shown in Figure 6.10(c). The -3-dB frequencies occur at a phase shift of $-\pi/4$ ($-45°$) for the two circuits.

For the nonrepetitive input voltage waveforms of cases (2) and (3), the output waveforms are given in Figure 6.11.

The output waveforms for the rectangular-wave input function represent a means of determining the time constant for differentiating and integrating circuits. This is called *square-wave testing*. The time constant RC for the differentiating circuit is obtained by using a square-wave input with a risetime much less than RC and a period much greater than RC. For times small compared with RC, the tilt of the top edge of the output as viewed with a fast-risetime oscilloscope (Figure 6.12) is directly related to RC. The fractional decrease in v_o in time t_1 is t_1/RC, which can be set equal to $(V-V')/V$ and solved for RC. Similarly, RC for the integrating circuit is obtained by measuring the risetime of the output waveform on a fast-risetime oscilloscope. Using the definition of the risetime t_r as the time between the 10% and 90% points on the leading edge of the output waveform, one has the relation

$$t_r = 2.2RC.$$

6.1.4 Resonant Circuits

The voltages and currents in circuits with capacitors, inductors, and resistors show oscillatory properties much like those of mechanical oscillators. Electronic circuits have natural frequencies of oscillation and can be criti-

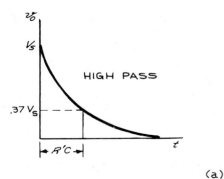

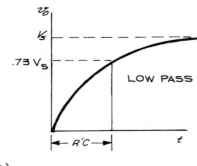

(a)

Figure 6.11 (a) Response of high-pass and low-pass circuits to a step voltage; (b) rectangular pulse of duration T.

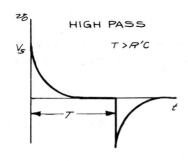

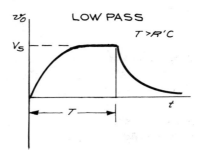

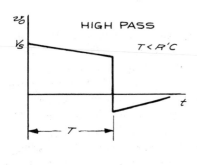

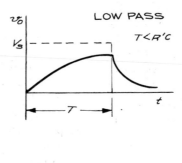

(b)

cally damped, underdamped, or overdamped, depending on the relations between the values of the circuit parameters. Resonant circuits with ideal capacitors and inductors are of the series or parallel type shown in Figure 6.13. When driven by a sinusoidal input source, the capacitative reactance in the series circuit will precisely cancel the inductive reactances at the resonant frequency ω_0, where $1/C\omega_0 = \omega_0 L$ and $\omega_0 = 1/\sqrt{LC}$.

At ω_0 the impedance of the series circuit is a minimum and the current through it is a maximum.

For the parallel resonant circuit at low frequencies, the L-branch will have a very low reactance, and the current drawn from the source will flow almost entirely through that branch. At high frequencies the current through the RC branch is limited by the value of R. The total impedance of the parallel circuit is therefore small

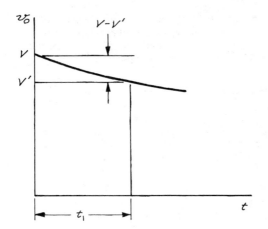

Figure 6.12 Square-wave testing of a high-pass circuit.

at low and high frequencies, passing through a maximum at the frequency $\omega_0 = 1/\sqrt{LC}$ provided that $R \ll \omega_0 L$. Graphs of the currents in the two circuits as a function of driving frequency are given in Figure 6.14. Real inductors have an associated resistance, which generally must also be taken into account when analyzing practical circuits.

One measure of the sharpness of the resonance in the series and parallel circuits is the Q or "quality" of the circuit. For practical purposes, $Q = \omega_0/\Delta\omega$, where $\Delta\omega$ is the full width at half maximum of the peak or valley. In terms of the circuit parameters, $1/Q = \omega_0 L/R$. This is the ratio of the energy stored (in the capacitor or inductor) to the energy dissipated in the resistor, per cycle, at resonance. Values of Q up to 100 can be attained in electrical circuits; mechanical oscillators can

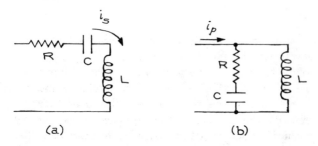

Figure 6.13 (*a*) Series resonant circuit; (*b*) parallel resonant circuit.

attain values as high as 10^6. The phase relationship between voltage and current is shown in Figure 6.15.

The behavior of an *RCL* circuit upon the application of a step input or rectangular wave is much like the response of a mechanical system to a sudden impulse. Critically damped, underdamped, and overdamped current flows are possible.

6.1.5 The Laplace-Transform Method

A very general technique for analyzing circuits for arbitrary input voltage waveforms is the method of Laplace transforms. With this method it is possible to use only algebra and tables of transforms such as those given in Table 6.2 for the solution of differential equations and the evaluation of boundary conditions. The results of the method will be presented without any proof. The vocabulary of Laplace transforms occurs very often in the discussion of circuits and is included in this chapter for that reason.

The method itself is based on an integral transform of the type

$$\bar{f}(s) = \int_0^\infty f(t)e^{-st}\,dt,$$

where $\bar{f}(s)$ is the Laplace transform of $f(t)$, often written as $L[f(t)]$. The function $f(t)$ can involve integrals and differentials. When it is applied to the second-order differential equations that arise in circuit analysis, rather important simplifications occur and results can often be written down by inspection. The Laplace transform of the output voltage $\bar{v}_o(s)$ of a circuit is the Laplace transform of the input voltage $\bar{v}_i(s)$ times the Laplace transform of the transfer function $\bar{T}(s)$, the *transfer function* being the function relating the output to input. To get $\bar{T}(s)$, the values of all elements in the circuit are replaced by their transform equivalents according to the recipe $R \rightarrow R, C \rightarrow 1/sC$, and $L \rightarrow sL$. In the simple case of the voltage divider, the transfer function is the ratio of the impedance of the output-circuit element to the total impedance of the circuit chain. $\bar{T}(s)$ is obtained in exactly the same way using the equivalences for R, C, and L. In general, $\bar{T}(s)$ is in the form of a ratio of two functions of

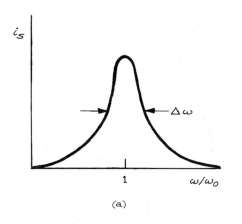

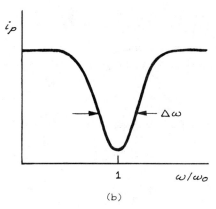

Figure 6.14 Current as a function of frequency for (*a*) the series resonant circuit and (*b*) the parallel resonant circuit.

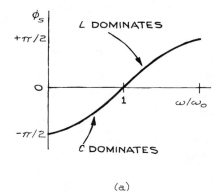

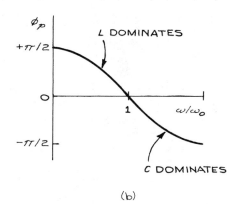

Figure 6.15 Phase relations between voltage and current in (*a*) the series resonant circuit and (*b*) the parallel resonant circuit.

s, $G(s)$ and $H(s)$, which are polynomials in s:

$$\bar{T}(s) = \frac{G(s)}{H(s)}.$$

The values of s for which $G(s)$ is 0 are called the *zeros* of $\bar{T}(s)$, while the values of s for which $H(s)$ is 0 are called the *poles* of $\bar{T}(s)$. In the most general case the zeros and poles are complex. The positions of the zeros and poles of $\bar{T}(s)$ in the complex plane give important information on the properties of the circuit under analysis. When the poles are complex, they occur in pairs, while real-valued poles can occur singly or in pairs. The values of the real and imaginary components of the pole coordinates, usually labeled σ and ω, have important physical meaning. The real component σ is a measure of the damping in the circuit while the imaginary part ω is

the natural frequency of oscillation. Negative values of σ give stable circuits in which transient signals all decay to zero with time. Circuits employing only passive elements behave in this way and are stable. Circuits with active elements can behave in such a way that the output increases with time in response to a transient input signal. Such circuits are unstable and have values of σ greater than zero. They are to be avoided except in the case of oscillators, which must have certain instabilities in order to function.

The Laplace-transform equivalents of the differentiating and integrating circuits are shown in Figure 6.16. For the high-pass circuit, the output voltage across the resistor R for the transformed circuit is

$$\bar{v}_o(s) = \bar{v}_s(s)\frac{sRC}{1 + sR'C},$$

where $R' = (R_s + R)$ and $\bar{T}(s)$, which is the transfer function, has a zero at $s = 0$ and a pole at $s = -1/R'C = -\omega_L$. For the low-pass circuit, the output voltage across the capacitor for the transformed circuit is given as follows:

$$\bar{v}_o(s) = \bar{v}_s(s)\frac{1}{1 + sR'C}.$$

Table 6.2   ELEMENTARY LAPLACE TRANSFORMS

$f(t)\ (t > 0)$	$\bar{f}(s)$
$\delta(t)$	1
1	$1/s$
$t^{n-1}/(n-1)!$	$1/s^n$ (n a positive integer)
e^{at}	$\dfrac{1}{s-a}$
$\sin at$	$\dfrac{a}{s^2 + a^2}$
$\cos at$	$\dfrac{s}{s^2 + a^2}$
$\sinh at$	$\dfrac{a}{s^2 - a^2}$
$\cosh at$	$\dfrac{s}{s^2 - a^2}$
$\dfrac{t}{2a}\sin at$	$\dfrac{s}{(s^2 + a^2)^2}$
$\dfrac{1}{2a^3}(\sin at - at\cos at)$	$\dfrac{1}{(s^2 + a^2)^2}$
$\dfrac{df(t)}{dt}$	$s\bar{f}(s) - f_0 \left[f_0 = \lim\limits_{t \to 0} f(t) \right]$
$\dfrac{d^2f(t)}{dt}$	$s^2\bar{f}(s) - sf_0 - f_1 \left[f_1 = \lim\limits_{t \to 0} \dfrac{df(t)}{dt} \right]$
$\displaystyle\int_0^t f(t')\, dt'$	$\dfrac{1}{s}\bar{f}(s)$
$\dfrac{1}{a-b}(e^{at} - e^{bt})$	$\dfrac{1}{(s-a)(s-b)}$
$\dfrac{1}{a-b}(ae^{at} - be^{bt})$	$\dfrac{s}{(s-a)(s-b)}$
$\dfrac{1}{a^2}(1 - \cos at)$	$\dfrac{1}{s(s^2 + a^2)}$
$\dfrac{1}{a^3}(at - \sin at)$	$\dfrac{1}{s^2(s^2 + a^2)}$
$\dfrac{1}{ab(b^2 - a^2)}(b\sin at - a\sin bt)$	$\dfrac{1}{(s^2 + a^2)(s^2 + b^2)}$
$\dfrac{1}{b^2 - a^2}(\cos at - \cos bt)$	$\dfrac{s}{(s^2 + a^2)(s^2 + b^2)}$
$\dfrac{1}{\alpha^2 + \beta^2} - \dfrac{e^{-\alpha t}}{\beta\sqrt{\alpha^2 + \beta^2}}\sin[\beta t + \arg(\alpha + i\beta)]$	$\dfrac{1}{s\left[(s + \alpha)^2 + \beta^2\right]}$

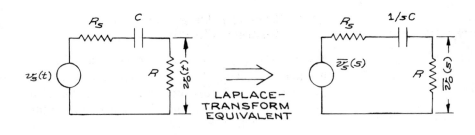

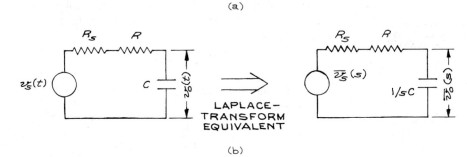

Figure 6.16 Laplace-transform equivalents of (a) the high-pass circuit and (b) the low-pass circuit.

$\bar{T}_o(s)$ has a pole at $s = -1/R'C = -\omega_H$. The steady-state frequency and phase response of the circuits are obtained from $\bar{T}(s)$ by replacing s with $j\omega$. The transfer functions are now, for the low-pass circuit,

$$T(\omega) = \frac{j\omega/\omega_L}{1 + j\omega/\omega_L}$$

and, for the high-pass circuit,

$$T(\omega) = \frac{1}{1 + j\omega/\omega_H}.$$

As seen previously, ω_L and ω_H are the corner frequencies of the circuits. By rationalizing the denominators of the transfer functions, one obtains the phase response. Since there are no inductive elements in the circuits, there is no natural frequency of oscillation. The poles lie on the negative real axis, since there are no active elements in the circuit.

6.1.6 *RCL* Circuits

Consider the equivalent circuit and the Laplace transform given in Figure 6.17. To analyze the circuit, consider the parallel combination of R and $1/sC$, which

is in series with sL in a voltage-divider configuration:

$$\frac{R/sC}{R + 1/sC} = \frac{R}{1 + sRC}.$$

Thus

$$\bar{v}_o(s) = \bar{v}_i(s) \frac{1}{1 + sL/R + s^2 LC}.$$

The poles of $\bar{T}(s)$ occur at

$$s = \frac{-L/R \pm \sqrt{(L/R)^2 - 4LC}}{2LC}.$$

Letting the natural frequency of oscillation of the circuit be $\omega_0 = 1/\sqrt{LC}$, we have $Q = R/\omega_0 L$, and the poles can be rewritten as

$$s = \frac{-\omega_0}{2Q} \pm \frac{\omega_0}{2}\sqrt{1/Q^2 - 4}.$$

There are three different possibilities for the roots of s:

1. $1/Q^2 = 4$: a single real root at $s = -\omega_0/2Q$.

2. $1/Q^2 - 4 = m^2$: two real roots at $s = -\omega_0/2Q \pm \omega_0 m/2$.

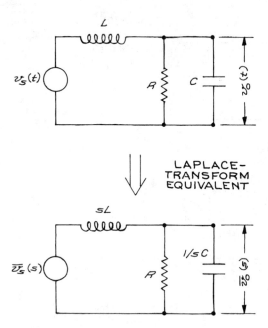

Figure 6.17   An *RLC* circuit and the Laplace-transform equivalent.

3. $1/Q^2 - 4 = -m^2$: two conjugate complex roots at $s = -\omega_0/2Q \pm j\omega_0 m/2$.

The magnitude of s in the $\sigma - j\omega$ plane is ω_0; in geometric terms this means that the roots of s are confined to a semicircle of radius ω_0 in the left half of the complex plane (Figure 6.18).

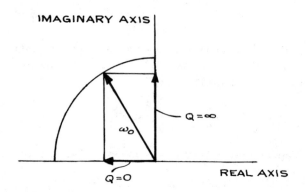

Figure 6.18   Pole trajectory for an *RLC* circuit.

6.1.7 Transient Response of Resonant Circuits

The stability of a linear system subjected to a driving function depends on the system itself rather than the driving function. For ease of analysis, consider the *RCL* circuit to be driven by a unit step input voltage of the form $v_i(t) = 1$ for $t > 0$. The Laplace transform of the output voltage is then the transform of the input voltage, $\bar{v}_i(s)$, multiplied by the transform of the transfer function, $\bar{T}(s)$:

$$\bar{v}_o(s) = \bar{v}_i(s)\bar{T}(s).$$

For the unit step input $\bar{v}_i(s) = 1/s$, from Table 6.2; therefore

$$\bar{v}_o(s) = \frac{1}{s(1 + sL/R + s^2LC)}.$$

To find $v_o(t)$ it is necessary to look up the inverse transform in Table 6.2. The result is

$$v_o(t) = 1 - \frac{e^{-k\omega_0 t}}{\sqrt{1 - k^2}}$$
$$\times \sin\left[\sqrt{1 - k^2}\,\omega_0 t + \arctan\left(\frac{\sqrt{1 - k^2}}{k}\right)\right],$$

where $k = 1/2Q$. Critical damping, overdamping, and underdamping correspond to $k = 1$, $k > 1$, and $k < 1$, respectively. The normalized response of the circuit for various values of k is shown in Figure 6.19.

As can be seen, the underdamped waveform overshoots the steady-state value of 1 and oscillates about it with decreasing amplitude (except for $k = 0$, no damping). This is sometimes called *ringing*. As damping is increased, the output waveform oscillations decrease in amplitude until for $k = 1$ (critical damping) the oscillations are completely gone. Increasing the damping beyond $k = 1$ only increases the time for the waveform to get to the steady-state value.

Overshoot, risetime, settling time, and *error band* are terms used to characterize such output waveforms. They are illustrated in Figure 6.20. For critical damping $t_r = 3.3/\omega_0$; by decreasing the damping so that $k < 1$, t_r

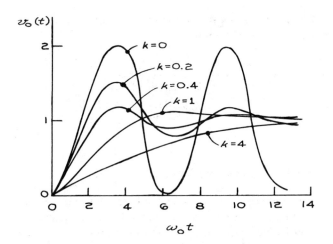

Figure 6.19   Response of a resonant circuit with no damping $(k = 0)$, underdamping $(k = 0.2, 0.4)$, critical damping $(k = 1)$, and overdamping $(k = 4)$.

can be further decreased, but at the expense of ringing. A suitable compromise that is often used is $k = 0.707$, under which circumstance the overshoot is 4.3% and t_r is reduced to $2.16/\omega_0$. Figure 6.21 is a graph of the percentage overshoot as a function of k.

The above analysis is not restricted to passive circuits.

Amplifiers with negative feedback can have response functions identical in form to those for *RCL* circuits. The response of such amplifiers to pulses and nonrepetitive waveforms is specified using the same terms as for the passive circuit above. In pulse amplifiers, the transient response is the fundamental limitation on the pulse repetition rate. For critically damped amplifiers, the width T of the impulse response and the risetime t_r are approximately related by $T = 1.5 t_r$, so that pulses cannot be sent at a rate greater than $1/T$ per second. T in this case is generally taken to be the time between the 50% points on the output waveform.

6.1.8   Transformers and Mutual Inductance

Transformers consist of primary and secondary windings coupled by a core, which can be of magnetic material or air. A voltage is induced in the secondary of the transformer when there is a change of current in the primary if mutual inductance is present. The reverse also occurs: induced voltage in the primary due to a current change in the secondary. A transformer circuit and the Laplace-transform equivalent are illustrated in

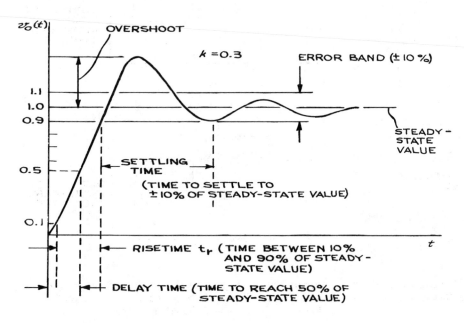

Figure 6.20   Output waveform of an underdamped resonant circuit.

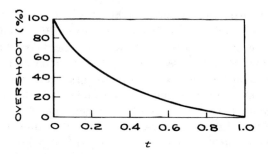

Figure 6.21 Percentage overshoot as a function of damping in a resonant circuit.

Figure 6.22. A dot is associated with each winding of the transformer. The convention is that M, the mutual inductance, is positive if the currents i_1 and i_2 both flow into or out of the dotted ends of the coils, and negative otherwise. The transfer function $\bar{T}(s)$ is obtained by applying Kirchhoff's laws to the two loops:

$$\bar{v}_i(s) = R\bar{i}_1(s) + \frac{\bar{i}_1(s)}{sC} + \bar{i}_1(s)sL_1 - sM\bar{i}_2(s),$$

$$0 = \bar{i}_2(s)sL_2 - \bar{i}_1(s)sM + \bar{i}_2(s)sL_3 + \bar{i}_2(s)R_2,$$

$$\bar{v}_0(s) = \bar{i}_2(s)R_2.$$

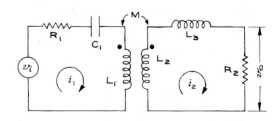

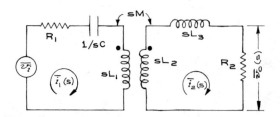

Figure 6.22 A transformer circuit and the Laplace-transform equivalent. M is positive if i_1 and i_2 both flow into or out of the dotted ends of the coils.

Eliminating $i_1(s)$ and substituting for $i_2(s)$ in the third equation gives the result

$$\bar{v}_o(s) = \bar{v}_i(s) \frac{R_2 MCs^2}{\begin{aligned} & \left[CL_1(L_2 + L_3) - M^2 \right] s^3 \\ & + \left[R_1(L_1 + L_2) + R_2 L_1 \right] Cs^2 \\ & + R_2 \left[L_2 + L_3 + R_1 R_2 C \right] s \end{aligned}}.$$

6.1.9 Compensation

It is often desired to modify the frequency response of a given circuit. Two common methods are the addition of a pole that is much smaller than the other poles of the transfer function and the simultaneous addition of a pole and zero to the transfer function. The addition of a pole is accomplished with a resistor and capacitor in a low-pass configuration at the output of the circuit to be modified, as illustrated in Figure 6.23(a). If the original

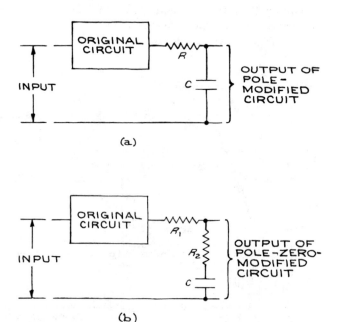

Figure 6.23 Compensation by (a) the addition of a pole and (b) the addition of a pole and a zero to cancel the lowest-frequency pole of the original circuit.

circuit had a pole ω_1 and the introduction of the RC circuit produced a pole at ω_2 where $\omega_2 \ll \omega_1$, the overall response of the circuit would be a 20-dB/decade decrease in gain from ω_2 to ω_1 followed by a 40-dB/decade decrease for frequencies greater than ω_1.

The addition of a pole and zero is accomplished with the circuit in Figure 6.23(b). The transfer function for the *pole-zero compensation* network is

$$\frac{R_2 + 1/sC}{R_1 + R_2 + 1/sC} = \frac{1 + sR_2C}{1 + s(R_1 + R_2)C},$$

with the zero at $s = -1/R_2C$ and the pole at $s = -1/(R_1 + R_2)C$. If the zero in the transfer function is chosen to cancel the smallest pole of the original circuit, the overall frequency response will be flat up to the frequency corresponding to the new pole. The gain will then decrease at 20 dB/decade with increasing frequency until the first pole of the original circuit. The gain will then decrease at 40 dB/decade until the next high-frequency pole, after which the decrease will be 60 dB/decade. The response of the pole-zero-compensated circuit is sharper than that of the single-pole-compensated one.

6.1.10 Filters

Filters are generally classed as *low-pass, high-pass, band-pass,* and *band-reject. RC* circuits are examples of the first two kinds, while the series and parallel *RCL* circuits illustrate the last two. A series of band-pass or band-reject filters with passbands at different but closely spaced frequencies is called a *comb filter.*

The ideal filter would pass unattenuated all frequencies in its passband while completely rejecting frequencies outside the passband. The simple circuits we have examined thus far are poor approximations to the ideal. Excellent approximations to the ideal filter are available however. They rely on judicious choices of the poles of the transfer functions of RC and RCL circuits. There are four classes of filter design, each of which has its advantages:

1. *Maximally flat* has the flattest amplitude response within the passband. The Butterworth filter is an example.

2. *Equal ripple* has fluctuations in the passband but greater attenuation in the stopband than the maximally flat filter.

3. *Elliptic* has the maximum rate of attenuation between the passband and stopband. The Chebyshev filter is an example of this type.

4. *Linear phase* has a much less sharp cutoff than the others but maintains an almost linear phase response in the passband. The Bessel filter is an example of this type.

Methods of filter design are well established, and numerous texts and handbooks provide tables for filter synthesis. Before designing or specifying a filter, it is important to know not only the cutoff and phase properties, but also the transient response and the input and output impedance.

As has already been noted, amplifiers with negative feedback provided by resistors and capacitors can have responses identical to RCL circuits. By an appropriate choice of the elements in a feedback network, these amplifiers can be made into filters. Especially at low frequencies, where large inductors with low resistance are heavy and expensive, these so-called *active filters* are very useful. There are numerous texts and handbooks on active filter design. Most designs are based on high-gain integrated-circuit operational amplifiers. With the frequency response of these amplifiers now extending into the megahertz region, active filters are finding increasing application to high-frequency circuits. An important factor in filter design, especially for high Q's, is stability. If drift is to be minimized, highly stable passive components are required.

6.2 PASSIVE COMPONENTS

In discussing passive components, emphasis will be placed on choosing the correct type for the function to be performed. It is rarely sufficient to specify the nominal value of a component. This can be seen by looking through the catalog of any electronics components supplier under "resistors" and finding perhaps 50 different resistor types, each with a range of resistance values. The following are some general considerations in the choice of passive components:

1. *Nominal value and tolerance.* An aspect of this is the coding applied to the component. Both color and number codes are used.

2. *Stability.* The nominal value as a function of temperature [the temperature coefficient ("tempco") is often expressed in percent or in parts per million (ppm)]. Age and environmental conditions such as humidity, vibration, or shock also effect component values. On the other hand, components can affect their environment—for example, the outgassing of a capacitor in vacuum or the heating of one component by another one.

3. *Size and shape.* Along with the reduction in size that solid-state electronics has brought, there has been a concomitant reduction in the size of passive components. Generally, one uses the smallest practical size; however, power-density considerations can limit the packing density of power-dissipating components. Also, components come in different shapes with different lead geometries depending on the intended mounting method.

4. *Power dissipation and voltage rating.* A resistor has both a power rating and a voltage rating. For very large values of resistance, it is quite possible to operate well within the resistor's power rating, yet exceed the voltage rating. The result is an electrical breakdown and destruction of the resistor. For moderate values of capacitance, capacitors dissipate very little power and power rating is of no importance. However, very high-capacitance capacitors with large effective surface areas have nonnegligible shunt conductances. The large currents that flow in such capacitors can cause considerable heat to be generated with potentially fatal (to the capacitor) results.

5. *Noise.* Passive components can introduce noise into an electronic circuit, and one must be particularly aware of this problem in low-level circuits and choose appropriate low-noise components.

6. *Frequency characteristics.* Pure resistors, capaci-

tors, and inductors are practically unrealizable, and the electrical properties of real passive components depend on frequency in a nonideal way. This is often expressed by an equivalent circuit. Inductors and transformers with magnetic cores, however, are inherently nonlinear and can be treated only approximately in this way.

7. *Derating.* This is an engineering term related to the safety factors to be used when operating components under a variety of conditions. Good design requires that components be operated at no more than 50% of the manufacturer's recommended maximum ratings, particularly with respect to power and voltage. At high ambient temperatures components should be derated by even larger amounts. The resulting decrease in stress greatly increases component lifetime. In laboratory equipment large safety factors should always be employed to enhance reliability.

8. *Cost.* For laboratory applications it is poor practice to economize on passive components by using the least expensive ones that will do the job. This is an important difference from accepted engineering practice. Laboratory work is labor intensive. The extra cost of top-grade components is greatly outweighed by the time saved in troubleshooting and repair.

6.2.1 Fixed Resistors and Capacitors

The more common types of fixed resistors and capacitors are listed in Tables 6.3, 6.4, and 6.5. In Table 6.5 the C range and V range refer to the minimum and maximum values available for all types of a given dielectric. It is not possible to have the maximum capacitance at the maximum voltage. The larger capacitances have lower *working voltages* (WV), while the smaller ones have the higher WV. The temperature coefficients generally apply only to a limited temperature range within the larger operating temperature range.

Tantalum is the most stable of all film-forming materials. Tantalum-foil electrolytics have similar characteristics to aluminum-foil ones, but offer superior stability and freedom from leaking. The wet-slug type has the highest volumetric efficiency of any capacitor.

Table 6.3 FIXED RESISTORS

Carbon Composition

R range:	$2.7\ \Omega\text{–}22\ M\Omega$
Power range:	$\frac{1}{10}\text{–}2\ W$
Tolerances:	$\pm 5, \pm 10, \pm 20\%$
Temperature range:	-55 to $+150°C$
Temperature coefficient:	$\pm 0.1\%/°C$
Voltage coefficient:	$\pm 0.03\%/V$ d.c.
Noise:	Highest
Typical capacitance:	0.25 pF, no inductance
Working voltage:	150 V ($\frac{1}{8}$ W) to 750 V (2 W)

Precision Carbon Film

R range:	$1\ \Omega\text{–}100\ M\Omega$
Power range:	$\frac{1}{10}\text{–}2\ W$
Tolerances:	$\pm 0.5, 1, 2\%$
Temperature range:	-55 to $165°C$
Temperature coefficient:	$0.02\text{–}0.05\%/°C$
Noise:	Low
Frequency response:	To 1 MHz for $R > 1\ k\Omega$ (spinal), to 100 MHz for $R < 1\ k\Omega$

Precision Metal Film

R range:	$10\ \Omega\text{–}158\ M\Omega$
Power range:	$\frac{1}{8}\text{–}\frac{1}{2}\ W$
Tolerances:	± 0.1 to $\pm 1\%$
Temperature range:	-55 to $165°C$
Temperature coefficient:	$0.0025\text{–}0.01\%/°C$
Noise:	Lowest

Precision Wire-Wound (bobbin)

R range:	$0.1\ \Omega\text{–}18\ M\Omega.$
Power range:	$\frac{1}{10}\text{–}2\ W$
Tolerances:	± 0.05 to $\pm 5\%$
Temperature range:	-55 to $145°C$
Temperature coefficient:	± 0.0001 to $\pm 0.002\%/°C$
Noise:	Low
Frequency response:	To 25 kHz for high-resistance values unless noninductively wound

Power Wire-Wound

R range:	$0.1\ \Omega\text{–}1.3\ M\Omega.$
Power range:	$2\text{–}225\ W$
Tolerances:	$\pm 5\%, \pm 10\%$
Operating temperature:	-75 to $+350°C$
Maximum rated-power temperature:	$25°C$
Temperature coefficient:	$\pm 0.026\%/°C$
Frequency response:	To 25 kHz for high-resistance values unless noninductively wound

Tantalum solid-electrolyte capacitors have much better stability, frequency, and temperature characteristics than liquid-electrolyte types.

The equivalent circuits for fixed resistors and capacitors are given in Figure 6.24. Figure 6.25 illustrates common capacitor types. Coding is given in Tables 6.6 and 6.7.

6.2.2 Variable Resistors

In electronic equipment variable resistors are generally called *potentiometers*. More specifically, they can be classed as potentiometers, trimmers, and rheostats. All are three-terminal devices with a fixed terminal at each end of the resistive element and the third terminal

Table 6.4 VARIABLE RESISTORS

Resistive-Element Material	R Range	Temperature Coefficient (ppm/°C)	Linearity (%)
Carbon composition	$50\ \Omega\text{–}10\ M\Omega$	± 1000	5
Resistance wire	$1\ \Omega\text{–}100\ K\Omega$	± 10	0.25
Conductive plastic	$100\ \Omega\text{–}5\ M\Omega$	± 100	0.25
Cermet [a]	$100\ \Omega\text{–}5\ M\Omega$	± 100	0.25

[a] Ceramic metal hybrid.

Table 6.5 FIXED CAPACITORS

Air

C range:	< 100 pF
Tolerances:	To 0.01%
Stability:	High
Temperature coefficient:	20 ppm/°C
Dissipation factor:	10^{-5}

Mica and Glass

C range:	1 pF–1 μF
Tolerance:	± 1 to ± 20%
Operating Temperature:	-55 to $+70$, $+85$, $+125$, or $+150$°C depending on construction
Temperature coefficient:	-20 to $+100$ ppm/°C
V range:	100 to 8000 WV d.c.
Dissipation factor:	$(1-7) \times 10^{-4}$

Ceramic Disc

C range:	1 pF–1 μF
Tolerance:	± 1 to ± 20%
V range:	100–7500 WV d.c.
Operating temperature:	-55 to $+150$°C
Temperature coefficient:	0 ppm/°C for NPO to ± 750 ppm/°C for N750
Dissipation factor:	5×10^{-4}

Film

Dielectric:	Polyester	Polycarbonate	Polystyrene
C range (μF):	0.0001–12 μF (metalized)	0.001–50 (metalized)	0.0001–1 (metalized)
V range (WV d.c.):	50–1600	50–600	50–600
Operating temp. (°C):	-55 to $+125$	-55 to $+125$	-55 to $+85$
Temp. coeff.:	$+250$ ppm/°C	Nonmonotonic, total ± 1%	-50 to -100 ppm/°C
Tolerances:	± 1 to ± 20%	± 1 to ± 10%	± 1 to ± 10%
Dissipation factor:	10^{-3}	3×10^{-3}	10^{-4}

Electrolytic (polarized or unpolarized)

Aluminum (general purpose)

C range:	1–50,000 μF
Tolerance:	-10 to $+75$%
V range:	3–450 WV d.c.
Maximum operating temperature:	85°C
Direct current leakage:	High
Low temperature stability:	Low
Internal inductance limits high-frequency performance	
Prone to leaking	

Tantalum (foil and wet slug)

Electrolyte:	Wet	Dry
C range (μF):	15–2200	0.0047–1000
V range (WV d.c.):	3–300	2–150
Tolerance:	± 10% to -15, $+75$%	± 10 to ± 40%

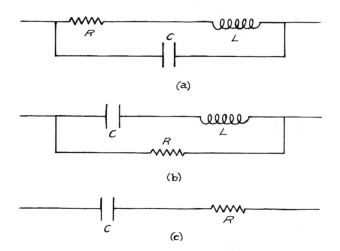

(a)

(b)

(c)

Figure 6.24 Equivalent circuits: (*a*) resistor; (*b*) capacitor (the power factor *D* is X_C/R or $1/\omega RC$ and is usually specified at 10^4 to 10^5 Hz, where $\omega L \ll 1/\omega C$); (*c*) electrolytic capacitor.

attached to the sliding contact or tap. Potentiometers are designed for regular movement, trimmers for occasional changes, and rheostats for current limiting in high-power circuits. Some common types of variable resistors are shown in Figure 6.26. There are a number of different resistive-element materials for potentiometers. They are listed in Table 6.4.

General-purpose, single-turn potentiometers come in power ratings from $\frac{1}{2}$ to 5 watts and a variety of sizes and tapers. (The *taper* is the change in resistance as a function of shaft angle. Nonlinear tapers are often used in volume controls, where an approximately logarithmic response is desired.) Some designs can be stacked on a common shaft, and it is often also possible to incorporate an on-off switch. Shafts with locking nuts are very useful when it is necessary to fix a given setting. Precision potentiometers can be single-turn or 3-, 5-, or 10-turn types. The resistive elements are wire, conductive plastic, or a ceramic-metal substance called *cermet*. The rotating shaft is usually supported in a bushing, which is mounted with a threaded collar. Power ratings are from $\frac{1}{4}$ to 2 watts with a maximum operating temperature of 125°C. Resistance values are from 50 Ω to 200 kΩ, depending on the model, and tolerances are ±1% to ±5%. Linearity is ±0.25% to ±1.0%, and

temperature coefficients are ±20 ppm/°C. For the 10-turn models, turns-counting dials are available.

Trimmers are used to compensate for the tolerance and variation of fixed components. They are not designed for continuous adjustment, 200 cycles being their design life. The resistance elements are carbon composition, cermet, and wire. Usually they are single-turn or have lead screws or worm gears.

6.2.3 Transmission Lines

The most common transmission lines are those with two conductors. A number of common configurations are shown in Figure 6.27. Common VHF flat television antenna lead-in cable is an example of a two-wire transmission line, and the RG/U cables found in almost all laboratories are examples of coaxial transmission line. Such coaxial cable is also used for UHF television leads. The wire-and-plane configuration is not common. However, the ribbon-and-plane configuration occurs very often in printed circuit boards (PCB), where the ribbon is the metal trace on one side of the board and the plane is a continuous conducting sheet on the opposite side,

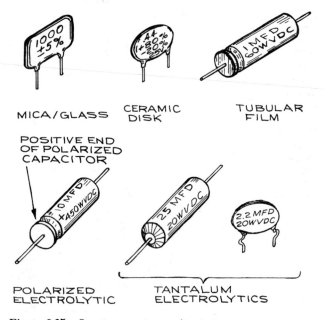

Figure 6.25   Some common capacitor types.

Table 6.6 RESISTOR CODING

Color Codes

For general-purpose industrial resistors:

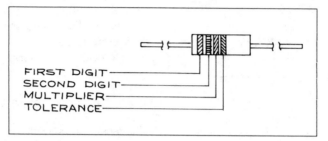

For general-purpose wire-wound resistors:

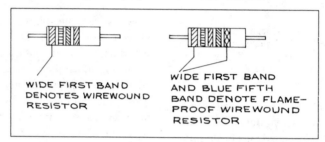

Color	Significant Digit	Multiplier	Tolerance
Black	0	1	—
Brown	1	10	—
Red	2	100	—
Orange	3	1000	—
Yellow	4	10,000	—
Green	5	100,000	—
Blue	6	1,000,000	—
Violet	7	10,000,000	—
Gray	8	—	—
White	9	—	—
Gold	—	—	±5%
Silver	—	—	±10%
No color	—	—	±20%

Example:

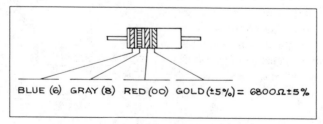

BLUE (6)   GRAY (8)   RED (00)   GOLD (±5%) = 6800 Ω ±5%

Table 6.6 (*continued*)

Number Code for Carbon Composition and Film Resistors

MIL-R-11 Resistor part numbers:

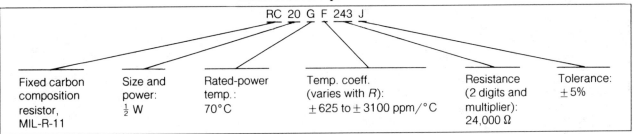

| Fixed carbon composition resistor, MIL-R-11 | Size and power: $\frac{1}{2}$ W | Rated-power temp.: 70°C | Temp. coeff. (varies with R): ± 625 to ± 3100 ppm/°C | Resistance (2 digits and multiplier): 24,000 Ω | Tolerance: $\pm 5\%$ |

MIL-R-22684 Resistor part numbers (all have ± 200-ppm/°C temp. coeff.):

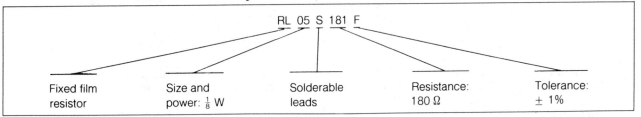

| Fixed film resistor | Size and power: $\frac{1}{8}$ W | Solderable leads | Resistance: 180 Ω | Tolerance: $\pm 1\%$ |

MIL-R-10509 Resistor part numbers:

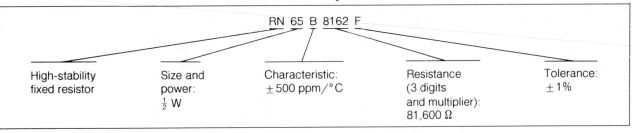

| High-stability fixed resistor | Size and power: $\frac{1}{2}$ W | Characteristic: ± 500 ppm/°C | Resistance (3 digits and multiplier): 81,600 Ω | Tolerance: $\pm 1\%$ |

Regular	Power (W) at 70°C	High Stability	Power (W) at 70°C	Tolerance
05	$\frac{1}{8}$	50	$\frac{1}{10}$	K $\pm 10\%$
07	$\frac{1}{4}$	55	$\frac{1}{8}$	J $\pm 5\%$
20	$\frac{1}{2}$	60	$\frac{1}{4}$	G $\pm 2\%$
32	1	65	$\frac{1}{2}$	F $\pm 1\%$
42	2	70	$\frac{3}{4}$	D $\pm 0.5\%$
		75	1	C $\pm 0.25\%$
		80	2	B $\pm 0.1\%$

Table 6.6 (*continued*)

Number Code for Established-Reliability
Carbon Composition, Film, and Wire-Wound Resistors

MIL-R-39008 Resistor part numbers:

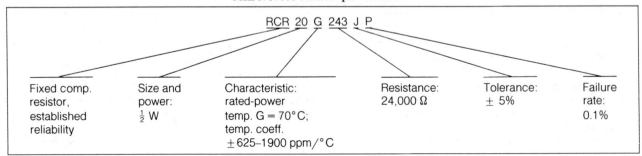

| Fixed comp. resistor, established reliability | Size and power: $\frac{1}{2}$ W | Characteristic: rated-power temp. G = 70°C; temp. coeff. ±625–1900 ppm/°C | Resistance: 24,000 Ω | Tolerance: ± 5% | Failure rate: 0.1% |

MIL-R-39017 Resistor part numbers:

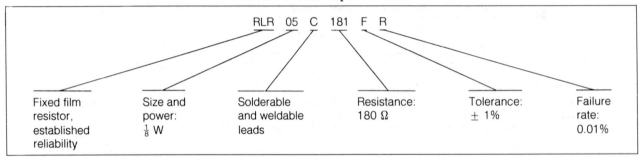

| Fixed film resistor, established reliability | Size and power: $\frac{1}{8}$ W | Solderable and weldable leads | Resistance: 180 Ω | Tolerance: ± 1% | Failure rate: 0.01% |

MIL-R-55182 Resistor part numbers:

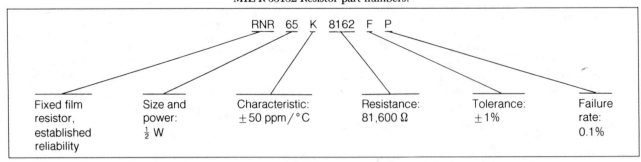

| Fixed film resistor, established reliability | Size and power: $\frac{1}{2}$ W | Characteristic: ±50 ppm/°C | Resistance: 81,600 Ω | Tolerance: ±1% | Failure rate: 0.1% |

Characteristic (RN) (ppm/°C)	*Characteristic (RNR)* (ppm/°C)	*Failure Rate per 1000 Hr* (60% *confidence*)
B ± 500	H ± 50	M 1.0%
C ± 50	J ± 25	P 0.1%
D ± 200 or ± 500	K ± 100	R 0.01%
E ± 25		S 0.001%
F ± 50		

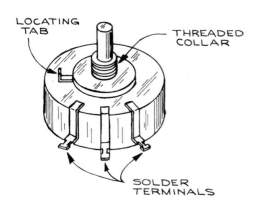

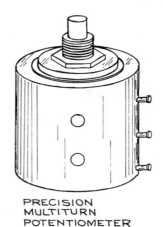

LOCATING TAB

THREADED COLLAR

SOLDER TERMINALS

PRECISION MULTITURN POTENTIOMETER

Figure 6.26 Some common types of variable resistors.

PRINTED CIRCUIT BOARD (PCB) MOUNTING TERMINALS (ON 0.100" CENTERS)

POTENTIOMETERS

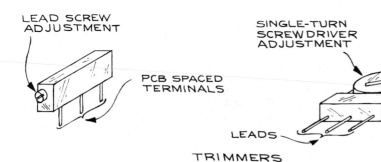

LEAD SCREW ADJUSTMENT

PCB SPACED TERMINALS

SINGLE-TURN SCREWDRIVER ADJUSTMENT

LEADS

TRIMMERS

often called the *ground plane*. A transmission line of this type is called a *microstrip line*. The formulae for the characteristic impedance Z_0 of the geometries in Figure 6.27 are given in Table 6.8.

The properties of a transmission line are given in terms of the series impedance and shunt admittance per unit length:

$$Z = R + j\omega L,$$
$$Y = G + j\omega C,$$

where $j = \sqrt{-1}$ and G is the shunt conductance, which is the reciprocal of the shunt resistance per unit length. The characteristic impedance Z_0 is $\sqrt{Z/Y}$. Voltage and current propagate along a transmission line as waves. The properties of the waves are determined by the propagation constant γ and phase velocity v_p. They are given by $\gamma = \sqrt{ZY} = \alpha + j\beta$ and $v_p = \omega/\beta$, where ω is the frequency of the wave. The wavelength in the transmission line is $\lambda = 2\pi v_p/\omega = 2\pi/\beta$. The attenuation of the wave is given by the factor $e^{-\alpha}$ per unit

Table 6.7 CAPACITOR CODING

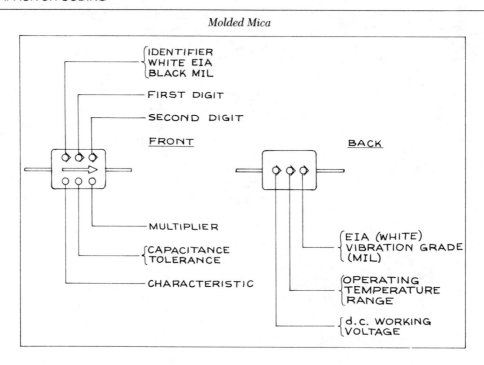

Molded Mica

Color	Significant Digits	Multiplier	Capacitance Tolerance (%)	Charac- teristic	D.c. Working Voltage	Operating Temperature (°C)	EIA/ Vibration
Black	0	1	± 20	—	—	− 55 to +70	10–55 Hz
Brown	1	10	± 1	B	100	—	—
Red	2	100	± 2	C	—	− 55 to +85	—
Orange	3	1,000	—	D	300	—	—
Yellow	4	10,000	—	E	—	− 55 to +125	10–2000 Hz
Green	5	—	± 5	F	500	—	—
Blue	6	—	—	—	—	− 55 to +150	—
Violet	7	—	—	—	—	—	—
Gray	8	—	—	—	—	—	—
White	9	—	—	—	—	—	EIA
Gold	—	—	±0.5	—	1000	—	—
Silver	—	—	±10	—	—	—	—

Equivalences:
10^{-6} F = 1 μF, 1 MFD.
10^{-9} F = 0.001 μF, 1000 pF.
10^{-12} F = 1 pF, 1 MMFD, 1 μμF.

Table 6.7 (*continued*)

Ceramic Capacitors[a]

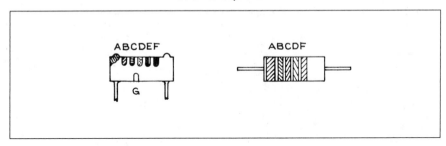

Six-Dot or -Band Code

A ⎫ Temperature coefficient

C ⎫
D ⎬ Capacitance
E ⎭
F   Capacitance tolerance
G   Military code number

Five-Dot or -Band Code

A   Temperature coefficient

C ⎫
D ⎬ Capacitance
E ⎭
F   Capacitance tolerance

Temperature Characteristics

A	B	Temp. Coeff.[b]	A	B	Temp. Coeff.[b]	Ppm/°C
Gray	Black	Gen. purpose	Black	—	NP0	0
Orange	Orange	N 1500	Brown	—	N030	− 30
Yellow	Orange	N 2200	Red	—	N080	− 80
Green	Orange	N 3300	Orange	—	N150	− 150
Blue	Orange	N 4700	Yellow	—	N220	− 220
Red	Violet	P100	Green	—	N330	− 330
Green	Blue	P030	Blue	—	N470	− 470
Gold	Orange	X5F	Violet	—	N750	− 750
Brown	Orange	Z5F	Gold	—	P100	+ 100
Gold	Yellow	X5P	White	—		
Brown	Yellow	Z5P	Gray	—		
Gold	Blue	X5S				
Brown	Blue	Z5S				
Gold	Gray	X5U				
Brown	Gray	Z5U				

[a]Nominal capacitance code is EIA-RS198. MIL-SPEC code not the same. Five- and six-digit codes are both used for radial-lead and axial-lead capacitors. Disc capacitors normally have typographical marking but may be color-coded.
[b]N designates a negative temperature coefficient; P designates a positive temperature coefficient.

Table 6.7 *(continued)*

		Capacitance		
			Nominal Capacitance	
Color	*Digit(C&D)*	*Multiplier(E)*	$\leqslant 10$ pF	> 10 pF
Black	0	1	± 2.0 pF	$\pm 20\%$
Brown	1	10	± 0.1 pF	$\pm\ 1\%$
Red	2	100	—	$\pm\ 2\%$
Orange	3	1,000	—	$\pm\ 3\%$
Yellow	4	10,000		$+100\%, -0\%$
Green	5	—	± 0.5 pF	$\pm\ 5\%$
Blue	6	—	—	—
Violet	7	—	—	—
Gray	8	0.01	± 0.25 pF	$+80\%, -20\%$
White	9	0.1	± 1.0 pF	$\pm 10\%$

length. These formulae are summarized in Table 6.9. For most practical transmission lines, the series resistance per unit length, R, is small compared to the inductive reactance ωL, and the shunt conductance G is small compared to the reciprocal of the capacitance reactance, $1/\omega C$. Under these conditions Z_0 is essentially $\sqrt{L/C}$, α becomes $\frac{1}{2}[R/Z_0 + GZ_0]$, and β becomes $\omega\sqrt{LC}$. The phase velocity is independent of frequency and is the ratio of the speed of light in vacuum to the square root of the dielectric constant of the material separating the conductors. For reasonable values of conductor geometry Z_0 lies approximately between 50 and 500 Ω. The attenuation α is obtained from the shunt conductance G and series resistance R

Figure 6.27 Fixed-geometry transmission-line configurations.

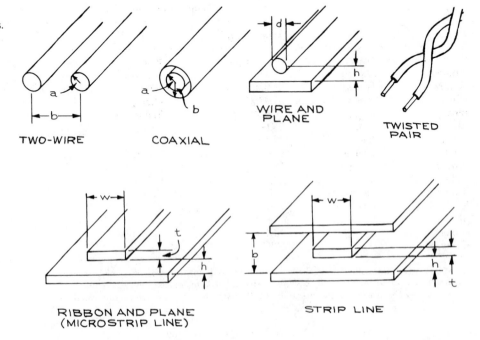

Table 6.8 CHARACTERISTIC IMPEDANCE OF TRANSMISSION-LINE GEOMETRIES

Geometry	Z_0 $(\Omega)^a$
Two-wire	$\dfrac{138}{\sqrt{\varepsilon}} \ln \dfrac{b}{a}$
Coaxial	$\dfrac{276}{\sqrt{\varepsilon}} \ln \dfrac{b}{a}$
Wire and plane	$\dfrac{60}{\sqrt{\varepsilon}} \ln \dfrac{4h}{d}$
Ribbon and plane or microstrip line[b]	$\dfrac{87}{\sqrt{\varepsilon + 1.41}} \ln \dfrac{5.98h}{0.8W + t}$
Twisted pair (AWG 24–28, 30 turns/inch)	110
Strip line[b]	$\dfrac{60}{\sqrt{\varepsilon}} \ln \dfrac{4b}{0.67\pi w(0.8 + t/w)}$

[a] $\varepsilon = 1.0$ for air and 5.0 for G-10 fiberglass epoxy circuit board.
[b] The thickness of the copper foil is 0.001 in. for 1-oz cladding and 0.002 in. for 2-oz cladding.

once Z_0 is known. For most dielectrics G is very small (for air G can be taken equal to zero) and R becomes the dominant factor. Because of the skin effect, R increases with frequency (see Table 6.1). The properties of common coaxial cable are given in Table 6.10. As is to be expected, the smaller cable exhibits greater at-

Table 6.9 GENERAL TRANSMISSION-LINE PROPERTIES

Quantity	Symbol	Relation to Fundamental Parameters
Shunt admittance	Y	$G + j\omega C$
Series impedance	Z	$R + j\omega L$
Characteristic impedance	Z_0	$\sqrt{Z/Y}$
Wavelength	λ	$2\pi/\beta$
Propagation velocity	v_p	$C/\sqrt{\varepsilon}$
Attenuation constant	α	
Propagation constant	γ	$\sqrt{ZY} = \alpha + j\beta$

tenuation due to the necessity of using a small-diameter inner conductor to maintain a reasonable ratio of b/a. The most common cable has a characteristic impedance of 50 Ω. Twisted pairs of single-conductor wire are used where immunity from noise is important, since unwanted signals induced in both conductors can be cancelled electronically by differential amplification.

As a general rule, if a transmission line is used for sending signals over distances greater than $\lambda/8$, at least one end should be terminated with a resistance equal to Z_0 to prevent ringing and undesirable reflections. For example, the transmission of pulses with nanosecond risetimes over distances of a few centimeters on a printed circuit board requires termination. In theory, termination of the source or load end of the line is sufficient; however, in practice it is usually the load end that is terminated. Since 50-Ω cable is so common, electronic instruments designed to operate at high frequencies often have 50-Ω input and output impedances to facilitate connections. Such low-impedance instruments are not suitable for use in combination with lower-frequency devices, which may be seriously overloaded by 50 Ω.

The velocity of propagation v_p of an electrical signal through a transmission line is $1/\sqrt{LC} = Z_0/L = 1/Z_0 C$. For 50-$\Omega$ cable with a polyethylene or Teflon insulation, v_p is of the order of 2×10^{10} m/sec, which gives a delay of 5 ns/m. For delays up to a few hundred nanoseconds, high-quality low-loss coaxial cable provides an excellent delay line, since when it is properly terminated, no distortion of the propagated signal occurs. The best cable to use for critical delay-line applications is rigid or semirigid air-core coaxial cable. With air as the dielectric, the shunt conductance is virtually zero. Semirigid construction assures that the coaxial geometry is maintained within tight tolerances, reducing distortion. Such cable is practical for use at frequencies in the microwave region.

For longer delays, lumped-element delays are the most practical. These delays consist of low-pass filter sections connected in series. For frequencies below their corner frequency they act as delay lines. The number of sections, n, required is related to the ratio of the delay time t_d to the risetime t_r, and is given by $n = 1.5 \, t_d/t_r$. The incorporation of multiple taps between the sections allows a choice of delays with a single unit.

Table 6.10 PROPERTIES OF COAXIAL CABLE

Military RG Number	Armor O.D. (in.)	Jacket O.D. (in.)	Jacket Type[a]	Dielectric O.D. (in.)	Dielectric Type[b]	Center Conductor[c]	V.P.[d] (%)	Capacitance (pF/ft)	Max. RMS Operating Voltage (V)	Nominal Impedance (Ω)
5B	—	0.328	IIa	.181	P	16 S	65.9	28.5	3,000	50
6A/U	—	0.332	IIa	.185	P	21 CW	65.9	20	2,700	75
7A/U	—	0.405	I	.285	P	7/21 C	65.9	29.5	5,000	52
8A/U	—	0.405	IIa	.285	P	7/21 C	65.9	29.5	5,000	52
9	—	0.420	II Grey	.280	P	7/21 S	65.9	30	5,000	51
9A	—	0.420	II Grey	.280	P	7/21 S	65.9	30	5,000	51
9B/U	—	0.420	IIa	.280	P	7/21 S	65.9	30	5,000	50
10A/U	0.475	0.405	IIa	.285	P	7/21 C	65.9	29.5	5,000	52
11A	—	0.405	I	.285	P	7/26 TC	65.9	20.5	5,000	75
11A/U	—	0.405	IIa	.285	P	7/26 TC	65.9	20.5	5,000	75
12A/U	0.475	0.405	IIa	.285	P	7/26 TC	65.9	20.5	5,000	75
13	—	0.420	I	.280	P	7/26 TC	65.9	20.5	5,000	74
13A/U	—	0.420	IIa	.280	P	7/26 T	65.9	20.5	5,000	74
14A/U	—	0.545	IIa	.370	P	10 C	65.9	29.5	7,000	52
17A	—	0.870	IIa	.680	P	.188 C	65.9	29.5	11,000	52
17A/U	—	0.870	IIa	.680	P	.188 C	65.9	29.5	11,000	52
18A/U	0.945	0.870	IIa	.680	P	.188 C	65.9	29.5	11,000	52
19A/U	—	1.120	IIa	.910	P	.250 C	65.9	29.5	14,000	52
20A/U	1.195	1.120	IIa	.910	P	.250 C	65.9	29.5	14,000	52
21A	—	0.332	IIa	.185	P	16 N	65.9	29	2,700	53
22	—	0.405	I	.285	P	Two 7/.0152 C	65.9	16	1,000	95
22B/U	—	0.420	IIa	.285	P	Two 7/.0152 C	65.9	16	1,000	95
34B/U	—	0.630	IIa	.460	P	7/.0249 C	65.9	21.5	6,500	75
35B/U	0.945	0.870	IIa	.680	P	.1045 C	65.9	21	10,000	75
55B/U	—	0.206	IIIa	.116	P	20 S	65.9	28.5	1,900	53.5
57A/U	—	0.625	IIa	.472	P	Two 7/21 C	65.9	16	3,000	95
58/U	—	0.195	I	.116	P	20 C	65.9	28.5	1,900	53.5
58A/U	—	0.195	I	.116	P	19/.0071 TC	65.9	30	1,900	50
58C/U	—	0.195	IIa	.116	P	19/.0071 TC	65.9	30	1,900	50
59/U	—	0.242	I	.146	P	22 CW	65.9	21.5	2,300	73
59B/U	—	0.242	IIa	.146	P	.023 CW	65.9	21	2,300	75
62/U	—	0.242	I	.146	SSP	22 CW	84	13.5	750	93
62A/U	—	0.242	IIa	.146	SSP	22 CW	84	13.5	750	93
62B/U	—	0.242	IIa	.146	SSP	7/32 CW	84	13.5	750	93
63/U	—	0.405	I	.285	SSP	22 CW	84	10	1,000	125
63B/U	—	0.405	IIa	.285	SSP	22 CW	84	10	1,000	125
71A	—	⩽ 0.250	I	.146	SSP	22 CW	84	13.5	750	93
71B/U	—	0.250	IIIa	.146	SSP	22 CW	84	13.5	750	93
74A/U	0.615	0.545	IIa	.370	P	10 C	65.9	29.5	7,000	52
79B/U	0.475	0.405	IIa	.285	SSP	22 CW	84	10	1,000	125

[a] Designation listed at end of this table.
[b] P: polyethylene; SSP: semisolid polyethylene; TF: Teflon; TT: Teflon tape; SST: semisolid Teflon.
[c] Number of strands, gauge (B&S) or O.D., and material are specified. A single solid wire results in the lowest cable attenuation. Stranding increases flexibility. S: silvered copper; C: copper; TC: tinned copper; N: Nichrome; CW: Copperweld; SCW: silvered Copperweld; K: Karma.
[d] 100% × (velocity of propagation)/(velocity in vacuum).

Table 6.10 (*continued*)

Military RG Number	Armor O.D. (in.)	Jacket O.D. (in.)	Jacket Type[a]	Dielectric O.D. (in.)	Dielectric Type[b]	Center Conductor[c]	V.P.[d] (%)	Capacitance (pF/ft)	Max. RMS Operating Voltage (V)	Nominal Impedance (Ω)
87A/U	—	0.425	V	.280	TF	7/20 S	69.5	29.5	5,000	50
108A/U	—	0.235	IIa	.079 ea.	P	Two 7/28 TC	68	23.5	1,000	78
111A/U	0.490	0.420	IIa	.285	P	Two 7/.0152 C	65.9	16	1,000	95
114/U	—	0.405	I	.285	SSP	33 CW	88	6.5	1,000	185
114A/U	—	0.405	IIa	.285	SSP	33 CW	88	6.5	1,000	185
115	—	0.375	V	.250	TT	7/21 S	70	29.5	5,000	50
115A/U	—	0.415	V	.250	TT	7/21 S	70	29.5	4,000	50
116/U	0.475	0.425	V	.280	TF	7/20 S	69.5	29.5	5,000	50
122/U	—	0.160	IIa	.096	P	27/37 TC	65.9	29.5	1,900	50
140/U	—	0.233	V	.146	TF	.025 SCW	69.5	21	2,300	75
141/U	—	0.190	V	.116	TF	.0359 SCW	69.5	28.5	1,900	50
141A/U	—	0.190	V	.116	TF	.039 SCW	69.5	28.5	1,900	50
142/U	—	0.206	V	.116	TF	.0359 SCW	69.5	28.5	1,900	50
142/B	—	0.195	IX	.116	TF	.039 SCW	69.5	28.5	1,900	50
143	—	0.325	V	.185	TF	.057 SCW	69.5	28.5	3,000	50
142A/U	—	0.206	V	.116	TF	.039 SCW	69.5	28.5	1,900	50
143A/U	—	0.325	V	.185	TF	.059 SCW	69.5	28.5	3,000	50
149/U	—	0.405	I	.285	P	7/26 TC	65.9	20.5	5,000	75
164/U	—	0.870	IIa	.680	P	.1045 C	65.9	21	10,000	75
174/U	—	0.100	I	.060	P	7/.0063 CW	65.9	30	1,500	50
178B/U	—	0.075	VII	.034	TF	7/38 SCW	69.5	29.0	1,000	50
179B/U	—	0.105	VII	.063	TF	7/38 SCW	69.5	19.5	1,200	75
180B/U	—	0.145	VII	.102	TF	7/38 SCW	69.5	15.0	1,500	95
187/U	—	0.110	VII	.063	TF	7/38 SCW	69.5	19.5	1,200	75
188/U	—	0.110	VII	.060	TF	7/.0067 SCW	69.5	29	1,200	50
195/U	—	0.155	VII	.102	TF	7/38 SCW	69.5	15	1,500	95
196/U	—	0.080	VII	.034	TF	7/38 SCW	69.5	29	1,000	50
209/U	—	0.745	VI	.500	SST	19/.0378 S	84	26.5	3,200	50
210/U	—	0.242	V	.146	SST	22 SCW	84	13.5	750	93
211A/U	—	0.730	V	.620	TF	.190 C	69.5	29.0	7,000	50
212/U	—	0.332	IIa	.185	P	.0556 S	65.9	28.5	3,000	50
213/U	—	0.405	IIa	.285	P	7/.0296 C	65.9	29.5	5,000	50
214/U	—	0.425	IIa	.285	P	7/.0296 S	65.9	30	5,000	50
215/U	0.475	0.405	IIa	.285	P	7/.0296 C	65.9	29.5	5,000	50
216/U	—	0.425	IIa	.285	P	7/26 TC	65.9	20.5	5,000	75
217/U	—	0.545	IIa	.370	P	.106 C	65.9	29.5	7,000	50
218/U	—	0.870	IIa	.680	P	.195 C	65.9	29.5	11,000	50
219/U	0.945	0.870	IIa	.680	P	.195 C	65.9	29.5	11,000	50
220/U	—	1.120	IIa	.910	P	.260 C	65.9	29.5	14,000	50
221/U	1.195	1.120	IIa	.910	P	.260 C	65.9	29.5	14,000	50
222/U	—	0.322	IIa	.185	P	.0556 N	65.9	29	2,700	50
223/U	—	0.216	IIa	.116	P	.035 S	65.9	29.5	1,900	50
225/U	—	0.430	V	.285	TF	7/.0312 S	69.5	29.5	5,000	50
227/U	0.490	0.430	V	.285	TF	7/.0312 S	69.5	29.5	5,000	50
228A/U	0.795	0.730	V	.620	TF	.190 C	69.5	29.0	7,000	50

Table 6.10 (continued)

Military RG Number	Armor O.D. (in.)	Jacket O.D. (in.)	Jacket Type[a]	Dielectric O.D. (in.)	Dielectric Type[b]	Center Conductor[c]	V.P.[d] (%)	Capacitance (pF/ft)	Max. RMS Operating Voltage (V)	Nominal Impedance (Ω)
264/U	—	0.750	PU	.176	P	Four 19/27 C	69.5	40.0	N.A.	40
280/U	—	0.480	IX	.327	TT	9 C	80	27	3,000	50
281/U	—	0.750	VI	.500	TT	19/.0378 S	80	27	3,200	50
301/U	—	0.245	IX	.185	TF	7/.0203 K	69.5	28.5	3,000	50
302/U	—	0.206	IX	.146	TF	22 SCW	69.5	21	2,300	75
303/U	—	0.170	IX	.116	TF	.038 SCW	69.5	28.5	1,900	50
304/U	—	0.280	IX	.185	TF	.059 SCW	69.5	28.5	3,000	50
307A/U	—	0.270	IIIa	.146	SSP	19/.0058 S	80.0	17	N.A.	75
316/U	—	0.102	IX	.060	TF	7/.0067 SCW	69.5	29	1,200	50

Attenuation Ratings for RG/U Cable

Frequency (MHz):	Nominal Attenuation [dB/(100 ft)]									
	1.0	10	50	100	200	400	1000	3000	5000	10,000
5, 5A, 5B, 6, 6A, 212	0.26	0.83	1.9	2.7	4.1	5.9	9.8	23.0	32.0	56.0
7	0.18	0.64	1.6	2.4	3.5	5.2	9.0	18.0	25.0	43.0
8, 8A, 10, 10A, 213, 215	0.15	0.55	1.3	1.9	2.7	4.1	8.0	16.0	27.0	> 100.0
9, 9A, 9B, 214	0.21	0.66	1.5	2.3	3.3	5.0	8.8	18.0	27.0	45.0
11, 11A, 12, 12A, 13, 13A, 216	0.19	0.66	1.6	2.3	3.3	4.8	7.8	16.5	26.5	> 100.0
14, 14A, 74, 74A, 217, 224	0.12	0.41	1.0	1.4	2.0	3.1	5.5	12.4	19.0	50.0
17, 17A, 18, 18A, 177, 218, 219	0.06	0.24	0.62	0.95	1.5	2.4	4.4	9.5	15.3	> 100.0
19, 19A, 20, 20A, 220, 221	0.04	0.17	0.45	0.69	1.12	1.85	3.6	7.7	11.5	> 100.0
21, 21A, 222	1.5	4.4	9.3	13.0	18.0	26.0	43.0	85.0	> 100.0	> 100.0
22, 22B, 111, 111A	0.24	0.80	2.0	3.0	4.5	6.8	12.0	25.0	> 100.0	> 100.0
29	0.32	1.20	2.95	4.4	6.5	9.6	16.2	30.0	44.0	> 100.0
34, 34A, 34B	0.08	0.32	0.85	1.4	2.1	3.3	5.8	16.0	28.0	> 100.0
35, 35A, 35B, 164	0.08	0.24	0.58	0.85	1.27	1.95	3.5	8.6	15.5	> 100.0
54, 54A	0.33	0.92	2.15	3.2	4.7	6.8	13.0	25.0	37.0	> 100.0
55, 55A, 55B, 223	0.30	1.2	3.2	4.8	7.0	10.0	16.5	30.5	46.0	> 100.0
57, 57A, 130, 131	0.18	0.65	1.6	2.4	3.5	5.4	9.8	21.0	> 100.0	> 100.0
58, 58B	0.33	1.25	3.15	4.6	6.9	10.5	17.5	37.5	60.0	> 100.0
58A, 58C	0.44	1.4	3.3	4.9	7.4	12.0	24.0	54.0	83.0	> 100.0
59, 59A, 59B	0.33	1.1	2.4	3.4	4.9	7.0	12.0	26.5	42.0	> 100.0
62, 62A, 71, 71A, 71B	0.25	0.85	1.9	2.7	3.8	5.3	8.7	18.5	30.0	83.0
62B	0.31	0.90	2.0	2.9	4.2	6.2	11.0	24.0	38.0	92.0
63, 63B, 79, 79B	0.19	0.52	1.1	1.5	2.3	3.4	5.8	12.0	20.5	44.0
87A, 116, 165, 166, 225, 227	0.18	0.60	1.4	2.1	3.0	4.5	7.6	15.0	21.5	36.5
94	0.15	0.60	1.6	2.2	3.3	5.0	7.0	16.0	25.0	60.0
94A, 226	0.15	0.55	1.2	1.7	2.5	3.5	6.6	15.0	23.0	50.0

Table 6.10   (*continued*)

Attenuation Ratings for RG/U Cable (continued)

Frequency (MHz):	Nominal Attenuation [dB/(100 ft)]									
	1.0	10	50	100	200	400	1000	3000	5000	10,000
108, 108A	0.70	2.3	5.2	7.5	11.0	16.0	26.0	54.0	86.0	> 100.0
114, 114A	0.95	1.3	2.1	2.9	4.4	6.7	11.6	26.0	40.0	65.0
115, 115A, 235	0.17	0.60	1.4	2.0	2.9	4.2	7.0	13.0	20.0	33.0
117, 118, 211, 228	0.09	0.24	0.60	0.90	1.35	2.0	3.5	7.5	12.0	37.0
119, 120	0.12	0.43	1.0	1.5	2.2	3.3	5.5	12.0	17.5	54.0
122	0.40	1.7	4.5	7.0	11.0	16.5	29.0	57.0	87.0	> 100.0
125	0.17	0.50	1.1	1.6	2.3	3.5	6.0	13.5	23.0	> 100.0
140, 141, 141A	0.30	0.90	2.1	3.3	4.7	6.9	13.0	26.0	40.0	90.0
142, 142A, 142B	0.34	1.1	2.7	3.9	5.6	8.0	13.5	27.0	39.0	70.0
143, 143A	0.25	0.85	1.9	2.8	4.0	5.8	9.5	18.0	25.5	52.0
144	0.19	0.60	1.3	1.8	2.6	3.9	7.0	14.0	22.0	50.0
149, 150	0.24	0.88	2.3	3.5	5.4	8.5	16.0	38.0	65.0	> 100.0
161, 174	2.3	3.9	6.6	8.9	12.0	17.5	30.0	64.0	99.0	> 100.0
178, 178A, 196	2.6	5.6	10.5	14.0	19.0	28.0	46.0	85.0	> 100.0	> 100.0
179, 179A, 187	3.0	5.3	8.5	10.0	12.5	16.0	24.0	44.0	64.0	> 100.0
180, 180A, 195	2.4	3.3	4.6	5.7	7.6	10.8	17.0	35.0	50.0	88.0
188, 188A	3.1	6.0	9.6	11.4	14.2	16.7	31.0	60.0	82.0	> 100.0
209	0.08	0.27	0.68	1.0	1.6	2.5	4.4	9.5	15.0	48.0
281	0.09	0.32	0.78	1.1	1.7	2.5	4.5	9.0	13.0	24.0

Jacket Type

Designation	Material	Temperature Limits (°C)
Type I	Black polyvinyl chloride	− 40 to +80
Type IIa	Black polyvinyl chloride - (noncontaminating):	
	$< \frac{1}{4}$ in.	− 55 to +80
	$> \frac{1}{4}$ in.	− 40 to +80
Type IIIa	Black polyethylene	− 55 to +85
Type IV	Black synthetic rubber:	
	$< \frac{1}{2}$ in.	− 55 to +80
	$> \frac{1}{2}$ in.	− 40 to +80
Type V	Fiber glass	− 55 to +250
Type VI	Silicone rubber	− 55 to +175
Type VII	Polytetrafluoroethylene (Teflon)	− 55 to +200
Type IX	Fluorinated Ethylene Propylene (FEP)	− 55 to +200
PU	Polyurethane (Estane)	− 60 to +180

Table 6.11   COMMON COAXIAL CONNECTOR TYPES

Size Classification	Connector Type	Coupling Method	Cable Size Range (in.)	Maximum Frequency (GHz)	RMS Working Voltage (V)
Medium	UHF	Screw	$\frac{3}{16} - \frac{7}{8}$	0.5	500
	N	Screw			1000
	C	Bayonet	$\frac{3}{8} - \frac{7}{8}$	11	1500
	SC	Screw			1500
	7-mm precision	Sexless screw	.141, .250, .325, semirigid; RG 214/U, RG 142B/U	18	1500
Miniature	BNC	Bayonet		4	500
	MHV[a]	Bayonet	$\frac{1}{8} - \frac{3}{8}$	4	5000
	TNC	Screw		11	500
	SHV	Bayonet	.080–.420	—	10000 d.c.[b]
Subminiature	SMB	Snap on		3	500
	SMC	Screw	$\frac{1}{16} - \frac{1}{8}$	10	500
	SMA	Screw		18	350
	LEMO[c]	Push on	.100–.195	4	1500

Source: Reference 2.
[a] Higher-voltage version of the BNC connector, not mateable with BNC connectors.
[b] NIM high-voltage connector.
[c] 00 shell size. Manufactured in the United States as K-Loc by Kings Electronics Co.

For a length of line less than a quarter wavelength, a transmission line behaves like a capacitor or an inductor depending on whether the end of the line is an open circuit or short circuit. A line that is an odd number of quarter-wavelengths long behaves like a parallel resonant *RCL* circuit when short-circuited at one end, and like a series resonant circuit when open-circuited. The Q's of such resonant circuits are much higher than that attainable with lumped elements. However, there are an infinite number of resonances, corresponding to frequencies where $v_p/f = (2n + 1)\lambda/4$, where n is an integer.

6.2.4 Coaxial Connectors

Probably more time is spent with connectors than with any other element of electronic hardware. When deciding on the method of connecting various pieces of electronic equipment, it is wise to be aware of all the kinds of connectors available and choose a family of connectors best suited to the task. In many cases the choice of connector is dictated by the connectors at the outputs and inputs of the existing equipment.

The most common connector is currently the BNC. It is a miniature bayonet type and replaced the bulkier RF screw-type connector in the 1960s. With the extensive use of integrated circuits, circuit density has increased enormously and the BNC connector is now in turn being replaced by a variety of subminiature connectors, among which the LEMO type has taken the lead for nuclear instrumentation. A list of common coaxial connectors is given in Table 6.11.

In addition to coaxial connector plugs, each series of connectors has a wide variety of jacks and adaptors (generally *plug* refers to the connector attached to the cable or wire, while *jack* refers to the connector to which the plug mates). Jacks for chassis mounting, bulkhead mounting, and printed-circuit-board mounting are available, as well as hermetically sealed jacks for

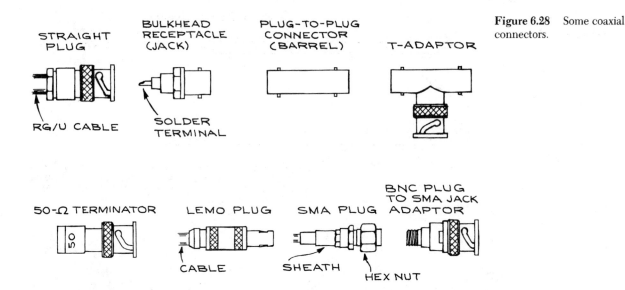

Figure 6.28 Some coaxial connectors.

making coaxial connections through a vacuum wall. *Tees* and *barrels* allow one to connect cables together and there are terminators which can be attached directly to input or output jacks. In addition, each series has a certain number of interseries adaptors that allow connection of the plugs or jacks of one series to those of another. Some of these different connectors are shown in Figure 6.28. It is advisable to reduce the number of adaptors to a minimum to maintain the transmission-line properties of the connections with as few reflection-producing discontinuities as possible.

A good connector should be easy to fit to the proper cable and should not introduce any discontinuities in the transmission-line properties of the cable. Needless to say, the characteristic impedance of the connector should be identical to that of the cable. The connector should be a strong mechanical fit to the cable, and it should be easy to connect and disconnect the connector from mating connectors.

Any given connector type comes in a variety of sizes to accommodate the different cable sizes. Different styles, such as straight and right-angle, are available, and often there are different choices for the mechanical and electrical connections of the connector to the cable, the three most common being a ferrule clamp and screw, crimp, and solder. The procedure for stripping coaxial cable and assembling a ferrule-clamp BNC/MHV con-

nector is shown in Figure 6.29. The crimp style requires the use of a special crimping tool but is stronger, less bulky, and electrically sounder than the clamp model. The solder type has the disadvantage that in soldering the shield the dielectric can be swelled or melted. When attaching connectors to cable it is highly recommended that the manufacturer's cable cutting procedures be followed, so that the inner conductor, insulator, outer conductor (braid or foil), and outer insulator are of the proper length. In this way no discontinuities in the electrical properties of the cable are introduced at the cable-connector interface. Special cutting jigs are available to assure the proper geometry when preparing the cable for insertion into the connector.

6.2.5 Relays

A relay is an electrically actuated switch. Relay action can be electromechanical or can be based on solid-state devices. A large number of contact configurations exist. Important parameters are the operating voltage and power of the relay and the contact rating. The speed at which mechanical relays operate is generally in the 10- to 100-ms range. General-purpose electromagnetic relays operate at 6, 12, 24, 48, and 110 V d.c. and 6, 12, 24, 48, 110, and 220 V a.c. with a coil power of around

Figure 6.29(a) Procedure for stripping coaxial cables.

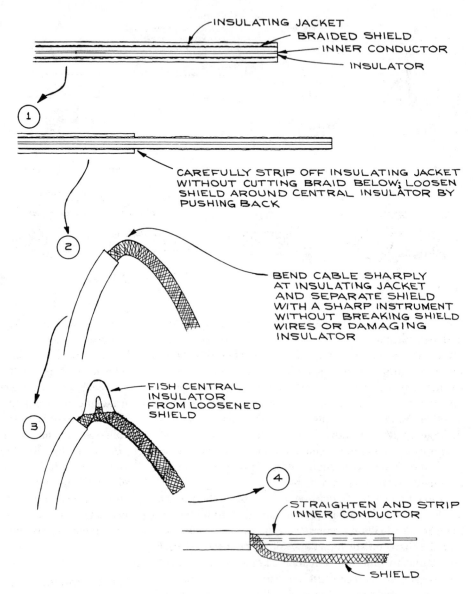

INSULATING JACKET
BRAIDED SHIELD
INNER CONDUCTOR
INSULATOR

① CAREFULLY STRIP OFF INSULATING JACKET WITHOUT CUTTING BRAID BELOW; LOOSEN SHIELD AROUND CENTRAL INSULATOR BY PUSHING BACK

② BEND CABLE SHARPLY AT INSULATING JACKET AND SEPARATE SHIELD WITH A SHARP INSTRUMENT WITHOUT BREAKING SHIELD WIRES OR DAMAGING INSULATOR

③ FISH CENTRAL INSULATOR FROM LOOSENED SHIELD

④ STRAIGHTEN AND STRIP INNER CONDUCTOR

SHIELD

1 watt [see Figure 6.30(a)]. Such relays have contacts capable of handling from 2 to 10 A. Sensitive mechanical relays use a large number of turns on the electromagnetic coil. They operate with 28 V d.c. and require only 1 to 40 mW of coil power. Contacts are rated from 0.5 to 2 A.

A *reed relay* is a glass-encapsulated switch having two flexible thin metal strips, or reeds, as contactors. The switch is placed inside an electromagnetic coil [Figure 6.30(b)]. The relay contacts can be dry or mercury-wetted. Standard coil voltages are 6, 12, and 24 V d.c., and the coil power is from 50 to 500 mW per

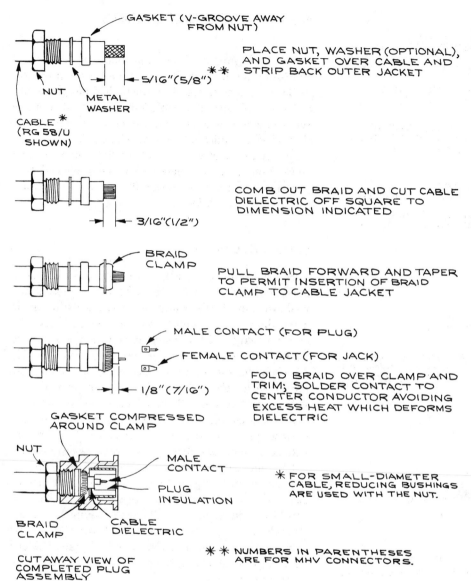

GASKET (V-GROOVE AWAY FROM NUT)

PLACE NUT, WASHER (OPTIONAL), AND GASKET OVER CABLE AND STRIP BACK OUTER JACKET **

5/16″ (5/8″)

NUT

METAL WASHER

CABLE * (RG 58/U SHOWN)

COMB OUT BRAID AND CUT CABLE DIELECTRIC OFF SQUARE TO DIMENSION INDICATED

3/16″ (1/2″)

BRAID CLAMP

PULL BRAID FORWARD AND TAPER TO PERMIT INSERTION OF BRAID CLAMP TO CABLE JACKET

MALE CONTACT (FOR PLUG)

FEMALE CONTACT (FOR JACK)

FOLD BRAID OVER CLAMP AND TRIM; SOLDER CONTACT TO CENTER CONDUCTOR AVOIDING EXCESS HEAT WHICH DEFORMS DIELECTRIC

1/8″ (7/16″)

GASKET COMPRESSED AROUND CLAMP

NUT

MALE CONTACT

PLUG INSULATION

* FOR SMALL-DIAMETER CABLE, REDUCING BUSHINGS ARE USED WITH THE NUT.

BRAID CLAMP

CABLE DIELECTRIC

CUTAWAY VIEW OF COMPLETED PLUG ASSEMBLY

** NUMBERS IN PARENTHESES ARE FOR MHV CONNECTORS.

Figure 6.29(b) Assembly procedure for BNC/MHV ferrule-clamp connector.

reed-switch capsule. Contacts are rated at 10 mA to 3 A, depending on size. There is also a maximum open-circuit voltage that can be sustained by the opened contacts. The low mass of the reeds makes operating times of 0.2 to 2 ms possible. Small reed relays are made for PCB mounting.

With inductive loads, the fast opening of relay contacts can cause a large voltage to appear across them. Some protective circuits for reducing this problem are illustrated in Figure 6.31.

Solid-state relays consist of a solid-state switching element driven by an appropriate amplifier. The load

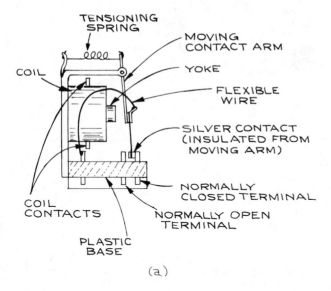

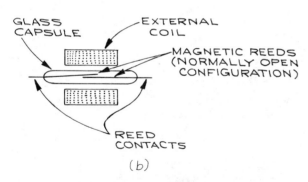

Figure 6.30 Mechanical relays: (*a*) medium-power relay; (*b*) reed relay.

current determines the switching element. Field-effect transistors (FETs) are used for low-level d.c., bipolar junction transistors (BJTs) for intermediate d.c. currents, and triacs and SCRs for large a.c. and d.c. currents. Section 6.3 contains a discussion of transistors and solid-state switches. Though there is no contact wear with solid-state relays, they have high "on" resistance compared with electromechanical devices and require carefully conditioned activating signals. They are much less resistant to overload than are electromechanical devices.

6.3 ACTIVE COMPONENTS

The overall properties of semiconductor diodes and transistors can be understood in terms of the properties of the *p-n* junctions that form them. Without going into solid-state physics, it is sufficient to know that *p*-type semiconductor material conducts electricity principally through the motion of positive charges (holes), while *n*-type material owes its conductivity mainly to electrons. Both *p*- and *n*-material can be produced from germanium or silicon by the addition of impurity atoms, a procedure called *doping*.

6.3.1 Diodes

Most semiconductor diodes consist of a *p-n* junction as illustrated in Figure 6.32. The schematic representation is shown next to the simplified drawing of the diode. The diode is a unidirectional device: its d.c. current-voltage (*I-V*) characteristics depend on the polarity of the *anode* with respect to the *cathode*. The *I-V* characteristics of a typical small-signal silicon diode are drawn in Figure 6.33. When the anode is positive with respect to the cathode (*forward bias*), the diode conducts a large current. When the polarity is reversed (*reverse bias*), the current is very small. If the reverse voltage is increased beyond the *peak reverse voltage* (PRV), breakdown occurs and the diode conducts strongly again. Except in the case of specially designed Zener diodes, this usually destroys the diode. The forward bias voltage at which the diode begins to conduct strongly is V_{cutin}. For germanium diodes V_{cutin} is 0.2 volts, while for silicon diodes it is 0.6 volts. The *I-V* characteristics shown in the graph are accurately represented by the formula

$$I = I_0 \left[e^{V/\eta V_T} - 1 \right],$$

where $I_0 = 10^{-9}$ A for silicon and 10^{-6} A for germanium. I_0 is temperature-dependent, increasing at a rate of 11%/°C for Ge and 8%/°C for Si. V_T has the value 0.026 volts at 300°K and is directly proportional to the absolute temperature, while η is a unitless constant equal to 1 for germanium and 2 for silicon.

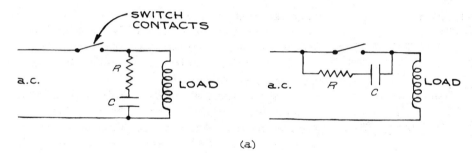

SWITCH CONTACTS

(a)

Figure 6.31 Contact-conditioning circuits for mechanical relays: (a) a.c. (for small loads driven from the power line, $R = 100\ \Omega$ and $C = 0.05\ \mu$F); (b) d.c. (the diodes D reduce the peak voltage from the inductive load; they should be able to handle the steady-state current through the inductor).

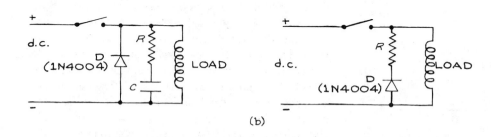

(b)

Diodes can be separated into two types according to the power they can handle. *Power diodes* are used in high-current applications, for example, as rectifiers in power supplies and as switches. Important specifications for power diodes are the *maximum forward current*, maximum PRV, and *effective forward bias resistance*. Low-power diodes or *signal diodes* are used to rectify small signals, mix two frequencies to produce sum and difference frequencies, and switch low voltages and currents. Besides the current and voltage ratings, the speed with which diodes can be changed from the forward-bias to the reverse-bias condition and back is

important. This is related to the effective capacitance of the *p-n* junction. Typical specifications for a power and signal diode are given in Tables 6.12 and 6.13.

Most diodes have a 1N-prefix designation. The two ends of a diode are usually distinguished from each other by a mark—the cathode end having a black band

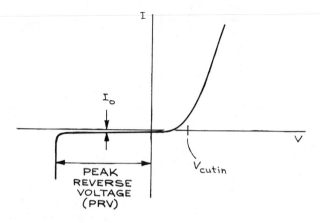

Figure 6.33 Typical current-voltage characteristic of a small-signal silicon diode.

Figure 6.32 A diode and the corresponding symbol.

Table 6.12 PROPERTIES OF THE 1N914 FAST-SWITCHING DIODE

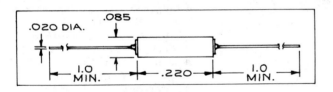

Maximum Ratings

Peak reverse voltage	75 V
Reverse current	25 nA
Average forward current	75 mA
Peak surge current (1 sec)	500 mA

Electrical Characteristics

Junction capacitance	4 pF
Reverse recovery time	4–8 ns

for glass-encapsulated diodes and a white band for black-plastic-encapsulated diodes. Power diodes can dissipate large amounts of heat and are often constructed to facilitate mounting to a heat sink for heat dissipation. Some diode configurations are shown in Figure 6.34. In the case of the high-power diode, the cathode is a stud which can be attached directly to a metal heat sink.

Generally, fiber or mica washers are used to isolate the cathode electrically from the metal heat sink, and thermal conductivity is enhanced with special silicone greases between the washers and heat sink. The mounting technique is shown in Figure 6.35.

For the purposes of circuit analysis, it is useful to simplify the *I-V* characteristic curve of a diode by a so-called *piecewise linear model*. With this model the forward- and reverse-bias regions are represented by straight lines with the transition occurring at the cutin voltage. The piecewise linear curve is shown in Figure 6.36. The slope of the curve at voltages above V_{cutin} is the reciprocal of the average forward resistance. With this model it is easy to calculate the effect of the diode in a circuit, since in the forward-bias condition it behaves like a resistor with resistance R_f, and in the reverse bias condition it acts essentially like an open circuit. A graphical representation of this can be obtained by what is called *load-line analysis*. Consider the circuit with an a.c. voltage source shown in Figure 6.37. The current through R_L can be determined in the following manner: given a value of v_S the voltage across the diode and the current through it will be determined by the intersection of the straight line with coordinates $(v_S, 0)$ and $(0, v_S/R_L)$, as in Figure 6.38. Different values of v_S will give a family of parallel straight lines. From the intersection points, one constructs a graph of v_L as a function of v_S. This *dynamic curve*, which is

Table 6.13 PROPERTIES OF 1N4001–1N4007 1-AMPERE SILICON RECTIFIERS

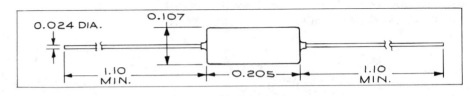

Maximum Ratings

	4001	4002	4003	4004	4005	4006	4007
Peak reverse voltage (V)	50	100	200	400	600	800	1000
Continuous reverse voltage (V)	50	100	200	400	600	800	1000
Average forward current (A)				1			
Peak surge current (1 cycle) (A)				30			
Operating temperature range (°C)				− 65 to + 175			

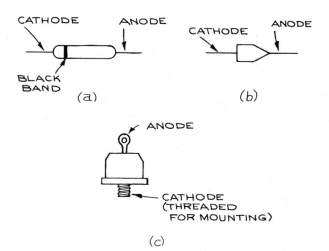

Figure 6.34   Diode case configurations: (*a*) glass-encapsulated signal diode; (*b*) plastic-encapsulated medium-power diode; (*c*) high-power diode.

shown in Figure 6.39, has much the same appearance as the original *static curve* from which it is derived. If the input waveform is drawn below the horizontal axis, the reflection of it on the dynamic curve will give the output waveform as shown.

6.3.2 Transistors

Transistors are three-terminal devices. Their *I-V* characteristics are therefore more complicated than for a two-terminal device. Of the three terminals one will be

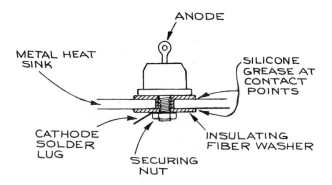

Figure 6.35   Proper power-diode mounting to provide electrical insulation with good heat dissipation.

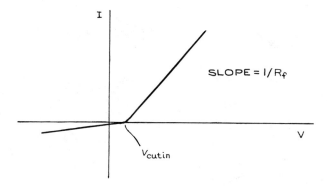

Figure 6.36   Linearized current-voltage characteristic of a diode.

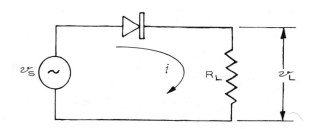

Figure 6.37   Circuit for load-line analysis of a diode.

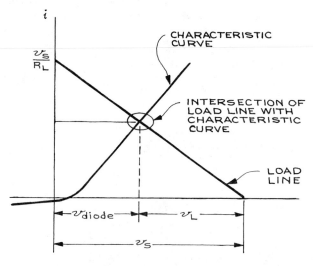

Figure 6.38   Graphical construction for determining the dynamic transfer curve from the load line and the static characteristic curve.

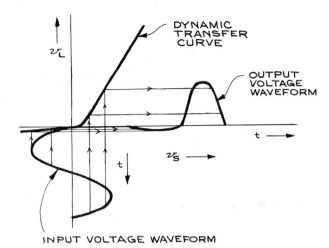

Figure 6.39 Graphical construction of output waveform from input waveform and the dynamic transfer curve.

an input, one will be an output, and the third will be common to the input and output circuits. *Bipolar junction transistors* (BJTs) consist of an *emitter*, *base*, and *collector*. They are designated *npn* or *pnp*, depending on the material used for the elements. As can be seen from Figure 6.40, a transistor is two back-to-back diodes with the base element in common. The major current flow in a transistor is from the emitter to the collector across the base. Relatively little current flows through the base lead. The arrow on the emitter shows the direction of the current flow.

Some common case styles are shown in Figure 6.41. To identify the leads of UHF transistors, manufacturers'

base diagrams should be consulted. With plastic cases, there is a flat on one side of the case to help identify the leads. Leads are sometimes identified by letters, or are in the order emitter, base, and collector starting at the flat or tab of sealed metal cans and proceeding clockwise as viewed from the lead side of the transistors. In power transistors, since most of the power is dissipated in the collector, the collector contact is the one with provision for attachment to a heat sink. Standard transistor types have code designations beginning with 2N. Several manufacturers often make the same transistors. Proprietary code designations are also used by manufacturers to identify their products.

Since transistors consist of two *p-n* junctions, a simple ohmmeter test is a good way to check them, once the transistor is isolated from the circuit. The ohmmeter must provide at least 0.6 volts, enough to forward-bias a silicon *p-n* junction. The resistance between the emitter and base should be low, of the order of a few hundred ohms or less, when the ohmmeter leads are attached so as to forward-bias the junction. When the leads are reversed, the resistance should be of the order of several megohms. The same should be true of the collector-base junction. The polarity of the ohmmeter leads can be checked with a separate voltmeter.

In normal operation the emitter-base junction of a transistor is forward-biased, while the base-collector junction is reverse-biased. The three most common configurations, *common emitter* (CE), *common base* (CB), and *common collector* (CC), are shown schematically in Figure 6.42. The *I-V* characteristics of each configuration consist of two sets of curves, one for the input and

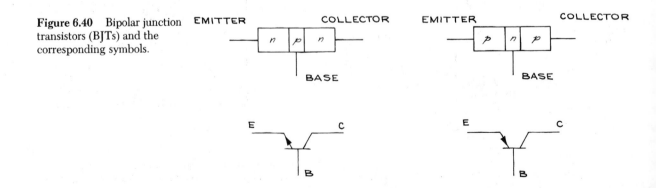

Figure 6.40 Bipolar junction transistors (BJTs) and the corresponding symbols.

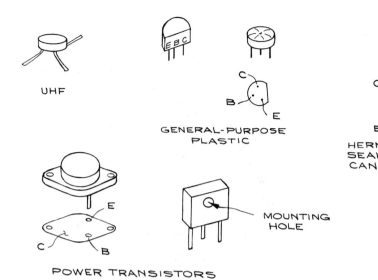

UHF

GENERAL-PURPOSE
PLASTIC

HERMETICALLY
SEALED METAL
CAN

MOUNTING
HOLE

POWER TRANSISTORS

Figure 6.41 Transistor case styles.

the other for the output. For example, the *I-V* curves for a typical general-purpose transistor in the CE configuration are shown in Figure 6.43. There are a set of curves rather than a single one for the input and output circuits because the two circuits are not isolated from each other. When measuring the *I-V* characteristics of one circuit, the condition of the other one must also therefore be specified. Figure 6.44 illustrates a circuit for establishing the conditions for which the curves can be obtained. Here the lowercase letters apply to instantaneous values of current and voltage. The uppercase subscript indicates that it is the total value of the current or voltage that is under consideration. As v_S increases, i_B increases as shown in the input curves, and i_C increases in accordance with output curves. This can be made more quantitative by using load-line analysis. Two points are sufficient to establish the load line on the output curves; they are points corresponding to $I_C = 0$ where $v_{CE} = V_{CC}$ and to $v_{CE} = 0$ where $I_C =$

V_{CC}/R_L. The output curves with a superimposed load line are shown in Figure 6.45. The intersections of the load line with the characteristic curves establish the operating voltages and currents. It is possible to predict quantitatively how the current through R_L varies with input base current from these curves. For linear operation, conditions are adjusted so that the transistor operates somewhere in the middle of the characteristic output curves, where they are equally spaced and parallel for equal base-current increments.

Another way of analyzing transistor operation is by constructing an equivalent circuit with passive elements and sources. Such a circuit is then amenable to analysis by standard methods. Because it contains only linear elements, it can represent the properties of the transistor only within a small region of the characteristic curves about given nominal values of the d.c., steady-state or quiescent voltages and currents. Four variables are sufficient to characterize a given configuration—in-

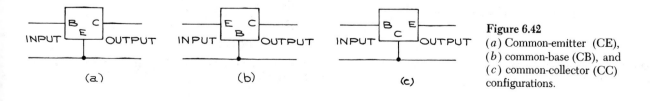

INPUT   OUTPUT   INPUT   OUTPUT   INPUT   OUTPUT

(a)   (b)   (c)

Figure 6.42
(*a*) Common-emitter (CE),
(*b*) common-base (CB), and
(*c*) common-collector (CC)
configurations.

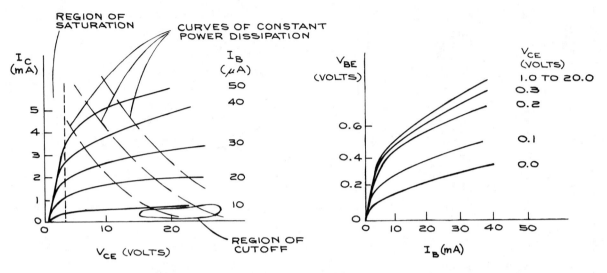

Figure 6.43   Current-voltage characteristics of a transistor in the common-emitter (CE) configuration.

put voltage, input current, output voltage, and output current. The normal convention is to take input voltage and output current as the dependent variables, and input current and output voltage as the independent variables. Mathematically this is expressed as

$$v_{\text{INPUT}} = v_{\text{INPUT}}(i_{\text{INPUT}}, v_{\text{OUTPUT}},)$$
$$i_{\text{OUTPUT}} = i_{\text{OUTPUT}}(i_{\text{INPUT}}, v_{\text{OUTPUT}}),$$

where the subscripts are in uppercase letters to indicate total instantaneous values. For the CE configuration these equations take the form

$$v_{BE} = v_{BE}(i_B, v_{CE}),$$
$$i_C = i_C(i_B, v_{CE}).$$

Expanding v_{BE} and i_C about the quiescent values I_B and

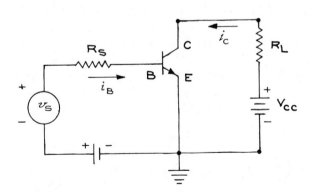

Figure 6.44   Biasing circuit for the common-emitter (CE) configuration.

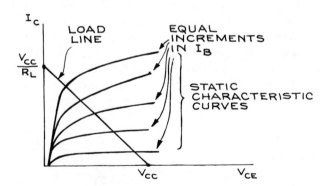

Figure 6.45   Relation between the static characteristic curves and the load line.

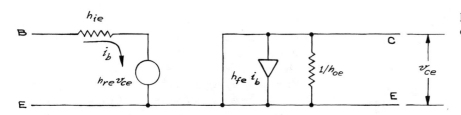

Figure 6.46 Hybrid-parameter equivalent circuit.

V_{CE} gives

$$\Delta v_{BE} = \left.\frac{\partial v_{BE}}{\partial i_B}\right|_{V_{CE}} \Delta i_B + \left.\frac{\partial v_{BE}}{\partial v_{CE}}\right|_{I_B} \Delta v_{CE},$$

$$\Delta i_C = \left.\frac{\partial i_C}{\partial i_B}\right|_{V_{CE}} \Delta i_B + \left.\frac{\partial i_C}{\partial v_{CE}}\right|_{I_B} \Delta v_{CE}.$$

The Δ's are now the small differences between the total instantaneous values and the d.c. quiescent values. By convention, the partial derivatives are represented by lowercase h's:

$$h_{ie} = \left.\frac{\partial v_{BE}}{\partial i_B}\right|_{V_{CE}},$$

$$h_{re} = \left.\frac{\partial v_{BE}}{\partial v_{CE}}\right|_{I_B},$$

$$h_{fe} = \left.\frac{\partial i_C}{\partial i_B}\right|_{V_{CE}},$$

$$h_{oe} = \left.\frac{\partial i_C}{\partial v_{CE}}\right|_{I_B},$$

so that

$$v_{be} = h_{ie}i_b + h_{re}v_{ce},$$
$$i_c = h_{fe}i_b + h_{oe}v_{ce}.$$

The h's are called small-signal *hybrid parameters*. The equations for v_{be} and i_c can be translated into the low-frequency equivalent circuit in Figure 6.46. It is necessary to keep in mind that this linear circuit is a good representation of the transistor only for small excursions of the input and output voltages and currents about specified quiescent d.c. values. The equivalent circuit does not show the biasing network necessary to establish the quiescent operating point. Practical circuits using separate power supplies for biasing voltages are shown in Figure 6.47. The output signal is developed across the load resistor R_L, while capacitors C_i and C_o isolate the input and output signals from the biasing levels. Transistor data sheets give the values of the hybrid parameters for different operating points. The hybrid parameters for the CB and CC configurations can be calculated from those of the CE with standard formulae.[3] The second letter in the subscript of the h's indicates whether the configuration is common *e*mitter, common *b*ase, or common *c*ollector. Lowercase subscripts indicate small-signal values while uppercase subscripts are for d.c., large-signal conditions. By applying the standard methods of circuit analysis to the equivalent circuit, the current, voltage, and power gains can be calculated, as well as the input and output impedances under small-signal conditions. The general properties of the three configurations are given in the graphs of Figure 6.48. Because the charge carriers in the *p*- and *n*-materials that make up the transistor cannot respond instantaneously to changes in voltage across the junctions, the hybrid parameter model is strictly valid only at low frequencies. The response of the transistor at high frequencies can be very different, and equivalent-circuit models exist for the high-frequency response.

If the emitter-base junction of a transistor is strongly forward-biased (0.8 V for silicon), the current through the transistor will be determined by the external resistance in the circuit. In other words, the resistance between emitter and collector will be low, and the voltage between emitter and collector will be of the order of 0.2 V. In this condition the transistor is said to be *saturated* or *on*, and the base current must be sufficient to sustain the current flow from emitter to

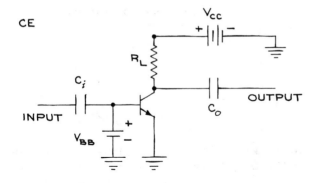

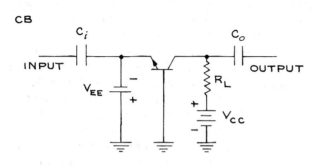

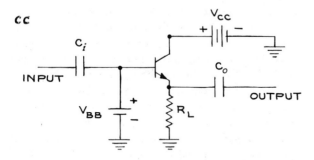

Figure 6.47 Practical transistor circuits using separate power supplies for biasing voltages.

frequencies, unlike mechanical switches, but the saturated resistance is never zero, and there is always at least a 0.2-V drop from the emitter to the collector.

The operating conditions for bipolar junction transistors are summarized in Table 6.14. It is easy to see from the table how to determine whether a transistor is operating in the linear region (*active*), is saturated, or cut off. When reading Tables 6.14 and 6.15, which give the specifications for a 2N3904 transistor, the following conventions are to be noted:

1. Voltages between two terminals are specified by either V or v with two or three subscripts. The first two subscripts indicate the terminals between which the voltage occurs, and the third subscript, when present, indicates the condition of the third terminal. Thus V_{CBO} indicates a d.c. voltage between collector and base with the emitter open. Subscripts with repeated letters indicate power-supply voltages. For example, V_{CC} is the power-supply voltage to the collector, and V_{BB} to the base.

2. By convention, all currents are taken to be positive when they flow into the leads of the transistor and negative when they flow out of the leads.

3. Currents are specified by I or i with a subscript indicating the terminal involved. I_B is therefore a d.c. base current.

Transistors are destroyed if their maximum power dissipation is exceeded, or if, when reverse-biasing the junctions, the maximum reverse voltages are exceeded, resulting in electrical breakdown. The maximum power dissipation depends on the temperature, and a transistor on a heat sink can dissipate considerably more power than one standing alone in free air. In pulse operation the maximum steady-state power levels can be momentarily exceeded without damage.

6.3.3 Field-Effect Transistors

The simplest *field-effect transistor* (FET) is the *junction FET* or JFET. It has three terminals—a *source*, a *gate*, and a *drain*. An *n*-channel device is shown in Figure

collector. If the emitter-base junction is reverse-biased, no base current will flow and no collector current will flow. Under this condition the transistor acts as an open circuit and is said to be *cut off*. In the saturated or the cut-off condition a transistor is similar to a closed or an open switch. Transistor switches can operate at high

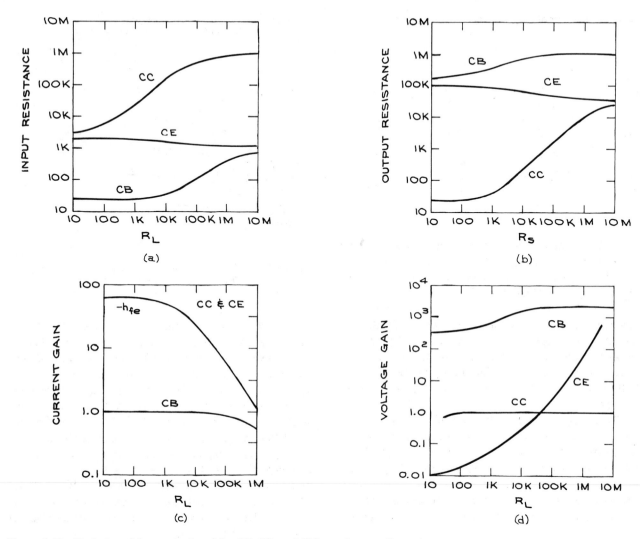

Figure 6.48 Variation of the properties of the CE, CB, and CC transistor configurations with various input parameters: (a) input resistance as a function of load resistance for a typical transistor; (b) output resistance as a function of source resistance; (c) current gain as a function of load resistance for a typical transistor; (d) voltage gain as a function of load resistance for a typical transistor.

6.49. Current flow is from source to drain with the gate reverse biased with respect to the channel. Because of this reverse bias, negligible current flows into the gate. The *I-V* characteristics of a small-signal JFET are shown in Figure 6.50. From the curves it can be seen that there is a region where the drain current decreases linearly with increasing reverse bias from the gate to source, independent of the drain-source voltage. This is the region where the transistor has a linear response. About the region $V_{DS} = 0$, I_D increases linearly with V_{DS} for

Table 6.14 2N3904 TRANSISTOR PARAMETERS

| Parameter | Condition of Transistor | | |
	Active	Saturated	Cut Off
V_{EB}	0.6 V	0.8 V	< 0.0 V
V_{BC}	5–10 V	0.6 V	$\simeq V_{CC}$
	(reverse bias)	(forward bias)	(reverse bias)
V_{EC}	5–10 V	0.2 V	V_{CC}
I_B	$\simeq \dfrac{I_C}{h_{FE}}$	$\dfrac{I_C}{h_{FE}}$	$\simeq 0$
I_C	$\simeq \dfrac{V_{CC} - V_{EC}}{R_L}$	$\dfrac{V_{CC}}{R_L}$	$\simeq 0$

fixed V_{GS}, with the slope equal to the reciprocal of the channel resistance R_{channel}. As V_{GS} becomes more negative, R_{channel} increases. Transistor operation in this region is that of a *voltage-controlled resistor*.

The FET can also be used as a switch, with the advantage of no offset voltage from source to drain and almost perfect electrical isolation of the gate signal from the output circuit. Disadvantages are rather high "on" resistances (of the order of hundreds of ohms) and low "off" resistances. Also, switching times are greater than for BJTs.

The simple structure of FETs results in their having low noise. As a consequence they are often used as the active element in amplifiers for low-level signals. FETs with gates insulated by an oxide layer from the conducting channel are called *MOSFETs* (metal-oxide-semiconductor FETs). The oxide gate insulation effectively prevents any gate current flow. These FETs can operate

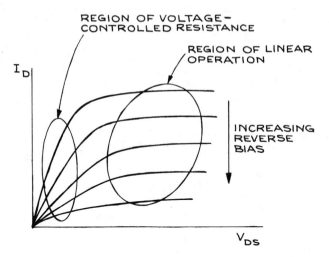

Figure 6.50 Current-voltage characteristics of a small-signal, junction-field-effect transistor (JFET).

in a *depletion mode*, where an effective reverse bias appears at the surface of the channel, or in an *enhancement mode*, where there is a forward bias. In addition, the channel can be of *p*- or *n*-material. The low power consumption, ease of fabrication, and high packing densities have made these devices extremely common in *large-scale integrated circuits* (LSI). Because of the gate insulation, static charge can accumulate on the gate and voltage levels sufficient to punch through the insulation attained. To avoid this, some MOSFET gates are protected with reverse-biased Zener diodes. MOSFET devices are normally packaged in conducting foam or with metal protecting rings, which are removed just prior to insertion into the circuit.

6.3.4 Silicon Controlled Rectifiers

The *silicon controlled rectifier* (SCR) is a switching device with only two states, on and off. In the on state the SCR has a low resistance, and some models are capable of passing over 100 A. In the off state the resistance is in the megohm range. The SCR has three terminals: an anode, cathode, and gate. The transition from off to on occurs when the voltage on the anode is positive with respect to the cathode and a pulse of the

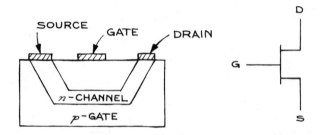

Figure 6.49 Field-effect transistor and the corresponding symbol.

Table 6.15   PROPERTIES OF THE 2N3904 N-P-N SILICON TRANSISTOR

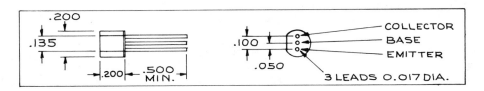

Parameter	Value	Conditions
V_{CBO}	60 V	$I_C = 10\ \mu\text{A}, I_E = 0$
V_{CEO}	40 V	$I_C = 1\ \text{mA}, I_B = 0$
V_{EBO}	6 V	$I_E = 10\ \mu\text{A}, I_C = 0$
h_{FE}	40	$V_{CE} = 1\ \text{V}, I_C = 100\ \mu\text{A}$
	70	$V_{CE} = 1\ \text{V}, I_C = 1\ \text{mA}$
V_{BE}	0.65–0.85 V	$I_B = 1\ \text{mA}, I_C = 10\ \text{mA}$
	0.95 V	$I_B = 5\ \text{mA}, I_C = 50\ \text{mA}$
V_{CE}[a]	0.2	$I_B = 1\ \text{mA}, I_C = 10\ \text{mA}$
	0.3	$I_B = 5\ \text{mA}, I_C = 50\ \text{mA}$
h_{ie}	$1\text{--}10 \times 10^4\ \Omega$	
h_{fe}	100–400	$V_{CE} = 10\ \text{V}, I_C = 1\ \text{mA},$
h_{re}	$(0.5\text{--}8 \times 10^{-4})$	$f = 1\ \text{Hz}$
h_{oe}	$1\text{--}40\ \mu\text{mho}$	
f_T[b]	300 MHz	$V_{CE} = 20\ \text{V}, I_C = 10\ \text{mA}$
C_{obo}[c]	4 pF	$V_{CB} = 5\ \text{V}, I_E = 0$
C_{ibo}[d]	8 pF	$V_{EB} = 0.5\ \text{V}, I_C = 0$

Maximum Ratings at 25°C in Free Air:
$$V_{CB} = 60\ \text{V}, \qquad V_{CE} = 40\ \text{V}, \qquad V_{EB} = 6\ \text{V},$$
$$I_C\ (\text{continuous}) = 200\ \text{mA}$$

Power Dissipation in Free Air:
310 mW at 25°

[a] Saturation.
[b] Transition frequency where $|h_{fe}| = 1$.
[c] Common-base open-circuit output capacitance.
[d] Common-base open-circuit input capacitance.

correct polarity and magnitude is applied to the gate. Once the SCR is on, the gate loses all control over the functioning of the device. The only way to turn it off is to reduce the anode-cathode voltage to zero. If a sinusoidal voltage is applied to the SCR anode, the device will be turned off once each half cycle. By controlling the point in each positive half cycle when the trigger pulse is applied, the average current through the SCR can be varied over wide limits (Figure 6.51). This way of regulating the power to a load is very

efficient because very little power is dissipated by the SCR. When it is off, there is no current through the SCR and its power dissipation is zero; when the current through the SCR is high, the voltage across it is low, resulting in low power dissipation. Bilateral triggering can be accomplished with a bilateral switch or *triac*. A common application of this kind of power control is the light dimmer used for incandescent lights. High-power switching devices like the SCR and triac produce *radiofrequency interference* (rfi) and should not be located

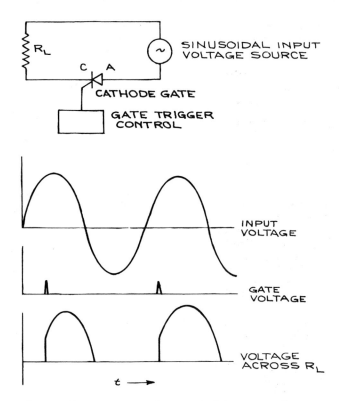

Figure 6.51   Current regulation with a silicon controlled rectifier (SCR).

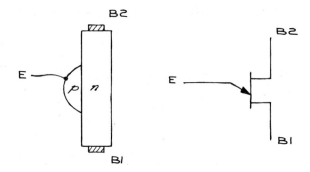

Figure 6.52   Unijunction transistor (UJT) and the corresponding symbol.

emitter is less positive than the bar at its point of attachment, the emitter junction will be reversed-biased and no current will flow through it. When the emitter potential is increased sufficiently to forward-bias the junction, current will flow in the E-$B1$ circuit and V_{EB_1} will fall. This is shown in the characteristic curve (Figure 6.53). Applications of the UJT include relaxation and sinusoidal oscillators, trigger circuits, pulse generators, and frequency dividers. The stable peak voltage, which is a fixed fraction of the interbase voltage, and high pulse-current capability make the UJT very useful in SCR control circuits.

near sensitive electronic circuits. Silicon controlled switches (SCSs) are similar to SCRs, but have the advantage of being able to be turned off with a pulse.

6.3.5   Unijunction Transistors

The *unijunction transistor* (UJT) is a single bar of n-material to the middle of which p-material is attached, forming a p-n junction. This is shown schematically in Figure 6.52. The two ends of the bar are bases 1 and 2 ($B1$ and $B2$), while the p-material is the emitter (E). When $B2$ is made positive with respect to $B1$, a potential gradient will exist along the bar. The potential at the point of attachment of the emitter will be a fixed fraction of the potential across the bar. So long as the

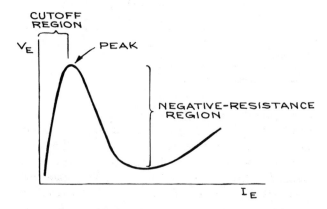

Figure 6.53   Current-voltage characteristic curve for a unijunction transistor showing the region of negative resistance.

6.3.6 Thyratrons

For fast switching of high currents at high voltages, gas-filled tubes called thyratrons are used. They consist of a heated cathode, an anode, and a grid sealed into a glass gas-filled envelope, and are analogs of SCRs. An arc can be struck between the cathode and anode, but the initiation of the arc can be controlled by the potential on the grid. The grid is usually a cylindrical structure surrounding the anode and cathode, from which a baffle or series of baffles with small holes extends between anode and cathode. Since the anode and cathode are almost completely shielded from each other, only a small grid potential is needed to overcome the field at the cathode resulting from the large anode potential. Once the arc has been initiated, the grid entirely loses control over the arc. Grid control is reestablished only when the anode potential falls below the level necessary to sustain the arc. Deuterium-filled thyratrons can switch hundreds of amperes at tens of kilovolts in microseconds or less, and are commonly used in the high-voltage discharge sources of lasers. SCRs are often used to supply the trigger pulse to the grid of the thyratron.

6.4 AMPLIFIERS AND PULSE ELECTRONICS

6.4.1 Definition of Terms

In laboratory applications, amplifiers are used to transform low-level signals to a level sufficient for observation and recording or for the operation of electronic or electromechanical devices. When choosing a particular amplifier for a given task, there are a number of considerations that necessarily require the definition and explanation of several special terms. Once again, the goal is to provide information for the experimentalist who must make decisions regarding the use and specifications of amplifiers.

Amplifiers can be classified in a number of ways, among them the following:

1. By input and output variables.

2. By frequency domain.

3. By power levels.

Each of these classifications will be treated in turn, and any given amplifier will have its place somewhere in each of the classifications. The two most common electrical input and output variables are current and voltage. Consequently, there are four different possible types of amplifier, corresponding to the two input-variable possibilities and the two output-variable possibilities. These are listed in Table 6.16. The *gain* of an amplifier is the ratio of the output variable to the input variable. This is the equivalent of the transfer function for the passive circuits already considered. For voltage and current amplifiers the gain is unitless, while for the transconductance and transresistance amplifiers it has units of siemens (ohm^{-1} or mho) and ohm respectively.

There are also amplifiers for which the input variable is the time integral of the current or the charge. Such *charge-sensitive* amplifiers have as an output a voltage, and the gain is expressed in volts per coulomb or inverse farads. Amplifiers with the time derivative of current as an input variable are also possible.

The amplifiers in Table 6.16 have very different input and output properties. If the input is a current, the amplifier should have a low input impedance so as to affect the source as little as possible; on the other hand, if the input is a voltage, the input impedance should be as large as possible to avoid drawing current from the source and affecting the source output voltage. The opposite considerations apply to the outputs of the amplifiers. If the output is a current, it is desirable that

Table 6.16   TYPES OF AMPLIFIER

Input Variable	Output Variable	Amplifier Type
Voltage	Voltage	Voltage
Voltage	Current	Transconductance
Current	Voltage	Transresistance
Current	Current	Current

Table 6.17   AMPLIFIER IMPEDANCES

Amplifier Type	Input Impedance	Output Impedance
Voltage	High	Low
Transconductance	High	High
Transresistance	Low	Low
Current	Low	High
Charge-sensitive	Low	Low

the output behave like an ideal current source, that is, have a very high output impedance. If the output is a voltage, the output impedance should be low to approximate as closely as possible an ideal voltage source. These considerations of input and output impedance (Table 6.17) are of fundamental importance when matching an amplifier to a detector or transducer at the input and a load at the output. If the input device is a current source, one should generally choose a current, transresistance, or charge-sensitive amplifier; if the input device is a voltage source, a voltage or transconductance amplifier is required. This assumes that amplification with minimum loading of the input device is desired rather than amplification with a maximum transfer of power. When efficient power transfer is desired, the impedance of the source and load should be identical to the input and output impedances of the amplifier.

Amplifiers can be divided into groups according to the frequency domain in which they are designed to operate (Table 6.18). Consider the Bode plots of gain and phase shift shown in Figure 6.54 for a capacitance-

Table 6.18   AMPLIFIER FREQUENCY RANGES

Amplifier Type	Frequency Range
D.c.	0–10 Hz
Audio	10 Hz–10 kHz
R.f.	100 kHz–1 MHz
Video	30–1000 MHz
VHF	30–300 MHz
UHF	300–1000 MHz
Microwave	1000 MHz–50 GHz

coupled inverting (output 180° out of phase with input) amplifier. Qualitatively, the curves can be understood in the following way: at low frequencies the coupling capacitors (capacitors in series with input and output to block d.c. levels) reduce the low-frequency gain just like a high-pass filter; at high frequencies the reduced gain of the active devices (transistors) has the effect of shunt capacitances across both the input and output circuits of the amplifier, resulting in behavior similar to that of a low-pass filter. The frequency difference between the corner frequencies (3-dB points of the gain curve) is the *bandwidth* of the amplifier. Since the low corner frequency is usually orders of magnitude smaller than the high corner frequency (except in the case of tuned amplifiers, which operate over a very small frequency range), the bandwidth can be taken equal to the upper corner frequency. An often used figure of merit for amplifiers is the *gain-bandwidth product* (GBWP), which is the midband gain times the bandwidth. A low-gain, wide-bandwidth amplifier is equivalent to a high-gain low-bandwidth amplifier by this criterion.

If it is possible for the output of an amplifier to be coupled back to the input (capacitance coupling is always present to some degree because of stray capacitances) in such a way as to be in phase with the signal at the input, a *positive feedback* or *regenerative* situation will occur, giving rise to oscillations and unstable behavior. This is the origin of the familiar squawking and whistling that occurs in public address systems when the output from the loudspeakers finds its way back to the microphone. The stability of an amplifier against such oscillations is expressed in terms of *gain* and *phase margins*. If the value of the gain of an amplifier (in dB) is negative at frequencies for which the phase of the output is equal to that of the input, oscillations cannot occur. Correspondingly, if the output is out of phase with the input at all frequencies for which the gain is positive, oscillations are prevented. The amount of negative gain at zero phase is the *gain margin*; the phase difference at 0-dB gain is the *phase margin* (Figure 6.55). The larger these margins, the more stable the amplifier. Wide-band video amplifiers are particularly susceptible to unstable behavior because even small stray capacitances are sufficient to provide a low-impedance path for high frequencies from the output to the

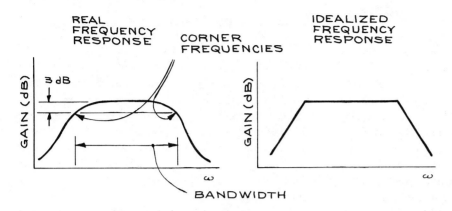

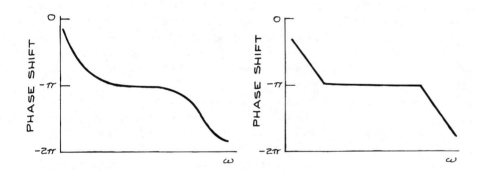

Figure 6.54   Bode plots for a capacitance-coupled inverting amplifier.

Figure 6.55   Gain and phase margins.

input. Because of the large power consumption in the positive feedback (regenerative) mode, prolonged unstable operation of an amplifier can destroy it. This is especially true in the case of fast pulse amplifiers, which must have a very large bandwidth in order to accurately amplify pulses with short risetimes. Compensation by the addition of poles and zeros to the transfer function can remedy this at the expense of reduced bandwidth.

In classifying amplifiers by power rating, the distinctions are rather arbitrary. Generally, *preamplifiers* are designed to operate with input voltage levels of millivolts and less, input currents of microamperes and less, and input charge of picocoulombs and less. Gain is not so important in preamplifiers as is noise, which limits ultimate sensitivity. In some applications, indeed, a unit-gain preamplifier is used as an impedance-matching device. *Power amplifiers* are designed to furnish anywhere from a few to several hundred watts to a matched load. Audio amplifiers are examples. Intermediate between preamplifiers and power amplifiers are amplifiers that operate with input signal levels from preamplifier outputs and produce outputs from a few to several hundred milliwatts. Gains of such amplifiers are high, and noise considerations are much less important than overload recovery. Frequency- and gain-compensating circuits are often incorporated in such amplifiers.

Thus far it has been assumed that the output of an amplifier is a linear function of the input at any frequency. Because the active elements (transistors) in solid-state amplifiers are actually nonlinear, it follows that amplifiers are linear only within a limited range of input conditions. For all amplifiers there is an input signal level beyond which the output signal is severely distorted. Clipping of the output waveform is an obvious form of distortion. Less obvious distortions can be quantified by a Fourier analysis of the output waveform for a sinusoidal input. The distortion in the output can be described in terms of the coefficients of the harmonics. This *harmonic distortion* is important in critical audio-amplifier applications, but not generally in the laboratory.

The properties of solid-state amplifiers may also be temperature dependent, and such factors as gain and distortion, though acceptable at room temperature, may become unacceptable at high or low temperatures.

6.4.2 General Transistor-Amplifier Operating Principles

The active elements in solid-state amplifiers can be n-p-n or p-n-p bipolar junction transistors or field-effect transistors, of which there are a very large variety. FETs can be n-channel or p-channel and can be operated in the depletion or the enhancement mode. Furthermore, they can have gate structures that are insulated from the conducting channel by an oxide layer (MOSFETs), or a direct contact or junction between the gate and conducting channel (JFETs).

To use an inherently nonlinear transistor to make a linear amplifier requires great ingenuity. Standard methods involve operating the transistors over a limited range of voltage and current where their nonlinear characteristics can be approximated by linear functions, using compensating circuits to cancel the nonlinearities with devices that have opposite characteristics, or applying negative feedback.

Transistors must be properly biased if they are to work at all. This means supplying, from external sources such as batteries or d.c. power-supply circuits, the correct potential differences across the terminals and the correct currents into them. Thus, even in the absence of an external signal, the transistors are dissipating power. The values of the d.c. bias, or quiescent, voltages and currents define the *operating point* of the transistor. Signals from an outside source or a previous stage are then superimposed on the quiescent values. So long as the signal does not represent too large a deviation from these values, the transistor will operate linearly. If the external signal is sufficiently large however, it can *cut off* the transistor (that is, reduce all the currents through it to zero) or *saturate* it (that is, cause the transistor to act as a short circuit). Clearly, when either of these two effects occurs, the transistor no longer operates as a linear circuit element.

Since the bias voltages and currents are d.c. and the signal voltages and currents are a.c., it is possible to separate them. Capacitors or transformers can be used to couple a.c. signals without disturbing the quiescent d.c. values. When capacitance coupling is used, the capacitors are called *blocking* capacitors because they block d.c. voltages and currents. Reasonable-size blocking capacitors usually limit the low-frequency response of amplifiers to ≥ 1 Hz. (Capacitor lead inductance and stray shunt capacitances to ground can also limit the high-frequency response.) In many applications the poor transient response of the high-pass circuits formed by the blocking capacitors cannot be tolerated. Direct-coupled amplifiers, where the bias levels of each transistor stage are designed to be compatible with the preceding and succeeding stages, provide high gain to 0 Hz. Because of the need to match active components and resistor ratios very precisely, such amplifiers are often *integrated-circuit* (IC) amplifiers, where all the elements of the circuit are manufactured simultaneously on a single chip. In fact, direct-coupled designs are particularly suited to the IC manufacturing process, which favors the production of transistors, resistors, and diodes but requires quite special techniques for the production of capacitors. *Chopping* is another method of achieving good low-frequency response in amplifiers. With this method, the input signal is chopped at a frequency much higher than the highest characteristic frequency of the input signal. The resulting a.c. signal can be amplified by conventional a.c. coupled amplifier stages. Rectification of the output of the final stage restores the input waveform. The advantages of chopper over direct-coupled amplifiers are lower uncompensated input currents and voltages, and comparative freedom from drifts with temperature. These advantages have been largely offset by IC direct-coupled amplifier designs with FET input stages.

Amplifier properties and specifications are best understood by considering as an example the general-purpose 741 IC operational amplifier. The name *operational amplifier* ("op amp") was originally applied to amplifiers used in analog-computer circuits, which performed a wide variety of mathematical operations. The term is now more generally applied to any high-gain, differential-input amplifier.

Differential-input amplifiers have two input terminals, which are electrically isolated from ground and each other. They are called the *noninverting* (+) and *inverting* (−) terminals. The output signal is proportional to the difference between the signals at the two input terminals. *Single-ended* inputs are those with one input terminal and a ground terminal. With regard to

outputs, most operational amplifiers are single-ended; that is, the output is developed between the single output terminal and ground. The ideal operational amplifier has the following characteristics:

1. Infinite input impedance.

2. Zero output impedance.

3. Infinite voltage gain, 180° out of phase with the input.

4. Infinite bandwidth.

5. Zero output voltage for identical input voltages at the two differential input terminals.

6. Properties independent of temperature, voltage levels, and frequency.

Real operational amplifiers depart from the ideal. Table 6.19 summarizes the properties of the common 741 integrated-circuit operational amplifier, whose symbol is shown in Figure 6.56. The supply voltages V_{CC+} and V_{CC-} are applied to the indicated terminals, not to be confused with the input terminals. The output is at the apex of the triangle and the terminals marked "null" are for an external potentiometer to cancel the offset voltage in critical applications.

Table 6.19   741 OPERATIONAL AMPLIFIER

CMRR	90 dB
Open-loop d.c. gain	200,000
GBWP	1.5 MHz
Slew rate	0.5 V/μs
Input resistance	2 MΩ
Output resistance	75 Ω
Input offset voltage	5.0 mV
Input offset current	20 nA
Input bias current	80 nA
Input voltage range	± 13 V
Maximum peak-to-peak output voltage swing	28 V
Input capacitance	1.4 pF
Power-supply rejection ratio	150 μV/V
Short-circuit output current	25 mA
Power consumption	50 mW

Note: Table shows typical characteristics at 25°C with power-supply voltages of ± 15 V.

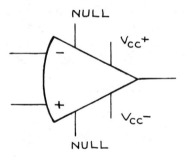

Figure 6.56 Schematic symbol for the operational amplifier, showing input, output, power supply, and null terminals.

Common operational-amplifier terms include:

Common-mode rejection ratio (CMRR). This is the ratio of the difference gain A_d to the common-mode gain A_c (often expressed in dB). The difference gain is the output voltage divided by the algebraic difference of the input voltages. The common mode gain is the output voltage divided by the algebraic sum of the input voltages. For the 741 the 90-dB value of the CMRR means that a one-volt signal at each input terminal will result in a 31-μV output signal. This is shown in the following calculation:

$$90 \text{ dB} = 20 \log \frac{A_d}{A_c},$$

$$\frac{A_d}{A_c} = 10^{4.5} = 3.2 \times 10^4.$$

Since the difference voltage is 0.0 V, the output voltage will be the common-mode voltage times the common mode gain A_c. For a common-mode voltage of 1.0 V, the output voltage is then $1.0/3.2 \times 10^4 = 31 \times 10^{-6}$ V.

Open-loop d.c. gain. This is the gain in the absence of a connection from output to the input, that is, in the absence of any *feedback*. The gain of the amplifier in the presence of a feedback network is called the *closed-loop gain*. With open-loop gains of the order of 200,000, it can be seen that the region of linear operation applies to only very small input signals. At a gain of 200,000, an

input signal of ± 75 μV is sufficient to drive the output to the maximum possible level, which in this case is ± 15 V from the power supply.

Frequency response. This is given in terms of either the GBWP or the frequency at which the gain equals 0 dB (unit gain). The two methods give nearly the same numbers for a single-pole transfer function. For the 741, the gain has fallen by over 5 orders of magnitude from the d.c. value at 1 MHz. It should be kept in mind that these frequency-response data are based on so called *small-signal* conditions, where the input is small enough so that the output stages are operating well within their linear region.

Slew rate. This gives the time response of the amplifier to large input signals. For this specification the amplifier is connected in a unit-gain (closed-loop) configuration with negative feedback, and an input voltage step is applied. The slew rate is then taken as the ratio of 10 V to time for the output voltage to go from 0 to 10 V. For the 741 this time is 20 μs, so the slew rate is 0.5 V/μs.

Input resistance. This is the resistance R_i between the input terminals with one terminal grounded.

Output resistance. This is the resistance R_o (seen by the load) between the output terminal and ground. It determines whether the amplifier may be considered as a voltage or a current source.

Input offset voltage. This is the d.c. voltage V_{IO} that must be applied across the input terminals to bring the d.c. output to zero. For the 741 it is 0.8 mV and can be compensated for, but one would not choose the 741 for the amplification of millivolt d.c. levels.

Input offset current (I_{IO}). This is the difference between the currents into the input terminals necessary to bring the output voltage to zero.

Input bias current (I_{IB}). This is the average of the currents into the input terminals with the output voltage at zero. It should be noted that these bias currents will flow through circuit elements (generally resistors) attached to the input terminals. If the resistances are not identical, voltage differences will be developed across the input terminals and amplified along with the input signal.

Input voltage range (V_I). This is the maximum

allowed input voltage for which proper operation can be maintained.

Maximum peak-to-peak output voltage swing.   This is the maximum peak-to-peak output voltage that can be obtained without clipping, about a quiescent output voltage of zero.

Input capacitance (C_i).   This is the capacitance between the input terminals with either input grounded.

Power-supply rejection ratio.   This is the change in input offset voltage for a given change in power-supply voltage. Operational amplifiers are particularly good in this regard and do not require highly regulated power supplies.

Short-circuit output current.   This is the maximum current the amplifier can deliver. This specification may seem to be at variance with the output impedance and maximum voltage output; however, it must be remembered that all the specifications apply only when the amplifier is operating within its acceptable range of parameters.

Temperature effects.   The various offset and bias currents and voltages are temperature-dependent. The variations of these parameters with temperature are specified by temperature coefficients.

The 741 is internally compensated with a capacitor at the high-gain second stage to produce a pole in the transfer function at 6 Hz. This ensures stable operation under even 100%-feedback conditions. (Of course, this gain in stability reduces the high-frequency response of the amplifier.) Two extra terminals are available for connection to an external 10 kΩ potentiometer, the wiper arm of which is connected to the negative power supply. By adjusting the potentiometer the input offset voltage can be eliminated. Additional features include short-circuit output protection and input overload protection. Single amplifiers come in at least four different packages: an 8-lead metal can, a 14-lead dual in-line package (DIP), an 8-lead mini-DIP, and a 10-lead *Flat Pak*. There is a commercial model, the 741C, which operates over a restricted temperature range, and a military model, the 741A, which operates over an extended temperature range. The 741 sells for about the same as a dozen $\frac{1}{4}$-watt composition resistors. Consider-

ing its complexity and performance, the 741 represents a triumph of electronic design, fabrication, and packaging.

Because of their high gain, wide bandwidth, and small offsets, operational amplifiers with suitable feedback networks can perform a wide range of electronic functions. There are many excellent texts especially devoted to operational-amplifier applications.[4] Here we give only a list of the basic configurations. In all of these it is important to keep in mind the following design rules:

1. The offset-voltage temperature coefficient must produce voltages much less than the input signal throughout the anticipated temperature range. At any given temperature, the offset voltage can be eliminated with an external nulling circuit, but variations with temperature cannot be easily accommodated.

2. The voltages created by the bias currents flowing through resistances attached to the input terminals must be less than the signal voltage. For high-output-impedance sources this may create a problem, but it can be solved by introducing an intermediate buffer stage with a high input impedance (so as not to load the source) and a low output impedance (to minimize the effects of input bias current). The temperature coefficient of the offset current must produce offset voltages much less than the signal voltage over the anticipated temperature range. To minimize bias-current effects, the resistances to ground from both input terminals should be the same.

3. The frequency response and compensation must be such that the amplifier will not break into oscillation. The 741 is internally compensated. High-frequency amplifiers lack internal compensation, but have extra terminals for external compensation.

4. The maximum rate of change of a sinusoidal output voltage $V_0 \sin \omega t$ is $V_0\omega$. If this exceeds the slew rate of the amplifier, distortion will result. Slew rate is directly related to frequency compensation, and high

slew rates are obtained with externally compensated amplifiers using the minimum compensation capacitance consistent with the particular configuration. Data sheets should be consulted.

6.4.3 Operational-Amplifier Circuit Analysis

If ideal behavior can be assumed (as is reasonable for circuits where the open-loop gain is much larger than the closed-loop gain), only two rules need be used to obtain the transfer function of an operational-amplifier circuit, as shown schematically in Figure 6.57:

1. The voltage difference across the input terminals is zero.

2. The currents into the input terminals are zero. (The inverting terminal is sometimes called the *summing point*.)

Applying these to the generalized circuit of Figure 6.57, we see that the voltage at the "+" input is

$v_{i2}Z_3/(Z_2 + Z_3)$, since there is no current flowing into the "+" input and Z_3, Z_2 form a voltage divider with no load. From rule 1, the voltage at the "−" input must also equal $v_{i2}Z_3/(Z_2 + Z_3)$. The current through Z_1 is then the potential difference across it divided by Z_1, or

$$\frac{1}{Z_1}\left[v_{i1} - v_{i2}\left(\frac{Z_3}{Z_2 + Z_3}\right)\right].$$

This current must be equal in magnitude and opposite in sign to the current from the output through the feedback element Z_f. This current i_f is given by the potential difference across Z_f divided by Z_f:

$$i_f = \frac{1}{Z_f}\left[v_{\text{out}} - v_{i2}\left(\frac{Z_3}{Z_2 + Z_3}\right)\right]$$

$$= -\frac{1}{Z_1}\left[v_{i1} - v_{i2}\left(\frac{Z_3}{Z_2 + Z_3}\right)\right].$$

Solving for v_{out}, we obtain

$$v_{\text{out}} = -v_{i1}\frac{Z_f}{Z_1} + v_{i2}\left(\frac{Z_3}{Z_2 + Z_3}\right)\left(1 + \frac{Z_f}{Z_1}\right).$$

Some useful configurations are illustrated in Figure 6.58.

The summer circuit [Figure 6.58(e)] is an elaboration of the inverting amplifier, with the current into the summing point coming from two external sources (v_A and v_B) and being canceled by the current from the feedback loop, v_{out}/R_f. If $R_A = R_B = R$, then $v_{\text{out}} = -(R_f/R)(v_A + v_B)$. If $R_A \ne R_B$, then v_{out} is equal to the negative of the weighted sum of voltages v_A and v_B, that is,

$$v_{\text{out}} = -\left[\left(\frac{R_f}{R_A}\right)v_A + \left(\frac{R_f}{R_B}\right)v_B\right].$$

Any number of external signal sources can be summed in this way.

A sinusoidal input to the *low-pass filter* [Figure 6.58(f)] gives $v_{\text{out}} = +(j/\omega RC)v_{i1}$. For a nonsinusoidal input, the current into the summing point from v_{i1} is

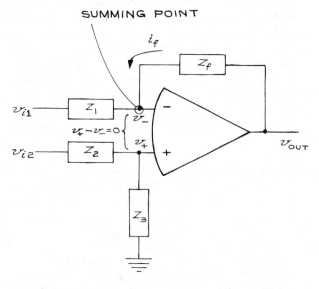

Figure 6.57 The general operational-amplifier circuit. Power supply and null terminals are not shown.

v_{i1}/R, since the inverting input is at ground (condition 1). Often one speaks of the inverting input being at *virtual ground*. There is no direct connection from the inverting input to ground, but the condition in which both inputs are at the same potential results in the inverting input assuming a potential of zero when the noninverting input is at ground potential. The current through the feedback capacitor of this circuit is i_f, and the voltage across the capacitor, v_{out}, is equal to $(1/C)\int i_f dt$. Since i_f and v_{i1}/R are equal in magnitude and opposite in sign,

$$v_{out} = \frac{1}{C}\int i_f dt = -\frac{1}{RC}\int v_{i1} dt$$

and the circuit acts as an *integrator* of the input voltage. Integration times of several minutes or even hours are possible with high-quality low-leakage capacitors and operational amplifiers with small offset voltages and currents.

For the *high-pass filter* [Figure 6.58(g)], $v_{out} = -RC\,dv_{i1}/dt$ and the circuit acts as a differentiator.

The integrator and differentiator circuits are simple forms of active filters. More complex networks involving only capacitors and resistors can be used to obtain poles in the left half of the complex s-plane, which in the past could only be obtained with inductors in passive filter networks. The operational amplifier furthermore allows the use of reasonable resistor and capacitor values even at frequencies as low as a few hertz. An additional benefit is the high isolation of input from output due to the low output impedance of most operational-amplifier circuits. The limitations of active filters are directly related to the properties of the operational amplifier. Inputs and outputs are usually single-ended and cannot be floated as passive filters can. The input and output voltage ranges are limited, as is the output current. Offset currents and voltages, bias currents, and temperature drifts can all affect active filter performance.

The circuits in Figure 6.58(b) through (g) are examples of the operational amplifier's use in analog computation. With these circuits, the mathematical operations of addition, subtraction, multiplication by a constant, integration, and differentiation can be performed. Mul-

tiplication or division of two voltages is accomplished by a logarithmic amplifier, an adder (for multiplication) or subtractor (for division), and an exponential amplifier.

Both the logarithmic and exponential amplifiers rely on the exponential *I-V* characteristics of a *p-n* junction. When this junction is the emitter-base junction of a transistor, the collector current I_C with zero collector-base voltage is

$$I_C = I_0\{\exp(qV_{EB}/kT) - 1\},$$

where I_0 is a constant for all transistors of a given type and V_{EB} is the emitter-base voltage. Typically, I_0 is 10–15 nA for silicon planar transistors.

For $I_C \gtrsim 10^{-8}$ A, the equation reduces to $I_C = I_0 \exp(qV_{EB}/kT)$. For the configuration shown in Figure 6.58(h), $i_1 = v_{i1}/R = I_0 \exp(qv_{out}/kT)$ and $v_{out} = -(kT/q)(\ln v_{i1} + \ln RI_0)$. Since k, T, q, R, and I_0 are constants, it can be seen that v_{out} will be proportional to $\ln v_{i1}$. Practical logarithmic amplifiers can operate over three decades of input voltages; they do, however, require temperature-compensating circuits, which generally involve the use of matched pairs of transistors. In addition to their arithmetic use, logarithmic amplifiers are frequently used for *data compression*. *Exponential* (antilogarithmic) amplifier circuits are logarithmic circuits with the input and feedback circuit elements interchanged.

Logarithmic and exponential amplifiers are examples of nonlinear circuits: the output is not linearly related to the input. Operational amplifiers are used extensively in nonlinear circuits. Since the cutin voltage of simple diodes is a few tenths of a volt (0.2 V for Ge and 0.6 V for Si), they cannot be used to rectify millivolt-level a.c. voltages. A diode in the feedback loop of an operational amplifier [Figure 6.58(i)] gives a rectifier with a cutin voltage equal to the diode cutin voltage divided by the open-loop gain of the amplifier. With a slight modification, the rectifier circuit can be converted to a *clamp* [Figure 6.58(j)] in which the output follows the input for voltages greater than a reference voltage V_R, but equals V_R for input voltages less than V_R.

Comparator circuits are used to compare an input signal with a reference voltage level. The output is

Figure 6.58   Operational-amplifier configurations: (a) follower; (b) follower with gain; (c) inverting amplifier; (d) subtractor; (e) summer.

(a)

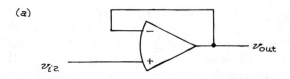

$$Z_1 = \infty \quad v_{i1} = 0$$
$$Z_2 = 0$$
$$Z_3 = \infty$$
$$Z_f = 0$$

$$v_{out} = v_{i2}$$
$$Z_{in} \simeq R_i \times A_{OL}$$
$$Z_{out} \simeq R_o / A_{OL}$$
$$A_{OL} = \text{OPEN LOOP GAIN}$$

(b)

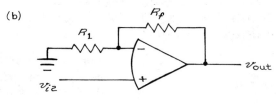

$$Z_1 = R_1$$
$$Z_f = R_f$$
$$Z_2 = 0$$
$$Z_3 = \infty$$

$$v_{out} = v_{i2} \underbrace{(1 + R_f / R_1)}_{A_{CL}}$$

$$Z_{in} = R_i \times A_{OL} / A_{CL}$$
$$Z_{out} = R_O / (A_{OL}/A_{CL})$$
$$A_{CL} = \text{CLOSED LOOP GAIN}$$

(c)

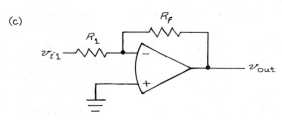

$$v_{i2} = 0 \quad Z_2 = \infty$$
$$Z_f = R_f \quad Z_3 = 0$$
$$Z_1 = R_1$$
$$v_{out} = -R_f / R_1$$
$$Z_{in} = R_1$$
$$Z_{out} \simeq R_O / (A_{OL}/A_{CL})$$

(d)

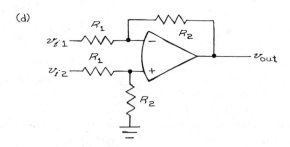

$$Z_1 = Z_2 = R_1$$
$$Z_3 = Z_f = R_2$$

$$v_{out} = R_2 / R_1 (v_{i2} - v_{i1})$$

(e)

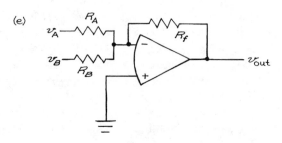

$$v_{out} = -R_f \left[\frac{v_A}{R_A} + \frac{v_B}{R_B} \right]$$

(f)

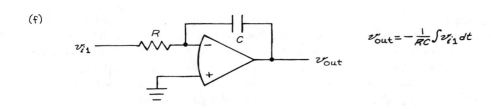

$$v_{out} = -\frac{1}{RC}\int v_{i_1}\,dt$$

Figure 6.58 (*continued*)   (*f*) low-pass filter (integrator); (*g*) high-pass filter (differentiator); (*h*) logarithmic amplifier; (*i*) precision rectifier; (*j*) clamp. The parameters are defined in Figure 6.57.

(g)

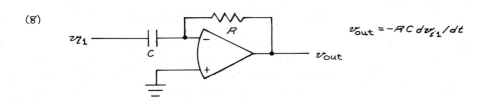

$$v_{out} = -RC\,dv_{i_1}/dt$$

(h)

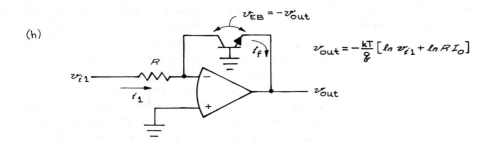

$$v_{out} = -\frac{kT}{q}\left[\ln v_{i_1} + \ln R I_o\right]$$

(i)

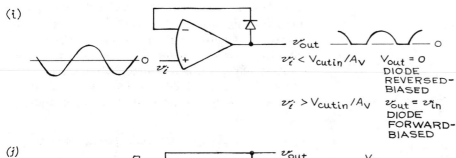

$v_i < V_{cutin}/A_v$ $V_{out} = 0$
DIODE REVERSED-BIASED

$v_i > V_{cutin}/A_v$ $v_{out} = v_{in}$
DIODE FORWARD-BIASED

(j)

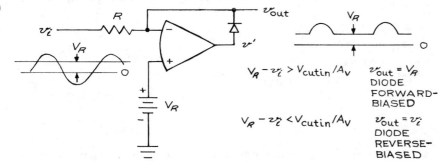

$V_R - v_i > V_{cutin}/A_v$ $v_{out} = V_R$
DIODE FORWARD-BIASED

$V_R - v_i < V_{cutin}/A_v$ $v_{out} = v_i$
DIODE REVERSE-BIASED

driven to V_{CC}^+ or V_{CC}^- depending on whether the input is less than or greater than the reference level [Figure 6.59(a)]. Using an amplifier in such a saturated mode is very poor practice and practical comparators with internally clamped outputs less than $|V_{CC}^+|$ and $|V_{CC}^-|$ are available. The transition between the two output states can be accelerated by the use of positive feedback [Figure 6.59(b)]. Such a circuit is called a *Schmitt*

trigger and finds wide application in signal conditioning when it is necessary to convert a slowly changing input voltage into an output waveform with a very steep edge. The price paid for the fast transition is *hysteresis*: the threshold value for a positive transition of the output is not the same as for a negative transition.

6.4.4 Stability and Oscillators

For circuits composed only of passive elements, the poles of the transfer function will always lie on the left-hand side of the complex $s = \sigma + j\omega$ plane, that is, σ will always be less than zero. For circuits with active elements this is not necessarily the case. If, for example, a fraction of the output signal of an amplifier is returned to the input in phase with the original input signal, an unstable situation will result, the output of the amplifier increasing with time. Analysis of such a circuit will show at least one pole to lie in the positive half of the complex *s*-plane. Unwanted oscillations in amplifiers are due to the regenerative coupling of a fraction of the output signal back to the input (Figure 6.60). Stray capacitances between input and output are sufficient, at high frequencies, to couple the output signal back to the input. This can be minimized by having the output and input physically as far apart as possible. A copper shield separating the output and input is also often effective, and high-frequency circuits built on printed circuit boards should have generous ground-plane areas.

While the lack of stability is undesirable in amplifiers, sinusoidal *oscillators* depend on such instabilities for

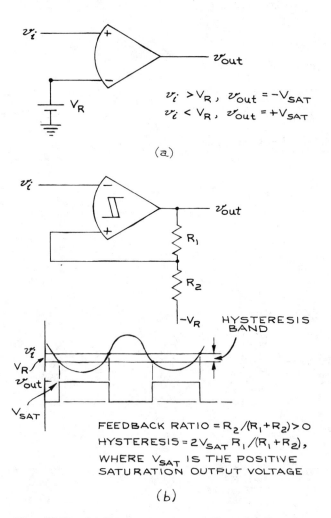

Figure 6.59 (a) Simple comparator ($-V_{sat}$ and $+V_{sat}$ are the minimum and maximum voltages the amplifier can deliver; V_R is a constant reference voltage); (b) Schmitt trigger (comparator with positive feedback).

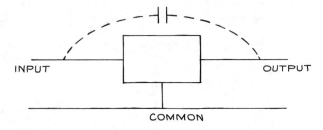

Figure 6.60 Coupling through stray capacitance causing unwanted feedback.

their operation. Consider an amplifier with gain A_0. It can be any one of the four already considered. If X_o represents the output signal (a current or voltage) and X_i the input signal (also a current or voltage), $X_o = A_0 X_i$. For a simple amplifier $X_i = X_s$ (the source signal). If a fraction β, the feedback ratio, of the output is sent back to the input in such a way as to cancel part of the source signal, the new signal X_i is $X_s - \beta X_0$, and the gain of the amplifier with feedback is the output signal divided by the source signal:

$$A_f = \frac{X_0}{X_s} = \frac{X_0}{X_i + \beta X_0},$$

$$A_f = \frac{A_0}{1 + \beta A_0}.$$

For $\beta > 0$ the feedback signal acts to cancel the signal from the source and $A_f < A_0$. This is called *negative* or *degenerative feedback* and is frequently used to stabilize amplifier gain and improve linearity and noise. For $\beta < 0$ the feedback signal adds to the source signal at the input and $A_f > A_0$. This is *positive* or *regenerative feedback* and is sometimes used to increase the gain of amplifiers at the expense of stability. If $\beta A_0 = -1$, $A_f = \infty$ and the amplifier is unstable, with an output that breaks into oscillation or saturates.

In practice, both A_0 and β are frequency-dependent. The response of such feedback amplifiers is straightfor-

ward to calculate; however, if only stability criteria are desired, one only needs to determine whether any poles lie in the right half of the complex plane.

If βA_0 can be made equal to -1 for only a single frequency by means of a frequency-selective feedback network, one has an *oscillator*. There are a large number of frequency-selective circuits based on *LC* and *RC* networks. Quartz crystal oscillators are particularly stable and free of temperature effects. These devices work by the application of a potential difference across the faces of the crystal, which then becomes deformed and oscillates at a resonant frequency determined by the size and orientation of the crystal planes with respect to the electrodes. Crystal oscillators with Q's of 10^4 to 10^6 are commercially available.

6.4.5 Detecting and Processing Pulses

There is a large class of experiments, especially in nuclear physics, that deals with the detection and analysis of single events. With the advent of detectors for low-energy photons, electrons, and ions these techniques have spread to the fields of atomic and molecular physics and physical chemistry. Commercial instrumentation is abundant, but to specify and use it to best advantage requires a knowledge of the basic properties of detectors, amplifiers, and associated processing circuits.

Table 6.20 summarizes the properties of the more

Table 6.20   PARTICLE DETECTORS

Particle Detector	Output Signal Level	Charge-Collection Time
Semiconductor:		
Ge	2.8 eV/electron-hole pair ⎫	0.1–10 ns[a]
Si	3.5 eV/electron-hole pair ⎭	
Photomultiplier	10^6–10^7 electrons/photoelectron	1 ns
Electron multiplier	10^6–10^8 electrons/incident electron	0.5–1 ns
Microchannel plate	10^3–10^4 electrons/incident electron	0.1 ns
Scintillator:		
Plastic ⎫		⎧ 0.1–10 ns
Alkali halide ⎭	3 keV/photon	⎩ 1 μs
Gas-filled tube	25–35 eV/ion pair	0.01–5 μs

[a] For narrow depletion depth.

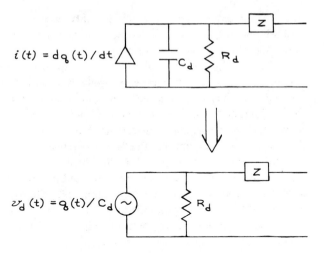

$$i(t) = dq(t)/dt$$

$$v_d(t) = q(t)/C_d$$

Figure 6.61   Equivalent circuit of a detector.

common types of detectors. The first part of the table allows one to estimate the output of a given detector for an arbitrary input. For example, a 5-MeV alpha particle depositing all its energy in a silicon detector will create 1.4×10^6 electron-hole pairs. With a suitable bias across the detector, the charge can be collected in 0.1 to 10 ns. Similar considerations allow one to calculate the outputs of other detectors. Electrically, the detectors can be represented by the equivalent circuit of Figure 6.61. The choice of preamplifier, amplifier, and shaping and analysis circuits now depends on the information to be obtained from the pulse. Figure 6.62 shows the arrangement of three general categories of experiments—counting, timing, and pulse-height analysis. Each one has different electronics requirements, but before consider-

Figure 6.62   Three main categories of pulse experiments: (*a*) counting, (*b*) timing, and (*c*) pulse-height analysis.

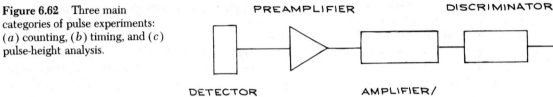

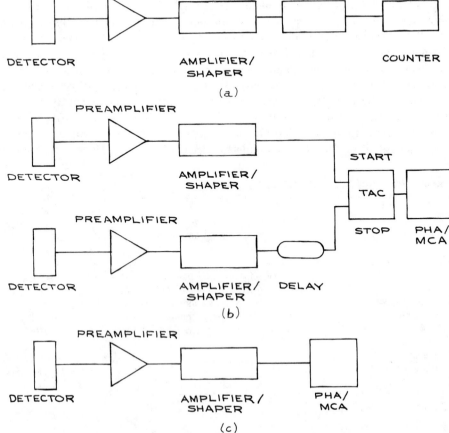

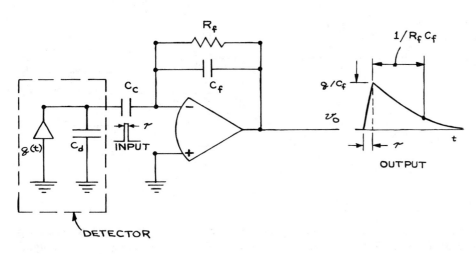

Figure 6.63 Combined detector and charge-sensitive preamplifier.

ing each of these systems in detail, it is worthwhile to examine the circuits commonly available for amplification and shaping.

Preamplifiers can be voltage-, current-, or charge-sensitive. Voltage preamplifiers are not generally used, because the voltage generated by common detectors depends on the capacitance of the detector (Figure 6.61), which can change with bias voltage and other parameters. Current preamplifiers have low input impedances and are designed to convert the fast current pulse from a photomultiplier or electron multiplier to a voltage pulse. Their sensitivity is given in output millivolts per input milliampere. Charge-sensitive preamplifiers are the most commonly used preamplifiers because their gain is independent of the capacitance of the detector. Their sensitivity is specified in output millivolts per unit input charge or in equivalent output millivolts per MeV deposited in a given solid-state detector.

A schematic diagram of a combined detector and charge-sensitive preamplifier is shown in Figure 6.63. Neglecting the effect of the coupling capacitor, the Laplace transform of the output voltage, $\bar{v}_0(s)$, is

$$\bar{v}_0(s) = \frac{\bar{Q}(s)}{C_f}\frac{1}{1+1/R_fC_fs}.$$

The function has a pole at $s = -1/R_fC_f$; the Bode plot is given in Figure 6.64. The high-frequency pole at ω_2 is a result of the preamplifier transfer function. The output

waveform for a rectangular input pulse of duration τ will be a *tail pulse* of risetime τ, amplitude Q/C_f, and decay time R_fC_f.

For very small values of τ, the risetime of the output pulse will be determined by the amplifier itself. Practical amplifiers use high-stability NPO capacitors for C_f with values from 0.1 to 5 pF; the feedback resistor R_f is made as large as is consistent with the signal rate and leakage current. Noise in charge-sensitive preamplifiers is specified in terms of the energy resolution, full width at half maximum (FWHM), in keV. This can be converted to an equivalent charge noise with the factors in Table 6.20. To test charge-sensitive preamplifiers, the circuit of Figure 6.65 is used. An input step voltage applied to the capacitor C_T results in an amount of

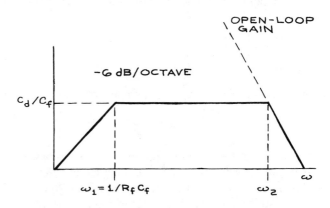

Figure 6.64 Bode plot for a charge-sensitive preamplifier.

Figure 6.65 Test circuit for charge-sensitive preamplifiers.

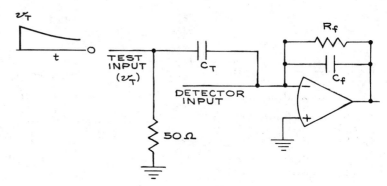

charge equal to $C_T V_T$ deposited on the input of the preamplifier. For 10^{-15} coulombs with $C_T = 10$ pF the amplitude of the input voltage pulse must be 10 mV. The 50-Ω resistor in the circuit is for termination purposes. The duration of the test input current will be equal to the risetime of the voltage input.

The output pulses from the preamplifier are generally very different in shape from the input signals. This is the case even with fast current preamplifiers. An important function of the amplifier and shaping circuits is to increase the amplitude of the preamplifier output signal and change its shape to minimize pulse overlap and increase the signal-to-noise ratio. To minimize pulse overlap, the duration of the pulses should be short compared with the average time between them. In more quantitative terms, the rms voltage level of the pulse train to be processed should be much less than the required voltage resolution of the system. If n is the pulse rate and $v(t)$ is the functional form of the pulse, then

$$v_{\text{rms}} = \left[n \int_0^\infty v(t)^2 \, dt \right]^{1/2}.$$

For rectangular pulses of unit height and width τ,

$$v_{\text{rms}} = n^{1/2} \tau^{1/2}.$$

The long-tail pulse from a charge-sensitive preamplifier can be shortened with a differentiating circuit [Figure 6.66(a)]. The circuit reduces the falltime from $R_f C_f$ to RC. A disadvantage of simple differentiation is under-

shoot, which appears on the trailing edge of the output pulse. This can be substantially eliminated by modifying the transfer function of the differentiating circuit to include a zero that exactly cancels the pole in the preamplifier transfer function [Figure 6.66(b)]. The

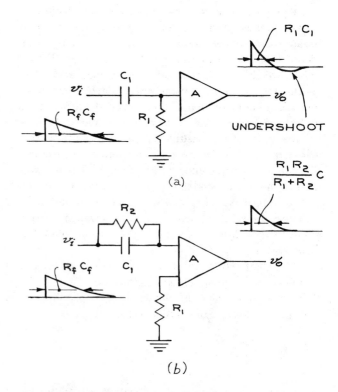

Figure 6.66 Differentiating circuits: (a) without pole-zero compensation; (b) with pole-zero compensation.

Laplace transform of the transfer function for Figure 6.66(b) is

$$\frac{s+1/R_2C_1}{s+(R_1+R_2)/R_1R_2C_1}.$$

If R_2C_1 is set equal to R_fC_f from the preamplifier, undershoot can be eliminated. This technique of *pole-zero compensation* is commonly used in pulse shaping.

For pulse-height analysis a Gaussian waveform has the best noise characteristics.[5] Such a shape can be obtained from a tail pulse by a single differentiation and an infinite number of cascaded integrations with integration time constants all equal to the differentiation constant. In practice, four integrations give a shape very close to Gaussian. Bipolar pulses can be obtained by double differentiation of a tail pulse with an intermediate integration (Figure 6.67). An important property of the bipolar pulse is that the zero crossing always occurs at a fixed time after its start, determined by the time constant, and independent of pulse height. When used with zero-crossing discriminators, such pulses permit reliable timing to be accomplished with long-decay-time pulses. Bipolar pulses have, however, poorer noise characteristics than unipolar Gaussian ones.

An alternative to RC shaping is delay-line shaping. The equivalent of unipolar and bipolar pulses can be obtained with single- and double-delay-line circuits (Figure 6.68). Advantages are the preservation of fast risetimes, sharp trailing-edge clipping, and greatly reduced base-line shifts. These advantages are offset by the poor noise properties of the pulses compared with RC shaping. Delay-line shaping is mostly confined to timing applications.

The shaping circuits discussed above are generally incorporated in the variable-gain main amplifier that follows the preamplifier. While noise is a consideration with such amplifiers, it is not so critical as with preamplifiers because of the larger signal levels involved. Important considerations are linearity and fast overload recovery. To reduce base-line shifts from pulse pileup, main amplifiers often incorporate a *base-line restorer circuit*. The principle of operation is illustrated in Figure 6.69. A noninverting amplifier with gain $H(s)$ is placed in the negative-feedback loop of an amplifier with gain $G(s)$. The Laplace transform of the transfer function for the system is

$$\frac{G(s)}{1+G(s)H(s)}.$$

If both $G(s)$ and $H(s)$ are themselves transforms of single-pole transfer functions, given by

$$G(s)=\frac{A}{1+T_1s} \quad \text{and} \quad H(s)=\frac{K}{1+T_2s},$$

the overall transfer function is

$$\frac{A(1+T_2s)}{AK+(1+T_1s)(1+T_2s)}.$$

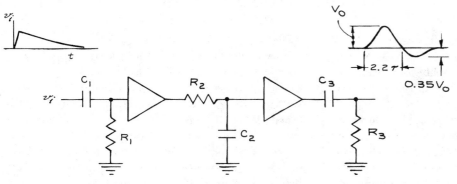

Figure 6.67 Bipolar pulse shaping with *CR-RC-CR* circuits.

$$R_1C_1 = R_2C_2 = R_3C_3 = \tau$$

Figure 6.68   (*a*) Unipolar and (*b*) bipolar pulse shaping with delay lines.

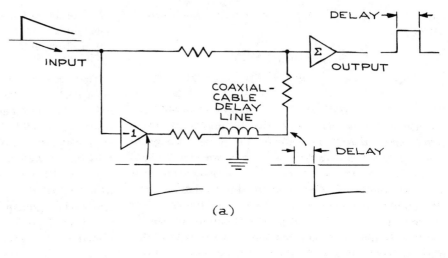

(a)

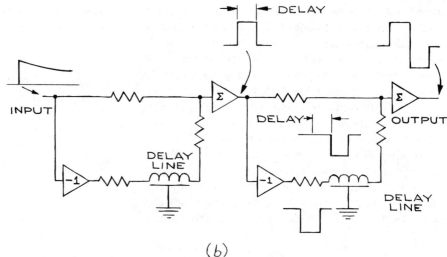

(*b*)

At high frequencies ($s \gg 1$) the transfer function reduces to $G(s)$, while for d.c. ($s = 0$) the function is $1/K$. The overall effect of the circuit is to amplify high-frequency-component pulses while attenuating low-frequency base-line shifts.

For counting applications, the principal considerations are the pulse rate and discriminator setting. The shaped pulses from the amplifier should be sufficiently short in duration to avoid pileup and base-line shifts. With a unipolar pulse a leading-edge discriminator can be used. Since such units are based on regenerative

(positive-feedback) comparator circuits, which have the property that the voltage level at which triggering occurs is greater than the level for recovery, the decay of the input pulse must not pass through the recovery voltage level until the triggered state is firmly established. When bipolar pulses are to be counted, zero-crossing discriminators are appropriate.

Timing systems require that the leading-edge information from the detector pulse be preserved as faithfully as possible. Timing errors due to time variations in the processed pulses relative to the input pulse, long-term

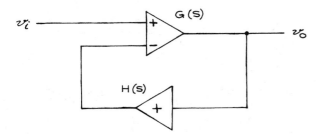

Figure 6.69 A base-line restorer circuit.

drift from component aging, and noise must be considered. Wide-band current preamplifiers and fast timing amplifiers with short integrating and differentiating shaping circuits or delay-line shaping should be used. When using wide-band amplifiers, precautions should be taken to electrically terminate all connections properly and avoid stray capacitances, which could form positive-feedback paths and produce instability. When the detector signal itself is of sufficient amplitude, as it is for the output of a photomultiplier tube used with a scintillation detector, a timing signal can be taken directly from the detector without further amplification or shaping. Leading-edge and constant-fraction discriminators are most commonly used for timing. The *constant-fraction discriminator* currently provides the best resolution times. With this discriminator the input signal is delayed and combined with a fixed fraction of the undelayed signal. A bipolar pulse with a zero crossing independent of risetime and amplitude results, and this can be used to activate a zero-crossing detector. In Figure 6.62(*b*) the delay inserted in the "stop" line of the system is to ensure that the stop pulse to the time-to-amplitude converter (TAC) will always follow the start pulse. The TAC converts the time difference between the start and stop input pulses to a single pulse of an amplitude proportional to the time difference. High-quality, low-attenuation coaxial cable provides an excellent delay line, and lumped-parameter delay lines are also available commercially for this purpose. Timing information is usually obtained by processing the output of the TAC with a pulse-height analyzer (PHA). A more direct method is with a time-to-digital converter which converts the time difference between the start and stop

pulses to a digital address for processing. Depending on detector characteristics, time resolution of a few hundred picoseconds can now be obtained with standard commercial units.

Particle spectroscopy experiments depend critically on the preservation of the pulse-amplitude information from the detector, since the amplitude is a direct measure of the particle energy. Optimum amplitude resolution is obtained by operating the preamplifier and amplifier in their linear ranges, reducing sources of noise and base-line shifts from pileup, and using semi-Gaussian pulse shapes as inputs to the PHA. Nonlinearities in the preamplifier and amplifier can be both differential and integral, illustrated in Figure 6.70. The *differential nonlinearity* is the ratio of amplifier gain at a specified input level to the gain at a reference input level. The *integral nonlinearity* is the maximum vertical deviation of the real gain curve from the ideal straight-line gain curve, expressed as a percentage of maximum output. Such nonlinearities also occur in the circuits of the PHA and must be taken into consideration for high-resolution work.

The manufacturers of pulse-processing equipment include applications notes and product guides with the

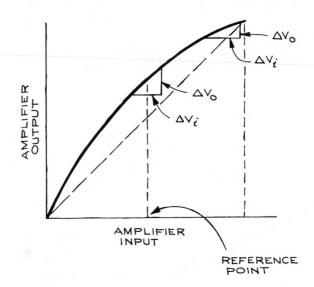

Figure 6.70 Amplifier linearity.

catalog descriptions of their instruments. These are generally very helpful and should be read thoroughly before deciding on a system and selecting components.

6.5   POWER SUPPLIES

The design of regulated power supplies follows a standard pattern (Figure 6.71). One starts with raw a.c. from the outlet and converts it, usually with a transformer, to the desired level. This transformed a.c. is then rectified to produce unregulated d.c. The unregulated d.c. is then the input to a regulator circuit, whose output is the desired regulated d.c. voltage.

6.5.1   Power-Supply Specifications

Power supplies are specified by the maximum current, maximum voltage, and maximum power they can deliver. It is not always the case that maximum current can be delivered over the complete voltage range of the supply, nor that maximum voltage can be obtained at all currents. This situation is illustrated in Figure 6.72. Maximum current can only be obtained over a limited range of voltage without exceeding the power rating of the supply.

The change in output voltage per unit change in input a.c. voltage is called the *line regulation*, while the

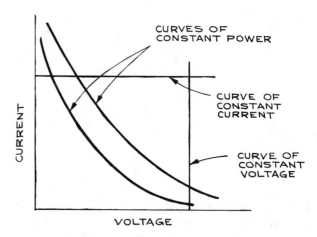

Figure 6.72   Current, voltage, and power relations for a power supply.

change in output voltage per unit change in load or output current is called the *load regulation*. Both are generally specified as a percentage. Many power supplies can be operated in a constant-current mode where the output is a current essentially independent of input and output voltage. Regulation specifications then of course apply to the output current. The load regulation can be translated into an equivalent dynamic-output resistance, since the relative change in load is the negative of the relative change in current ($\Delta R / R = - \Delta I / I$):

$$load regulation = \frac{\Delta V / V}{\Delta R / R} = -\frac{\Delta V / V}{\Delta I / I}$$

$$= \left| \frac{\Delta V}{\Delta I} \right| \frac{I}{V} = \frac{R_{\text{dynamic output}}}{R_{\text{load}}} .$$

As an example, if a filament power supply produces 3 A at 5 V, and the load regulation is 0.1%, then the dynamic output resistance is

$$R_{\text{dynamic output}} = \frac{1 \times 10^{-3} \times 5 \text{ V}}{3 \text{ A}} = 1.67 \times 10^{-3} \ \Omega .$$

Transient response and *recovery time* relate to the power supply's ability to recover from sudden changes in load or line voltage at the stated operating point. If

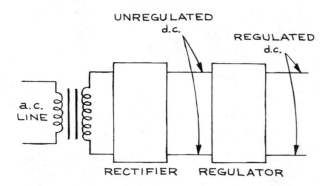

Figure 6.71   Block diagram of a regulated power supply.

the line and load fluctuations are rapid, the power-supply regulating circuits may not be able to keep up with them and the resulting regulation will be considerably worse than specified. Generally, the better the regulation, the slower the supply responds to changes of line and load conditions, since high-gain feedback circuits must be used. The regulation may also depend on the output power level, a point worth checking in the specifications.

As well as maintaining a constant average d.c. level, instantaneous values of the output voltage should not deviate appreciably from the average. Deviations come from ripple at twice the line frequency for dissipative regulators, at the switching frequency for switching regulators, and from transients originating in the regulating circuit. Specification of the rms ripple gives an indication of the effectiveness of the filtering circuits; however, it gives no indication of the presence of short-duration large-amplitude voltage spikes on the output. A maximum peak-to-peak noise specification is used to describe this kind of instantaneous voltage deviation. As with regulation, ripple and noise may depend upon the power level.

Even with a constant line voltage and constant load, the output voltage can change if the temperature changes. This is generally indicated by a *temperature-coefficient* specification, which gives the percentage change in output voltage per degree change in temperature about a specified temperature. For laboratory applications, this is usually not a critical specification; however, when working in extreme environments it may be significant. (For highly compact units that depend on forced-air cooling or free-convection cooling, adequate space must be provided in the mounting to prevent overheating.)

Though often not included in the specification list, r.f. noise can be an important consideration for supplies used in the vicinity of sensitive wide-band amplifiers and when trying to extract a weak signal from noise. Some types of power supplies produce more r.f. noise than others because of the design of the regulating circuit. Switching regulators with SCRs are particularly poor in this regard.

Other factors to consider in a power supply are *bipolar operation* (can both the negative and positive terminals be grounded to give positive and negative output voltages, respectively?) and insulation from ground if floating the supply is anticipated. Operation is simplified by front-panel meters indicating both output voltage and current, overload reset switches, and convenient output terminal configuration and location.

A single output power supply can never be used to supply two voltages, one negative and one positive with respect to ground. Often it is necessary to have such voltages for operational-amplifier circuits, and two supplies must be used.

6.5.2 Regulator Circuits and Programmable Power Supplies

A block diagram of a dissipative regulator circuit in a power supply is illustrated in Figure 6.73. The sensing resistors R_1 and R_2 are across the output of the power supply. A fraction $R_2/(R_1 + R_2)$ of the output voltage is sent to a difference amplifier, where it is compared with a reference voltage, and the amplified difference used to control the *pass element*, which in this case is a transistor in the common-emitter configuration. Depending on the output of the difference amplifier, the pass-element output voltage will adjust itself to a value just sufficient to sustain the required input from the difference amplifier. If for some reason the output voltage rises, the output of the difference amplifier will decrease and the pass-element output will decrease. The opposite occurs for a falling output. The series dissipative regulator is inefficient because the difference in voltage between the raw d.c. input and regulated d.c. output falls across the pass element, which must then dissipate power equal to this voltage drop times the output current.

More efficient power supplies use circuits to control the duty cycle of the pass element. In these circuits the pass element (a transistor, SCR or other solid state device) acts as a switch and is either on (no voltage drop across it) or off (no current through it). The output voltage from such a regulator is

$$V_{\text{out}} = V_{\text{in}} \frac{t_{\text{on}}}{t_{\text{on}} + t_{\text{off}}},$$

Figure 6.73 Block diagram of the regulator circuit in a power supply.

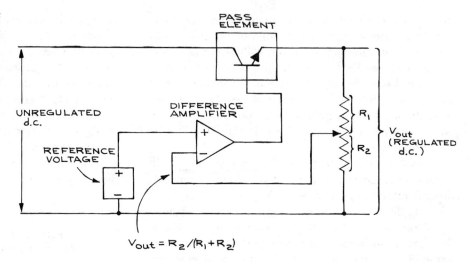

$$V_{out} = R_2/(R_1 + R_2)$$

where t_{on} is the time for which the pass element is on and t_{off} is the time for which it is off, in each cycle. The optimum switching frequency for switching regulators is between 20 and 100 kHz. At these high frequencies, rfi can be a problem.

In practice, regulator circuits can be very complex, with high-gain difference amplifiers or comparator preregulators, multiple pass elements, overload and overvoltage protection, and temperature compensation. The above description, however, is sufficient for an understanding of remote programmable power supplies. If R_1 and R_2 in Figure 6.73 are replaced by a potentiometer, a change in its setting will result in a change in the regulated output voltage. This is what occurs when one changes the dial settings on the front panel of a variable power supply. Clearly, the internal potentiometer can be replaced by an external potentiometer (*resistance programming*)—or even more directly, an external voltage source (*voltage programming*). The addition of such features requires very little modification to the basic power-supply regulator circuit, and they are often available at no cost or as low-cost options. Remote programming can be particularly useful in control operations, where a digital code from a computer is changed to a voltage by a digital-to-analog-converter (DAC) and this voltage is used to set the output of a power supply.

Overvoltage protection is an important feature in a power supply. With such protection, one is assured that no failure will result in a high voltage across the output terminals of the supply. *Crowbar* circuits using SCR switches are often used for overvoltage protection. Such circuits short the output terminals of the supply when the output voltage exceeds a preset value. The resulting short circuit current causes a fuse to blow or a circuit breaker to open, shutting down the power supply until the fault that caused the overvoltage is corrected. Since it is essential that the protection circuit respond quickly, the speed of response should be included in the specifications.

Another built-in safety feature found on power supplies is *foldback current limiting*. When the output current exceeds a specified limit, it is automatically reduced by such circuits to a low level and maintained there for a specified time before being reset. Current-limiting circuits are triggered by temperature-sensitive or power-dissipating elements.

In addition to variable power supplies there are a vast number of fixed-voltage power supplies on the market. When working with logic circuits and operational amplifier circuits, only fixed power-supply voltages are required, and the additional expense of a variable supply can be avoided.

Most fixed-voltage supplies are designed to operate from the a.c. line, though there are a number of d.c.-to-d.c. converters. The least expensive configuration is the card power supply [Figure 6.74(*a*)]. All components are

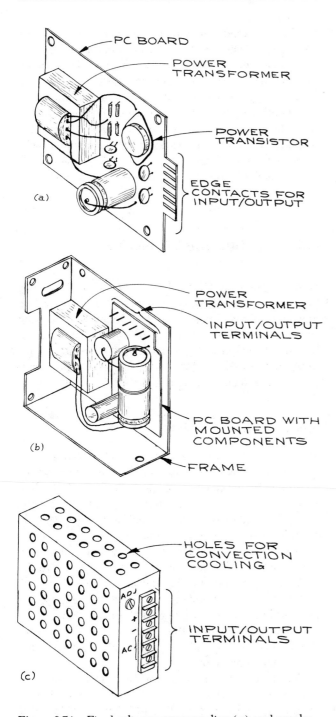

PC BOARD

POWER TRANSFORMER

POWER TRANSISTOR

EDGE CONTACTS FOR INPUT/OUTPUT

(a)

POWER TRANSFORMER

INPUT/OUTPUT TERMINALS

PC BOARD WITH MOUNTED COMPONENTS

(b)

FRAME

HOLES FOR CONVECTION COOLING

ADJ

AC

INPUT/OUTPUT TERMINALS

(c)

Figure 6.74 Fixed-voltage power supplies: (*a*) card supply; (*b*) frame supply; (*c*) enclosed case.

on a single printed circuit board (PCB) with edge contacts. The a.c. input must be brought to one set of contacts and the d.c. output taken from another. For routine use, the card supply should be mounted in a case or on a rack panel with an on-off switch, pilot light, fuse, and output terminals. This requires additional time and expense, which may cancel the initial cost savings from purchasing such a supply.

Frame-mounted supplies [Figure 6.74(*b*)] are similar to the card supplies in that all external connections, switches, and output terminals are missing. However, the metal frame facilitates mounting and affords some protection to the components. Rather than PCB contacts, frame supplies usually have screw-type barrier connectors.

Fixed-voltage supplies also come in enclosed cases and in encapsulated units suitable for direct connection to a printed circuit board [Figure 6.74(*c*)]. Many of the so-called fixed-voltage supplies can be adjusted over a 5–10% range about the nominal output voltage value. Multiple fixed-voltage units with two and even three voltages are also common. It is often useful to have ± 15 V and $+5$ V available from a single power supply when both linear and digital circuits are used together.

Manufacturers produce a wide range of supplies with different voltages, power ratings, and mechanical configurations. Thus it is rarely worthwhile to construct one's own power supply. A possible exception lies in the use of integrated-circuit modules. There are two general types: fixed and variable output. The fixed-output units are three-terminal devices; unregulated d.c. is applied across the input and common terminals, and regulated d.c. at the appropriate voltage appears between the output and common terminals. A representative unit of this type is the LM109 5-V, 1.5-A monolithic regulator. The regulator combines on-chip thermal shutdown with current limiting, an on-chip series pass transistor, and a stable internal voltage reference. Further details about the LM109 and its use in other applications are available from the manufacturer.[6]

The LM105 is an example of a variable-voltage integrated-circuit regulator. For this unit the output is set by the ratio of two external resistances. The maximum output voltage is 40 V, and the maximum output current is 20 mA. When an external pass transistor is added

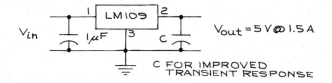

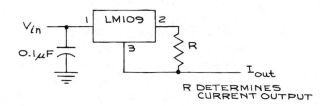

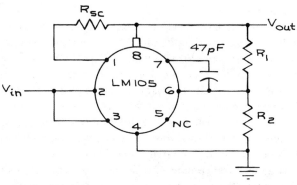

$$\frac{R_2}{R_1 + R_2} V_{out} = 1.8$$

$$\frac{R_1 R_2}{R_1 + R_2} = 2\,k\Omega$$

$$R_{sc} = \frac{325}{I_{sc}}$$

1 CURRENT LIMIT
2 BOOSTER OUTPUT
3 UNREGULATED
   INPUT
4 GROUND
5 REFERENCE
   BYPASS
6 FEEDBACK
7 COMPENSATION
   SHUTDOWN
8 REGULATED
   OUTPUT

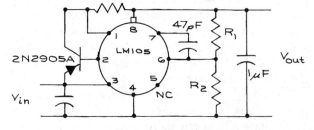

Figure 6.75 Examples of circuits with the LM109 fixed-voltage regulator and the LM105 variable-voltage regulator.

to the basic regulator circuit, the load current can be increased to 500 mA. The LM105 can also be used as a switching regulator. Data and application sheets should be consulted when using such regulators.

Typical power-supply circuits employing integrated regulators are illustrated in Figure 6.75. These circuits show how simple it is to use the voltage regulators. Fixed-voltage units are made by several manufacturers at several fixed voltages from 4 to 20 V. Variable units are available for both positive and negative regulation.

6.6 DIGITAL ELECTRONICS

Digital systems are based on circuit elements (usually transistors) operated in a way such that they exist in only one of two states. This is in contrast to analog systems where the outputs are continuous functions of the input variables. Combinations of two-state, or *binary*, devices can perform arithmetic and logic operations of any degree of complexity.

6.6.1 Binary Counting

With a binary system it is usual to call the states 0 and 1. Combinations of 0's and 1's can then be used for counting. Some schemes are given in Table 6.21. Note that the *least significant bit* (LSB) or column is 2^0, the next 2^1, and the last or *most significant bit* (MSB) is 2^4. In other words, the binary number system is merely a base-2 system, while the decimal system is base 10. Other systems derived from the binary system are the octal (base 8) and the hexadecimal (base 16) (Table 6.21). In the binary system the digit 2 does not appear, and in the octal system 8 does not appear. The hexadecimal system substitutes the letters A, B, C, D, E, F for the decimal numbers 10, 11, 12, 13, 14, 15, for which a single symbol does not exist.

The logic state of the output of a logic device can depend on the voltage level or the presence or absence of a pulse within a given time window. A voltage-level system in which the most positive voltage level corresponds to logical 1 is a *positive-logic* system. One in

Table 6.21 SYSTEMS OF NUMERATION

Decimal	Binary	Octal	Hexadecimal
0	0000	00	0
1	0001	01	1
2	0010	02	2
3	0011	03	3
4	0100	04	4
5	0101	05	5
6	0110	06	6
7	0111	07	7
8	1000	10	8
9	1001	11	9
10	1010	12	A
11	1011	13	B
12	1100	14	C
13	1101	15	D
14	1110	16	E
15	1111	17	F

↗ ↖
MSB LSB

which the least positive level corresponds to logical 1 is a *negative-logic* system.

6.6.2 Elementary Functions

The basic logic gates, along with the corresponding relationships between input and output variables, are given in Table 6.22 in *truth-table* form.

6.6.3 Boolean Algebra

The rules for manipulating logic expressions based on binary logic were formulated in the nineteenth century by George Boole. The laws and identities are given in Table 6.23. Boolean algebra is useful for implementing complex truth tables by means of the fewest possible gates. Consider for example the three-variable (A, B, C), two-function (X, Y) truth table at the top of Figure 6.76. The Boolean algebra expression for X and Y is obtained by looking at the state of each of the variables (A, B, C) for which the function under consideration (X or Y) is in the 1 state. For example, $X = 1$ when $A = 0$, $B = 0$, and $C = 1$. By convention, a bar over a variable symbol indicates negation, so that if $A = 0$, $\overline{A} = 1$. The

A	B	C	X	Y
0	0	0	0	1
0	0	1	1	0
0	1	0	1	1
0	1	1	0	1
1	0	0	1	0
1	0	1	0	0
1	1	0	0	1
1	1	1	1	0

$$X = (\overline{A} \cdot \overline{B} \cdot C) + (\overline{A} \cdot B \cdot \overline{C}) + (A \cdot \overline{B} \cdot \overline{C}) + (A \cdot B \cdot C)$$
$$Y = (\overline{A} \cdot \overline{B} \cdot \overline{C}) + (\overline{A} \cdot B \cdot \overline{C}) + (\overline{A} \cdot B \cdot C) + (A \cdot B \cdot \overline{C}),$$

WHICH SIMPLIFY TO

EXCLUSIVE OR

$$X = (\overline{A} \cdot \overline{B} + A \cdot B) \cdot C + (\overline{A} \cdot B + A \cdot \overline{B}) \cdot \overline{C}$$
$$Y = \overline{A} \cdot B + (\overline{A} \cdot \overline{B} + A \cdot B) \cdot \overline{C}$$

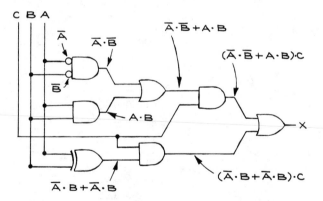

Figure 6.76 Example of a truth table, the equivalent Boolean algebraic expression, and implementation with elementary gates.

coincidence of 1's at $\overline{A}$, $\overline{B}$, and C to give $X = 1$ requires an AND operation, which is the first term in the complete expression for X in Figure 6.76. The other terms are derived from the truth table in the same way, and they are all connected together by OR operations. After simplification by application of the laws in Table 6.23,

Table 6.22   LOGIC GATES

Name	Symbol	Truth Table	Boolean Expression
OR		A B Y 0 0 0 0 1 1 1 0 1 1 1 1	$Y = A + B$
AND (coincidence)		A B Y 0 0 0 0 1 0 1 0 0 1 1 1	$Y = A \cdot B$
NOT (inverter)		A Y 0 1 1 0	$Y = \bar{A}$
NOR		A B Y 0 0 1 0 1 0 1 0 0 1 1 0	$Y = \overline{A + B}$
NAND		A B Y 0 0 1 0 1 1 1 0 1 1 1 0	$Y = \overline{A \cdot B}$
XOR (exclusive OR, anticoincidence)		A B Y 0 0 0 0 1 1 1 0 1 1 1 0	$Y = (B \cdot \bar{A}) + (\bar{B} \cdot A)$
XNOR (exclusive NOR, equivalence)		A B Y 0 0 1 0 1 0 1 0 0 1 1 1	$Y = \bar{A} \cdot \bar{B} + AB$

the expressions can be implemented with elementary gates. The gates for the function X are shown at the bottom of Figure 6.76.

Another way of simplifying truth tables is by making a logic map called a Karnaugh map. The map for the truth table of Figure 6.76 is shown in Figure 6.77. The squares corresponding to the variables for which X and Y are 1 are shaded. Where there are contiguous shaded squares corresponding to both values of one variable, say C, with the other variables constant, the value of the function is independent of the state of C. Thus, in the map for Y, it can be seen that the squares for $\bar{A} \cdot B$ are

Table 6.23   BOOLEAN ALGEBRA

Laws	*Identities*
Commutative	$A + A = A$
$\quad A + B = B + A$	$A \cdot A = A$
$\quad A \cdot B = B \cdot A$	$A + 1 = 1$
	$A \cdot 1 = A$
Distributive	$A + 0 = A$
$\quad A \cdot (B + C) = A \cdot B + A \cdot C$	$A \cdot 0 = 0$
De Morgan	
$\quad \overline{A \cdot B \cdot C} = \bar{A} + \bar{B} + \bar{C}$	
$\quad \overline{A + B + C} = \bar{A} \cdot \bar{B} \cdot \bar{C}$	

shaded for C and $\bar{C}$; therefore, it follows that the state of Y is not dependent on the variable C for these cases.

6.6.4   Arithmetic Units

After elementary gates, the next simplest logic units are those that perform elementary binary arithmetic operations. Some common arithmetic units are given in Table 6.24. The truth table for the *half adder* follows the rules for binary addition with C the low-order bit of the sum and D the high-order bit. The *full adder* is an elaboration of the half adder with an extra input. When adding two binary numbers of more than one bit each, provision must be made for the carry operation. This is the function of the C_{n-1} terminal, which accepts the carry bit from the result of adding the preceding lower-order bits in the string (thus the designation C_{n-1}); the results

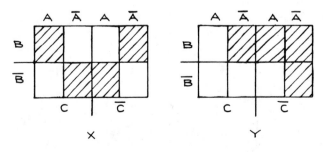

Figure 6.77   Karnaugh map of the truth table of Figure 6.76.

from the addition of the n order bits appear at terminals C_n and S_n. The *digital comparator* compares the magnitude of the digital signals at input A_n and B_n and activates output C_n, D_n, or E_n depending on whether $A_n > B_n$, $A_n < B_n$, or $A_n = B_n$. Such elements can be connected together to compare two n-bit binary numbers. The parity of a binary number is defined as even (0) if there are an even number of 1's and odd (1) if there are an odd number of 1's. The parity code is often used to check for transmission errors in bit strings. By sending a parity bit with an n-bit word and testing the received word against the parity bit, single bit errors can be detected. This is illustrated for a four-bit word with the *parity generator/checker* in Table 6.24.

6.6.5   Data Units

For data acquisition and transmission, *decoders*, *encoders*, *multiplexers*, and *demultiplexers* are commonly used. The *read-only memory* (ROM) is a combination of decoder and encoder and is frequently used to implement complex truth tables. *Programmable* and *erasable* ROMs, called PROMs and EPROMs, can have their internal logic modified for the particular application of the user. Table 6.25 gives the truth tables and input-output terminal arrangements for some simple units.

6.6.6   Dynamic Systems

The logic units described so far are considered to be static because the output levels depend only on the input levels at the time of observation. The previous history of the signal levels does not matter. With dynamic systems, output signal levels depend on the history of the signal levels at the input terminals. For this reason the truth tables for dynamic systems must specify previous states of the inputs and outputs in order for the state of a particular output to be completely determined. Subscripts are used to differentiate the time sequence of the output states. Table 6.26 lists a number of *flip-flops*, which are the basic building blocks in dynamic systems. Flip-flops can be strung together to

Table 6.24   ARITHMETIC UNITS

Name	Symbol	Truth Table

Half adder

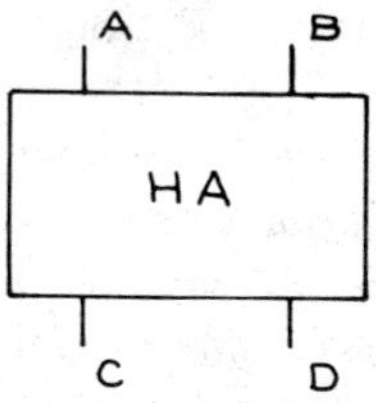

A	B	Sum	C	D
0	0	00	0	0
0	1	01	0	1
1	0	01	0	1
1	1	10	1	0

Full adder

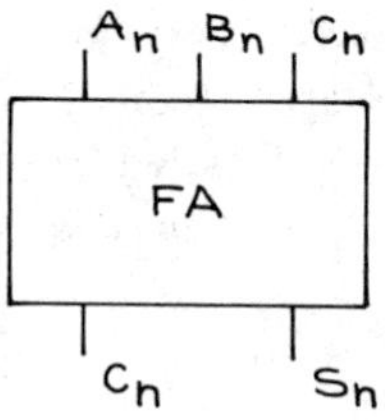

A_n	B_n	C_{n-1}	S_n (sum)	C_n (carry)
0	0	0	0	0
0	1	0	1	0
1	0	0	1	0
1	1	0	0	1
0	0	1	1	0
0	1	1	0	1
1	0	1	0	1
1	1	1	1	1

Digital comparator

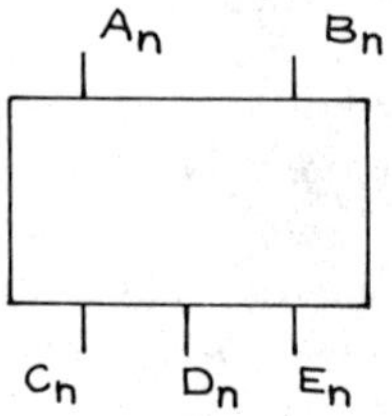

A_n	B_n	$C_n\,(A_n > B_n)$	$D_n\,(B_n > A_n)$	$E_n\,(A_n = B_n)$
0	0	0	0	1
0	1	0	1	0
1	0	1	0	0
1	1	0	0	1

Parity generator/checker

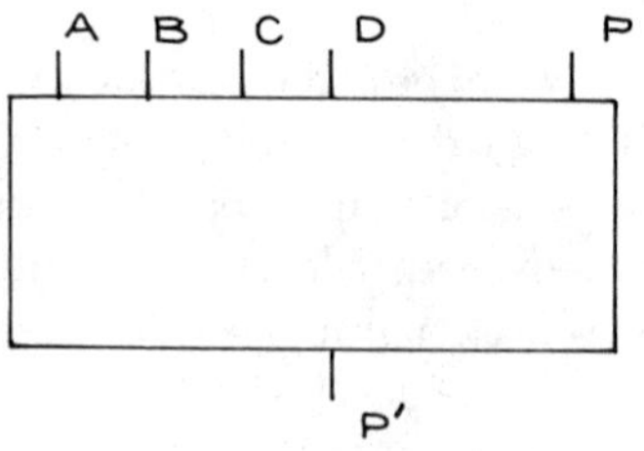

4-bit word				Parity bit	
A	B	C	D	P	P'
0	0	0	0	0	0
0	0	0	1	1	0
0	0	1	0	1	0
0	0	1	1	0	0
0	1	0	0	1	0
0	1	0	1	0	0
0	1	1	0	0	0
0	1	1	1	1	0
1	0	0	0	1	0
1	0	0	1	0	0
1	0	1	0	0	0
1	0	1	1	1	0
1	1	0	0	0	0
1	1	0	1	1	0
1	1	1	0	1	0
1	1	1	1	0	0

Table 6.25 DATA UNITS

Unit	Truth Table

Decoder: Activates one of 2^n outputs according to an n-bit code.

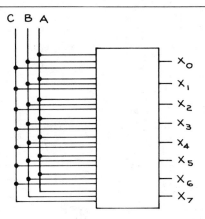

C	B	A	X_0	X_1	X_2	X_3	X_4	X_5	X_6	X_7
0	0	0	1	0	0	0	0	0	0	0
0	0	1	0	1	0	0	0	0	0	0
0	1	0	0	0	1	0	0	0	0	0
0	1	1	0	0	0	1	0	0	0	0
1	0	0	0	0	0	0	1	0	0	0
1	0	1	0	0	0	0	0	1	0	0
1	1	0	0	0	0	0	0	0	1	0
1	1	1	0	0	0	0	0	0	0	1

Encoder: Upon the activation of one of 2^n lines, an n-bit code is generated at the n-output terminals.

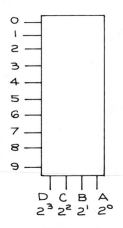

Input	D	C	B	A
0	0	0	0	0
1	0	0	0	1
2	0	0	1	0
3	0	0	1	1
4	0	1	0	0
5	0	1	0	1
6	0	1	1	0
7	0	1	1	1
8	1	0	0	0
9	1	0	0	1

Multiplexer: According to an n-bit code, one of 2^n signal input lines is connected to a single output line.

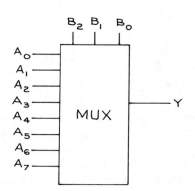

8-to-1 line multiplexer

B_2	B_1	B_0	Y
0	0	0	A_0
0	0	1	A_1
0	1	0	A_2
0	1	1	A_3
1	0	0	A_4
1	0	1	A_5
1	1	0	A_6
1	1	1	A_7

Table 6.25   (*continued*)

Unit		Truth Table

Demultiplexer: According to an *n*-bit code, a single input line is routed to one of 2^n output lines.

C	B	A	Y connected to
0	0	0	X_0
0	0	1	X_1
0	1	0	X_2
0	1	1	X_3
1	0	0	X_4
1	0	1	X_5
1	1	0	X_6
1	1	1	X_7

Read-only memory (ROM): Combination of a decoder and encoder to convert an *n*-bit code to an *m*-bit code, where *n* and *m* are not necessarily equal.

make *shift registers* and *counters*. A shift register is a series of J/K *flip-flops* with the Q and $\overline{Q}$ outputs of each stage connected to the J and K inputs of the next stage (Figure 6.78). Data are entered serially in the first unit, connected as a D *flip-flop*, and at each clock transition the data move from unit to unit. Shift registers can be used as counters, for serial-to-parallel data conversion using the parallel outputs, and for parallel-to-serial data conversion using the parallel inputs.

The basic serial or *ripple* binary counter is shown in Figure 6.79. Three J/K flip-flops can count to 2^3 in this arrangement. In the shift-register configuration eight flip-flops are needed. Modulo-*n* counters ($n \neq 2$) can be made, by decoding the output with gates and resetting

the flip-flops after *n* counts. Shift registers and counters are manufactured in a variety of configurations in single 14- and 16-pin DIP packages.

6.6.7 Digital-to-Analog Conversion

Whenever it is necessary to connect a digital system to an external device that operates in an analog mode, such as a thermocouple (temperature proportional to output voltage), pressure gauge (pressure proportional to output current), or chart recorder (deflection proportional to input voltage), converters are needed, either *digital-to-analog* (DAC) or *analog-to-digital* (ADC). Of the two,

Table 6.26   FLIP-FLOPS

Unit		Truth Tables

R/S (reset-set) flip-flop: The basic digital memory unit.

S	R	Q	$\overline{Q}$
0	1	0	1
1	0	1	0
0	0	Indeterminate	
1	1	Indeterminate	

J/K flip-flop: An elaboration of the R/S avoiding the indeterminate 1,1 and 0,0 states and with separate preset (Pr) and clear (Cr) inputs, and a clock (Ck) input that must be activated for the outputs to follow the input conditions.

Upon the application of a Ck pulse: [a]

J	K	Q_{n+1}	$\overline{Q}_{n+1}$
0	0	Q_n	$\overline{Q}_n$
0	1	0	1
1	0	1	0
1	1	$\overline{Q}_n$	Q_n

Direct inputs:

Pr	Cr	Q	$\overline{Q}$
0	0	Disallowed	
0	1	1	0
1	0	0	1
1	1	Normal clocked operation	

D (delay) flip-flop: Whatever is on the input appears at the output after a clock pulse. Can be constructed from a *J/K* flip-flop by putting an inverter from *J* to *K*, with the signal applied to *J*.

T (toggle) flip-flop: Output changes state at each clock transition. Used as a binary divider. Constructed from a J/K flip-flop by fixing J and K at 1.

[a] The subscript $n+1$ designates the $(n+1)$th state as distinct from the previous nth state. When R and S are 0, the output of Q remains constant, equal to its previous value.

Figure 6.78 *J/K* flip-flops
arranged as a 3-bit shift register.

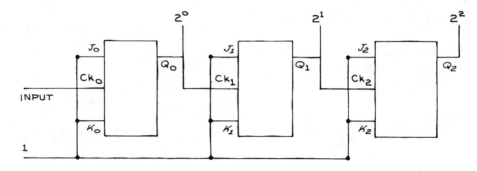

Figure 6.79 *J/K* flip-flops
arranged as an 8-bit ripple binary
counter.

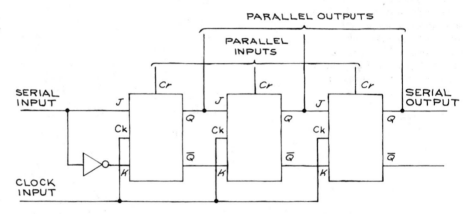

the DAC is the more fundamental, since many ADCs use DACs in their construction.

A DAC produces an output current or voltage proportional to the magnitude of the digital number at its input terminals (Figure 6.80). DACs operate by having the digital input activate FET analog switches, which connect the properly weighted current or voltage from the reference source to the output. In the case of voltage DACs, an operational amplifier in the summing mode produces the output voltage. Electrical specifications to consider with a DAC are *resolution*, the number of input bits or steps (an 8-bit DAC has 256 steps); *linearity*, the maximum deviation from the best straight line drawn through the graph of output versus digital input; and *accuracy*, the deviation of the output from that computed on the basis of the digital input. Linearity primarily depends on the resistors in the DAC circuit and the voltage drops across the internal transistor switches, so it is temperature dependent. Accuracy depends on the factors affecting linearity and also on the

reference voltage, either internal or external. In applications where speed of conversion is important, the *settling time* must be considered. This is the time necessary for the output to stabilize within $\pm \frac{1}{2}$ of the least significant bit of the final steady-state value. Other

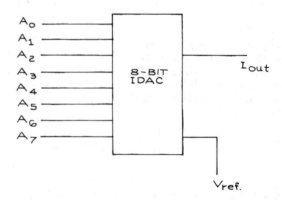

Figure 6.80 An 8-bit current digital-to-analog converter (IDAC).

considerations are the power-supply sensitivity, and the converter's compatibility with input logic and output levels. If a current DAC is used in an application where the output must be a voltage, a current-to-voltage converter is needed. For low-voltage applications this can be simply a resistor. When higher voltages are required, the operational-amplifier circuits shown in Figure 6.81 can be used. The characteristics of the amplifier should not degrade the expected performance of the DAC with respect to linearity, accuracy, and settling time.

A variation of the DAC is the *multiplying DAC*. Since the output of a DAC is directly proportional to the digital input and reference voltage, varying the reference voltage has the effect of multiplying the output by a constant. Digitally modulated waveforms can be easily produced with such devices.

There is a larger variety of ADCs than DACs because the complexity of the task lends itself to different techniques. Three ADC schemes are shown in Figure 6.82.

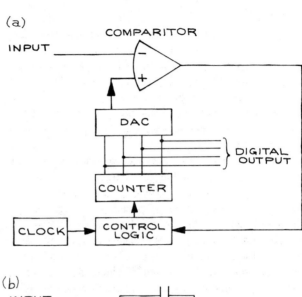

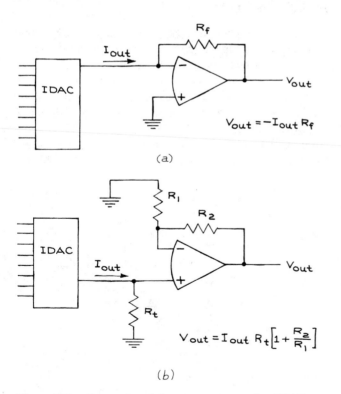

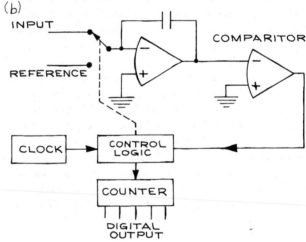

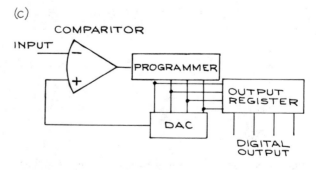

Figure 6.81 Conversion of the current output of an IDAC to a voltage output: (*a*) inverting; (*b*) noninverting.

Figure 6.82 Three analog-to-digital converters: (*a*) counter type; (*b*) dual slope; (*c*) successive approximation.

With the *counter ADC*, pulses from an oscillator or *clock* are routed to a counter through control logic, which is activated by the output of a comparator. The counter output goes to a DAC, the output of which is compared with the input signal. When the DAC output equals the input voltage, the comparator changes state and disconnects the clock from the counter. The digital number at the counter outputs is then proportional to the input voltage.

The *dual-slope ADC* converts voltage to time by using first the input voltage to charge a capacitor to a predetermined voltage and then discharging the capacitor by connecting it to a reference voltage of the opposite polarity. The input voltage is proportional to the ratio of the charge to the discharge time, which is recorded by a counter driven by a clock through control logic. So long as the capacitor, clock, and reference voltage are stable during the conversion time, the resolution will depend only on the resolution of the capacitor circuit. This method is the one most commonly used in digital multimeters.

The *successive-approximation ADC* works like the counter type except that a programmer, rather than a clock and counter, drives the DAC. The programmer starts by activating the DAC's most significant bit. If the DAC output is greater than the input, the programmer turns the MSB off and the next MSB on, and the comparison is done again. The process of comparison continues to the LSB, after which the conversion is complete. This method is fast and has high resolution.

As with DACs, resolution, linearity, accuracy, and settling time are important parameters used for characterizing the operation of a unit.

Modular data-acquisition systems or *data loggers* with an analog multiplexer (MUX), *sample-and-hold* (S/H) circuit, and ADC are available from many sources. A block diagram of such a system is shown in Figure 6.83. Each of the transducer-amplifier-filter circuits is connected to the sample-and-hold circuit for a fixed time through the multiplexer. The output of the sample-and-hold circuit is then digitally encoded by the ADC. The frequency at which the various inputs is sampled depends on the rate at which the signals are varying. An important theorem of sampling theory states that if a sampled signal contains no Fourier components higher than f_{max}, the signal can be recovered with no distortion if sampled at a rate of $2f_{max}$.

6.6.8 Memories

Large-scale integration (LSI) has made possible the production of high-density solid-state memories, which have almost completely displaced other types in many applications. The typical one-bit memory cell is an R/S flip-flop with supplementary gates shown in Figure 6.84. When the "address" and "write enable" lines are activated, the signal on the "write" line is registered on the "Q" line. It will remain there as long as the "write enable" line is not subsequently activated. To read the

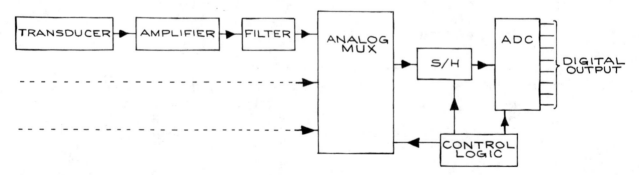

Figure 6.83 Multiple-input data-acquisition system with an analog multiplexer (MUX), sample-and-hold (S/H) circuit, and ADC.

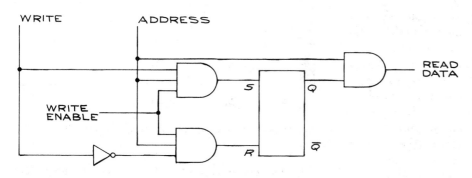

Figure 6.84 Basic memory cell.

contents of the memory cell, the "address" line has only to be activated. Large-scale memories are composed of hundreds to thousands of such elementary cells. Because the contents of any single cell can be read by activating the appropriate address line, such memories are called random-access memories (RAMs). There are several different ways of arranging the basic memory cells for addressing purposes. For example, a 1024-bit RAM can be arranged as 1024 one-bit words or 256 four-bit words. The addressing scheme in the first case requires 10 address lines, one "data out" line, and one "data in" line. For the 256×4 memory, eight address lines are needed, as well as four "data in" and four "data out" lines. In addition to the address and data lines a "read/write" (R/W) line and a "chip enable" (CE) line are needed. The state of the R/W line determines whether data are to be read into or out of the memory, and the CE line is activated when reading or writing data. *Dynamic* RAMs require an additional input, called the *refresh*, to hold the contents of the memory. Refreshing is required continuously; the signal for it is usually supplied from a clock. *Static* RAMs require no such refreshing. Solid-state RAMs are volatile, meaning that their contents are destroyed when the power is interrupted. Battery backup can be provided in applications where the memory contents must be saved in the event of a power failure.

Important parameters to consider for RAMs are power dissipation, access time, and cycle time. The *access time* is the time required after the activation of the address lines to obtain valid data at the output. The *cycle times* are the times required to complete a read- or write-data procedure. The access and cycle times are related to the

power dissipation: generally, the lower the power dissipation, the slower the memory.

Other considerations are power-supply voltages and input and output signal levels. The RAM must be able to accept data from external devices and in turn produce output levels capable of driving the circuits connected to its outputs.

6.6.9 Logic and Function

To illustrate the ideas in the preceding sections on digital logic, we now discuss a method for designing an interlock circuit for a vacuum system. High-vacuum systems often have a vacuum chamber connected to an oil diffusion pump through a gate valve (see Section 3.6.1). The diffusion pump has its high-pressure outlet connected to a mechanical rotary pump via a foreline and foreline valve. A bypass roughing line with a roughing valve is commonly used to initially evacuate the chamber. Before this is done, the foreline valve is closed. When the pressure in the chamber is sufficiently low, the roughing line is closed, the foreline is opened, and the chamber is pumped with the diffusion pump through the gate valve.

To guard against accidents it is worthwhile having a system of interlocks that activates valves in the proper sequence and turns the diffusion pump off in case of an accident. The interlock system consists, first, of *sensors* that continuously sample the variables in the vacuum system, such as the cooling-water temperature, cooling-water pressure, foreline pressure, vacuum-chamber pressure, and roughing-line pressure. The security of the system depends on whether the physical property sampled is above or below a predetermined threshold. The second section of the interlock system is the *logic*, which, with combinations of elementary gates, establishes the relationships between the input variables from the sensors and the output functions that control the various vacuum-system operations, such as the opening and closing of the various valves and the application of voltage to the diffusion-pump heaters. These output functions also only assume one of two values. The interlock system is thus amenable to binary-logic analysis either using Boolean algebra or Karnaugh maps. Examples of increasing complexity are given below. In addition, a way of implementing the system with a *data selector* is also discussed.

The simplest practical interlock system monitors the cooling-water temperature and pressure and turns off the voltage to the diffusion-pump heaters if the temperature is too high, the pressure too low, or both. A truth table relating the variables to the function is shown in Table 6.27. The letters and numbers in parentheses are logic symbols assigned to the variables (A, B), function (X), and states of the variables $(0, 1)$. The Boolean logic expression for the truth table is

$$X = \bar{A} \cdot \bar{B} = \overline{A + B} \, .$$

This is obtained by noting the condition under which X is in the logical 1 state. Had there been more than one occurrence of 1 for X, the expression would have additional OR terms. The Boolean expression can be implemented with the inverter-and-gate arrangement of Figure 6.85. A bimetallic-switch temperature sensor can provide the temperature variable, and a diaphragm-type pressure sensor the pressure variable. Power to the

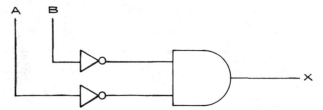

Figure 6.85   Gate arrangement corresponding to Table 6.27.

diffusion pump is controlled by a heavy-duty relay actuated by a sensitive relay driven by the output of the logic circuit. A practical circuit is shown in Figure 6.86. It is assumed that logical 0 is less than 0.8 volts and logical 1 is greater than 2.4 volts, consistent with TTL logic.

A more useful interlock system using three variables and two functions has the truth table of Table 6.28. The variables monitored are water pressure and temperature and diffusion-pump foreline pressure. The functions are the diffusion-pump heater (on or off) and the gate valve (open or closed). The Boolean expression and logic circuit are also given in the table, while the Karnaugh map is shown in Figure 6.87. In the map for Y the 1's are adjacent, resulting in the simplification that Y is independent of the state of C.

As a final example of logic design, consider increasing the number of variables to five by including the roughing-line pressure and chamber pressure and adding the roughing valve and foreline valve to the functions. With five variables there will be 2^5 or 32 combinations. Systematic design of the interlock circuit will ensure that none of them is left out. The truth table is given in Table 6.29. Rather than obtain the Boolean expressions for the truth table or make a map, we can use a *data selector* to establish the relationships among the variables and functions. A 5-bit data selector has 5-input address lines, 1 output line, and 32 data lines. Depending on the 5-bit input address, the data on the appropriate input line are transferred to the output line. Implementation of the expression for Y is shown in Figure 6.88(a). A 4-bit data selector [Figure 6.88(b)] can also be used by employing a foldback arrangement. Looking at the A, B, C, and D bits, one sees that there are 16 identical pairs corresponding to E being 1 or 0. For each of those pairs, the function is independent of

Table 6.27   TRUTH TABLE

Variables		Function
Water Pressure (A)	Water Temperature (B)	Diffusion Pump Heater (X)
Low (1)	Low (0)	Off (0)
Low (1)	High (1)	Off (0)
High (0)	Low (0)	On (1)
High (0)	High (1)	Off (0)

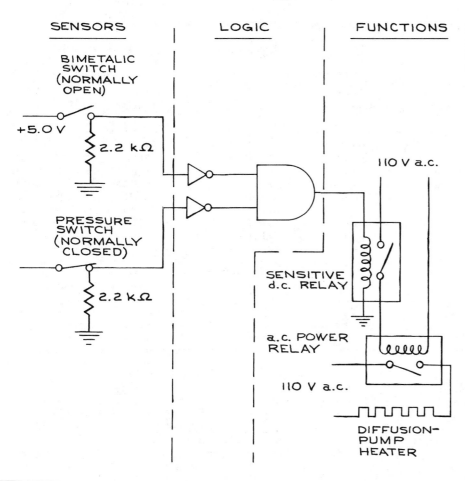

Figure 6.86 Practical circuit for a vacuum-system interlock according to the truth table of Table 6.27.

the state of E, equals E, or equals $\bar{E}$. Correspondingly, the 16 input lines of the 4-bit data selector will each have 0, 1, E, or $\bar{E}$ attached to them.

Even more complexity can be built into the circuit by

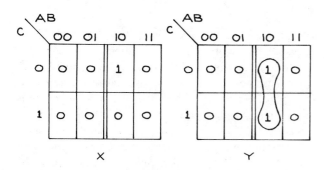

Figure 6.87 Karnaugh map of the truth table of Table 6.28.

having time delays on the functions so that one valve is not opening while the other is closing. Extra functions such as turning on and off and venting the mechanical pump can also be added. This systematic method of analysis ensures that nothing is overlooked. By noting when two functions change state upon the change of state of a variable, decisions about sequencing can be made.

6.6.10 Implementing Logic Functions

Implementing functions with transistor gates requires that the 0- and 1-states be associated with voltage or current levels. There are a large number of logic families, all of which have internally consistent 0- and 1-levels,

Table 6.28   THREE-VARIABLE, TWO-FUNCTION INTERLOCK

Truth Table

Variables			Function	
Water Pressure (A)	Water Temperature (B)	Foreline Pressure (C)	Diffusion-Pump Heater (X)	Gate Valve (Y)
low (0)	low (0)	low (0)	off (0)	closed (0)
low (0)	low (0)	high (1)	off (0)	closed (0)
low (0)	high (1)	low (0)	off (0)	closed (0)
low (0)	high (1)	high (1)	off (0)	closed (0)
high (1)	low (0)	low (0)	on (1)	open (1)
high (1)	low (0)	high (1)	off (0)	open (1)
high (1)	high (1)	low (0)	off (0)	closed (0)
high (1)	high (1)	high (1)	off (0)	closed (0)

Boolean Expressions

$$X = A \cdot \bar{B} \cdot \bar{C}$$
$$Y = A \cdot \bar{B} \cdot \bar{C} + A \cdot \bar{B} \cdot C$$
$$Y = A \cdot \bar{B}(C + \bar{C})$$
$$Y = A \cdot \bar{B}$$

Implementation

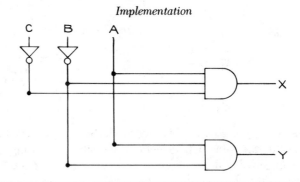

but which generally are not mutually compatible. The choice of a logic family for the implementation of a given function depends on a number of factors; among them are speed, power dissipation, immunity to noise, number of available functions, cost, and compatibility. A summary of the properties of the most common logic families is given in Table 6.30. As can be seen from the table, there is a tradeoff between speed and power dissipation, the fastest logic dissipating the most power

per gate and the slowest dissipating the least. *Noise immunity* is measured by *noise margin*, the meaning of which is illustrated in Figure 6.89. In general, logic families are not directly compatible; however, translator chips exist for interfamily conversion, and some simple conversion circuits are given in Fig. 6.90. The abbreviations SSI, MSI, LSI, and VLSI stand for small-scale integration, medium-scale integration, large-scale integration, and very large-scale integration and are

Table 6.29   FIVE-VARIABLE, FOUR-FUNCTION INTERLOCK TRUTH TABLE

T_{H_2O} (A)	P_{H_2O} (B)	$P_{foreline}$ (C)	$P_{roughing}$ (D)	$P_{chamber}$ (E)	Diffusion Pump (X)	Gate Valve (Y)	Roughing Valve (R)	Foreline Valve (W)
0 (low)	0 (low)	0 (low)	0 (low)	0 (low)	0 (off)	0 (closed)	0 (closed)	0 (closed)
0	0	0	0	1 (high)	0	0	0	0
0	0	0	1 (high)	0	0	0	0	0
0	0	0	1	1	0	0	0	0
0	0	1 (high)	0	0	0	0	0	0
0	0	1	0	1	0	0	0	0
0	0	1	1	0	0	0	0	0
0	0	1	1	1	0	0	0	0
0	1 (high)	0	0	0	1 (on)	1 (open)	0	1 (open)
0	1	0	0	1	0	0	1 (open)	0
0	1	0	1	0	1	1	0	1
0	1	0	1	1	0	0	1	0
0	1	1	0	0	0	0	0	0
0	1	1	0	1	0	0	1	0
0	1	1	1	0	0	0	0	1
0	1	1	1	1	0	0	1	0
1 (high)	0	0	0	0	0	0	0	0
1	0	0	0	1	0	0	0	0
1	0	0	1	0	0	0	0	0
1	0	0	1	1	0	0	0	0
1	0	1	0	0	0	0	0	0
1	0	1	0	1	0	0	0	0
1	0	1	1	0	0	0	0	0
1	0	1	1	1	0	0	0	0
1	1	0	0	0	0	0	0	0
1	1	0	0	1	0	0	0	0
1	1	0	1	0	0	0	0	0
1	1	0	1	1	0	0	0	0
1	1	1	0	0	0	0	0	0
1	1	1	0	1	0	0	0	0
1	1	1	1	0	0	0	0	0
1	1	1	1	1	0	0	0	0

used to specify the number of gates on a chip. SSI is less than 12 gates per chip, MSI is between 12 and 100, LSI is more than 100, and VLSI is more than 1000. Simple logic functions employ SSI and MSI.

The most common logic family is TTL (transistor-transistor logic) though CMOS (complementary-symmetry metal-oxide-semiconductor) finds wide application because of its low power consumption and wide range of usable power-supply voltages. The TTL line has expanded from the basic 74/54 line to a high-performance, high-speed series, the 74H/54S, while the newest 74LS/54LS series is very popular with designers because it has the performance of the older 74/54 series with one-fourth the power consumption. Emitter-coupled logic (ECL) is only used when the highest speeds are required. The ECL 10,000 series is a slower version of the ECL II series and is not so sensitive to the quality of the interconnections. With the II series, a circuit

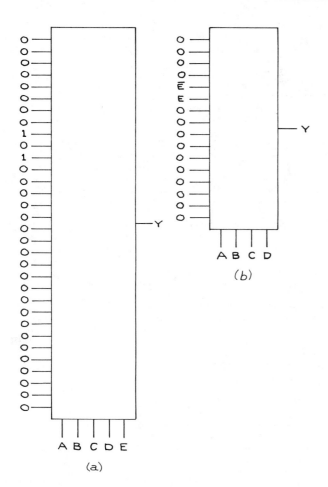

Figure 6.88 Implementation of the Y-function of the truth table of Table 6.28 with (*a*) a 5-bit data selector and (*b*) a 4-bit data selector using a foldback arrangement.

board with a ground plane must be used and the interconnectors must be carefully laid out to minimize cross coupling.

Because TTL is so common, some simple TTL circuits are given in Figure 6.91. The mechanical contact conditioner ensures that the output from a mechanical switch will be a single transition rather than the series of voltage spikes normally resulting from the mechanical bouncing of the contacts. TTL gates are either *edge-* or *level-triggered*. Edge triggering refers to the transition between the 0- and 1-levels (positive edge) or 1- and 0-levels (negative edge). Level triggering occurs when

the 0 or 1 is attained. Data sheets specify which type of triggering each logic unit requires.

With TTL, as with most logic families, there is the tendency to generate current spikes during switching, thus creating power-supply noise. To reduce this noise, the power supply must be capacitatively decoupled from the logic system. Generally, a $0.01\text{-}\mu\text{F}$ capacitor for each gate and an additional $0.1\text{-}\mu\text{F}$ capacitor for each 20 gates are required. Counters and shift registers are especially sensitive to power-supply noise and should be decoupled with a $0.1\text{-}\mu\text{F}$ capacitor for each two devices. For optimum performance, unused inputs should be set to 0 by direct connection to ground or 1 by connection to the power-supply voltage through a 1- to 10-KΩ resistor.

Some additional features of TTL are three-state outputs and open-collector outputs. A *three-state output* has the standard 0- and 1-levels and also a high-impedance nonactive state. In this state the output will assume the level of any active output connected to it. The three-state outputs are designed to be connected together in a common line or bus structure so that when any single output is active, all the others assume its level. The *open-collector output* requires a collector resistor to the power supply for operation. Open-collector outputs when connected together allow one to implement an output AND function (often incorrectly termed *wired*-OR); that is, when any output is 0, all become 0. Open-collector outputs have the disadvantage of reduced speed and high output impedance in the 1-state. They are usually used as line drivers.

6.7 DATA ACQUISITION

When designing an experiment, one important consideration is the rate at which data are collected. This is necessary because of the need to store and display the data in a form that can be used for subsequent reduction and analysis. Usually it is useful to think of an experiment in terms of the sampling rate, the sampling time, and the desired precision of the measurement.

Table 6.30   LOGIC FAMILIES

	TTL				ECL		CMOS		
					MECL[a]	MECL[a]	$V_{DD}=$	$V_{DD}=$	$V_{DD}=$
	54/74	54/74H	54S/74S	54LS/74LS	10,000	III	5.0 V	10.0 V	15.0 V
Propagation delay (ns)	10	6	3	10	2	1	115	55	40
Power dissipation (mW/gate)	10	22	19	2	25	60	.005	.020	.060
0-level (V)	0.2	0.2	0.2	0.2	−1.75	−1.75	.05	.05	.05
1-level (V)	3.9	3.9	3.9	3.9	−0.90	−0.90	4.95	9.95	14.95
Noise margin (V)	0.4	0.4	0.4	0.4	0.2	0.2	1.0	2.0	2.5
Fanout	10	10	10	20	10	6	50	50	50
Power supply (V)	+5	+5	+5	+5	−5.2	−5.2 (−2.0)	5.0	10.0	15.0

[a]Trademark of Motorola, Inc.

Consider a spectroscopy experiment where the output of a monochromator is incident on a photomultiplier, the output of which is pulses, which are counted with a counter. The monochromator is stepped at a regular rate at a predetermined step size over the wavelength interval of interest. One important parameter is the minimum time required for the measurement. The time spent at each step and the number of steps determine the total number of counts recorded. If the total number of signal counts in any interval is C_S and the number of noise counts is C_N, the total number of counts C_T is $C_S + C_N$. The statistical error in C_T is $\sqrt{C_T}$, and that in

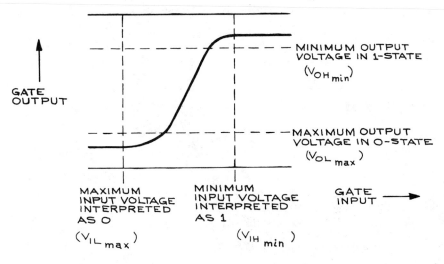

Figure 6.89   Graphical illustration of "noise margin." Noise margin for 0-state $= V_{OH_{min}} - V_{IH_{min}}$; noise margin for 1-state $= V_{IL_{max}} - V_{OL_{max}}$.

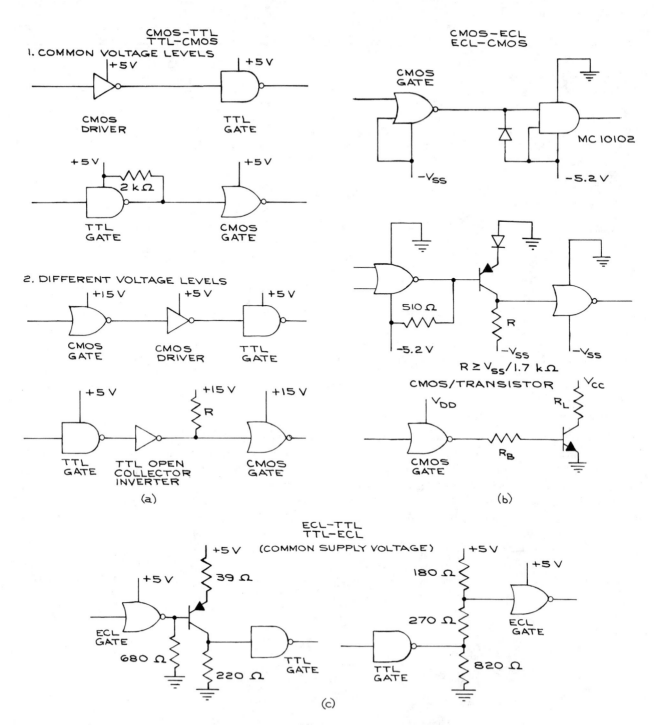

Figure 6.90 TTL, CMOS, ECL, and discrete transistor interface circuits: (*a*) CMOS-TTL and TTL-CMOS (*R* is determined by risetime and loading considerations) (from *McMOS Integrated Circuits Data Book*, Motorola, Inc., 1973); (*b*) CMOS-ECL and ECL-CMOS (*R_B* is determined by V_{DD}: 0 Ω for V_{DD} = 5 V, 500 Ω for 10 V, 1 kΩ for 15 V); (*c*) ECL-TTL and TTL-ECL (from *MECL System Design Handbook*, W. R. Blood, Jr., and E. C. Tynan, Eds., 2nd edition, Motorola Semiconductor Products, 1972).

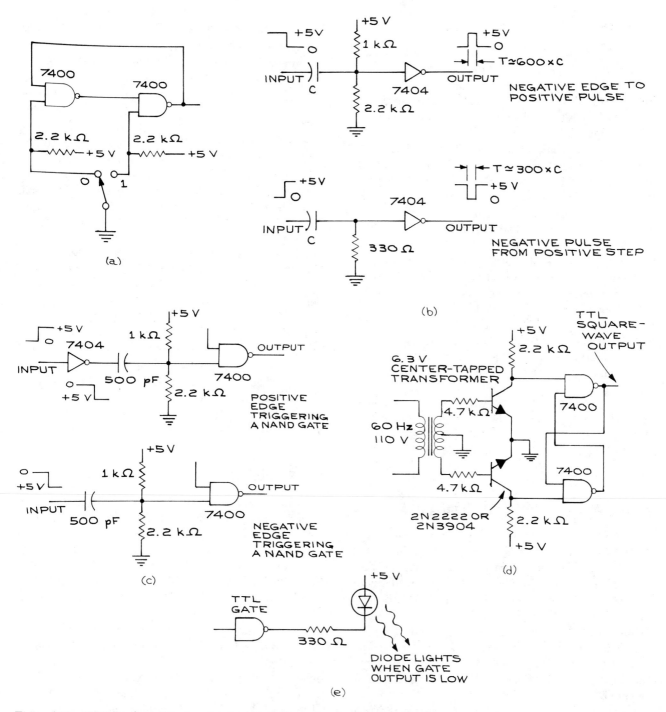

Figure 6.91 TTL signal-conditioning circuits: (*a*) mechanical-contact conditioner; (*b*) pulses from steps (input-step duration must exceed output-pulse width); (*c*) edge triggering; (*d*) 60-Hz TTL square wave from the power line; (*e*) TTL to LED.

C_N is $\sqrt{C_N}$. The error in determining C_S is therefore $\sqrt{C_T + C_N}$, so that the relative error is $\sqrt{C_T + C_N}/(C_T - C_N)$. The counts increase linearly with the time of measurement, but the relative error only decreases as the inverse square root of the time of the measurement. The *data rate* is the number of counts in a measurement interval divided by the interval time.

Bits, bytes, words, and *characters* are all used to specify the amount of information. The number of *bits* necessary to specify a given integer decimal D is given by $N = 3.32\log_{10} D$. If N is noninteger, it is to be rounded to the next higher integer. If the decimal number is to be signed (that is, plus or minus), another bit should be added. Eight bits make a *byte*, and 2 bytes (16 bits) commonly, but not always, make a *word*.

A *character* is the equivalent of one word. If data are to be transferred from the counter to a printer or other display or storage device, the data acceptance rate of the device must match the data output rate of the counter. Data rates are often specified as a *baud rate*, which is the number of times per second there is a transition on the signal line from a 0 to a 1 or from a 1 to a 0.

For compatibility between two devices, a number of parameters must match besides the rate. These include voltage levels, timing, format, code, and hand-shaking protocols and housekeeping procedures.

6.7.1 Voltage Levels and Timing

The most positive and most negative voltages corresponding to the two logic states must match. For example, the TTL logic levels are $+3.2$ and $+0.2$ V, while the ECL levels are -0.75 and -1.55 V. Furthermore, the sign of the logic must be identical, that is, positive or negative. *Positive logic* means that the more positive voltage level is taken as a logical 1, and the more negative as logical 0. The reverse is true for *negative logic*. If the signs do not match and everything else is satisfactory, only inversion of the input signal will be required.

Often a digital input or output circuit is compatible with a logic family of which it is not a member. For example, MOSFET logic exists that is *TTL-compatible*, which means that the voltage levels are compatible. The sinking and sourcing capabilities may be very different, however, for so-called compatible logic families. *Sinking* has to do with the number of gates that can be attached to the output of a single output gate and still maintain reliable operation. With TTL logic, when an output is at logical 0 it must be able to hold all inputs connected to it at logical 0. Since this condition is a consequence of the input transistors being in saturation, the output must be able to sink all the saturation currents and still keep the inputs at their required levels. A TTL output at logical 1 attached to a TTL input will turn off the input transistor, and very little current will flow in the input circuit. This is the *sourcing* specification. *Fan-in* and *fan-out* are directly related to the sinking and sourcing capabilities of a gate. Fan-in is the number of logic inputs to a gate, while fan-out is the number of logic outputs it can drive. These specifications should be carefully considered when connecting two different logic families. Timing is also important: the required rise and fall times as well as pulse widths must be compatible. Timing diagrams, which show the relationship between signals, are invaluable in this regard. Converter chips that convert the signals from one logic family directly to another are readily available.

6.7.2 Format

A *serial format* has all data on a single line in a sequential pattern. This is economical as far as wiring is concerned, but slow. A teletype works with a serial format. Both input and output data are transferred over a single line. With a *parallel format* there are as many data lines as bits, and all the data are transferred simultaneously. This is the fastest way to transmit data, but the least economical.

There are also *parallel-serial* formats where the data are transmitted as a series of parallel codes. For example, an eight-bit byte could be transmitted with four lines as two four-bit codes. A synchronization code is necessary with such an arrangement to be able to distinguish between the beginning and end of a single byte string. It is possible to convert from a serial code to a parallel code and vice versa with converters based on shift registers. For serial-to-parallel conversion, data are injected into the registers from one end and then read from the outputs of each stage. For parallel-to-serial conversion, data are put into each register simultaneously and then extracted from the last register a bit at a time. Data transmission in one direction over a single line is called *simplex*; in two directions over a single line, *half duplex;* and in two directions simultaneously, with a single line and two coding frequencies or two separate lines, *full duplex.* Complete ICs made for data conversion are called *UARTs* (universal asynchronous receiver-transmitters). They take a wide variety of data formats and convert them to standard codes, one of which is ASCII (American Standard Computer Information Interchange). The ASCII computer code is given in Table 6.31.

A major difficulty in setting up data-handling equipment is the interface. In the days when electromechanical teletypes were the most common form of terminal, the 20-mA current loop was the standard interface. Data transfer rates are limited to less than 100 bps (bits per second) with this interface. It has now been superseded by the RS232C serial interface, which can handle data rates to 20 kbps. This interface has a high impedance (3–5 kΩ), uses high voltage (± 25 V), and is unbalanced and therefore noise-susceptible. The pin assignments for the standard 25-pin connector used with this interface are given in Figure 6.92. More advanced interfaces such as the RS-449 use balanced circuitry and can handle data rates well in excess of 20 kbps.

Modems (modulator-demodulators) are devices for converting a binary code to audio frequencies (300–3300 Hz) for transmission over telephone lines, and reconverting the audio frequencies to binary. Commonly, they are separate units to which both a telephone receiver and a data terminal are attached, but some modems are incorporated directly in the telephone base.

6.7.3 System Overhead

In all data-transmission schemes there are additional signals that are necessary to assure that the data transfer occurs correctly. These are sometimes called *housekeep-*

Table 6.31   ASCII COMPUTER CODE

b_4 ↓	b_3 ↓	b_2 ↓	b_1 ↓	0 0 0 ← b_7	0 0 1 ← b_6	0 1 0 ← b_5	0 1 1	1 0 0	1 0 1	1 1 0	1 1 1
0	0	0	0	NUL	DLE	SP	0	@	P	\	p
0	0	0	1	SOH	DC1	!	1	A	Q	a	q
0	0	1	0	STX	DC2	"	2	B	R	b	r
0	0	1	1	ETX	DC3	#	3	C	S	c	s
0	1	0	0	EOT	DC4	$	4	D	T	d	t
0	1	0	1	ENQ	NAK	%	5	E	U	e	u
0	1	1	0	ACK	SYN	&	6	F	V	f	v
0	1	1	1	BEL	ETB	'	7	G	W	g	w
1	0	0	0	BS	CAN	(	8	H	X	h	x
1	0	0	1	HT	EM	)	9	I	Y	i	y
1	0	1	0	LF	SUB	*	:	J	Z	j	z
1	0	1	1	VT	ESC	+	;	K	[	k	⟨
1	1	0	0	FF	FS	,	⟨	L	\	l	|
1	1	0	1	CR	GS	–	=	M	]	m	⟩
1	1	1	0	SO	RS	·	⟩	N	∧	n	~
1	1	1	1	SI	US	/	?	O	___	o	DEL

Note: This is a 7-bit code to give $2^7 = 128$ different words. Columns 1 and 2 are machine commands such as carriage return (CR), line feed (LF), escape (ESC), etc.

ing signals, or *hand-shaking* in the case where such information is exchanged back and forth between the sending unit and the receiving unit. These additional signals contribute to what is called the *system overhead*, that is, additional data capacity which must exist regardless of the useful information transmitted. When data are transferred at a regular rate between two units, they must be kept in step. A *sync* bit is often used to assure synchronization between the two units. When data are transferred at a variable rate that depends on other independent parameters of the system, they are said to be asynchronous. In this case the sending unit must alert the receiving unit of the coming of data, and the receiving unit then should give a "ready to receive" signal, after which the data are transferred. When data transfer is complete an "end of data" signal is also usually sent. Without these signals no reliable data transfer can be effected.

To ensure the reliability of data transfer and detect errors in the transmission, an extra *parity bit* can be sent. One scheme requires the parity bit to be 0 when

Figure 6.92 RS232C serial interface: 25-contact connector pin assignment.

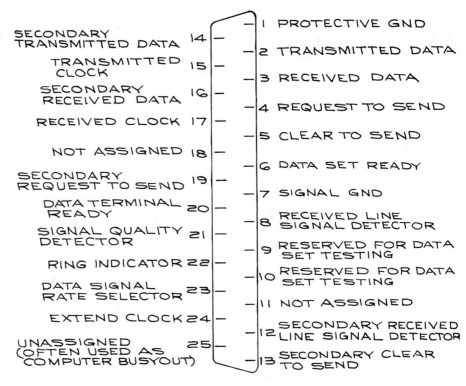

the sum of the data bits is even and 1 when the sum is odd. At the receiving unit data bits are summed and compared with the parity bit, and if there is consistency the data are accepted; if not, rejected. A more elaborate way of ensuring integrity is to have the receiving unit echo the received data back to the sender, where they are compared with the original data. Any differences cause the data to be rejected. This system minimizes transmission errors at the cost of additional complexity and lower speed.

Returning to the example of the spectroscopy experiment, if the expected signal produces 10^4 counts per second and the background is 10 counts per second, the statistical error after one second is $\sqrt{10^4 + 10}\,/10^4 \simeq 1\%$. If the counting time is one second, at least 13 bits are needed for the data alone. To be safe, 16 bits should be used, with the extra bits reserved for housekeeping functions. This data transfer rate of (2 bytes per second) is very modest. However, it is necessary to consider the

true rate of data transfer. If the data are held on the output lines of the counter for only 10 ms, the effective rate is 200 bytes per second.

To circumvent this problem, a *latch* can be used on the output lines to hold the data for a predetermined amount of time regardless of the subsequent status of the lines. The operation of a 4-bit latch is illustrated in Figure 6.93. Data at the input lines to the latch (A_0, A_1, A_2, and A_3) are transferred to the output lines (B_0, B_1, B_2, and B_3) when the control line is activated. In the absence of a start signal on the control line, the output levels cannot change even though the input levels do. In this way data can be held on the output lines for a time sufficient for them to be processed by the subsequent data-receiving unit. Obviously, the timing relations among the various signals are important for proper operation of the latch. The control signal must arrive when there are valid data on the input lines.

A data *buffer* serves a purpose similar to a latch. It

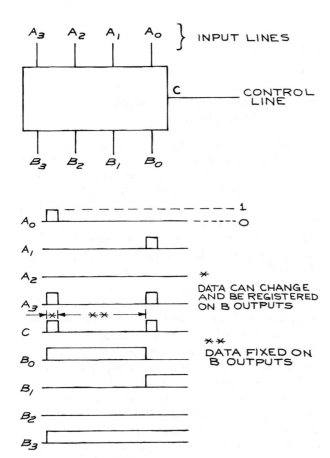

Figure 6.93   A 4-bit latch.

holds received data for processing by subsequent units or circuits. Printers, which are inherently slow because they are mechanical, often have data buffers that hold the input data until the printing mechanisms can transfer them to paper. Then the buffer is ready to accept another batch of data.

6.7.4 Analog Input Signals

The spectroscopy experiment used as an example would be more complicated if the output of the detector, instead of being pulses, were a current proportional to the incident light intensity. To convert the current to a digital signal requires an analog-to-digital converter (ADC). The resolution of the ADC is determined by the number of output bits, and the analog signal must be compatible both in magnitude and in sign with the input requirements of the ADC. If the ADC accepts only a voltage with a full-scale maximum value of 1.0 V and the analog signal is a current of 10^{-8} A maximum, we may clearly use a current-to-voltage converter using an operational amplifier with a variable-gain element. Response and sampling times are again a consideration. The analog input to the ADC must remain at its input terminals long enough for its outputs to respond. This is called the *settling time*. At times less than the settling time the ADC output will not be a valid reflection of the analog input. A block diagram of a possible arrangement is shown in Figure 6.94.

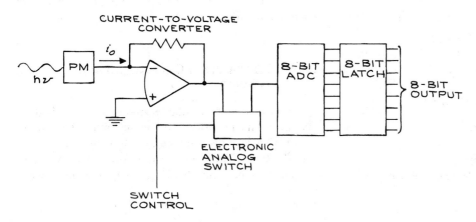

Figure 6.94 Example of a light-measuring system using an ADC with a sample-and-hold circuit at the input.

For the systems under consideration the output current must not be changing faster than the ADC can respond. When the analog signal is changing too rapidly, a sample-and-hold (S/H) circuit can be used ahead of the ADC. Important parameters of a S/H circuit are the sampling time and the output decay time. S/H circuits use a capacitor as a memory element, and the sampling time is proportional to its capacitance. Because of leakage, capacitors cannot permanently maintain their voltage; fast sampling circuits with small capacitors maintain their output voltages for much less time than slower circuits. The time for the output to fall to a fixed fraction of the original value is the *decay time*. A figure of merit for S/H circuits is the ratio of sampling time to decay time. For the simple S/H circuit shown in Figure 6.95, a control signal closes the MOSFET switch, and the 1-μF capacitor is charged with a time constant $R_{on}C$, where R_{on} is the "on" resistance of the switch. Typical values of R_{on} are 10^2 to 10^4 Ω, giving a time constant of 10^{-4} to 10^{-2} sec for the sampling time. The decay time depends on leakage of the capacitor, the "off" resistance of the MOSFET and the bias currents of the operational amplifier. When high-quality capacitors with polyethylene, polystyrene, polycarbonate, or Teflon dielectrics are used, most of the decay will be due to current through the switch ($\approx 10^{-9}$ A) and bias currents (10^{-9}–10^{-8} A for an average low-bias-current operational amplifier). In this example the decay will be of the order of 2 to 20 mV/sec. Special MOSFET switches and picoampere-bias-current operational amplifiers can reduce these rates even further.

6.7.5 Multiple Signal Sources: Data Loggers

A multiple-transducer data-acquisition system is a straightforward extension of the single-transducer system already described. Figure 6.96 shows one possible arrangement. Such systems are useful when it is necessary to monitor and record several different signals at virtually the same time. If the transducers all have analog outputs, the most convenient arrangement is an *analog multiplexer*. The function of the multiplexer is to alternately connect each of the analog signal lines to an ADC circuit and the recording instrument. The switching between the various signal lines is accomplished by separate control lines. Multiplexing reduces the number of circuit components at the expense of more complex control and synchronization circuits. With the system in Figure 6.96, the data rate is potentially four times that of the simple system.

A number of complete data-acquisition units specifically designed for the laboratory are commercially available. So long as the input signals are compatible and the data rates are within the limits of the unit, they offer an excellent solution to data-acquisition problems. The problem of data acquisition from several transducers occurs often in industry. Industrial data loggers, however, are designed for reliability rather than flexibility and are generally not suited to laboratory situations, where many different kinds of transducers are used.

6.7.6 Standardized Data-Acquisition Systems

There have been two notable attempts to standardize laboratory data acquisition and manipulation. The earlier is NIM (Nuclear Instrumentation Module) which dates from the 1950s. The more recent CAMAC uses a digital computer to perform the logic that NIM does with cables.

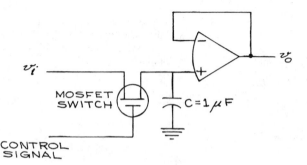

Figure 6.95 A sample-and-hold circuit using a capacitor as a memory unit.

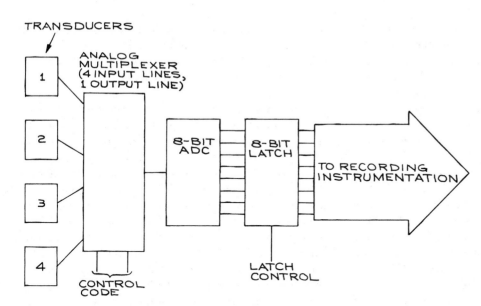

Figure 6.96 A multiple-transducer data-acquisition system using a multiplexer.

The NIM system is based upon an instrument mounting frame or *bin*, which incorporates a central power supply. The NIM bin is capable of powering several units plugged into it. The supply has ± 24 V, ± 12 V, and optionally ± 6 V. Each bin has 12 slots in which instrument module units can be placed. Units can occupy anywhere from 1 to 6 slots. The power supply and individual units have been designed so that a bin can accept the full complement of modules without overloading the supply. A short list of standard NIM modules is given in Table 6.32. Interconnections between

modules are made with coaxial cable. BNC connections are most common, but in some units higher-density LEMO-type coaxial connectors are used. Input and output specifications are standardized so that NIM modules can be connected together without worrying about compatibility. NIM specifications are given in Table 6.33.

Originally the NIM standard voltages were ± 12 and ± 24 V. As integrated circuits came into increasing use, ± 6 V were included in the standard; however, modules operating from 12-24-V supplies are still the most flexi-

Table 6.32   MODULES AVAILABLE IN NIM

Power supplies
Amplifiers
Discriminators
Delays
Signal generators, pulse generators
Counters/timers
Gates
DACs and ADCs
Analog adders, subtractors, dividers, multipliers
Printers
Signal averagers
Multichannel analyzers

Table 6.33   NIM LOGIC LEVELS

	Output (*must deliver*)	Input (*must accept*)
	Digital Data	
Logic 1	+ 4 to + 12 V	+ 3 to + 12 V
Logic 0	+ 1 to − 2 V	+ 1.5 to − 2 V
	Fast Logic (50 Ω impedance)	
Logic 1	− 14 to − 18 mA	− 12 to − 36 mA
Logic 0	− 1 to + 1 mA	− 4 to + 20 mA

ble, since all bins have such voltages. Where 6 V is required in a 12-24-V bin, it is best to derive it within the module from the available 12 V. Use of high-current 6-V modules in a 12-24-V bin should be avoided, since the regulation of the 12- and 24-V supplies will be degraded by the high current in the common return line. Precise reference voltages should be produced in the modules themselves, since the tolerances in the voltages are $24.00(\pm 0.7\%)$, $12.00(\pm 1.0\%)$, and $6.00(\pm 3.0\%)$.[7] The NIM power-connector pin assignments and other details are given in Figure 6.97. The NIM standard logic levels are not compatible directly with common logic families such as TTL or ECL; also, the power-supply voltages are different from common logic supplies ($+5$ V) and operational-amplifier supplies (± 15 V). Some conversion circuits between NIM, ECL, and TTL levels are given in Figure 6.98.

NIM systems are limited in complexity to hard-wired logic, which must be used to interconnect individual modules. For complex experiments this is a disadvantage when there are several data-transfer as well as control functions to be performed. To circumvent the limitations of NIM systems, the CAMAC system was developed. This system retains the use of modular function units that can be plugged into a central power supply unit—in this case a *crate*. The CAMAC crate, however, has a large number of additional internal connections besides the power connections to each module.

The connections that tie each unit separately to a central crate controller form the *data bus* and are used both for data transfer and the control of each of the units. The bus has 24 read and 24 write lines, 4 station subaddress lines, and 5 station function lines; each station has its own identification line and *look-at-me* (LAM) line. Common to all stations are 2 timing lines, 3 status lines, and 3 control lines. These lines all go to the control unit. Also going to the controller are the lines from the computer, generally divided into a set of data bits and address bits. The controller interprets the signals from the computer and sends appropriate signals to the CAMAC stations. This is shown schematically in Figure 6.99, and the line designations are given in Table 6.34. With each of the units under digital control, it is possible, through the use of a computer connected to

the crate controller, to produce systems capable of extremely complex logic and control functions. Data can be collected by the computer and stored, displayed, and manipulated in the same manner as for off-line data reduction.

The CAMAC system offers complete hardware compatibility and reduces the work of the experimenter to producing appropriate computer programs to perform the desired functions. To the uninitiated this can be a formidable task, because the information transferred between the crate controller and the computer is a binary bit code which has no relation to the high-level languages most scientists use for calculational purposes. Computers are, of course, able to generate the appropriate code, but the language used for individual bit manipulation must be learned. A less painful alternative is to write the necessary logic in a high-level language, such as FORTRAN, and at appropriate points in the program call a machine-language subroutine that converts the FORTRAN variables to the bit code necessary for control of the crate modules. When done in this way, the machine-language programming can be very simple.

The investment in a CAMAC system can be considerable, both in money and in the time necessary to learn the required programming techniques. The need for a computer to run the system properly and the computer peripherals, such as terminals, printers, and mass storage units, are an additional expense, which can sometimes be written off if there are other uses for them. The overwhelming advantage of the CAMAC system is the complete flexibility which it provides. As in NIM systems, the individual modules are not costly, because there is no need for internal power supplies or complex logic circuits. There are now several manufacturers who offer an extensive line of CAMAC crates, controllers, and individual function modules. Additionally, there are NIM-CAMAC adaptors that permit one to run certain NIM modules under CAMAC control.

A third type of data-acquisition system can be arranged around the IEEE-488 bus configuration, also known as the GPIB (general-purpose interface bus) or HP-IB (Hewlett-Packard interface bus). Input-output data paths are 8 bits (1 byte) wide, and a complete data-transmission consists of two serial 8-bit bytes. Unlike the CAMAC system, each unit connected to the bus

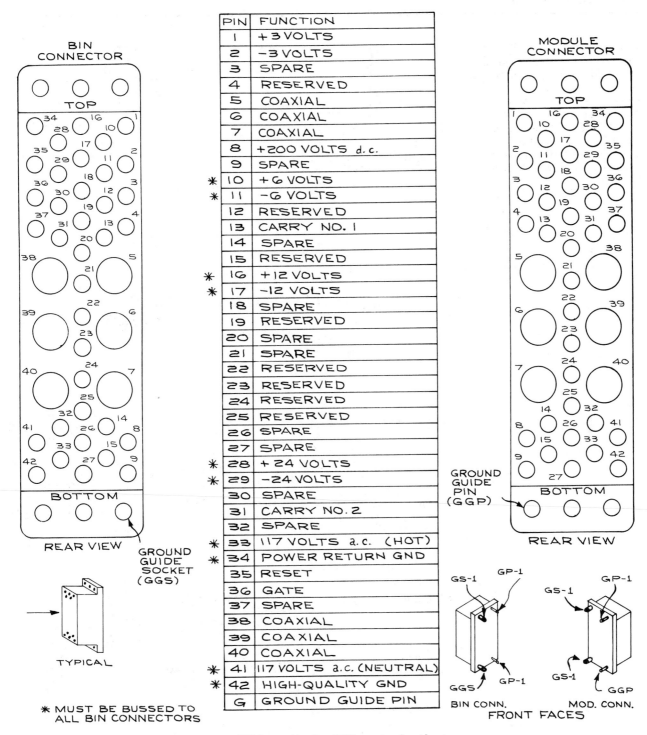

PIN	FUNCTION
1	+ 3 VOLTS
2	− 3 VOLTS
3	SPARE
4	RESERVED
5	COAXIAL
6	COAXIAL
7	COAXIAL
8	+ 200 VOLTS d.c.
9	SPARE
* 10	+ 6 VOLTS
* 11	− 6 VOLTS
12	RESERVED
13	CARRY NO. 1
14	SPARE
15	RESERVED
* 16	+ 12 VOLTS
* 17	− 12 VOLTS
18	SPARE
19	RESERVED
20	SPARE
21	SPARE
22	RESERVED
23	RESERVED
24	RESERVED
25	RESERVED
26	SPARE
27	SPARE
* 28	+ 24 VOLTS
* 29	− 24 VOLTS
30	SPARE
31	CARRY NO. 2
32	SPARE
* 33	117 VOLTS a.c. (HOT)
* 34	POWER RETURN GND
35	RESET
36	GATE
37	SPARE
38	COAXIAL
39	COAXIAL
40	COAXIAL
* 41	117 VOLTS a.c. (NEUTRAL)
* 42	HIGH-QUALITY GND
G	GROUND GUIDE PIN

* MUST BE BUSSED TO ALL BIN CONNECTORS

Figure 6.97 NIM connector pin assignment (GP-1 = guide pin; GGP = ground guide pin; GS-1 = guide socket; GGS = ground guide socket).

429

Figure 6.98 (*a*) NIM-TTL and (*b*) NIM-ECL conversion circuits (fast NIM levels: −0.200 V maximum high, −0.600 V minimum low, 17 mA; MECL threshold at −1.45 V). (From R. F. Althaus and L. W. Nagel, "NIM Fast Logic Modules Utilizing MECL III Integrated Circuits," IEEE Trans. Nucl. Sci., NS-19, 520, 1972.)

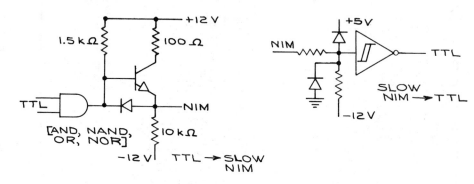

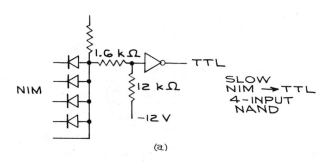

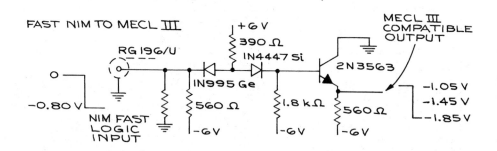

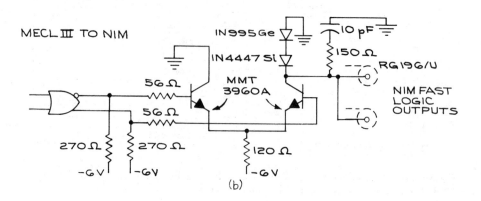

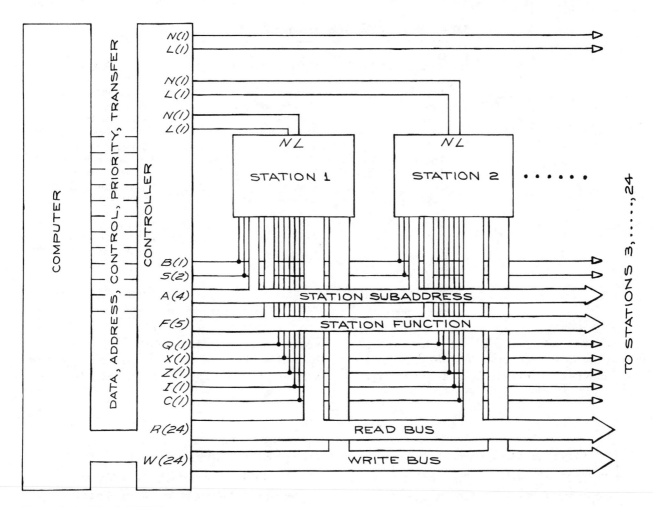

Figure 6.99 The CAMAC bus.

is automomous, with its own power supply. Connections to the bus are through standard 24-pin connectors (Figure 6.100) arranged so that they can be stacked one on top of the other. There can be up to 15 interconnected devices on a single bus, and the total transmission path length cannot exceed 20 meters. Instruments connected to the bus are designated as *talkers, listeners,* and *controllers.* For example, a tape reader is a talker, while a signal generator is a listener. Some instruments may be able to talk and listen, or even (like a variable-range frequency counter) talk, listen, and control. The bus itself is passive, and all the circuitry enabling an instrument to talk, listen, or control is contained within the instrument itself. Simple systems built around the bus need not even use a controller, but can have just one talker and one listener, so long as the instruments have the built-in interface circuitry that allows their functions to be assigned under some form of local control. A more usual configuration incorporates a controller, which uses interface messages to assign the other instruments on the bus the function of talker or listener. A simple controller-based system is shown in Figure 6.101.

What makes the IEEE-488 bus arrangement so attractive is the increasing availability of instruments and

Table 6.34   CAMAC LINE DESIGNATIONS

Line Code (no. of lines used)	Command	Function
$N(1)$	Station number	Selects station
$A(4)$	Subaddress	Selects section of station
$F(5)$	Function	Defines function
$S(2)$	Strobe	Controls phase of operation
$R(24)$	Read	Parallel data to module
$W(24)$	Write	Parallel data from module
$L(1)$	Look-at-me	Request for service
$B(1)$	Busy	Operation in progress
$X(1)$	Command accepted	Module ready
$Q(1)$	Response	Status
$Z(1)$	Initialize	Sets module
$I(1)$	Inhibit	Disables module
$C(1)$	Clear	Clears registers
$P(5)$	Extra lines	
± 24 V $\rbrace$		Power
± 6 V		

microcomputers with the IEEE-488 interface logic. Adaptors for conversion from the 488 bus to the RS232C interface (see Table 6.32) are also readily available. As with the CAMAC system, the logic operations for a controller-based system require an appropriate computer program. The necessity to write and debug

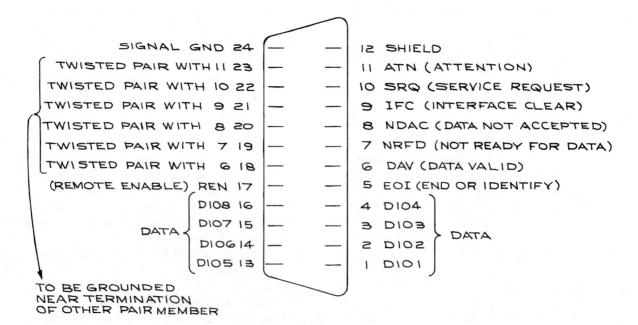

Figure 6.100   IEEE-488 interface-bus connector pin assignment. Logic levels are TTL-compatible negative logic (logic 1, 0–0.4 V; logic 0, 2.5–5.0 V).

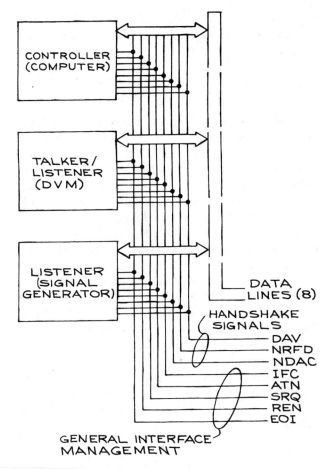

Figure 6.101 A simple controller-based IEEE-488 bus system.

such a program so the system can work takes time, which must be justified by a gain in efficiency.

6.8 EXTRACTION OF SIGNAL FROM NOISE

6.8.1 Signal-to-Noise Ratio

With many experiments, an important consideration is the minimum detectable signal that can be recovered. This depends on the *signal-to-noise ratio*[8] (SNR or

S/N) rather than the absolute value of the signal. In counting experiments where the minimum signal is one event, the uncertainty associated with C_S signal counts is $\sqrt{C_S}$, so that the relative uncertainty is $\sqrt{C_S}/C_S$, or $1/\sqrt{C_S}$. In principle, one could attain arbitrarily high precision by counting for long times, but this is often not practical because of changes in the experimental system.

With analog systems, where the signal is a voltage, current, or charge, the SNR depends on the source resistance and capacitance, shot noise, $1/f$ or flicker noise, and amplifier noise. In this discussion we assume that sources of noise such as rfi do not exist. The noise will be assumed to be confined to the source and preamplifier and to represent the theoretical limit for such a combination.

Consider a detector that is a source of voltage with output impedance R_S, connected to an amplifier with a gain A_0 as shown in Figure 6.102. *Johnson noise* is associated with the random motion of electrons in R_S. Its rms value over a frequency bandwidth Δf in Hz is

$$v_{\text{Johnson}} = (4kTR_S\Delta f)^{1/2},$$

where k is Boltzman's constant (1.38×10^{-23} J/K), and T the absolute temperature of the resistor. Johnson noise is frequency-independent; that is, the rms noise per unit bandwidth (noise density) is the same at all frequencies. For this reason, Johnson noise is often called *white noise*. For the purposes of circuit analysis, R_S can be represented by a voltage source giving a voltage v_{Johnson} in series with a noiseless resistor R. The noise current from this source is $i_{\text{Johnson}} = v_{\text{Johnson}}/R$.

Shot noise has its origin in the discrete nature of the

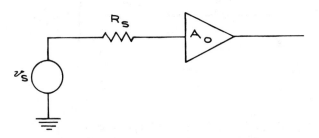

Figure 6.102 Voltage source v_s with output impedance R_S connected to an amplifier of gain A_0.

electronic charge. The shot current noise accompanying a d.c. current I is

$$i_{\text{SHOT}} = (2qI\Delta f)^{1/2},$$

where q is the electronic charge in coulombs. Like Johnson noise, shot noise is frequency-independent.

Flicker or $1/f$ *noise* tends to dominate Johnson and shot noise below 100 Hz. The frequency dependence of this noise is of the form $1/f^n$, where n is usually between 0.9 and 1.35. Flicker noise imposes an important limitation on the SNR at low frequencies. For this reason measurements should, if at all possible, be made at frequencies where white noise dominates.

6.8.2 Optimizing the Signal-to-Noise Ratio

A real amplifier can be represented by an ideal amplifier of gain A_0 with current and voltage noise generators at the input terminal. Usually it is assumed that the noise sources are frequency-independent. Consider the system in Figure 6.103, where v_n and i_n represent the rms voltage and current noise per $\text{Hz}^{1/2}$. The SNR is then the output voltage due to the signal alone divided by the total output voltage, including noise:

$$\text{SNR} = \frac{v_s}{\left[4kTR_S + v_n^2 + (i_n R_S)^2\right]^{1/2}\Delta f^{1/2}}.$$

From this formula it is clear that the SNR can be

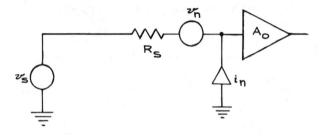

Figure 6.103   Real amplifier and signal source v_s with noise sources explicitly represented by voltage and current generators v_n and i_n.

improved by reducing R_S, Δf, and T. When characterizing noise in an amplifier, the *noise figure* (NF) is often used:

$$\text{NF(dB)} = 20\log_{10}\frac{\text{input voltage SNR without amplifier}}{\text{output SNR from amplifier}}.$$

Since it is impossible for an amplifier to do any better than not reduce the SNR at its input terminal, the smallest possible NF is 0. An NF of 3 dB means that the amplifier has reduced the SNR by $\sqrt{2}$ at its output terminals from what it was at the input terminals. Exceedingly good commercial amplifiers can have NFs as low as 0.05 dB over a limited range of frequency and source resistance. To evaluate the importance of the NF it is necessary to know the input signal level, source resistance, and frequency bandwidth. For example, if the SNR at the input terminals of an amplifier is 10^3, an amplifier with a NF of 20 will reduce it to 10^2 at the output terminals, and this may still be entirely satisfactory for the intended application. With their products, manufacturers often supply contour plots of constant NF in frequency–source-resistance space.

For optimum SNR it is important to match the source resistance to the amplifier. The optimum value of the source resistor, R_{SO}, is v_n/i_n. This results in a minimum NF given by

$$\text{NF}_{\text{min}} = 10\log_{10}\left[1 + \frac{2v_n^2}{4kTR_{SO}}\right],$$

where R_{SO} is the optimum value of the source resistance. Generally the source resistance is inherent to the source. To transform the source's intrinsic resistance R_S to the optimum resistance R_{SO}, a transformer is used (never an additional series or shunt resistor). If the ratio of secondary to primary turns is α, then $\alpha^2 R_S = R_{SO}$. Though it was stated that the maximum SNR occurs when $R_S = 0$, in practice all sources have finite resistance and the SNR can be optimized by transformer matching. Photomultipliers, photodiodes, and electron multipliers are best represented (Figure 6.104) as a noiseless constant current source in parallel with a shot-noise source and a Johnson-noise source resulting from the addition of a load resistor R_L to the circuit. In

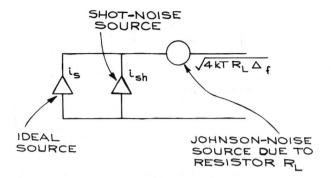

Figure 6.104   Explicit representation of noise sources in a detector.

this case the SNR is optimized by using as large a value of R_L as possible and an amplifier with high input impedance and low v_n and i_n. Amplifiers with FET input stages are most appropriate in this case. Maximum practical values of R_L are limited by the capacitance C of the input circuit (including cables) and the input impedance of the amplifier. There is no benefit in having R_L greater than the input impedance of the amplifier or so large that the time constant $R_L C$ requires one to work at low frequencies where $1/f$ noise dominates. The justification for using large values of R_L comes from the fact that the voltage signal from a current source is proportional to R_L, while the Johnson noise is proportional to $\sqrt{R_L}$.

An alternative way of specifying amplifier performance is with the *equivalent noise temperature, Te,* defined as the increase in temperature of the source resistor necessary to produce the observed noise at the amplifier output, the amplifier being considered noiseless for this purpose. We have

$$T_e = T(10^{\mathrm{NF}/10} - 1),$$

where T is the absolute temperature of the source resistor. Using the example of the amplifier with an NF of 3 dB and a source resistor at 300 K,

$$T_e = (300 \text{ K})(10^{0.3} - 1) = 300 \text{ K}.$$

In this case the amplifier introduces an amount of noise equal to the noise from the source resistor.

6.8.3   The Lock-in Amplifier and Gated Integrator or Boxcar

The proper matching of a signal source to an amplifier is the first step to be taken in the recovery of a signal accompanied by noise. Once this has been done correctly, a number of signal-enhancing techniques can be used to extract the signal.

Because of the difficulty in making d.c. measurements due to zero drifts, amplifier instabilities, and flicker noise, the signal of interest should, if at all possible, be at a frequency sufficiently high that the dominant noise is white noise. Often choppers are used to convert a d.c. or low-frequency signal to a higher-frequency one. Clearly it is wise to choose a frequency far from the power-line frequency and its harmonics. Also to be avoided are frequencies near known noise frequencies.

Signal-enhancing techniques are mainly based on bandwidth reduction. Since the white-noise power per unit bandwidth is constant, reducing the bandwidth will reduce the noise power proportionally. Of course, in the limit as the bandwidth goes to zero, the noise power goes to zero and the signal power as well. Common bandwidth-narrowing circuits are the resonant filter and low-pass filter. For resonant filters, the sharpness of the resonance is given in terms of Q, which is approximately equal to $f_0/\Delta f$, where f_0 is the frequency to which the filter is tuned and Δf is the bandwidth. Practical values of Q for electronic resonant filters vary from 10 to 100. A simple detection system might take the form shown in Figure 6.105. The rectifier converts the a.c. signal back to d.c., so it can be recorded on a chart recorder or read from a meter. With this arrangement the noise passed by the filter is rectified along with the signal and adds to the recorded d.c. level.

The effective bandwidth can be substantially narrowed if in the above arrangement the ordinary rectifier is replaced by a synchronous rectifier. This kind of rectifier acts like a switch that is opened and closed in synchronization with the chopper. Since the phase relations for each of the noise components are random (a requirement of white noise), they will tend to cancel each other when averaged, and a low-pass filter at the output of the rectifier can be used to do the averaging. The RC time constant of the filter is related to the

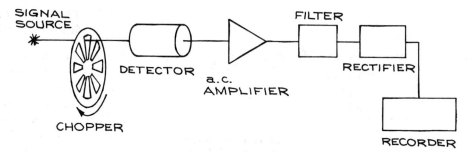

Figure 6.105 Simple detection and signal-enhancement system.

effective passband of the system, Δf, by

$$\Delta f = \frac{1}{4RC}$$

for a single-section low-pass filter, and

$$\Delta f = \frac{1}{4nRC}$$

for n concatenated low-pass filters, each with time constant RC. Quite small values of Δf are possible with such a system, the only limitation being the length of time necessary for the measurement. If Δf is the passband, the measurement time is approximately $1/\Delta f$, so that it is necessary for the source signal to remain stable over times comparable to $1/\Delta f$. Normally the chopping frequency is chosen to be ten times the highest component frequency to be recovered in the source signal.

The system described above is often called a *lock-in amplifier* or *phase-sensitive detector*. The details of the operation of such devices are well documented.[9] However, as far as signal enhancement is concerned, the formulas given above are sufficient for estimating the benefits of such devices. It is well to remember that the SNR is proportional to $1/\sqrt{\Delta f}$, so that narrowing the passband by a factor of 4 increases SNR by a factor of 2, but increases the time of measurement by a factor of 4.

When the signal to be detected has the form of a repetitive low-duty-cycle train of pulses, the lock-in amplifier may not be the best method for signal enhancement. Here the *duty cycle* is defined as the fraction of the time during which the signal of interest is present. With low-duty cycles, signal information is available for only a fraction of the total time, while noise

is always on the line. With timing and gating circuits it is possible to connect the signal line to an RC integrating circuit only during those times when the signal is present. The time constant of the integrator is then chosen to be very much larger than the period of the pulse train. The time required for the capacitor to charge to 99% of the final voltage level is $4.6RC$, so that 5 time constants after the first gate opening the capacitor should be charged to within 1% of its final steady-state value, provided the signal is continuously present. If the signal is present for a fraction γ of the time, the time constant of the integrator must be increased to $5RC/\gamma$. With this system, known as a *gated* or *boxcar integrator*, the SNR is increased only by increasing the time of the measurement. The effective bandwidth of the instrument is $\gamma/4RC$. Table 6.35 compares the important parameters associated with lock-in amplifiers and gated integrators.

6.8.4 Signal Averaging

With a repetitive signal, improvement in the SNR can be effected by merely averaging the signal over many cycles. Let SNR_0 represent the SNR in one cycle, and N the number of cycles over which one averages. The SNR improvement is proportional to $\sqrt{N}$. If each cycle lasts for a time τ, the time T necessary to arrive at a specified SNR is given by

$$T = \left(\frac{SNR_0}{SNR} \right)^2 \tau.$$

Lock-in amplifiers and gated integrators are inherently superior to simple signal averaging because of the

Table 6.35   COMPARISON OF LOCK-IN AMPLIFIER AND GATED INTEGRATOR

	Lock-in Amplifier	*Gated Integrator*
Duty cycle δ	> 0.05	< 0.50
Bandwidth	$1/8RC$	$\gamma/4RC$
Minimum measurement time	$5RC$	$5RC/\gamma$
Highest recoverable signal frequency	$1/20\pi RC$ for chopping frequency $\geqslant 1/2\pi RC$	$\gamma/20\pi RC$ for repetition frequency $\geqslant 1/2\pi RC$
Design notes	1. Determine f_{max}, the highest frequency component of the signal to be recovered. 2. Choose f_s, the chopping frequency, where $f_s = 10f_{max}$. 3. Choose low-pass filter constants $1/RC = 2\pi f_s$ so as to pass frequency f_{max}. 4. The bandwidth Δf is $1/8RC$, and a tuned amplifier with a Q of 10 is sufficient to pass all frequency components of the signal to f_{max}.	1. Required measurement time is $5RC/\gamma$. 2. Bandwidth is $\gamma/4RC$. 3. Preferred to a lock-in amplifier for signal repetition rates $\leqslant 10$ Hz and low duty cycle.

band-narrowing functions they perform; however, the bandwidth improvement is only an advantage when the highest component frequency in the signal to be recovered permits a long integrating time.

6.8.5 Waveform Recovery

The gated integrator can be converted to an instrument for waveform recovery by the inclusion of variable delay and variable gate-width functions. A separate RC network is required for each delay time in such a scheme. A schematic representation of such a system is shown in Figure 6.106. Each separate delay corresponds to a different part of the waveform. The duty cycle for each gate opening is $\tau/T = \gamma$, so that the effective bandwidth is $\gamma/4RC$. An integration time of $5RC/\gamma$ is needed for the charge on each capacitor to reach a steady-state

value. If the waveform has been divided into N parts, the total time of measurement is $5NRC/\gamma$.

Digital schemes for recovering waveforms use ADCs to convert the analog signal level to a digital number, which is then recorded in the appropriate time channel with the aid of delay and gating signals. With such instruments, called *digital signal analyzers*, SNRs are only limited by the length of time of the measurement; however, the long-term stability of the various components limits the use of very long measuring times. If the absolute value of the waveform is needed, one must record the number of cycles treated by the instrument, and data rates are limited by the conversion speed of the ADC that encodes the amplitude information. An advantage of digital signal analysis is the increased versatility in data manipulation and the production of permanent records on paper or magnetic media. Such data are then available for numerical analysis.

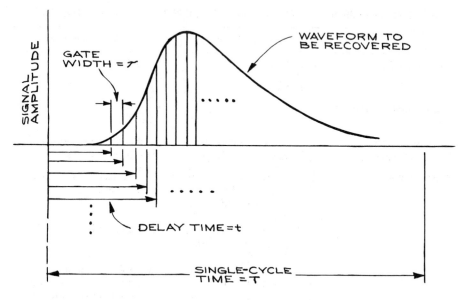

Figure 6.106 Relation between gate width τ, delay time t, and single-cycle time T in waveform recovery. The duty cycle γ is given by τ/T.

Instruments related to the digital waveform recorders are *transient recorders* and *memory oscilloscopes*. While these offer no gain in SNR over that in the original signal, they do provide outputs that are a record of a single waveform. An inefficient, but commonly used, method of SNR enhancement with these instruments is the separate recording of many waveforms and subsequent digital averaging by a computer.

6.9   GROUNDS AND GROUNDING

6.9.1   Electrical Grounds and Safety

A battery or power supply produces a potential difference between its two output terminals. Only by defining in some absolute way the potential of one of the terminals can the potential of the other terminal be given an absolute value. By convention, the electrical potential of the earth is defined to be zero. When one output terminal of a battery or power supply is connected to the earth, the absolute potential of the other terminal is then equal to the potential difference. Since the surface of the earth is a fairly good conductor of

electricity, all points will be at nearly the same potential. A good connection to the earth via a copper rod buried in the ground is the most convenient way of establishing zero potential. Since such connections are not always readily available in the laboratory, one often uses connections to metal water pipes, which go underground, to define zero potential.

Most connections to the a.c. line are now made with three-prong plugs and three-conductor wire. The wire is usually color coded in North America: black corresponding to the high-voltage or *hot* side of the line, white to the low-voltage or *neutral*, and green to the earth or *ground*. The terminals on a.c. plugs and receptacles are also coded: the hot terminal is brass, the neutral terminal is chrome-plated, and the ground terminal is dyed green. Switches and fuses for use with a.c. lines should always be placed in the hot line; otherwise the full line voltage will be present at the input even with an open switch or a blown fuse.

The three-wire system is designed for safety. Consider a piece of electrical equipment with a metal case operating from the a.c. line. Normal practice requires that the case be connected to the green ground wire of the line cord. Should, for some reason, the hot wire come in contact with the case, a very large current would flow in the hot line and blow a fuse or open a circuit breaker. Were the case not connected to ground, it would assume the potential of the hot line and could kill anyone touching it. Of course, if the case were not a conductor, even if the hot line made a connection to it there would be no danger. There are a large number of appliances and hand power tools with so-called *double insulated* cases, which therefore do not require a separate ground connection.

From this discussion it should be clear that fuses and circuit breakers are important safety elements in electrical circuits. Fuses are generally of either the standard quick-response type or the slow-blow type. The former will open whenever its maximum rated current is exceeded, even momentarily, while the *slow-blow* fuse is designed to sustain momentary current surges in excess of the maximum rated continuous value. The two types are not interchangeable even though they may have identical maximum continuous current ratings. Circuit breakers are generally of the electromagnetic-relay or the bimetallic-strip type. They can be reset after opening from an overload.

It is sometimes necessary to *float* or remove all ground connections from a piece of electrical equipment. This is necessary with measuring instruments when a signal between two points in a circuit, neither of which is at ground, is to be measured, amplified, or processed in some way. Power supplies are floated when it is necessary to produce a well-defined potential difference with respect to a nonzero potential. Floating introduces the possibility of additional noise, electrical shock, and electrical breakdown within the circuit with destruction of components.

At this point it is worthwhile to make the distinction between d.c. and a.c. grounds. A *d.c. ground* connection is a direct low-resistance connection to ground while an *a.c. ground* connection is one that does not permit d.c. currents to flow while offering low impedance to a.c. currents. In Figure 6.107 power supply 1 has the negative terminal and case (G terminal) at d.c. ground. Power supply 2 is floating on power supply 1, and in the absence of any external connections no current will flow through the terminals of the supplies. Since high-quality power supplies generally have very low output impedances, the a.c. impedances of the terminals marked

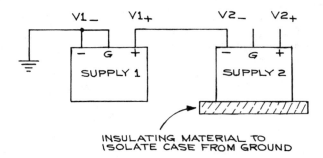

Figure 6.107 Power supply 2 floating on power supply 1. The $+$, $-$, and G terminals of supply 2 are assumed to have no direct d.c. connection to ground.

$V1_+$, $V2_-$, and $V2_+$ with respect to ground are low, though the potentials at the terminals are not zero. It is well to keep in mind that the impedances to ground are frequency-dependent because of the characteristics of the output circuits. At high frequencies the impedances can become large through the use of components with poor high-frequency characteristics (such as electrolytic capacitors). As a result, transient voltage spikes may not be effectively shunted to ground and may adversely affect the circuits to which the power supplies are connected. A remedy is the use of good high-frequency capacitors of an adequate voltage rating between the terminals $V2_+$ and $V1_+$ and ground.

Figure 6.108 illustrates a typical line-operated power supply connected to the a.c. outlet via a three-wire line cord. Under normal operation, depending on whether a positive or a negative voltage with respect to ground is desired, either the negative or the positive terminal is connected to the ground terminal. When floating the power supply, the ground connection is removed and the potential at which the supply is to be floated is applied to either the positive or the negative output terminal. This may be satisfactory for levels of as much as a few hundred volts, but has its dangers because the internal circuitry is floating while the case is at ground through the green wire in the three-wire line plug. Potentials at least equal to the floating potential plus the potential difference at the output terminals now exist at the terminals of several components in the internal circuitry. Components are often mechanically fixed

Figure 6.108 Mounting a line-operated power supply so that both output terminals are independent of the a.c. line ground.

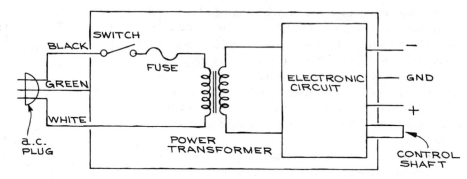

to the grounded chassis; the electrical insulation between the components and the chassis, as well as that between the primary and secondary of the transformer, must be good enough to sustain the d.c. voltage difference. The maximum for floating line-powered equipment in this way is generally 600 volts. In the event of electrical breakdown, components will be destroyed, and the shafts of the controls may attain a potential that is equal to the floating potential and thus present a shock hazard.

When floating line-operated equipment at voltages above the rated maximum voltages, it is necessary to completely remove the ground connection to the case. The easiest method is to use a two-terminal-to-three-terminal a.c. line-plug converter and not connect the third terminal to ground at the a.c. outlet. If the case of the instrument is then electrically isolated from all conducting surfaces by placing it on a sheet of Plexiglas, it can then be electrically floated at the desired potential without risk of electrical breakdown between the components of the circuit and the case. With the case at a nonzero potential, there is a shock hazard associated with touching it and the controls so that *access must be provided via auxiliary insulated knobs and shafts*. A further difficulty is the possibility of electrical breakdown between the primary and secondary windings of the input transformer. At the primary, the peak voltage values are ± 155 volts, corresponding to the normal peak values of the a.c. line voltage. The secondary voltage, however, will have a d.c. component equal to the floating voltage superimposed on the normal a.c. value. If the resulting d.c. plus peak-a.c. secondary

voltage exceeds the voltage rating of the transformer insulation, electrical breakdown can occur, destroying the transformer and possibly the rest of the circuit. *Isolation transformers* are commonly used to solve this problem. These transformers have a 1 : 1 turns ratio and high-voltage insulation between the primary and secondary windings. The primary is connected to the a.c. line, while the secondary is connected to the primary of the input transformer of the instrument to be floated. In this way the floating voltage appears across the isolation-transformer windings. Of course the case and controls are still at the floating potential, and precautions similar to those in the former situation must be observed. When specifying an isolation transformer, the power-handling capacity in watts or kilovolt-amperes (kVA) as well as the maximum isolation voltage must be considered.

A substitute for an isolation transformer is two high-voltage filament transformers of the same turns ratio connected back to back. Such transformers are often used in oscilloscope power supplies. This will produce the necessary 1 : 1 overall voltage ratio with isolation. Often such transformers are more readily available and less expensive than high-voltage, high-power single-isolation transformers.

When floating instruments that are not directly powered from the a.c. line, the external power supply, whether it is a battery or line-operated, must be floated too. The considerations that apply to the floating of line-operated instruments apply here also. Table 6.36 summarizes the important considerations involved when floating power supplies.

Table 6.36 FLOATING POWER SUPPLIES

A. Normal operation
1. Input-line cord connected to three-terminal a.c. outlet
2. Chassis at ground potential
3. + or − output terminal to grounded center terminal for a − or + output voltage
B. Low-voltage floating operation
1. Input-line cord to three-terminal a.c. outlet
2. Chassis at ground potential
3. + or − terminal to floating potential
4. Grounded output terminal unconnected
C. Medium-voltage floating operation
1. Input-line cord attached to 3-2 plug adaptor with ground terminal unconnected
2. Chassis at floating potential
3. + or − output terminal connected to chassis output terminal (at floating potential)
4. Case and chassis isolated from all conductors
5. Insulating shafts and knobs on all switches and dials
D. High-voltage floating operation
1. A.c. input from isolation transformer
2. Instrument case at floating potential
3. Case and chassis electrically isolated from all conducting surfaces
4. Insulated shafts and knobs on all switches and controls

6.9.2 Electrical Pickup: Capacitive Effects

It is often necessary to detect and measure electrical signals in the millivolt and even microvolt range. If the signals are accompanied by noise generated by external sources, they can be so degraded as to render the measurement useless. The elimination of these external sources—or, if this is not possible, the elimination of their effects—is an important element in the design and construction of electronic measuring, control, and detection systems.

Consider a wire at a potential V_1 in the vicinity of a second wire that is insulated from all conducting surfaces [Figure 6.109(a)]. Coulomb forces between the charges on the surfaces of the wires will cause a polarization of the free charges in wire 2. Because wire 2 is electrically isolated, it will remain electrically neutral. If now wire 2 is connected to ground through a resistance R, a current

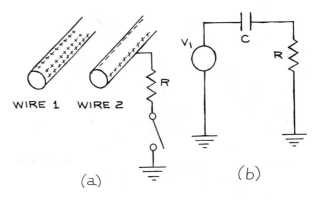

Figure 6.109 (a) Capacitive coupling between two wires; (b) the equivalent electrical circuit with the effect of wire 1 on wire 2 represented by C.

will momentarily flow due to the presence of the positive charges on wire attracting negative charge to wire 2 from the ground. The equivalent circuit is shown in Figure 6.109(b). Typical values of the coupling capacitance are 1 to 100 pF. For example, two twisted #22 cloth-insulated wires will have a capacitance of 20 pF/ft while the center-wire-to-shield capacitance of RG/58 coaxial cable is 33 pF/ft. If we assume that V_1 is varying sinusoidally at frequency ω, the rms potential across R is given by

$$V_{1\,\text{rms}} \frac{R}{R + 1/\omega C} = V_{1\text{rms}} \frac{\omega RC}{1 + \omega RC}$$

(the voltage-divider formula). Thus the rms potential across R increases with R, ω, and C.

In the average laboratory, there are many sources of a.c. voltage that can couple capacitively to signal lines; open-line sockets, lighting fixtures, and line cords are a few examples. An estimate of the capacitance C_{min} necessary to induce a 1-mV rms voltage across 1 MΩ from a power-line source can easily be made by setting $V_{1\,\text{rms}}$ equal to 120 V, $\omega = 2\pi f = 188$ rad/sec, and neglecting ωRC with respect to 1:

$$C_{\text{min}} = \frac{1 \times 10^{-3}\,\text{V}}{1.20 \times 10^2\,\text{V} \times 1.88 \times 10^2\,\text{sec}^{-1} \times 10^6\,\Omega}$$
$$= 0.044\,\text{pF},$$

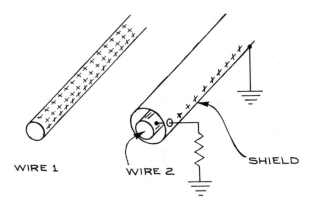

WIRE 1

WIRE 2

SHIELD

Figure 6.110 Shielding to prevent capacitive coupling.

a value easily attained. Such induced voltages can be substantially reduced by the use of shielding. If a grounded shield is placed around wire 2, the situation illustrated in Figure 6.110 will occur. The conducting shield around wire 2 will have a current induced in it when connected to ground, due to the potential on wire 1. However, this current flows through a low-resistance connection. In addition, and most importantly, any charge on the shield will exist on its surface, and the interior will be field-free. In practice, common coaxial braided shield is about 90% effective against capacitively coupled voltages from external sources. Foil shields are even more effective.

6.9.3 Electrical Pickup: Inductive Effects

A changing magnetic field can induce a current in any loop it cuts. The magnitude of the induced current depends on the area of the loop and the time rate of change of the magnetic field. High-resistance circuits are not greatly affected by such inductive effects, but they are important in low-resistance circuits such as the input circuits of current and pulse amplifiers. Common sources of magnetic-field pickup are transformers, inductors, and wires carrying large a.c. currents. Effective shielding against low-frequency magnetic fields is accomplished with ferromagnetic enclosures, which act

to concentrate the stray magnetic fields within the shield material. The usual practice is to enclose the source of the magnetic fields within such shields. Higher-frequency magnetic fields are best shielded with a copper enclosure. Eddy currents induced in the copper produce counter magnetic fields. A rapidly changing current is a source of time-varying magnetic fields, which can themselves induce currents in other circuits. Thus, fast-risetime pulses in the low-impedance output circuits of pulse amplifiers can generate rapidly changing magnetic fields, which can induce unwanted currents in any nearby low-impedance input circuit. Once again the most effective shielding is copper sheet or foil.

6.9.4 Electromagnetic Interference and RFI

The high-frequency fluctuation of current or charge in a conductor results in the radiation of part of the energy in the conductor in the form of an electromagnetic wave. Such a wave propagates through space at the speed of light and, when not wanted, is called *radio-frequency interference* (rfi). Sources of such radiation are automobile ignition systems, microwave ovens, electrical discharges, electric motors, electromechanical switches and relays, and electronic switches such as thyratrons, rectifiers, SCRs, and triacs. Power supplies with switching regulators are often sources of rfi.

This interference can be reduced by the use of zero-crossing switching regulators in which the switching only occurs when the voltage across the switch is zero. The most effective shield against rfi is a grounded enclosure of a conducting material, such as copper or aluminum. On the surfaces of such a shield the electric component of any incident electromagnetic wave is zero and further propagation is not possible. Conducting screens rather than solid sheet are often used for rfi shielding because of the savings in weight. When using such screen it is important that the mesh size be small compared to the wavelength of the highest-frequency component of the rfi. Standards have now been established for rfi emission from certain classes of electrical and electronic equipment. Sensitive, high-frequency measuring equipment must have good immunity to rfi, and additional rfi shielding can be often purchased as an option for equipment that must operate in an especially noisy environment.

6.9.5 Power-Line-Coupled Noise

Consider the case illustrated in Figure 6.111, where three circuits are powered by a single supply line. If circuit 1 suddenly requires a large amount of current, circuits 2 and 3 may be affected, because of the resistance and inductance of the power supply line. A large current drawn by circuit 1 can cause a momentary voltage drop at circuits 2 and 3, which can be propagated as noise throughout the circuit. The larger the time rate of change of the current pulse through circuit 1, the greater the effect on the other circuits. This problem is very common with logic circuits, particularly TTL circuits, which have saturating output stages. If a number of gates switch simultaneously, very large transient current spikes can appear on the power line. These transients are then interpreted as a change in logic level by other gates, and transitions occur that are entirely spurious. Standard practice is to use low-resistance, low-inductance power-supply lines to the logic gates and decouple the gates from the power supply with high-frequency capacitors by connecting the power-supply terminal of the gate to ground through a capacitor. For TTL, every group of 20 gates must be decoupled from the power supply with a 0.1-μF capacitor, and every group of 100 gates must be decoupled with an addi-

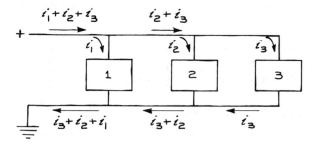

Figure 6.111 Three circuits on a common power line.

tional 0.1-μF tantalum capacitor with good high-frequency characteristics.

Similar situations arise with switched high-current devices such as temperature-controlled furnaces operated directly from the a.c. line. Large current transients may affect all equipment plugged into the same line. Here, the only remedy is to operate the high-current device from a completely separate line circuit.

As an example, consider a series of logic gates to be connected to a power supply by 1 foot of #22 wire, which has a resistance of 16 Ω/1000 ft and an inductance of 640 μH/1000 ft. If a few gates switch simultaneously, resulting in a current spike of 10 mA with a risetime t_r of 10 ns, the momentary voltage drop along the line will be approximately iZ, where Z is $\sqrt{R^2 + (\omega L)^2}$. In this case

$$Z = \left[(16 \times 10^{-3}\,\Omega)^2 + (2\pi \times 3.3 \times 10^{+7}\,\text{sec}^{-1} \times 6.4 \times 10^{-7}\,\text{H})^2 \right]^{1/2},$$

where $\omega = (2\pi/3)\,t_r$ has been used in the estimate of the inductive reactance of the wire. The voltage drop is then

$$10^{-2}\,\text{A} \times 133\,\Omega = 1.3\ V.$$

For the TTL logic this drop could be enough to cause some marginally substandard gates to change state momentarily.

6.9.6 Ground Loops

Instruments are connected to ground in order to have the potentials of those terminals be at 0 V for reference purposes. This, of course, assumes that every ground connection is made directly to the zero potential of the earth. This is often far from the case, since most ground connections are through the third wire of the a.c. line. Since this line has finite resistance, currents flowing in it will cause potential drops and, depending on the point on the line where a ground connection is made, the potential can be very different from 0 V. When the 0 V references of two instruments that are connected together differ, there is danger of introducing noise into the system. The problem is made all the worse by the fact that the reference points can change their potentials with respect to each other in a way independent of the other system parameters. The solution to this problem is to have only a single ground point in the system, to which all the zero reference points and shields are connected. This means removing the ground connections in the power-line cords of all line-operated instruments, and isolating all shields and cases from conducting surfaces that may be connected to ground at points other than the single system ground. All connections to this single system ground should be made as short as possible, with the lowest-resistance conductors available. By having the 0 V reference potential connection and shield connection at the same point in the system, no currents can be induced in the reference line.

To illustrate these ideas, two different ways of connecting a system composed of a signal source, amplifier, and recorder are shown in Figure 6.112. In the top arrangement each case is separately grounded and additionally connected to the other cases through the outer conductor of the coaxial cable between them. If the three grounds are at different potentials, currents will flow through the coaxial outer shields connecting the devices together. Since the reference-potential line is the shield, it will take on different potentials at different points in the circuit. As these potentials change, the signal levels will change—the result being seen as noise on the signal line.

In the second arrangement, the common reference line and the instrument shields are all attached to ground at a single point, the reference ground. No ground loops can exist, and the reference line remains at the same potential everywhere in the system. Clearly this is a superior arrangement. However, the implementation may be difficult, since it is necessary to connect all the cases and shields separately to a single ground point. If all the instruments are mounted in a single rack, each must be electrically isolated from the conducting rack structure. The coaxial-shield connections must also be isolated from the cases, and the ground connection in the a.c. line cord must be removed from each of the instruments. It is not always necessary to take such precautions with all stages of a system: it may be sufficient to eliminate ground loops

Figure 6.112 System grounding: (*a*) multiple grounds; (*b*) single ground to eliminate ground loops.

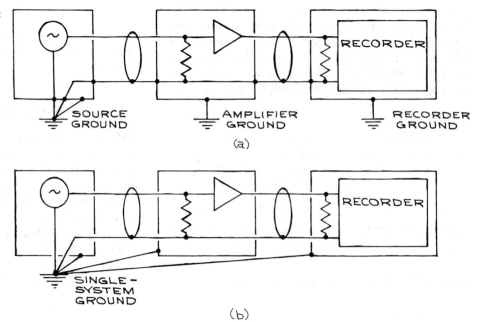

from the most sensitive element to which the signal source is connected.

It is worthwhile understanding the origins of ground-loop and shielding problems so as to be able to solve them when they arise. A much more extensive treatment is given by Morrison.[10]

The problem of ground loops also occurs when measurements must be made across two electrical terminals, neither of which is at ground potential.[11] Bridge measurements are a common example of this. For floating measurements, instruments that have the reference input terminal electrically separate from the ground or common terminal must be used; otherwise the floating potential (common-mode potential) will be short-circuited.

In Figure 6.113, v_S is the source voltage to be measured, v_C is the common-mode voltage, Z_1 is the input impedance of the measuring system, and Z_2 is the impedance from the common of the measuring circuit to the low terminal, normally 10^8 to 10^{10} Ω in parallel with 10^3 to 10^5 pF. The presence of v_C results in currents i_{C_1} and i_{C_2} flowing through Z_1 and Z_2. This presents no difficulties so long as the resistance of the lines is zero, but if they are not zero, the common-mode currents will result in potential differences at the low input terminal and add to the error of the measurement. Generally $Z_1 \ll Z_2$; therefore the critical parameter is the ratio of the resistance of the low signal line to Z_2. This ratio is frequency-dependent because Z_2 is complex, so that if the source and instrument grounds are not at the same potential, the difference will add to v_C.

The *common-mode rejection* (CMR) is a quantity for specifying how well an instrument rejects a common-mode voltage superimposed on the voltage to be measured (called the *normal mode*):

$$\text{CMR(dB)} = -20 \log \frac{V_{NM}}{V_{CM}},$$

where V_{NM} is the common-mode voltage, which appears as the normal-mode voltage and therefore is a source of error. Floatable meters have CMRs from 80 to 120 dB at d.c. and 60 to 100 dB at line frequency. When higher values are required, special guarded voltmeters having an additional guard terminal must be used. This terminal, when connected to the low terminal of the source

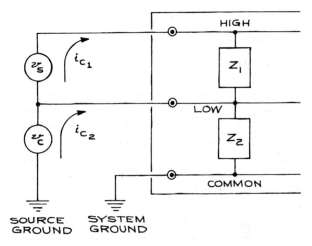

Figure 6.113 Measurement of v_S across two terminals, one of which is not at ground. Current flow through Z_2, the impedance from the low terminal of the measuring instrument to ground, will result in an error voltage v_C.

by a low-resistance connection, shunts the common-mode current from the measuring terminals. Such meters offer CMRs of 160 dB at d.c. and 140 dB at line frequency.

6.10   HARDWARE AND CONSTRUCTION

It is almost invariably more practical to purchase a piece of electronic equipment than to design and construct it oneself. This is certainly the case for power supplies, amplifiers, signal generators, and similar equipment, which are mass-produced by a large number of companies in a wide variety of models.

When one does, however, decide to construct a piece of electronic equipment because of cost or the unavailability of commercial units, there are a number of well-defined steps to follow:

1. Design of the circuit.

2. Selection of components.

3. Construction and testing of a breadboard model.

4. Construction of the final circuit.

5. Mounting.

6. Final testing.

Except for the simplest circuits, one is well advised to use proven designs, which can be found in manufacturers' application books and in books and articles on circuit design. A few such publications are given in the list of references.[12] However, there are some simple procedures, well known to electronics engineers, but not often included in textbook examples of circuit design, that can make the difference between a circuit that works and one that does not. These include bandwidth limiting, power-supply decoupling, and signal conditioning, discussed above.

6.10.1   Circuit Diagrams

Two kinds of circuit diagrams are the block diagram and the schematic wiring diagram. The *block diagram* shows the logical layout of the circuit by grouping all elements necessary for single function in a single block. Though simple, such diagrams when well done are very useful in understanding the operation of an electronic circuit and isolating the cause of any malfunction to a well-defined group of components. The block diagram is much like the flowchart of a computer program.

The *schematic wiring diagram* shows all the components of the circuit and the connections between them. Figure 6.114 lists symbols used in schematic diagrams. The more complete the schematic, the more useful it is for troubleshooting. A schematic will have the values and ratings of all the components, as well as numbers and color codes. Good schematics include the values of voltages at critical points in the circuit and drawings of waveforms where appropriate. Generally, one reads schematic diagrams starting at the upper left with the input and proceeding to the lower right to the output.

A useful addition to the schematic diagram is the *chassis layout diagram*, which shows the physical location of all parts. For ease of drawing, components that appear next to each other on the schematic may be far away from each other on the actual chassis or circuit board.

With increasing use of integrated circuits, it is more and more difficult to identify the functions of different parts of a schematic. This is because the integrated circuits are often only represented by rectangles with pin numbers. To identify their function it is necessary to consult a manufacturer's catalog. Certain abbreviations common with ICs are given in Tables 6.24–6.26.

Schematic diagrams also help identify specific components. If there is an operational amplifier on a circuit board with a balance control near it for nulling purposes, it is useful to know where the amplifier is in the circuit and what comes before and after before proceeding with adjustments. The location of damaged components on a schematic can be useful when deciding on the probable cause of the damage and other components that may also have been affected.

6.10.2   Component Selection and Construction Techniques

When selecting components the considerations that an electronic engineer finds important may not be the same as those of a laboratory scientist building only a single example of a circuit. It is always wise to over-specify components, that is, use components of higher ratings and better quality than a critical cost-effective analysis of the circuit would show to be necessary. Usually the cost of components is a small fraction of total project cost when time is considered.

Once the circuit design has been decided upon and the components assembled, a *breadboard* model can be constructed. This may seem an unnecessary and time-consuming step, but care at this stage will save a great deal of time later. The breadboard circuit should be constructed from a complete, detailed schematic circuit diagram.

There are a number of prototype circuit boards and aids for making breadboard models. Prototype boards use sockets to which components and connecting wires can be connected in a temporary way. The spacing of the sockets is usually on 0.100-in centers to accommodate the usual 14- and 16-pin dual in-line integrated circuits. Usually the sockets accommodate only a limited range of solid-wire sizes. Outside of this range, either the wire is not held securely or the socket is bent out of

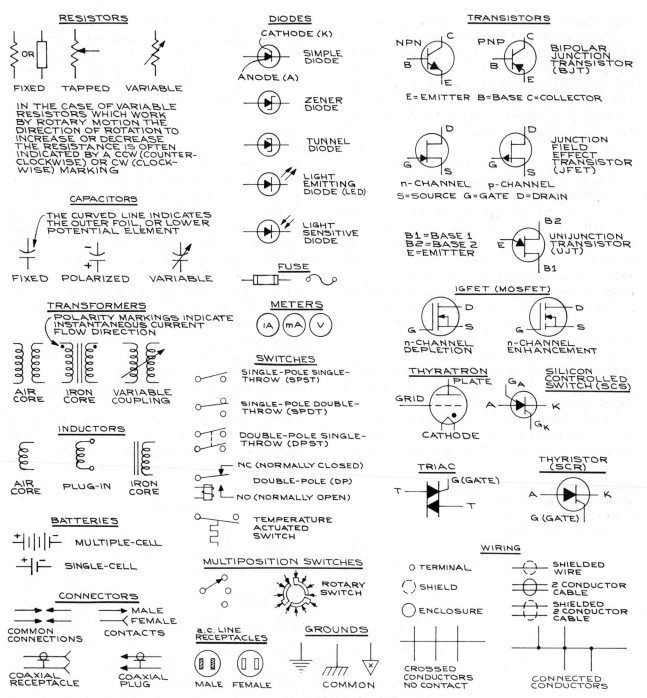

Figure 6.114 Symbols for schematic circuit diagrams. (From *Electrical and Electronics Graphics Symbols*, 76-ANSI/IEEE Y3 ZE, Institute of Electrical and Electronic Engineers, New York.)

shape. Boards of this type are satisfactory for low-frequency circuits and logic circuits using TTL and CMOS. For sensitive high-frequency circuits and ECL logic the stray capacitance between sockets is a problem, and such circuits must be built on a circuit board with a ground plane if they are to work properly.[13] At this stage it is necessary that the breadboard circuit follow the schematic in every detail. The mechanical layout should be neat, with interconnections between components as short as possible. This simplifies troubleshooting and avoids pickup problems.

A common difficulty with breadboards is getting signals in and out and applying d.c. power. Dangling wires and alligator-clip connections invite short circuits and damaged components. Whenever possible the breadboard should be mounted on a larger structure with terminals and connectors to which semipermanent signal and power-supply connections can be made (Figure 6.115). Connections from these terminals to the breadboard can then be made with short pieces of hookup wire.

The breadboard circuit is fully tested before proceeding to the final construction stage. During testing, the circuit should be operated with the same input and output connections the final circuit will have.

Once the breadboard circuit is functioning properly, the final circuit can be constructed. Depending on the complexity of the circuit, there are a number of con-

struction techniques available. Generally, all components are mounted on circuit boards and the boards are fitted into a case or attached to a panel. The power supply can be incorporated into the circuit, or power-supply voltages can be brought in from the outside through connectors. For simple circuits involving only a few components and a moderate number of connections, perforated circuit board and push-in terminals can be used [Figure 6.116(a)]. For such circuits all connections are individually made and soldered. When possible, active components such as ICs and transistors should be used with sockets, and all soldering done with the components removed from the sockets. Perforated circuit board with printed circuit wiring is also a convenient medium for making final circuits [Figure 6.116(b)]. With this type, all soldering is done on the board and there is no need for separate terminals. Power-supply and ground-bus structures are often printed on such boards to facilitate bringing power to the components.

For more complex circuits custom-printed circuit boards (Section 6.10.3) or Wire Wrap boards (Section 6.10.4) can be used.

Typically, the circuit board containing all the electrical components and a power supply, if required, is mounted on a standard 19-in.-aluminum rack panel on standoffs. When there is the need for more than two or three circuit boards, the means of mounting and of making the interconnections must be considered more

Figure 6.115 A circuit breadboard mounted to allow proper connection of power supply and input and output leads.

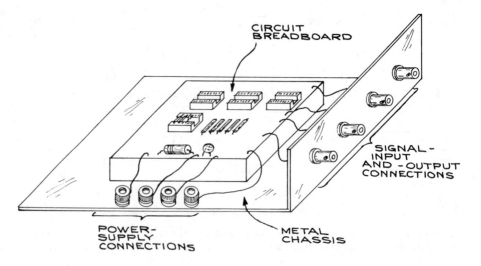

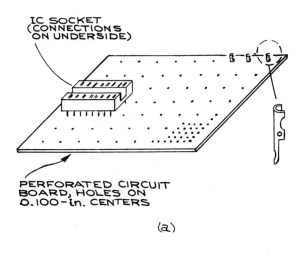

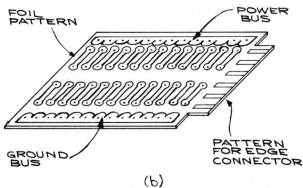

Figure 6.116 Perforated circuit boards: for use with (*a*) pin connectors and (*b*) direct solder connectors.

carefully. For multiple circuit boards, *card-cage* mounting, using guides and edge connectors as shown in Figure 6.117, is a common technique. Interconnections between boards are made between the edge connectors on the rear panel of the chassis. A variation of this method uses a *mother board*, to which all the connections from the other boards are brought. In this method, the edge connectors provide power to the separate boards and help to retain the boards mechanically in the chassis. Signal leads are brought to the mother board via multiple-pin connectors and flat multiple-conductor cable, using mass termination hardware for increased reliability. A variation of the mother-board configuration

includes the edge connectors on the mother board. An on-off switch, pilot light, and fuse are put on the front panel, as well as input and output connectors and controls. When labeling the panel it is wise to indicate schematically with standard symbols the electronic functions performed by each section of the circuit. Labels can be applied with the silk-screen technique or with dry transfer letters, India ink, or embossed pressure-sensitive tape. A protective cover should be placed over the circuit (Figure 6.118).

Construction can be simple with the proper tools. A portable electric drill or drill press is necessary for drilling proper-size holes in the panel and cover. A list of electronic-circuit construction tools is given in Table 6.37 and a list of useful hardware is given in Table 6.38. Flathead screws should be used on exterior surfaces, and lock washers under nuts. Standard $\frac{1}{16}$-in. circuit board can be cut easily with a hacksaw and drilled with standard high-speed drills. A nibbling tool is useful for making irregular-shaped holes in sheet metal, and a set of chassis punches from $\frac{1}{2}$ to $1\frac{1}{2}$ in. speeds up work and makes accurate holes in sheet metal. A tapered hand-reamer is very useful for enlarging holes in sheet metal.

6.10.3 Printed Circuit Boards

The use of printed circuit boards for laboratory-constructed electronics is rarely justified. It is, however, when many identical circuit boards are needed and when the electrical properties of the circuit demand the kind of controlled geometry that printed boards can offer. The boards consist of an insulated substrate covered with a metal (usually copper) foil. Some of the common rigid substrates used for copper-clad boards are listed in Table 6.39 along with their codes. For laboratory purposes the G-10 grade of board is usually used. The copper-foil cladding is usually a cathode-quality electrolytic copper. The most common thickness is 1.0 oz/ft^2 (35 μm). Thinner foils do, however, exist.

To make a PCB, one first produces the artwork, which is an enlarged version of the final traces to appear on the board. Usually the artwork master is a $2\times$ or $4\times$ enlargement of the final board. The master is produced manually using drafting techniques. To increase speed

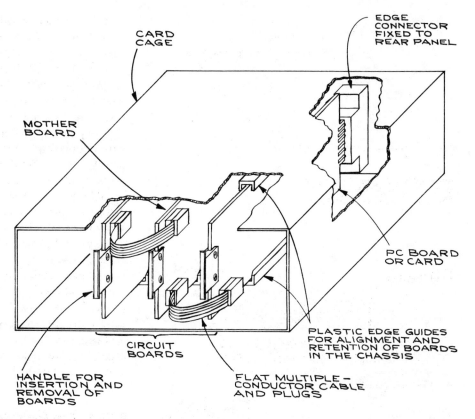

Figure 6.117 Circuit boards mounted in a card cage.

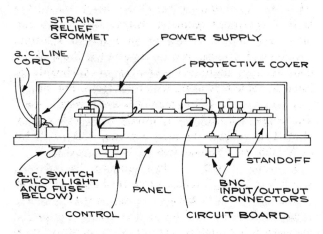

Figure 6.118 Mounting of a circuit board on a rack panel.

and accuracy, PCB dry-transfer drafting tapes, socket-hole patterns, and terminal pads should be used. These are available from a number of manufacturers. The only tools needed are a T-square, triangles, an X-acto knife to cut the tape to length, and tweezers for removing the dry transfers from the backing sheet. To avoid dimensional changes with temperature and humidity, the artwork can be prepared on a Mylar sheet rather than ordinary drafting paper or film. The completed master is then reduced photographically to the required size.

The pattern on the photographic negative is transferred to the circuit board by the use of a *photoresist*, which coats the copper cladding and is sensitive to light (Figure 6.119). When exposed through the photographic negative, it undergoes chemical and physical changes, making it resistant to a solvent. The unexposed portions are then dissolved away, uncovering the copper cladding beneath. When the board is placed in a suitable etching bath, unprotected copper is removed while the resist-covered copper remains. Alternatively, the desired pattern can be made by a silk-screen printing technique. Because bare copper oxidizes easily, it is usually covered

Table 6.37 ELECTRONIC-CIRCUIT CONSTRUCTION TOOLS

Nutdrivers
Sheet-metal nibblers
Universal electrician's tool
Needle-nose pliers
Slip-joint pliers
$\frac{1}{2}$-in. tapered reamer
Wire cutters
Wire strippers
Soldering gun, pencil
Knife
Single-edge razor blades
Solder sucker
Solder wick
Heat gun
Clip-on heat sinks
Chassis punches ($\frac{5}{8}$, $\frac{3}{4}$, $1\frac{1}{8}$, $1\frac{1}{4}$ in.)
Circuit-board holder

Note: These tools supplement those listed in Table 1.1.

Table 6.38 HARDWARE AND ELECTRONIC EQUIPMENT

Hardware

Screws (flathead and roundhead), nuts, flat washers, lock washers (4-40, 6-32, 10-24)
Solder lugs and terminals
Binding posts
Grommets
Standoffs
Cable clamps
Line cord
Hookup wire
Spaghetti (flexible thin insulating tubing) in assorted sizes
Shrink tubing in assorted sizes
Eyelets
Fuse holders
Indicator lamps
Switches

Electronic Equipment

Signal sources:
  Signal generator
  Pulse generator
  Logic pulse source
VOM/DVM
Oscilloscope
Logic probe
Logic clips
Test leads for VOM and DVM
Oscilloscope probes:
  ×1 probe for low frequencies
  ×10 compensated probe
    for frequencies above 10 MHz
Radio-frequency probe
    for demodulating r.f. signals.

with solder plate or electrolytically plated with tin, nickel, or gold. Finally, the board is cut to size and holes are drilled for mounting the components to the board.

Of all of these procedures, only the production of the artwork is practical in a laboratory. The other procedures require highly specialized equipment. In addition, the production of PCBs is a very competitive business, and there are many firms that can produce them from artwork at a moderate price. Hobbyist-type kits do not give results that justify the time and effort required to produce a satisfactory PCB.

Single-sided circuit boards do not permit very high component density. Double-sided boards are superior in this respect. However, when producing the artwork for such boards, accurate registration of the two sides is an absolute necessity. An ingenious solution is to use two colors of drafting tape and terminal pads. Commonly, blue indicates the component side of the board and red the other side. Both colors can be used on a single sheet, and when photographed through blue and red filters the blue and red patterns are selectively filtered out. In this way perfect registration of the two sides is assured. Connections between the two sides of the board are routinely made with *through-plated holes*. These are holes extending through the circuit board, which are

plated with metal on the sides of the hole and both sides of the board.

When making artwork masters, time can be saved in assembly if the positions and values of the components to be inserted into the board are included on the master. Table 6.40 summarizes the steps necessary to produce a PCB master, and Table 6.41 lists the instructions normally supplied to the PCB maker along with the master drawings. Except for the drilling of holes, the time to make several boards is only a little more than for a single one. Usually a photoresist is used rather than the high-production silk-screen printing method. The photoresist is capable of much higher resolution than the

Table 6.39 CIRCUIT-BOARD SUBSTRATES

Type of board material	Designation	Application
Paper-base phenolic	XXXP	Hot punching
	XXXPC	Room-temperature punching
Paper-base phenolic	FR-2	Flame-resistant
Paper-base epoxy resin	FR-3	Flame-resistant
Glass-fabric-base epoxy resin	FR-4	General-purpose flame-resistant
	FR-5	Temperature- and flame-resistant
	G-10	General-purpose
	G-11	Temperature-resistant
Glass-fabric-base polytetrafluoroethylene resin	GT and GX	Controlled dielectric constant

Note: Standard thicknesses, including the copper cladding, range from $\frac{1}{32}$ (0.031) to $\frac{1}{4}$ (0.250) in., with $\frac{1}{16}$ (0.062) in. the most common.

printing technique but is more expensive when many boards are required, so that the making of a screen may then be economical.

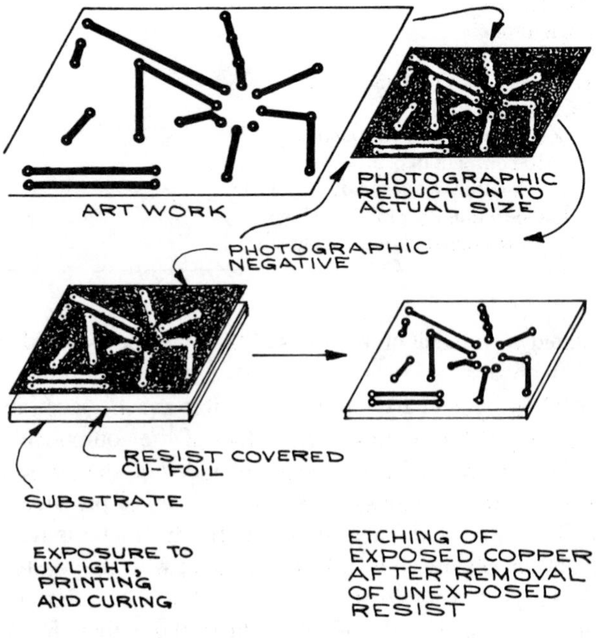

Figure 6.119 Preparation of a printed circuit board from artwork.

Components are fixed to PCBs by soldering. After inserting the leads through the appropriate holes in the board, they should be bent outward at 45° so that the

Table 6.40 PCB CONSTRUCTION CONSIDERATIONS

1. Board size, shape, material [$\frac{1}{16}$-in. G10 is most common with 1-oz/ft^2 (35-μm) copper cladding]
2. External connections
   a. Solder or wrapped to terminals
   b. Plugs
   c. Edge connectors
3. Mounting
   a. Screws
   b. Guide slots and edge connector
4. Component layout
   a. Mechanical stress
   b. Thermal stress
   c. Inputs and outputs separated to avoid positive feedback
5. Lead layout
   a. $\frac{1}{16}$ or 0.050 in. preferred width; no sharp corners
   b. 0.015-in. minimum lead clearance; larger in the vicinity of solder connections
   c. Crossovers
      (1) Insulated wire bridges when few crossovers remain
      (2) Double-sided board and plated-through holes when many crossovers remain

Table 6.41 INSTRUCTIONS TO THE PCB MAKER

1. Board material, size, cladding
2. Size of submitted master with respect to the completed board (i.e., 2X, 4X, etc.)
3. Component side (for double-sided boards, there should be enough alignment marks to make positioning easy and positive)
4. Plating
5. Size and location of holes
6. Number of boards required

component will not fall out of the board when it is turned over for soldering. After soldering, the leads are clipped short.

Most solder is an alloy of tin and lead and is specified by the ratio of weight percent tin to lead. Thus 40/60 solder is 40% tin and 60% lead by weight. An eutectic alloy with the sharp melting point of 183°C (361°F) is formed when the mixture is 63/37. As one departs from this ratio, the melting point becomes less sharp, extending over a larger temperature range. Fluxes, whose purpose it is to dissolve the oxides on the surfaces of the metals to be joined, are almost always used in soldering. Rosin fluxes are commonly used for electrical and electronic work. Under no circumstances should acid fluxes be used. These are extremely corrosive and can severely damage components, insulation, and the circuit board itself. There are three types of rosin fluxes: R (low activity), RMA (mild activity), and RA (active). Normally the R type is used. Solder wire for electronic use generally has flux incorporated in it, though pure flux is also available. The solder should be maintained at 35–65°C (60–120°F) above its melting point for a time sufficient to completely wet the surface to be joined. Too little heat will result in insufficient wetting of the surfaces and a so called "cold-solder joint," which is mechanically and electrically unsatisfactory. Too much heat can destroy components, melt plastic insulation, and burn the circuit board. When soldering heat-sensitive components, a heat sink can be used near the component to protect it. This can be in the form of a pair of long-nose pliers or special tweezers, which are clamped on the lead between the source of heat and component to be protected (Figure 6.120). To ensure a

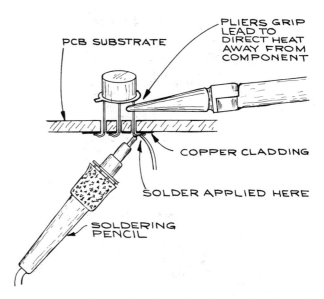

Figure 6.120 Pliers used as heat sink. A heat sink is required when soldering to the leads of sensitive components.

good joint, the tip of the soldering iron should have a thin bright coat of molten solder covering it, for thermal contact. With time, this coating becomes contaminated with oxides and should be renewed by wiping the surface and reapplying fresh solder. This is called *tinning* the iron. The properly tinned hot tip is then brought in contact with the surfaces to be soldered, and only when they have reached the temperature necessary to melt the solder should solder be applied to them. A common mistake is to melt the solder on the surface of the iron and hope it will flow on to the surfaces to be soldered. Since the molten solder flows toward the hottest point, this method leads to wasted solder and cold-solder joints. Poor solder joints also result from insufficient heat and from contaminated surfaces, which resist being wetted by the solder even in the presence of flux. When soldering is complete, the flux should be removed with a commercial flux solvent or isopropyl alcohol.

Soldering irons are generally of two types, the soldering gun [Figure 6.121(*a*)] and the single-element soldering pencil [Figure 6.121(*b*)]. Guns come in single- and dual-wattage models from 100 to 325 watts. They have the advantage of being able to be turned on and

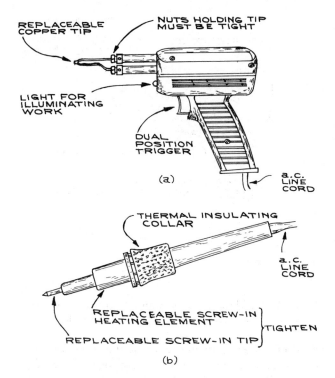

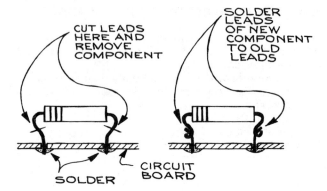

Figure 6.122 Replacing a component on a printed circuit board.

Figure 6.121 (*a*) Soldering gun; (*b*) single-element soldering pencil.

off quickly, but are generally too powerful and bulky for PCB work. Soldering pencils come in wattages from 12 to 75 watts with a large variety of tip shapes. Thermostatic control and stands with tip cleaning sponges are often used with them. For best results, soldering-iron tips should be cleaned and tinned regularly, and the heating elements should be checked for good electrical, mechanical, and thermal contact to the tips.

Often it is necessary to remove a component from a PCB and replace it. One method that does not require desoldering the old component is pictured in Figure 6.122. This technique does not work with ICs and multilead components, and desoldering from the board is necessary. Three common desoldering methods are solder wick, solder aspiration, and heat and pull. Solder wick is copper braid that withdraws solder from a joint by capillary action when heated and placed on top of the solder to be removed. Solder can also be removed by

using a sucking tool to aspirate the solder after melting it with a soldering iron. The tool can be a rubber bulb with a heat-resistant nonmetallic tip, or a triggered spring-loaded syringe. Special heat-and-pull soldering irons are needed to remove multilead components such as ICs. These irons have tips shaped to heat all leads simultaneously. After removal of the component, the holes must be cleaned of residual solder before proceeding with the insertion of another component. Solder removal by heating followed by shaking or blowing should be avoided. It results in solder splashes, which can cause short circuits.

PCB tracks that have been damaged by overheating or mechanical stresses can be removed with a knife and replaced with new sections soldered directly to undamaged foil. When pads become detached from the substrate, they can be replaced with new ones anchored to the board with swaged eyelets.

In the preceding discussion of PCBs, low-frequency applications have been assumed. If the board is made for high-frequency applications, a great deal more care is required in component placement and lead geometry. For fast logic circuits using ECL ICs, microstrip line geometries are recommended.[12]

A technique related to PCBs and based on thick-film technology is used to produce high-density circuit patterns on an alumina (Al_2O_3) substrate. A screen with 200–300 lines/inch transfers the desired circuit pattern with a special ink containing suspended metal particles and a binder. When the inked substrate is fired, the metal particles fuse to each other and the substrate, producing a tightly bonded conducting pattern. Components are then soldered to the appropriate pads on the substrate with microsoldering irons. Thick-film rather than discrete resistors are often used with this technique, since they can be produced by the same printing technique, merely using a different ink. Small, high-capacitance chip capacitors replace the disc and tubular capacitors used with PCBs. With the thick-film technique it is often quite easy to incorporate ICs and discrete components on the same substrate to produce a self-contained hybrid circuit. Such a circuit can be hermetically sealed and has the appearance of a large IC. The advantage of such circuits are small size, excellent electrical properties, and good mechanical strength. There are a number of companies that produce hybrid circuits starting from schematic diagrams. Because of its specialized nature, thick-film work is best left to such firms.

6.10.4   Wire Wrap Boards

These boards, made by the Gardner-Denver Company under the trade name Wire Wrap, have sockets spaced at intervals corresponding to the spacing of the leads on integrated circuits. The sockets have long, square-cross-section posts, so that when the socket is swaged into a substrate material the post extends on the back side. The length of the post determines how many separate connections are to be made to it. Three- and four-layer length posts are the most common (Figure 6.123). To make connections to the pins a special wrapping tool is used. The tool can be manual, line-operated, battery operated, or air-operated. The hand tool is entirely adequate for circuits with up to 10 ICs. As the wire is wrapped around the post, the high pressures generated between the wire and the sharp corners of the post form a kind of cold weld, which has good electrical and mechanical properties. The advantages of such a system are as follows:

1. No solder connections to cause heat damage to components and circuit board.

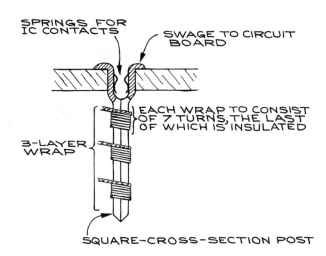

Figure 6.123   Wire Wrap post.

2. Electrical components that plug into sockets and are therefore easily removed for replacement or testing.

3. Higher component density than obtainable with double-sided circuit boards, because wire crossings are possible without short circuits.

4. Ease of removing and remaking connections. (An unwrapping tool that is a companion to the wrapping tool allows one to remove connections. However, if the connection that one wants to remove is not the highest of a stack, one must remove the upper ones to get to it.)

5. No artwork, photographic reproduction, masking, or etching required.

6. Reliability.

7. Speed of fabrication.

In practice one uses single-conductor wire specially made for wrapping. The three most common gauges are AWG 30 (0.25 mm), 28 (0.32 mm), and 26 (0.40 mm). Different wrap and unwrap tools are used for each gauge. One can obtain the wire in rolls or precut and prestripped lengths. If one uses rolls, it is necessary to have a wire stripper to remove 1 in. of insulation at either end of the length of wire to be wrapped. Thermal

Figure 6.124 (*a*) A Wire Wrap board; (*b*) an example of a Wire Wrap list.

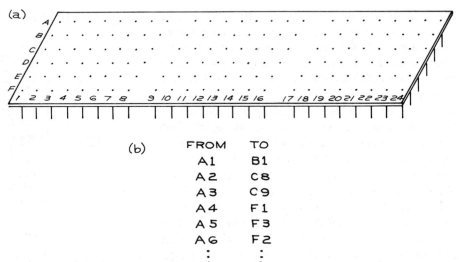

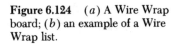

FROM	TO
A1	B1
A2	C8
A3	C9
A4	F1
A5	F3
A6	F2
⋮	⋮

strippers ensure that the conductor will not be nicked when stripped, but special mechanical strippers that do an adequate job can also be obtained inexpensively.

The usual Wire Wrap board consists of a matrix of sockets with spacing corresponding to the spacing of the terminals on the normal 14- or 16-pin IC. After the placement of ICs on the board has been decided upon, a Wire Wrap list is made, which indicates the pins between which connections are to be made. Each pin is identified by a letter and number code identifying the column and row of the pin (Figure 6.124). Once the list is complete, all the wrapping can be done at one time. It is common for Wire Wrap boards to have a ground plane and a power-supply voltage plane or bus. The Wire Wrap posts can be connected to these planes at intervals of 14 to 16 pins. Sometimes a solder bridge can be made between the appropriate pin and the exposed ground plane. Connections to the plane from outside the board can be made with a screw-and-lug connection to tie points on the board. Signal connections are usually via ribbon connectors to DIP sockets.

Discrete components are used with mounting platforms, or if sufficient space is available, they can be wrapped directly to the posts. The platforms have pins that fit the socket holes in the board, and the discrete components are soldered to the posts or forked terminals on the top of the platform (Figure 6.125). When a

circuit with only a few ICs must be constructed, an inexpensive method is to use individual Wire Wrap sockets and glue them to perforated circuit board with 0.100-in. hole spacing. This is much less expensive than commercial boards and entirely adequate for low-frequency applications. A disadvantage of Wire Wrap boards is their high initial cost; however, they may be reused.

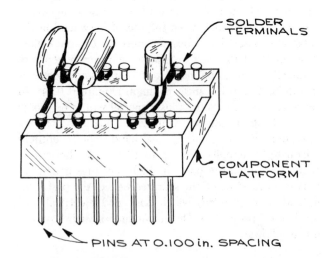

Figure 6.125 Platform for mounting discrete components. The platform then plugs into the Wire Wrap board.

Initial verification of the connections on a board is done by removing all components, attaching the ground and power-supply leads, and measuring the voltage at each pin. They should all be consistent with the Wire Wrap list.

6.10.5 Wires and Cables

There are many types of electrical wires used in the laboratory. Selection of the correct type is important for correctly functioning equipment. Wire should be selected according to its voltage and current rating for routine low-frequency operation. For applications involving high frequencies and low signal levels more care must be taken. The use of multiple-conductor cables can simplify wiring, and some knowledge of the kinds of multiple conductor configurations is useful. Table 6.42 lists the diameter, allowable current, and resistance per 1000 feet of B&S-gauge insulated copper wire. Table 6.43 lists the electrical properties of common thermoplastics used for insulation.

The most common wire is a.c.-line cord, the type used for connection to the a.c. outlet. Twin-conductor "zip cord" should be avoided in laboratory applications because of its limited resistance to mechanical stresses. Three-conductor color-coded line cord is best suited for the laboratory. The double insulation of the wires provides extra mechanical and electrical protection. The standard code is black for hot, white for neutral, and green for ground, and should be observed at both the plug and the chassis end of the cord. To minimize damage to the line cord at the chassis end, strain relief and mechanical protection of the insulation should be provided. This is often accomplished with a single strain-relief grommet or combination plastic grommet and clamp as in Figure 6.126. Line cord is almost always stranded to enhance flexibility. If the wire is to be used in high-current-carrying applications, care should be taken not to cut the strands when stripping, since that will limit the capacity of the wire and cause heating at the stripped end. Sometimes the individual strands are insulated with varnish. When making electrical connections with wire of this kind the insulation must be removed with fine emery paper or steel wool.

Table 6.42 CURRENT-CARRYING CAPACITIES OF COPPER-INSULATED WIRE

B&S Gauge	Diameter (in.)	Allowable Current[a] (A)	Resistance per 1000 ft[b] (Ω)
8	.128	50	0.628
10	.102	30	0.999
12	.081	25	1.588
14	.064	20	2.525
16	.051	10	4.016
18	.040	5	6.385
20	.032	3.2	10.15
22	.025	2.0	16.14
24	.020	1.25	25.67
26	.016	0.80	40.81
28	.013	0.53	64.90
30	.010	0.31	103.2

[a] For rubber-insulated wires the allowable current should be reduced by 30%.
[b] At 20°C (68°F).

Another type of wire used for a.c. lines is the solid-conductor #12 or #14 gauge wire for power distribution. The most common are two- and three-conductor Romex, which is PVC-insulated, and two-conductor BX,

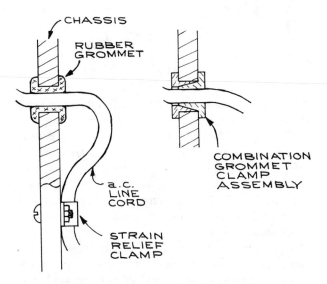

Figure 6.126 Mounting line cord at the chassis to relieve strain on the electrical connection.

Table 6.43 ELECTRICAL PROPERTIES OF THERMOPLASTICS

Material	Trade Name	Volume Resistivity (Ω cm)	Dielectric Strength (V/mil)	Power Factor at 60 Hz	Characteristics
ABS	Lustran	10^{15}–10^{17}	300–450	.003–.007	Tough, with average overall electrical properties
Acetals	Delrin	10^{14}	500	.004–.005	Strong with good electrical properties to 125°C
Acrylics	Lucite, Plexiglas, Perspex	$> 10^{14}$	450–480	.04–.05	Resistant to arcing
Fluorocarbons:					
CTFE	Kel-F	10^{18}	450	.015	Excellent electrical properties, some cold flow
FEP	Teflon FEP	$> 10^{18}$	500	.0002	Properties similar to TFE, good to 400°F
TFE	Teflon TFE	$> 10^{18}$	400	< .0001	One of the best electrical materials to 300 to 500°F; cold flow
Polyamides	Nylon	10^{14}–10^{15}	300–400	.04–.6	Good general electrical properties; absorbs water
Polyamide-imides and polyimides	Vespel, Kapton	10^{16}–10^{17}	400	.002–.003	Useful operating temperatures from 400 to 700°F; excellent electrical properties
Polycarbonates	Lexan	10^{16}	410	.0001–.0005	Good electrical, excellent mechanical properties; low water absorption
Polyethylene and polypropylenes	—		450–1000	.0001–.006	Good electrical, weak mechanical and thermal properties
Polyethylene terephthalates	Mylar	$> 10^{16}$	500–710	.0003	Tough, excellent dielectric properties
PVC	Saran	10^{11}–10^{16}	300–1100	.01–.15	Low cost; general purpose; average electrical properties

which has a metal-armored outer covering. Both types can be routed within walls or outside, in either metal or PVC conduit. Because of the heavy gauge of the conductors, connections are made to screw terminals or with *wire nuts* when splices are involved. The installation of a.c. lines generally requires a professional electrician who knows the conventions, regulations (codes), and procedures for fusing and connection to existing power lines.

For electronic circuits where the voltages are less than a few hundred volts and the currents are a few amperes, one uses hookup wire. This can be solid conductor or multistranded. The solid-conductor wire is easier to use when making connections because it does not fray, but it lacks the flexibility of multistranded wire. Hookup wire is generally *tinned*, that is, coated with a tin-lead alloy to enhance solderability. The insulation is usually PVC, polyethylene, or Teflon.

When stripping the wire, care should be taken to avoid nicking the conductor of the solid wire and cutting strands of the multistranded wire. There are a large variety of wire strippers on the market (Figure 6.127), but none is foolproof and all require a certain amount of skill if the conductor is to remain undamaged. Thermal strippers that melt the insulation locally before it is withdrawn from the wire work very well but are not

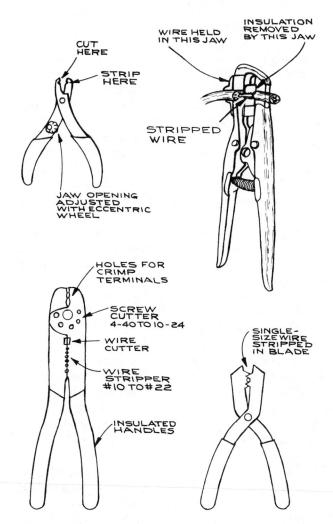

Figure 6.127 Wire strippers.

portable. For Teflon-insulated wire, thermal strippers are very useful because the slippery material is difficult to grip.

When working with stranded wire, it is recommended that the ends be tinned. It should be remembered, however, that solder will destroy the flexibility of multistrand wire so that the tinning should be confined to a short length at the end. For coding purposes, the insulation of hookup wire comes in many colors and combinations of colors. Color coding is a very useful way to avoid wiring errors and makes subsequent troubleshooting easier.

Test-prod wire is highly flexible multistrand wire with rubber insulation rated at a few thousand volts. Such wire comes with red or black insulation and is used with multimeters and in high-voltage circuitry.

For voltages from a few kilovolts to several tens of kilovolts, special high-voltage cable must be used. The most common jacket materials are silicone rubber, Teflon, and Kapton. Silicone-rubber insulation is very flexible but has only modest dielectric breakdown strength and volume resistivity, so that high-voltage cables made with it have large diameters. Kapton, a polyimide, has high breakdown strength and volume resistivity, and cables made with it are relatively small in diameter. The principal disadvantage of Kapton is its stiffness. The properties of Teflon are intermediate between those of silicone rubber and Kapton, and it is usually the best compromise. Typical properties of high-voltage Teflon cable are given in Table 6.44. One source of high-voltage cable is the ignition cables used in automobiles. The silicone-rubber-insulated stainless-steel or copper stranded-conductor type found in high-performance cars can be used to 50 kV. The more common carbon-filament conductor cable used to minimize rfi is not suitable.

Magnet wire is varnish-insulated solid-conductor wire used for winding electromagnets, transformers, and inductors. The thin insulation means that one can pack a large number of turns in a given volume. To make

Table 6.44 TYPICAL HIGH-VOLTAGE
TEFLON-INSULATED CABLE

D.c. Voltage Rating[a] (kV)	Conductor	Cable diameter (in.)
10	19/36[b]	0.077
15		0.097
20		0.117
25		0.137
40		0.197
50		0.237

[a]A.c. voltage rating is typically 20–25% of d.c.
[b]19 strands of #36 wire.

electrical connections with such wire the varnish must be removed with emery paper or steel wool. This wire is not a suitable substitute for hookup wire. When using it for electromagnets, one should be aware of the possible buildup of heat around the inner windings, which can destroy the insulation and create short circuits. The varnish insulation is rated for voltage breakdown and heat resistance.

For high-current electromagnets, rectangular-cross-section wire is used because the high surface-area-to-volume ratio facilitates convection cooling. In applications that require conduction cooling, hollow-core wire is used, through which cooling water can be circulated.

In addition to two- and three-conductor line cord, there are many other multiconductor round cable configurations. The conductors can be solid or stranded. The wires are color-coded for identification. For low-level, low-frequency signals, the wires can be enclosed in a single shield, as is done for microphone cable. Configurations where there is individual shielding of single wires also are available. Shielding can be in the form of a braid or a foil; if foil, a *drain wire* is also included, which is electrically connected to the shield and to which connections can be made. For high-frequency applications there is coaxial cable or 300-Ω parallel-conductor cable with transmission-line properties. These are discussed in the sections on coaxial cable and connectors (6.2.3 and 6.2.4).

Flat multiconductor cable is very useful when a large number of wires are needed as with computer and data interfaces. There are many different arrangements of the wires within the cable. The simplest uses round parallel conductors. There are also flat conductors and alternating round and flat. Flat conductors minimize interconductor capacitance and interference among signals. Twisted pairs are used for transmission lines, and an alternating twisted-pair–straight geometry is often used for data transmission. The great advantage of flat cable is that mass termination connectors can be used. These connectors are designed so that the cable is clamped and electrical contact is made to each of the wires in a single operation with a special tool. No stripping, soldering, or crimping is required. The reliability of such connections is excellent.

6.10.6   Connectors

Probably the best-known connector combination is the *binding post* and *banana plug* or *tip plug* (Figure 6.128). Binding posts have the advantage of permitting a wide variety of conductors and conductor terminations to be made to them in a rapid, reliable semipermanent way. They have the disadvantage of being bulky and highly susceptible to the pickup and radiation of electromagnetic energy. When mounting binding posts on a chassis, the mounting hole should be made large enough to accommodate the shoulders on the insulating sleeves. In this way the central conductor is kept well away from the chassis. The standard spacing between posts is 0.75 in. This corresponds to the spacing of twin-conductor banana plugs (Figure 6.129), which are quite common and useful, especially when it is necessary to connect coaxial cable to binding posts. The twin plug usually has one terminal marked "ground," to which the shield of coaxial cable is attached.

Alligator clips (Figure 6.130), though useful with test leads, should not be used in any permanent or even semipermanent installation. The type of clip into which one inserts a banana plug is more useful than a clip requiring solder or screw connections. Flexible insulating sleeves that fit over the clips prevent short circuits and are always used when the clip is at the end of a test lead.

Phone plugs (Figure 6.131) come in several different sizes and provide one means for the termination and connection of coaxial cable carrying low-frequency signals. Because this type of plug is polarized (that is, the connections can only be made in one way), it is sometimes used for low-voltage, low-current d.c. power-supply connections. There are many variations on the basic plug-jack design. There are plugs that can accommodate three or more separate connections, and there are jacks that remain shorted until the insertion of the plug.

Insulated *barrier strips* (Figure 6.132) with screw terminals are a good way of making semipermanent connections for d.c. applications and are often found on the rear panels of power supplies.

For terminating the ends of wires, there are a variety of terminals which are attached to the wire by soldering

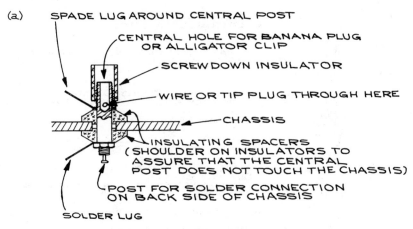

(a)

SPADE LUG AROUND CENTRAL POST

CENTRAL HOLE FOR BANANA PLUG OR ALLIGATOR CLIP

SCREW DOWN INSULATOR

WIRE OR TIP PLUG THROUGH HERE

CHASSIS

INSULATING SPACERS (SHOULDER ON INSULATORS TO ASSURE THAT THE CENTRAL POST DOES NOT TOUCH THE CHASSIS)

POST FOR SOLDER CONNECTION ON BACK SIDE OF CHASSIS

SOLDER LUG

Figure 6.128   (*a*) Binding post; (*b*) banana plug.

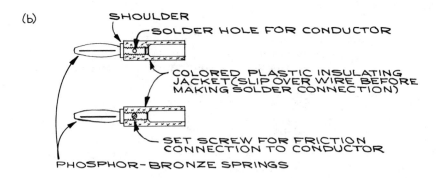

(b)

SHOULDER

SOLDER HOLE FOR CONDUCTOR

COLORED PLASTIC INSULATING JACKET (SLIP OVER WIRE BEFORE MAKING SOLDER CONNECTION)

SET SCREW FOR FRICTION CONNECTION TO CONDUCTOR

PHOSPHOR-BRONZE SPRINGS

or crimping (Figure 6.133). The terminal used should be of the correct size with respect to both the wire used and the opening at the end. Quick-connect, friction-type, push-on terminals are very useful in d.c. applications when the connection must be made and broken repeatedly. Such terminals are often also found on mechanical relays, circuit breakers, and mechanical switches. Crimp connectors are color-coded according to the wire-size range they can accommodate. It is important to use the proper crimp-tool opening, which is also color coded on many tools, to avoid too loose a crimp (too large a hole) or a severed wire (too small a hole).

When soldering connectors, only enough heat should

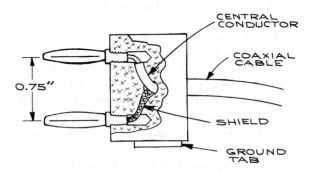

CENTRAL CONDUCTOR

COAXIAL CABLE

0.75"

SHIELD

GROUND TAB

Figure 6.129   Twin-conductor banana plug.

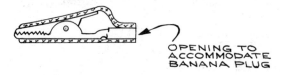

OPENING TO ACCOMMODATE BANANA PLUG

Figure 6.130   Alligator clip.

Figure 6.131   Phone plug.

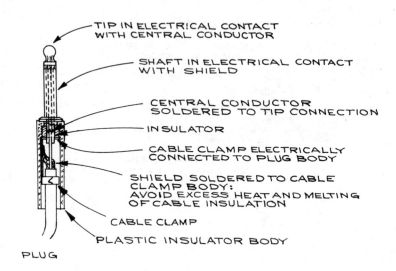

TIP IN ELECTRICAL CONTACT WITH CENTRAL CONDUCTOR

SHAFT IN ELECTRICAL CONTACT WITH SHIELD

CENTRAL CONDUCTOR SOLDERED TO TIP CONNECTION

INSULATOR

CABLE CLAMP ELECTRICALLY CONNECTED TO PLUG BODY

SHIELD SOLDERED TO CABLE CLAMP BODY: AVOID EXCESS HEAT AND MELTING OF CABLE INSULATION

CABLE CLAMP

PLASTIC INSULATOR BODY

PLUG

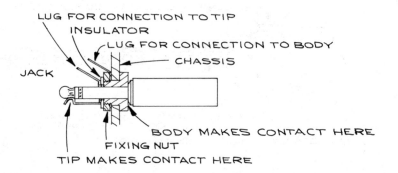

LUG FOR CONNECTION TO TIP

INSULATOR

LUG FOR CONNECTION TO BODY

CHASSIS

JACK

BODY MAKES CONTACT HERE

FIXING NUT

TIP MAKES CONTACT HERE

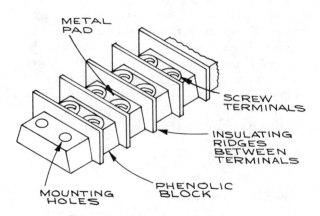

METAL PAD

SCREW TERMINALS

INSULATING RIDGES BETWEEN TERMINALS

PHENOLIC BLOCK

MOUNTING HOLES

Figure 6.132   Barrier strip.

be used to make a good joint, since too much heat will melt the insulation. Some type of "third hand" is helpful in this regard—a small vice or an alligator clip at the end of a heavy piece of solid copper wire anchored to a metal base.

A very large part of the electronics-hardware industry is devoted to the manufacture of connectors. For laboratory applications only a few of the most common ones will be described, and some guidelines for connector selection will be given. Electrical contact between the wires and the connector pins can be made by soldering or crimping. For some connectors the pins must be removed to make the electrical connection and then inserted into the connector block. Removal of the pin

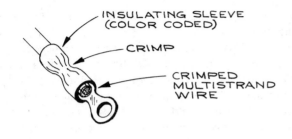

INSULATING SLEEVE
(COLOR CODED)

CRIMP

CRIMPED
MULTISTRAND
WIRE

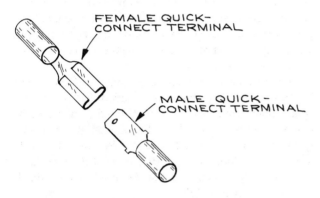

FEMALE QUICK-
CONNECT TERMINAL

MALE QUICK-
CONNECT TERMINAL

Figure 6.133 Wire terminals.

then requires a special tool. A common, older multiple-pin connector, which is a holdover from vacuum-tube days, is the *octal* connector (Figure 6.134).

D-connectors derive their name from the cross-sectional shape of the connector body (Figure 6.135). The connector can have up to 50 contacts and connection is made by soldering the conductor into a hollow recess at the back of the contact (solder pot). Such connectors can be attached directly to a panel with the correct cutout. When fitted with a shell (either plastic or metal) incorporating a cable clamp, they become plugs. Locking accessories are also available for securing the plug to the mating jack, and the D-shape assures correct mating. Double-density connectors of this type are also made, and there exist a wide variety of configurations with high-voltage, high-current, and coaxial contacts in addition to the standard single pins. Because pins are very closely spaced, *shrink tubing* is routinely used over the solder connections. The tubing serves a strain-relief as well as an insulating function in this case. Packages of shrink tubing with an assortment of sizes are convenient. If the tubing is too small in diameter it will split upon being shrunk; if too large, it will not fit the enclosed wire tightly. Shrinking should be done with a

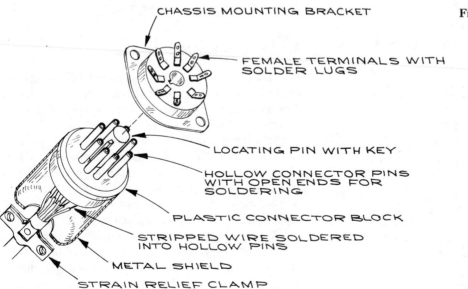

CHASSIS MOUNTING BRACKET

Figure 6.134 Octal plug.

FEMALE TERMINALS WITH
SOLDER LUGS

LOCATING PIN WITH KEY

HOLLOW CONNECTOR PINS
WITH OPEN ENDS FOR
SOLDERING

PLASTIC CONNECTOR BLOCK

STRIPPED WIRE SOLDERED
INTO HOLLOW PINS

METAL SHIELD

STRAIN RELIEF CLAMP

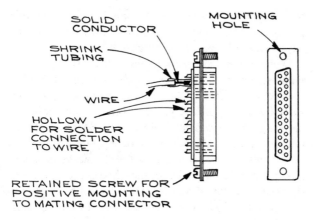

Figure 6.135   D-connector.

heat gun rather than a soldering iron. The usual RS232C interface uses a 25-pin subminiature D-connector.

Threaded circular connectors (Figure 6.136) are based on designs originally used by the military for aircraft. They employ separate removable pins and a threaded mating collar to make a mechanically secure union between male and female connectors. There is a broad range of sizes, styles, and pin arrangements, and for this reason these connectors are used on a variety of laboratory electronic equipment. There are various ways of retaining the pins in the insulator block. Some of them require special insertion and removal tools, while others

use a second backup block held in place inside the connector.

Rack and panel connectors are used to connect chassis-mounted equipment to a stationary rack or panel. NIM and CAMAC equipment are examples of such arrangements. The basic connector is a rectangular insulator block, which holds contact pins, maintained in place by one-way spring action. Connection to the pins is by crimping. The pins can be readily inserted into the block, but an extraction tool is required to remove them. When the connector has a large number of pins, considerable force is required for mating, and threaded screw jacks are often used. Extra pins can be purchased separately, and shrouds are available to convert the connector to a plug.

No high-voltage connector is entirely satisfactory, and for this reason they should be avoided and permanent connections with ceramic feedthroughs or standoffs made whenever possible. European-type phenolic spark-plug connectors make acceptable connectors when used with ignition cable; a drilled-out Teflon-insulated r.f. coaxial connector can also be used. These are illustrated in Figure 6.137.

6.11   TROUBLESHOOTING

6.11.1   General Procedures

When a piece of electronic equipment fails to operate properly there are a number of steps that can be taken, depending on the complexity of the equipment, and the availability of test equipment and diagrams.

The worst symptoms are usually the easiest to treat. An experienced TV repairman once noted that 95% of all malfunctions in TV sets could be found and corrected in less than one-half hour. For the other 5% the best solution was to sell the customer a new set. This also applies to laboratory equipment. With increasing use of highly reliable integrated circuits and resistance and capacitance modules, the most unreliable elements of a circuit are the mechanical parts (such as switches, dials, and fans) and the connectors. Thus, when an

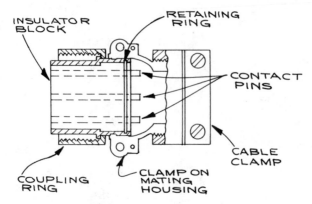

Figure 6.136   Threaded military-type connector.

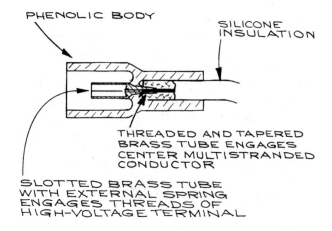

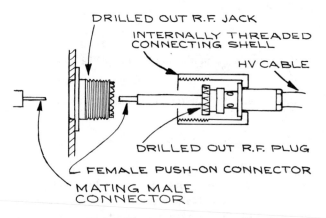

Figure 6.137 High-voltage connector made from a spark-plug and r.f. coaxial connector.

amplifier or power supply ceases to operate, one's first impulse should not be to take out screwdriver and wrench and begin disassembling the chassis, but to be sure that it is plugged in, that the on-off switch is functioning, that all fuses are intact, that the cooling fans are operating, and that the air filters are clean and not blocked. Switch contacts can be cleaned with spray cleaners made for the purpose.

If these measures produce no results, the various dial and switch settings of the instrument should be checked. Often what is interpreted as a malfunction is merely an incorrect setting, causing the instrument to operate in an unexpected mode.

If these simple actions are not effective, the next course of action depends on the complexity of the equipment and the availability of circuit diagrams. To repair a complex circuit without circuit diagrams is almost impossible. If they are not at hand, write to the manufacturer for the necessary diagrams and for troubleshooting procedures.

There are of course certain visual checks that one can perform without knowledge of the circuit. Charred resistors and an overheated circuit board are readily apparent, as are loose and dangling wires. One difficulty with merely replacing a damaged component, however, is that only the symptom may be treated. The charred resistor may be due to a failure in another component.

Troubleshooting is most effective and efficient when a complete set of circuit diagrams, schematic drawings, chassis diagrams, and functional block diagrams are available. These are often accompanied by a troubleshooting guide, which lists the most frequently encountered malfunctions, the symptoms, and the remedies. Often by studying the functional block diagram of the unit, the area in which the malfunction is occurring can be localized. As an example, consider a counter-timer that responds normally to input signals but gives erratic readings when counting for preset time intervals, and erratic times when in the preset-counts mode. In this case, obviously the gating circuit and the time base are to be suspected. One would begin by looking at that part of the circuit.

Good schematic diagrams include waveforms and voltage levels at all critical points in the circuit. These should be checked first. Well-designed circuit boards often have the d.c. voltage levels printed directly at the appropriate test points and components identified by a printed letter-and-number code that corresponds to an identical code on the schematic diagram.

A common difficulty is gaining access to components and connections in the restricted space between boards in a multiple-circuit-board assembly. *Extender boards* (Figure 6.138) are circuit boards arranged in such a way that they mate to the circuit-board connector at one end and the circuit board itself at the other end. The ex-

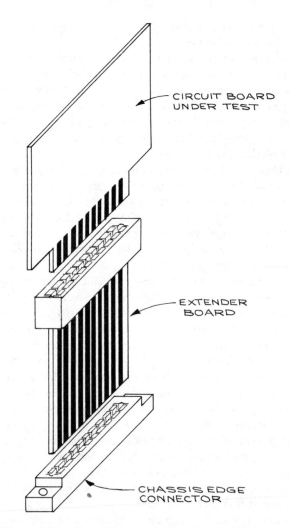

Figure 6.138 Use of the extender board for troubleshooting.

tender board is of sufficient length to place the board in a position where the components and circuit interconnections are easily accessible.

Signal injection is a common method for isolating a circuit fault. With this technique, an appropriate signal is injected at the input of the suspected circuit element, and the output monitored. This works well for linear circuits such as amplifiers, but becomes quite complex for digital circuits, where it may be necessary to stimulate several input terminals and monitor several output

terminals simultaneously and often in synchronization with a clock signal. Logic pulsers and probes as well as clips are useful for small-scale testing. They are shown in Figure 6.139. For more complex circuits it is necessary to capture the signals and store them for subsequent analysis. There are logic analyzers made expressly for this purpose; however, they are expensive and only worthwhile in situations where a large amount of digital circuit troubleshooting is done—more the domain of the electronic technician than the experimental scientist. Often it is much more economical to replace all the suspected ICs in a given section (especially if they are easily removable and in sockets) than to test each one individually.

Some common sources of faults in electronic equipment are as follows:

1. Electrolytic capacitors in general.

2. Pass transistors in the output circuit of power supplies.

3. Input transistors in the input circuits of amplifiers and preamplifiers.

4. Mechanical switches, potentiometers.

Repeated failure of input transistors in low-level preamplifier circuits can often be cured by placing two diodes such as 1N914s across the input, as shown in Figure 6.140. So long as the input signal does not exceed a few tenths of a volt, both diodes remain nonconducting, allowing the unattenuated signal to pass. Should the signal exceed 0.6 V, positive or negative, one or the other diode will conduct, shunting the signal to ground.

6.11.2 Identifying parts

Often one can localize the malfunction to a circuit component but does not know enough about the component to be able to replace it. The parts list may give the equipment manufacturer's component code rather than the standard code of the manufacturer of the component. When this is the case, it is necessary to identify the component from the code printed on it.

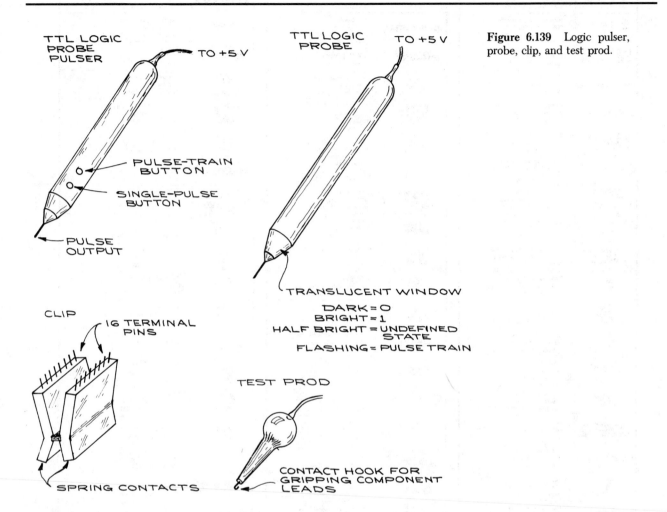

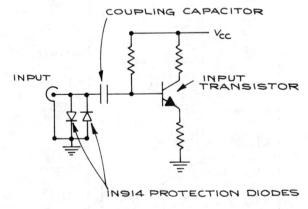

Figure 6.139 Logic pulser, probe, clip, and test prod.

Most ICs, transistors, and other components have a four-digit date code giving the date of manufacture. The first two digits are the year, and the next two are the week of the year. Thus, 7712 is the 12th week (last week in April) of 1977. The other codes are the manufacturer's component designation. To help identify the manufacturer, a list of logos is given in Figure 6.141. A list of the prefix codes of semiconductor manufacturers is given in Table 6.45. Once the manufacturer is known, the appropriate data book can be consulted. One should replace components with caution, paying attention to the package type and temperature range. The three standard ranges are commercial (0 to 70°C), industrial (-25 to $+85$°C), and military (-55 to $+125$°C).

Figure 6.140   Use of protection diodes at a low-level input.

Logo	Manufacturer
AM	Advanced Micro Devices 901 Thompson Place Sunnyvale, CA 94086
AML	American Microsystems, Inc. 3800 Homestead Road Santa Clara, CA 95051
ANALOG DEVICES	Analog Devices Route 1 Industrial Park Norwood, MA 02062
ANALOGIC	Analogic Audubon Road Wakefield, MA 01880
BECKMAN	Beckman Instruments 2500 Harbor Boulevard Fullerton, CA 92634
BURR-BROWN BB	Burr-Brown International Airport Park Tucson, AZ 85734
CS MCC	Cherry Semiconductor (Micro Components) 99 Bald Hill Road Cranston, RI 02920
	Consumer Microcircuits 114 E. Simmons Street Galesburg, IL 61401
CY	Cybernetic Micro Devices 2378B Walsh Avenue Santa Clara, CA 95050
DDC	Data Device Corporation Airport International Plaza Bohemia, NY 11716
	Data General 15 Turnpike Road Westboro, MA 01587
D	Datel Systems 11 Cabot Boulevard Mansfield, MA 02048
	Electronic Arrays 550 E. Middlefield Road Mountain View, CA 94043
EMM SEMI, INC.	EMM Semi 2000 W. 14th Street Tempe, AZ 85281
	EXAR Integrated Systems 750 Palomar Avenue Sunnyvale, CA 94086
FAIRCHILD	Fairchild 464 Ellis Street Mountain View, CA 94042
FERRANTI	Ferranti 87 Modula Avenue Commack, NY 11725
F	Fujitsu 2945 Oakmead Village Court Santa Clara, CA 95051
GI	General Instrument 600 W. John Street Hicksville, NY 11802
	Harris P.O. Box 883 Melbourne, FL 32901
(O)	Hitachi 707 W. Algonquin Road Arlington Heights, IL 60005
HUGHES	Hughes Aircraft 500 Superior Avenue Newport Beach, CA 92663
Hybrid Systems	Hybrid Systems Crosby Drive Bedford, MA 01730
	Intech/FMI 282 Brokaw Road Santa Clara, CA 95050
	Integrated Photomatrix 1101 Bristol Road Mountainside, NJ 07092
intel i	Intel 3065 Bowers Avenue Santa Clara, CA 95051

Logo	Manufacturer
interdesign	Interdesign 1255 Reamwood Avenue Sunnyvale, CA 94086
INTERSIL	Intersil 10900 N. Tantau Avenue Cupertino, CA 95014
	ITT Semiconductor 500 Broadway Lawrence, MA 01841
LSI	LSI Computer Systems 1235 Walt Whitman Road Melville, NY 11746
	Lambda 515 Broad Hollow Road Melville, NY 11746
ML	Master Logic 716 E. Evelyn Sunnyvale, CA 94086
MN	Micro Networks 324 Clark Street Worcester, MA 01606
Mii	Micropac Industries 905 E. Walnut Street Garland, TX 75040
M	Micro Power Systems 3100 Alfred Street Santa Clara, CA 95050
	Mitel Semiconductor P.O. Box 13089, Kanata, Ottawa, Ontario, Canada K2K IX3
	Mitsubishi Marunouchi, Tokyo Post Code 100, Japan
MMI	Monolithic Memories, Inc. 1165 E. Arques Avenue Sunnyvale, CA 94086
MOS	MOS Technology 950 Rittenhouse Road Norristown, PA 19401
MOSTEK	Mostek 1215 W. Crosby Road Carrollton, TX 75006
M	Motorola 5005 E. McDowell Road Phoenix, AZ 85008
NCR	NCR 8181 Byers Road Miamisburg, OH 45342
NEC	NEC 3120 Central Expressway Santa Clara, CA 95051
	National Semiconductor 2900 Semiconductor Drive Santa Clara, CA 95051
Nitron	Nitron 10420 Bubb Road Cupertino, CA 95014
NORTEC	Nortec 3697 Tahoe Way Santa Clara, CA 95051
NPC	Nucleonic Products Co. 6660 Variel Avenue Canoga Park, CA 91303
OEi	Optical Electronics P.O. Box 11140 Tucson, AZ 85734
MATSUSHITA	Panasonic (Matsushita) 1 Panasonic Way Secaucus, NJ 07094
PLESSEY	Plessey 1641 Kaiser Irvine, CA 92714
PMI	Precision Monolithics 1500 Space Park Drive Santa Clara, CA 95050

Logo	Manufacturer
RCA Solid State	RCA Box 3200 Somerville, NJ 08876
RAY	Raytheon Semiconductor 350 Ellis Street Mountain View, CA 94042
RETICON	Reticon 345 Potrero Sunnyvale, CA 94086
Rockwell	Rockwell (Collins) 3310 Miraloma Avenue Anaheim, CA 92803
	SGS-ATES Semiconductor 240 Bear Hill Road Waltham, MA 02154
	Sanyo 291 S. Van Brunt Street Englewood, NJ 07631
signetics	Signetics 811 E. Arques Avenue Sunnyvale, CA 94086
SILICON GENERAL	Silicon General 11651 Monarch Street Garden Grove, CA 92641
B	Siliconix 2201 Laurelwood Road Santa Clara, CA 95054
SSi	Silicon Systems 14351 Myford Road Tustin, CA 92680
SSS	Solid State Scientific Industrial Center Montgomeryville, PA 18936
S	Solitron 8808 Balboa Avenue San Diego, CA 92123
(2) SPRAGUE	Sprague Electric 115 N.E. Cutoff Worcester, MA 01606
	Standard Microsystems 35 Marcus Boulevard Hauppauge, NY 11787
	Supertex 1225 Bordeaux Drive Sunnyvale, CA 94086
S	Synertek 3001 Stender Way Santa Clara, CA 95051
	Texas Instruments P.O. Box 225012 Dallas, TX 75222
TMX	TMX 1100 Glendon Avenue Los Angeles, CA 90024
TRW	TRW P.O. Box 1125 Redondo Beach, CA 90278
TELEDYNE CRYSTALONICS	Teledyne Crystalonics 147 Sherman Street Cambridge, MA 02140
TELEDYNE PHILBRICK	Teledyne Philbrick Allied Drive Dedham, MA 02026
	Teledyne Semiconductor 1300 Terra Bella Avenue Mountain View, CA 94043
TFK	Telefunken Route 22 at Orr Street Somerville, NJ 08876
	Toshiba 2900 MacArthur Boulevard Northbrook, IL 60062
	Western Digital 3128 Red Hill Avenue Newport Beach, CA 92663
Zilog	Zilog 10340 Bubb Road Cupertino, CA 95014

Figure 6.141 Manufacturers' logos.

Table 6.45   SEMICONDUCTOR INTEGRATED-CIRCUIT CODE PREFIXES

Company	Prefix[a]
Analog Devices	AD
Advanced Micro Devices	Am
General Instrument	AY, GIC, GP
Intel	C, I
RCA	CA, CD, CDP
TRW	CA, TDC, MPY, CMP, DAC, MAT, OP
Precision Monolithics	PM, REF, SSS
National Semiconductor	DM, LF, LFT, LH, LM, NH
Fairchild	F, μA, μL, Unx
Ferranti	FSS, ZLD
GE	GEL
Harris	HA
Motorola	HEP, MC, MCC, MCM, MFC, MM, MWM
Intersil	ICH, ICL, ICM, IM
ITT	ITT, MIC
Siliconix	L, LD
Fugitsu	MB
Mostek	MK
Plessey	MN, SL, SP
Signetics	N, NE, S, SE, SP
Raytheon	R, RAY, RC, RM
Texas Instruments	SN, TMS
Sprague	ULN, ULS
Westinghouse	WC, WM
Hewlett-Packard	5082-$nnnn$

[a] x = number; n = letter.

Sometimes specially selected or matched components are used. Replacement with off-the-shelf units may not work in this case.

CITED REFERENCES

1. H. W. Bode, *Network Analysis and Feedback Amplifier Design*, Van Nostrand, Princeton, N.J., 1945.
2. Electronic Design, 24, 63, 1976.
3. J. Millman and C. C. Halkias, *Integrated Electronics: Analog and Digital Circuits and Systems*, McGraw-Hill, New York, 1972, pp. 244–245.
4. J. G. Graeme, *Operational Amplifiers, Design and Application*, G. E. Tobey and L. P. Huelsman, Eds., McGraw-Hill, New York, 1971; *Designing with Operational Amplifiers*, McGraw-Hill, New York, 1977.
5. E. Fairstein and J. Hahn, "Nuclear Pulse Amplifiers—Fundamentals and Design Practice," Nucleonics, 23, No. 7, 56, 1965; ibid., No. 9, 81, 1965; ibid., No. 11, 50, 1965; ibid., 24, No. 1, 54, 1966; ibid., No. 3, 68, 1966.
6. *Voltage Regulator Handbook*, National Semiconductor Corporation, Santa Clara, Calif.
7. *Standard Nuclear Instrument Modules*, adopted by AEC Committee on Nuclear Instrument Modules, U.S. Government Publication TID-20893 (Rev. 3).
8. S. Letzter and N. Webster, "Noise in Amplifiers," IEEE Spectrum, August 1970, pp. 67–75.
9. John C. Fisher, "Lock in the Devil, Educe Him or Take Him for the Last Ride in a Boxcar?" Tek Talk, Princeton Applied Research, 6, No. 1.
10. R. Morrison, *Grounding and Shielding Techniques in Instrumentation*, Wiley, New York, 1967.
11. *Floating Measurements and Guarding*, Application Note 123, Hewlett-Packard, 1970.
12. *Circuits for Electronics Engineers*, S. Weber, Ed., Electronics Book Series, McGraw-Hill, New York, 1977; *Circuit Design Idea Handbook*, W. Furlow, Ed., Cahner's Books, Boston, 1974; *Electronics Circuit Designer's Casebook*, Electronics, New York; *Signetics Analog Manual, Applications, Specifications*, Signetics Corporation, Sunnyvale, Calif.; *Linear Applications Handbook*, Vols. 1 and 2, National Semiconductor Corporation, Santa Clara, Calif.
13. W. R. Blood, Jr., *MECL Applications Handbook*, 2nd edition, Motorola Semiconductor Products, 1972.

GENERAL REFERENCES

CAMAC and IEEE-488 (GPIB or HP-IB)

CAMAC: A Modular Instrumentation System for Data Handling, ESONE Committee, Report EUR 4100, 1972, Chapters 4–6.
CAMAC Tutorial Issue, IEEE Trans. Nucl. Sci., NS-20, No. 2, April 1973.
D. Horelick and R. S. Larsen, CAMAC: "A Modular Standard," IEEE Spectrum, April 1976, p. 50.

HP-IB General Information, Hewlett-Packard Publication No. 5952-0058.

IEEE Standard 488, available from the IEEE Standards Office, 345 E. 47th St., New York, NY 10017.

Specifications for the CAMAC Serial Highway and Serial Crate Controller Type 62, Report EUR 6100e, Commission of the European Communities, Greel, Belgium, 1976, Chapter 14.

Circuit Theory

P. Grivet, *The Physics of Transmission Lines at High and Very High Frequencies*, Academic Press, New York, 1970.

H. V. Malmstadt, C. G. Enke, and S. R. Crouch with G. Horlick, *Electronic Measurements for Scientists*, Benjamin, Menlo Park, Calif., 1974.

J. Millman, *Microelectronics*, McGraw-Hill, New York, 1979.

C. J. Savant, Jr., *Fundamentals of the Laplace Transform*, McGraw-Hill, New York, 1962.

A. I. Zverev, *Handbook of Filter Synthesis*, Wiley, New York, 1967.

Components

M. Grossman, "Focus on rf Connectors," Electronic Design, 24, No. 11, 60, 1976.

C. A. Harper, "To Compare Electrical Insulators," Electronic Design, 24, No. 11, 72, 1976.

C. A. Harper, Ed., *Handbook of Components for Electronics*, McGraw-Hill, New York, 1977.

T. H. Jones, *Electronic Components Handbook*, Reston Publishing, Reston, Va., 1978.

"Selecting Capacitors Properly," Electronic Design, 13, 66, 1977.

Data Books

D.A.T.A. Books, D.A.T.A., San Diego, Calif. Volumes include *Optoelectronics, Digital I.C., Linear I.C., Transistor, Diode, Thyristor, Power Semiconductor, Interface I.C.*

MECL System Design Handbook, W. R. Blood, Jr., and E. C. Tynan, Jr., Eds., 2nd edition, Motorola Semiconductor Products, 1972.

Motorola Semiconductor Data Library: Vol. 1, *EIA Type Numbers to 1N5000 and 2N5000*; Vol. 2, *Discrete Products, EIA Type Numbers 1N5000 and Up, 2N5000 and Up, 3N...*

and 4N...; Vol. 3, *Discrete Products, Motorola Non-Registered Type Numbers*; Vol. 4, *MECL Integrated Circuits*.

National Semiconductor Data Books: Linear, TTL, CMOS, MOSFET, Memory, Discrete, National Semiconductor Corporation, Santa Clara, Calif.

Signetics Data Manual: Logic, Memories, Interface, Analog, Microprocessor, Military, Signetics, Sunnyvale, Calif.

The TTL Data Book for Design Engineers, 1st edition, Texas Instruments, 1973.

Voltage Regulator Handbook, National Semiconductor Corporation, Santa Clara, Calif.

Handbooks

Electronics Designers' Handbook, L. J. Giacoletto, Ed., 2nd edition, McGraw-Hill, New York, 1977.

T. D. S. Hamilton, *Handbook of Linear Integrated Electronics for Research*, McGraw-Hill, New York, 1977.

ITT Engineering Staff, *Reference Data for Radio Engineers*, 6th edition, Sams, Indianapolis, 1979.

Radio Engineering Handbook, K. Henney, Ed., McGraw-Hill, New York, 1959.

The Radio Amateur's Handbook, 58th edition, American Radio Relay League, Hartford, Conn., 1981.

Master Catalogs

Electronic Design's Gold Book: Master Catalog and Directory of Suppliers to Electronics Manufacturers, 7th edition, Hayden, Rochelle Park, N.J., 1980.

Electronic Engineers Master, Vols. 1 and 2, 23rd edition, United Technical Publications, Garden City, N.J., 1980.

Noise

S. Letzter and N. Webster, "Noise in Amplifiers," IEEE Spectrum, August 1970, pp. 67–75.

A. V. D. Ziel, "Noise in Solid-State Devices and Lasers," Proc. IEEE, 58, No. 8, 1178, 1970.

Particle and Radiation Detection

Electronics for Nuclear Particle Analysis, L. J. Herbst, Ed., Oxford University Press, London, 1970.

Electro-Optics Handbook: A Compendium of Useful Informa-

tion and Technical Data, RCA Defense Electronics Products, Aerospace Systems Division, Burlington, Mass., 1968.

Glenn F. Knoll, *Radiation Detection and Measurement*, Wiley, New York, 1979.

Optoelectronics Designer's Catalog, Hewlett-Packard, 1980.

Practical Electronics

Design Techniques for Electronics Engineers, Electronics Book Series, McGraw-Hill, New York, 1977.

P. Horowitz and W. Hill, *The Art of Electronics*, Cambridge University Press, Cambridge, 1980.

D. Lancaster, *CMOS Cookbook*, Sams, Indianapolis, 1977.

D. Lancaster, *TTL Cookbook*, Sams, Indianapolis, 1974.

Trade Publications

Digital Design, Benwill, Boston: computers, peripherals, systems; monthly.

Electronic Component News, Chilton, Radnor, Pa.; monthly.

Electronic Engineering Times, CMP Publications, Manhasset, N.Y.; biweekly tabloid.

Instrument and Apparatus News (IAN), Chilton, Radnor, Pa.: instruments, industrial controls, digital systems; monthly.

Troubleshooting

R. E. Gasperini, *Digital Trouble Shooting*, Hayden, Rochelle Park, N.J., 1976.

Techniques of Digital Trouble Shooting, Hewlett-Packard Application Note 163-1.

MANUFACTURERS AND SUPPLIERS

Breadboards, and Circuit Boards

Continental Specialties Corp.
44 Kendall St.
New Haven, CT 06512

E. L. Instruments
61 First St.
Derby, CT

Vector Electronic Co.
12460 Gladstone Ave.
Sylmar, CA 91342

Coaxial Connectors

AMP Inc.
449 Eisenhower Blvd.
Harrisburg, PA 17105

Bunker-Ramo, RF Division
33 E. Franklin Street
Danbury, CT 06810

ITT Cannon Electric
666 E. Dyer Road
Santa Ana, CA 92702

Kings Electronics Co., Inc.
40 Marbledale Rd.
Tuckahoe, NY 10707

Omni Spectra, Inc.
21 Continental Blvd.
Merrimack, NH 03057

TRW—Cinch Connectors
1501 Morse Avenue
Elk Grove, IL 60007

Lock-in Amplifiers

EG and G Princeton Applied Research
P.O. Box 2565
Princeton, NJ 08540

Ithaco
735 W. Clinton St.
Box 818
Ithaca, NY 14850

Multiple-Pin Connectors

AMP Special Industries
Valley Forge, PA 19482

ITT Cannon Electric
666 E. Dyer Rd.
Santa Ana, CA 92702

Nuclear Electronics (NIM)

The Aston Company
P.O. Box 49123
Atlanta, GA 30359

Berkeley Nucleonics Corp.
1188 Tenth St.
Berkeley, CA 94710

Canberra Industries, Inc.
45 Gracey Ave.
Meriden, CT

EG and G Ortec
100 Midland Rd.
Oak Ridge, TN 37830

Le Croy
700 S. Main St.
Spring Valley, NY 10877

Mech-Tronics
430A Kay Ave.
Addison, IL 60101

Tennelec, Inc.
601 Turnpike
Oak Ridge, TN 37830

Tracor Northern
2551 W. Beltline Hwy.
Middleton, WI 53562

Operational Amplifiers and Devices Derived from Them

Analog Devices, Inc.
Norwood, MA 02062

Burr-Brown Research Corp.
International Airport Industrial Park
Tucson, AZ 85734

Datel Systems, Inc.
11 Cabot Blvd.
Mansfield, MA 02048

Hybrid Systems Corp.
95 Terrace Hall Ave.
Burlington, MA 01803

Power Supplies

Deltron, Inc.
Wissahickon Avenue
North Wales, PA 18454

John Fluke Mfg. Co., Inc.
P.O. Box 43210
Mountlake Terrace, WA 98043

Kepco, Inc.
131–38 Sanford Avenue
Flushing, NY 11352

Lambda Electronics
515 Broad Hollow Rd.
Melville, NY 11747

Power/Mate Corp.
514 S. River St.
Hackensack, NJ 07601

Sola
1717 Busse Rd.
Elk Grove, IL 60007

Sorenson Co.
676 Island Pond Rd.
Manchester, NH 03103

Printed Circuit Board Graphics Materials

Bishop Graphics, Inc.
Chatsworth, CA 91311

Wire Wrap Boards

Augat, Inc.
33 Perry Ave.
P.O. Box 779
Attleboro, MA 02703

Wire Wrap Tools

Gardner-Denver Co.
Pneutronics Division
P.O. Box 88
Reed City, MI 49677

INDEX